STUDENT'S SOLUTIONS MANUAL TO ACCOMPANY

HOFFMANN/BRADLEY

CALCULUS

For Business, Economics, and the Social and Life Sciences

STUDENT'S SOLUTIONS MANUAL TO ACCOMPANY

HOFFMANN/BRADLEY

CALCULUS

For Business, Economics, and the Social and Life Sciences

FIFTH EDITION

Stanley M. Lukawecki
CLEMSON UNIVERSITY

Henri Feiner
WEST LOS ANGELES COLLEGE
CALIFORNIA ACADEMY OF
MATHEMATICS AND SCIENCE

McGRAW-HILL, INC.

NEW YORK ST. LOUIS SAN FRANCISCO AUCKLAND BOGOTÁ CARACAS
LISBON LONDON MADRID MEXICO MILAN MONTREAL NEW DELHI
PARIS SAN JUAN SINGAPORE SYDNEY TOKYO TORONTO

Student's Solutions Manual to Accompany Hoffmann/Bradley:
CALCULUS for Business, Economics, and the Social and Life Sciences

2 3 4 5 6 7 8 9 0 SEM SEM 9 0 9 8 7 6 5 4 3 2

ISBN 0-07-029346-5

The editor was Maggie Lanzillo;
the production supervisor was Friederich W. Schulte.
Semline, Inc., was printer and binder.

TABLE OF CONTENTS

STUDENT'S SOLUTIONS MANUAL TO ACCOMPANY

HOFFMANN/BRADLEY

CALCULUS

For Business, Economics, and the Social and Life Sciences

Chapter 1

Functions, Graphs, and Limits

1.1 Functions

1. $f(x) = 3x^2 + 5x - 2$, $f(1) = 3(1^2) + 5(1) - 2 = 6$, $f(0) = 3(0^2) + 5(0) - 2 = -2$, $f(-2) = 3(-2)^2 + 5(-2) - 2 = 0$.

3. $g(x) = 1 + \frac{1}{x}$, $g(-1) = -1 + \frac{1}{-1} = -2$, $g(1) = 1 + \frac{1}{1} = 2$, $g(2) = 2 + \frac{1}{2} = \frac{5}{2}$.

5. $h(t) = \sqrt{t^2 + 2t + 4}$, $h(2) = \sqrt{2^2 + 2(2) + 4} = \sqrt{12} = 2\sqrt{3} \approx 3.46$, $h(0) = \sqrt{0^2 + 2(0) + 4} = \sqrt{4} = 2$, $h(-4) = \sqrt{(-4)^2 + 2(-4) + 4} = \sqrt{12} = 2\sqrt{3} \approx 3.46$

7. $f(t) = (2t - 1)^{-3/2} = \frac{1}{(\sqrt{2t-1})^3}$, $f(1) = \frac{1}{[\sqrt{2(1)-1}]^3} = 1$, $f(5) = \frac{1}{[\sqrt{2(5)-1}]^3} = \frac{1}{[\sqrt{9}]^3} = \frac{1}{27}$, $f(13) = \frac{1}{[\sqrt{2(13)-1}]^3} = \frac{1}{[\sqrt{25}]^3} = \frac{1}{125}$.

9. $f(x) = x - |x - 2|$, $f(1) = 1 - |1 - 2| = 1 - |-1| = 1 - 1 = 0$, $f(2) = 2 - |2 - 2| = 2 - |0| = 2$, $f(3) = 3 - |3 - 2| = 3 - |1| = 3 - 1 = 2$.

11. $$f(t) = \begin{cases} 3 & \text{if } t < -5 \\ t + 1 & \text{if } -1 \le t \le 5 \\ \sqrt{t} & \text{if } t > 5 \end{cases}$$
$f(-6) = 3$, $f(-5) = -5 + 1 = -4$, $f(16) = \sqrt{16} = 4$.

13. $g(x) = \frac{x^2+5}{x+2}$. Since $x + 2 \neq 0$, the domain consists of all real numbers such that $x \neq -2$.

15. $y = \sqrt{x-5}$. Radicands cannot be negative (in the real number system), so $x-5 \geq 0$, or $x \geq 5$

17. $g(t) = \sqrt{t^2+9}$. Since the radicand is not negative, any real number t is valid.

19. $f(t) = (\sqrt{2t-4})^3$. Radicands cannot be negative (in the real number system), so $2t-4 \geq 0$, or $t \geq 2$

21. $f(x) = (\sqrt{x^2-9})^{-1/2} = \frac{1}{\sqrt{x^2-9}}$. Radicands cannot be negative, denominators must not vanish (become 0), so $x^2 > 9$, or $(x+3)(x-3) > 0$. $x = -3$ and $x = 3$ are of interest, because $x^2 - 9$ becomes 0 at those values. If $x < -3$ or $x > 3$, $x^2 - 9 > 0$, else $x^2 - 9 < 0$, ($0^2 - 9 < 0$ for instance). The domain consists of real numbers $|x| > 3$.

23. $g(t) = \frac{1}{|t-1|}$. Since denominators cannot become 0, $|t-1| \neq 0$, $t - 1 \neq 0$, or $t \neq 1$.

25. a) $C(q) = q^3 - 30q^2 + 400q + 500$, where q is the number of units. Thus $C(20) = (20)^3 - 30(20)^2 + 400(20) + 500 = \$4,500$.
b) The cost of manufacturing the 20th unit is $C(20) - C(19) = 4,500 - [(19)^3 - 30(19)^2 + 400(19) + 500] = 4,500 - 4,129 = \$371$

27. a) $C(t) = -\frac{t^2}{6} + 4t + 10$ degrees Celsius, where t represents the number of hours after midnight. Thus $t = 14$ at 2:00 p.m. and $C(14) = -\frac{(14)^2}{6} + 4(14) + 10 = 33\frac{1}{3}$.
b) The difference in temperature between 6:00 p.m. ($t = 18$) and 9:00 p.m. $t = 21$ is $C(21) - C(18) = [-\frac{(21)^2}{6} + 4(21) + 10] - [-\frac{(18)^2}{6} + 4(18) + 10] = 20\frac{1}{2} - 28 = -7\frac{1}{2}$.

29. a) $f(n) = 3 + \frac{12}{n}$. The domain consists of all real numbers $n \neq 0$ (because of the denominator).
b) Since n represents the number of trials, n is a positive integer, like $n = 1, 2, 3, \cdots$.
c) For the third trial $n = 3$, thus $f(3) = 3 + \frac{12}{3} = 7$ minutes.
d) $f(n) \leq 4$, so $3 + \frac{12}{n} \leq 4$, $\frac{12}{n} \leq 1$ or $n = 12$.
e) $\frac{12}{n}$ gets smaller and smaller as n increases. Thus $\frac{12}{n} \to 0$ as $n \to \infty$ and $f(n)$ gets closer and closer to 3.

31. a) $f(x) = \frac{600x}{300-x}$. The domain consists of all $x \neq 300$, since denominators must not go to 0.
b) x represents a percentage, so $0 \leq x \leq 300$, so that $f(x) \geq 0$, or better, $0 \leq x \leq 100$, since books need not be distributed to more than the rural population.
c) If $x = 50$, $f(50) = \frac{600(50)}{300-50} = 120$.
d) With $x = 100$, $f(100) = \frac{600(100)}{300-100} = 300$.

7. $f(x) = \frac{1}{x^3}$. Note that if $x > 0$ then $f(x) > 0$ and if $x < 0$, then $f(x) < 0$, so the curve will only appear in the first and third quadrants. Since $\frac{1}{x^3}$ and $\frac{1}{(-x)^3}$ have the same absolute value, only their signs are opposites, the curve will be symmetric WRT the origin. Note that the denominators get smaller and smaller as the x-values shrink, so the values of $f(x)$ get larger and larger. $f(0)$ is not defined. Similarly, $f(x) \to 0$ as $x \to \infty$.

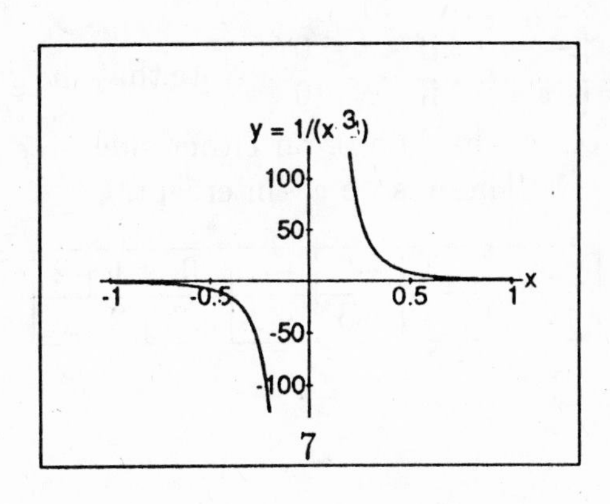

x	0.1	1	2	4	10
$f(x)$	1000	1	0.125	$\frac{1}{64}$	0.001

9. $f(x) = 2x - 1$. A function of the form $y = f(x) = ax + b$ is a linear function, that is its graph is a straight line. Two points are sufficient to draw that line. Note that $f(0) = -1$ and $f(\frac{1}{2}) = 0$.

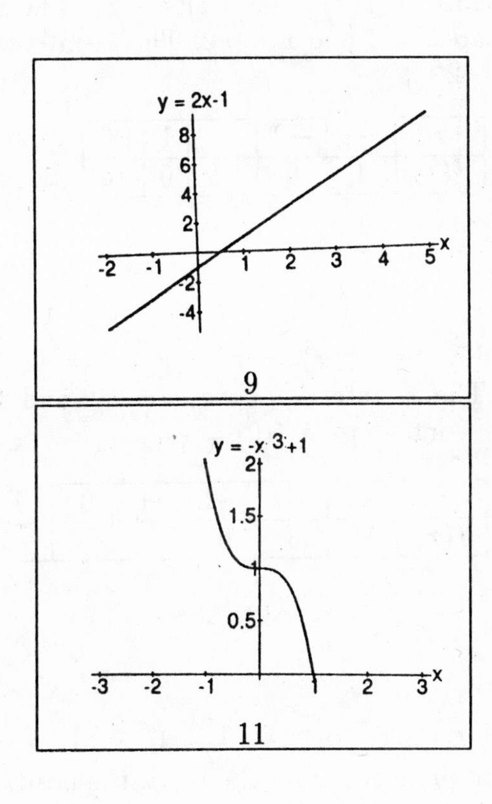

11. $f(x) = -x^3 + 1$. Note the similarities between this graph and the one in exercise 3. The y-values here are the negatives of those in 3 and the curve is translated (moved up) by 1 unit.

x	0	1	2	3	5
$f(x)$	1	0	-7	-26	-124

13. $f(x) = \begin{cases} x-1 & \text{if } x \le 0 \\ x+1 & \text{if } x > 0 \end{cases}$ Note that the graph consists of two half lines, on either side of $x = 0$. $f(0) = -1$. There is no x–intercept.

x	-3	-2	-1	0	1	2	3
$f(x)$	-4	-3	-2	-1	2	3	4

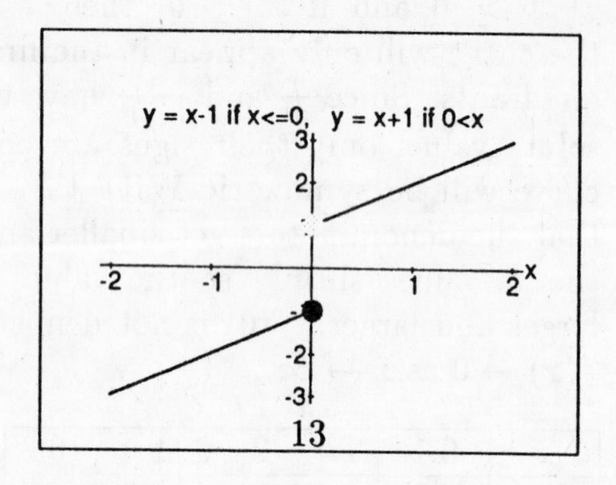

13

15. $f(x) = (x-1)(x+2)$. The x–intercepts are $x = 1$ and $x = -2$. The y–intercept is $f(0) = -2$.

x	-3	-2	0	1	3
$f(x)$	4	0	-2	0	10

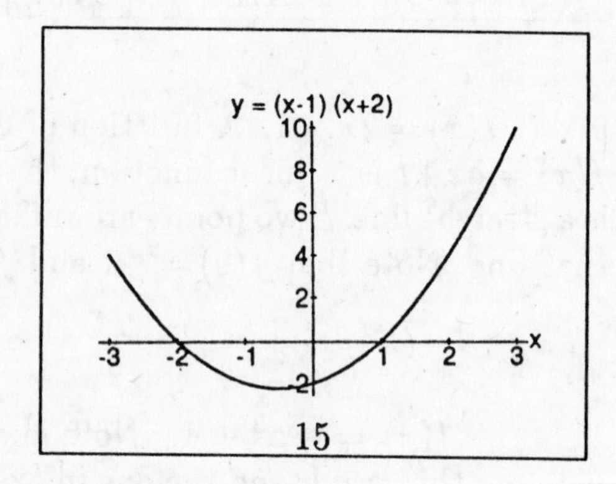

15

17. $f(x) = x^2 - x - 6 = (x-3)(x+2)$. The x–intercepts are $x = -2$ and $x = 3$. The y–intercept is $f(0) = -6$.

x	-4	-3	-2	-1	0	1	2	3	4	5
$f(x)$	14	6	0	-4	-6	-6	-4	0	6	14

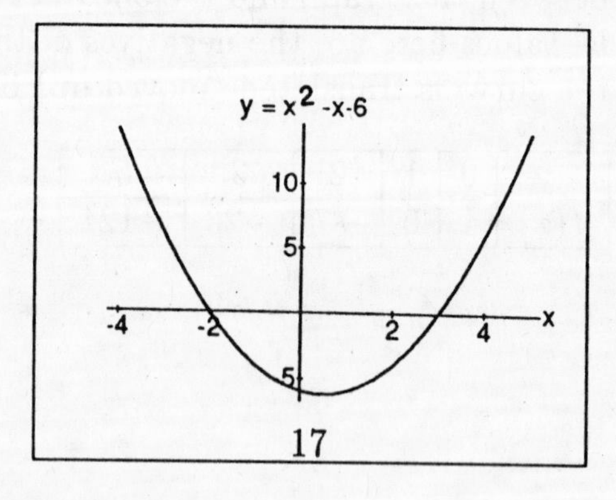

17

19. $f(x) = \frac{1}{x-2}$. If $x \to 2^+$, that is 2 is approached from the right, the positive side, $y \to +\infty$, while if $x \to 2^-$, $y \to -\infty$. Also, as $x \to \infty$, $y \to 0$. The lines $x = 2$ and $y = 0$ are called asymptotes. There is no x–intercept.

x	0	1	$\frac{3}{2}$	$\frac{3}{4}$	3	$\frac{5}{2}$	$\frac{5}{4}$
$f(x)$	$-\frac{1}{2}$	-1	-2	-4	1	2	4

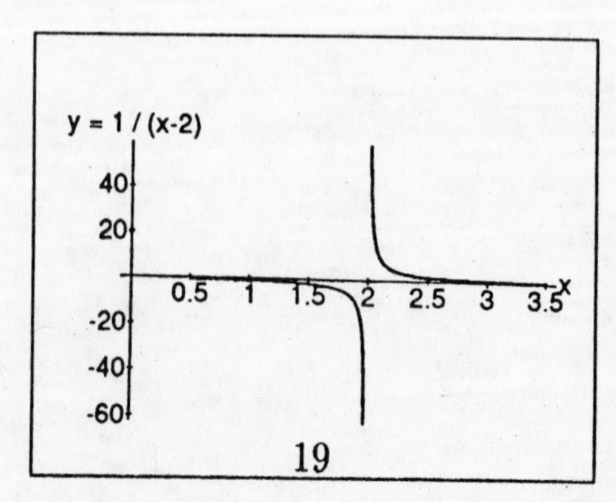

19

21. $f(x) = \frac{x^2-2x}{x+1}$. There is a discontinuity at $x = -1$. Since $f(x) = \frac{x(x-2)}{x+1}$, the intercepts are at $(0,0)$ and $(2,0)$.

x	-10	-3	$-\frac{3}{2}$	$-\frac{1}{2}$	0	1	2	10
$f(x)$	-13.3	-7.5	-10.5	2.5	0	-0.5	0	7.3

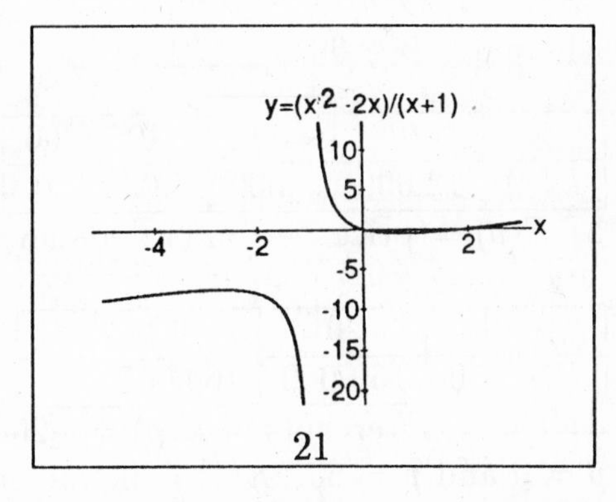

21

23. $f(x) = x + \frac{1}{x}$. There is a discontinuity at $x = 0$. Note that $f(x) \to \pm\infty$ as $x \to 0$. $f(x)$ and x both tend to ∞. Also observe the symmetry WRT the origin.

x	$\frac{1}{10}$	$\frac{1}{2}$	1	2	10
$f(x)$	10.1	2.5	2	2.5	10.1

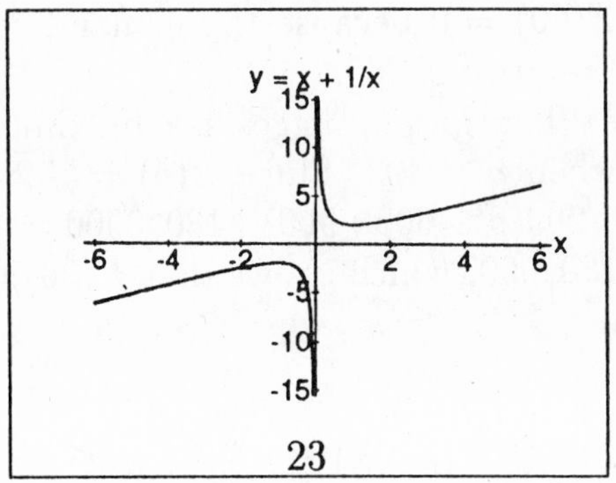

23

25. The monthly profit is $P(x) =$ (number of recorders sold)(price − cost) $= (120-x)(x-20)$. Thus the intercepts are at $(20,0)$, $(120,0)$, and $(0,-2400)$. The graph suggests a maximum profit when $x \approx 70$, that is when 70 recorders are sold. Since $P(x) = -x^2+140x-2400 = -(x^2-140x+70^2)+4900-2400 = -(x-70)^2+2,500$, which shows that $P(70) = \$2,500$.

x	0	20	30	60	90	120
$P(x)$	-2400	0	900	2400	2100	0

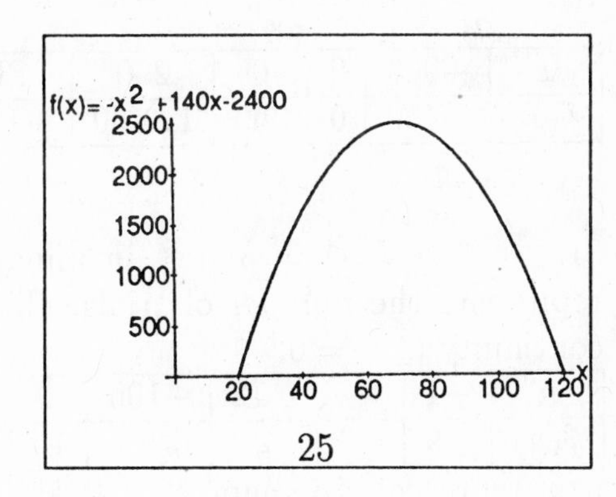

25

27. a) $D(p) = -200p+12,000 = -200(p-60)$ where $0 \leq p \leq 60$.

p	0	20	40	60
$D(p)$	12,000	8,000	4,000	0

b) $E(p)$ =(price per unit)(demand)= $-200p(p-60)$

p	0	20	40	60
$E(p)$	0	160,000	160,000	0

d) The p–intercepts of $E(p) = -200p(p-60)$ are $p = 0$ and $p = 60$. $E(0) = 0$ since the price is 0, $E(60) = 0$ because the demand drops to 0 when the price is \$60.

e) The graph suggests a maximum expenditure when $p \approx 30$. Since $E(p) = -200(p^2 - 60p) = -200(p^2-60p+900)+180,0000 = -200(p-30)^2+180,000$, which shows that $E(30) = \$180,000$.

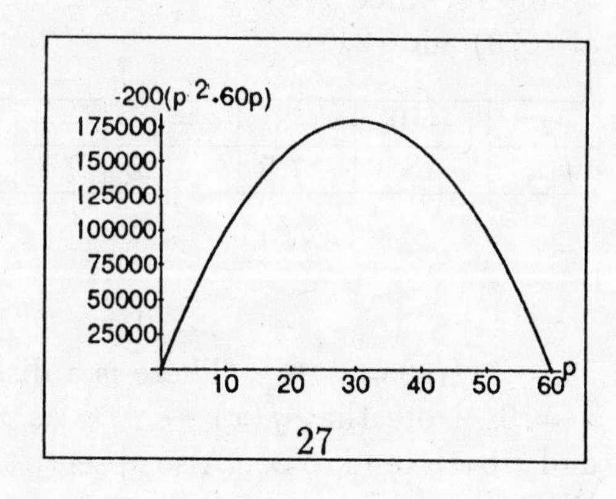

27

29. a) $f(x) = \frac{600x}{300-x}$, where x represents the percent of homes. There is a discontinuity at $x = 300$. The practical values of x here are $0 \leq x \leq 100$.

x	−900	0	100	290	310	1,100
$f(x)$	−450	0	300	17,400	−18,600	−825

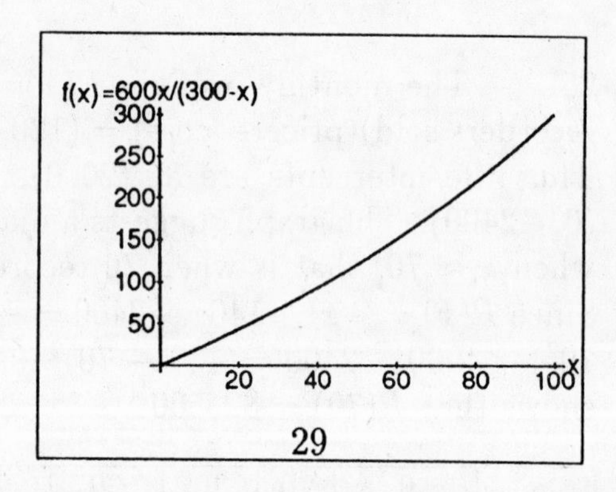

29

31. a) $f(n) = 3 + \frac{12}{n}$ in minutes, where n represents the number of trials. There is a discontinuity at $n = 0$.

n	−1	−2	−12	−100	1	2	12	100
$f(n)$	−9	−3	2	2.88	15	9	4	3.12

b) n represents the number of trials, so it is practical only if we use a positive integer, namely 1, 2, 3, ...

c) $\frac{12}{n}$ gets closer and closer to 0 as n gets larger and larger. Thus $f(n)$ tends to 3 as n goes to ∞. The more often the rat traverses the maze, the closer the required time will be to 3 minutes.

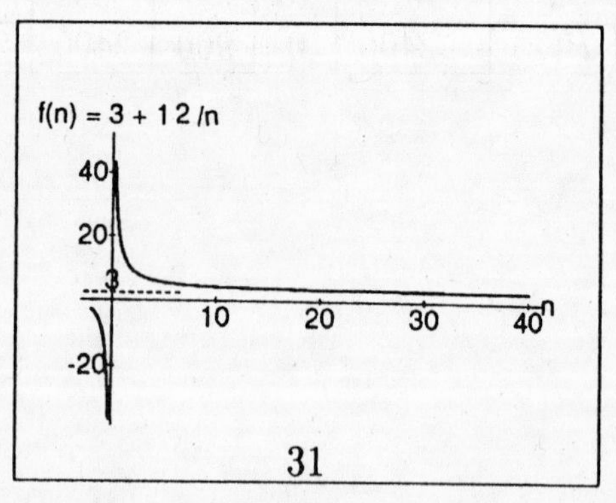

31

33. a) $C(x) = 20x + \frac{2{,}000}{x}$. There is a discontinuity at $x = 0$. The graph suggests that the cost is minimal in the neighborhood of $x = 10$, that is when 10 matches are made.

x	1	2	5	8	10	20
$C(x)$	2,020	1,040	500	410	400	500

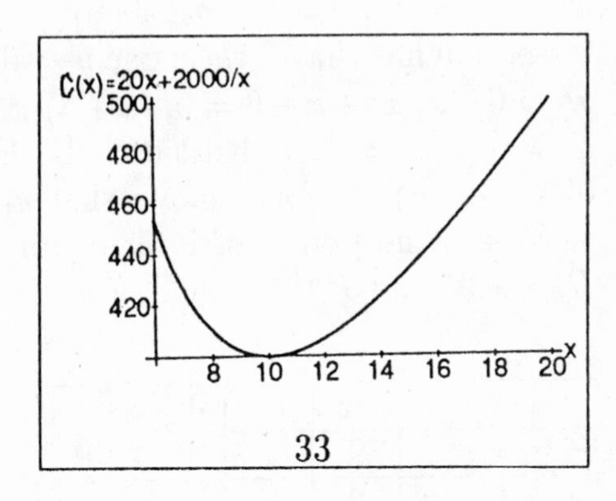

33

35. $C(x) = x^2 + 6x + 19$. The average cost is $A(x) = x + 6 + \frac{19}{x}$.

x	1	2	5	8	10	20
$A(x)$	26	17.5	14.8	16.4	17.9	27

x	1	2	5	8	10	20
$C(x)$	26	35	74	131	179	539

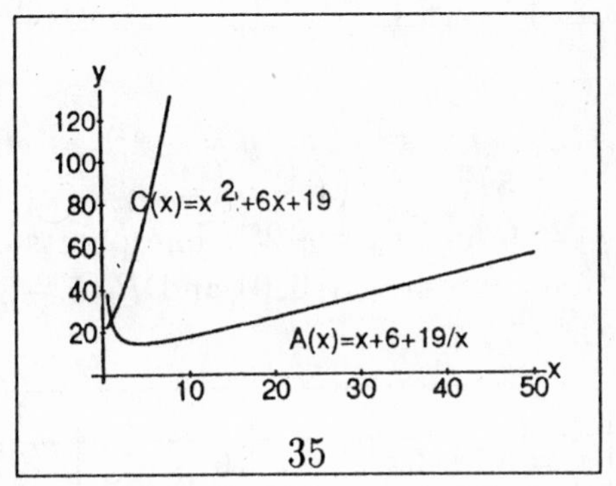

35

37. $y = 5x - 14$, $y = 4 - x$. The point(s) of intersection has (have) the same y-value. Thus $5x - 14 = 4 - x$, $6x = 18$, or $x = 3$. Substituting back into one of the equations for y shows that $y = 1$. The point of intersection then is $P(3, 1)$.

x	0	1	3	4
$y = 5x - 14$	-14	-9	1	6
$y = 4 - x$	4	3	1	0

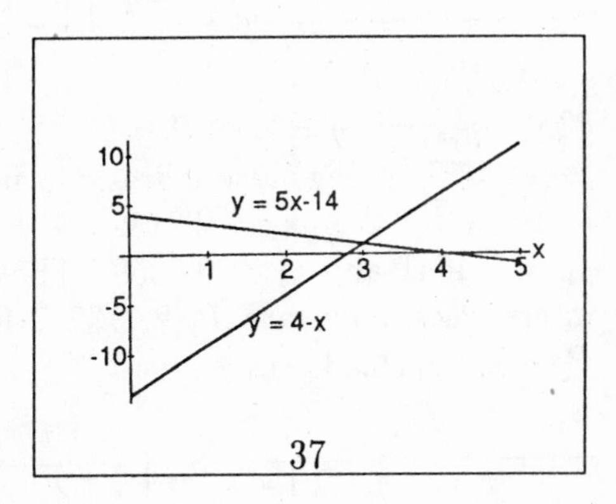

37

39. $y = x^2$, $y = 6 - x$. The point(s) of intersection has (have) the same y–value. Thus $x^2 = 6 - x$, $x^2 + x - 6 = 0$, $(x+3)(x-2) = 0$, $x_1 = -3$, $x_2 = 2$. Substituting back into one of the equations for y shows that $y_1 = 9$ and $y_2 = 4$. The points of intersection then are $P_1(-3, 9)$ and $P_2(2, 4)$.

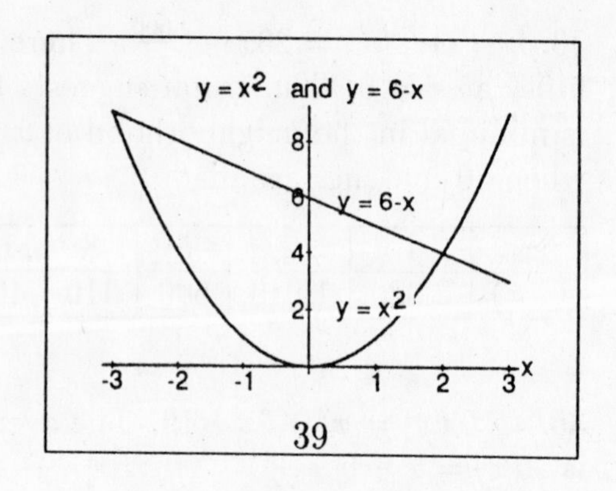

x	-3	-1	0	2	3
$y = x^2$	9	1	0	4	9
$y = 6 - x$	9	7	6	4	3

41. $y = x^3 - 6x^2$, $y = -x^2$. $x^3 - 6x^2 = -x^2$, $x^3 - 5x^2 = x^2(x - 5) = 0$, $x_1 = 0$, $x_2 = 5$. $y_1 = 0$ and $y_2 = -25$. The points of intersection then are $P_1(0, 0)$ and $P_2(5, -25)$.

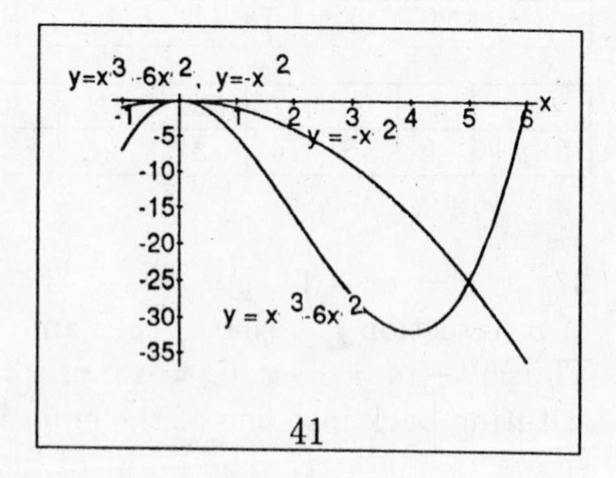

x	-1	0	2	4	5	6
$y = x^3 - 6x^2$	-7	0	-16	-32	-25	0
$y = -x^2$	-1	0	-4	-16	-25	-36

43. $y = x^2$, $y = 2x + 2$. $x^2 = 2x + 2$, $x^2 - 2x - 2 = 0$. Using the quadratic formula leads to $1 \pm \sqrt{1+2}$ or $x_1 = 2.7132$, $x_2 = -0.7132$. $y_1 = 7.4641$ and $y_2 = 0.5359$. The points of intersection then are $P_1(2.7132, 7.4641)$ and $P_2(-0.7132, 0.5359)$.

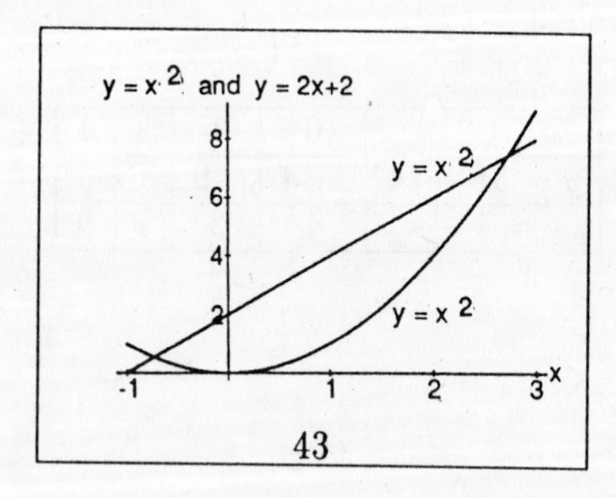

x	-2	-1	0	1	2	3
$y = x^2$	4	1	0	1	4	9
$y = 2x + 2$	-2	0	2	4	6	8

45. $3y-2x=5$, $y+3x=9$. Let's multiply the second equation by -3 and add it to the first one. Then $-2x-9x=5-27$, $x=2$, $y=9-3(2)=3$. The point of intersection then is $P(2,3)$.

x	0	1	2	3
$y=\frac{2x+5}{2}$	1.6667	2.3333	3	3.6667
$y=9-3x$	9	6	3	0

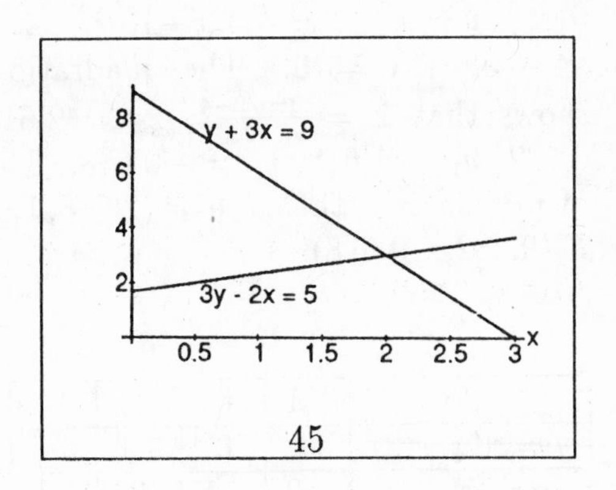

45

47. $y=\frac{1}{x}$, $y=x^2$. $\frac{1}{x}=x^2$, $x^3=1$, $x=1$, $y=1$. The point of intersection then is $P(1,1)$.

x	-1	0	1	2
$y=\frac{1}{x}$	-1	∞	1	0.5
$y=x^2$	1	0	1	4

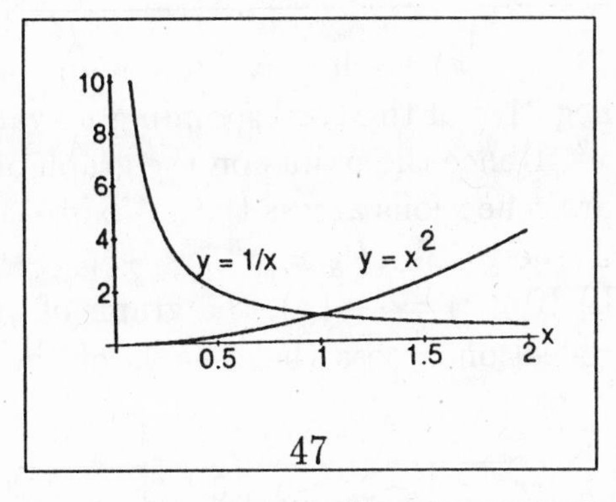

47

49. $y=\frac{1}{x}$, $y=\frac{1}{x^2}$. $\frac{1}{x}=\frac{1}{x^2}$, $x\neq 0$, $x=1$, $y=1$. The point of intersection then is $P(1,1)$.

x	-1	0	1	2
$y=\frac{1}{x}$	-1	∞	1	0.5
$y=\frac{1}{x^2}$	1	∞	1	0.25

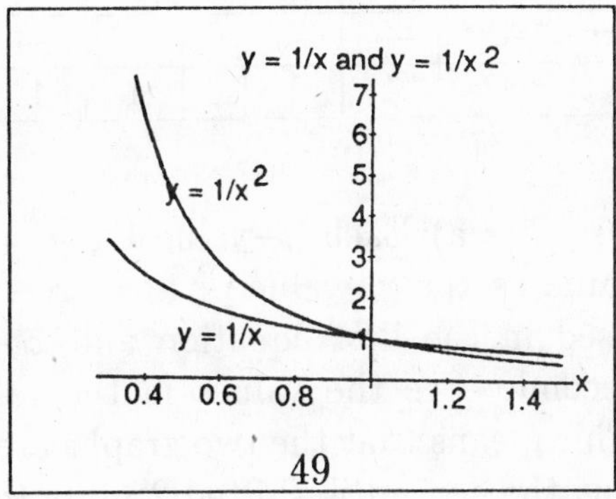

49

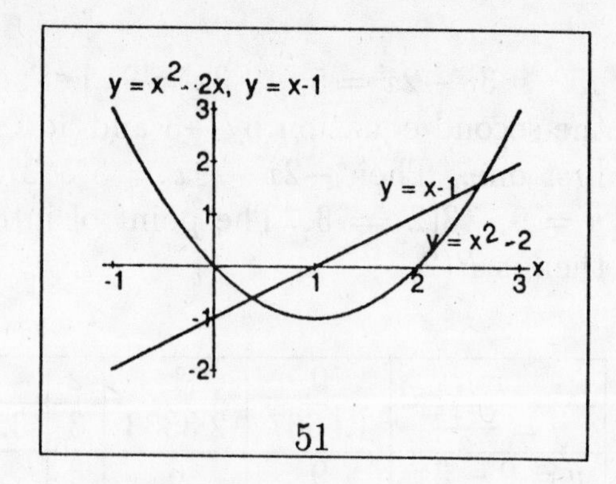

51

51. $y = x^2 - 2x$, $y = x - 1$. $x^2 - 2x = x - 1$, $x^2 - 3x + 1 = 0$. The quadratic formula shows that $x = \frac{3 \pm \sqrt{9-4}}{2}$, $x_1 = 2.618$, $x_2 = 0.382$, $y_1 = 1.618$, $y_2 = -0.618$. The points of intersection then are $P_1(2.618, 1.618)$ and $P_2(0.382, -0.618)$.

x	-1	0	1	2	3
$y = x(x-2)$	3	0	-1	0	3
$y = x - 1$	-2	-1	0	1	2

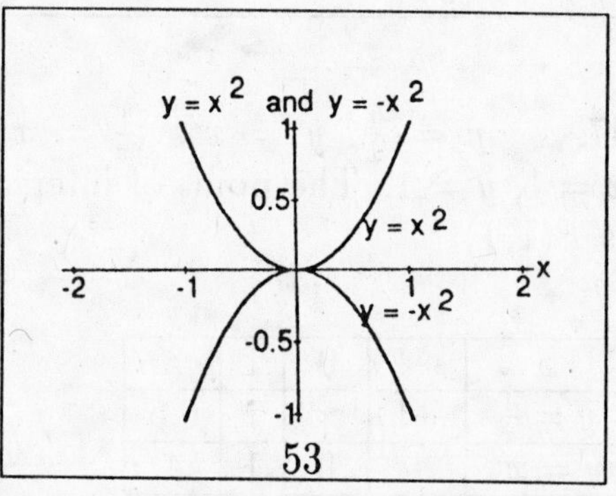

53

53. a) Each y-value for $y = -x^2$ is the negative of the corresponding y-value of $y = x^2$. Hence the points on the graph of $y = -x^2$ are reflections across the x-axis of the points of the graph of $y = x^2$.
b) If $g(x) = -f(x)$, the graph of $g(x)$ is the reflection across the x-axis of the graph of $f(x)$.

x	-2	-1	0	1	2
$y = -x^2$	-4	-1	0	-1	-4
$y = x^2$	4	1	0	1	4

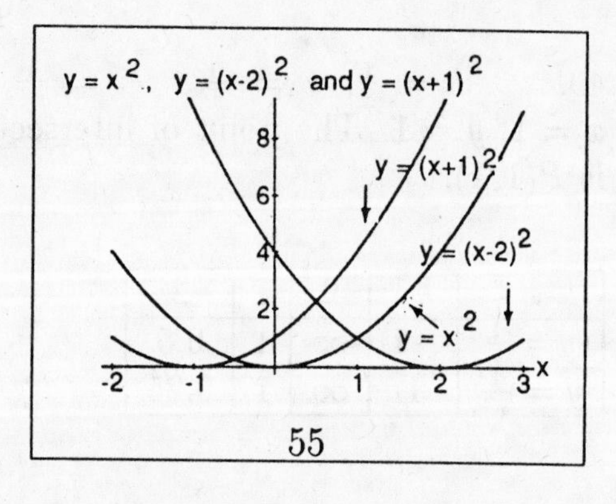

55

55. a) Each y-value for $y = x^2$ is the same as the y-value of $y = (x-2)^2$ if x is used in the first equation and $x + 2$ in the second. See the values in the table below. This means that the two graphs are identical, but the second is shifted 2 units to the right of the first one.
b) Similarly, the graph of $y = (x+1)^2$ is shifted 1 unit to the left of the graph of $y = x^2$.
c) If $g(x) = f(x - c)$, the graph of $g(x)$ is the graph of $f(x)$ shifted c units to the right if c is positive and c units to the left if c is negative.

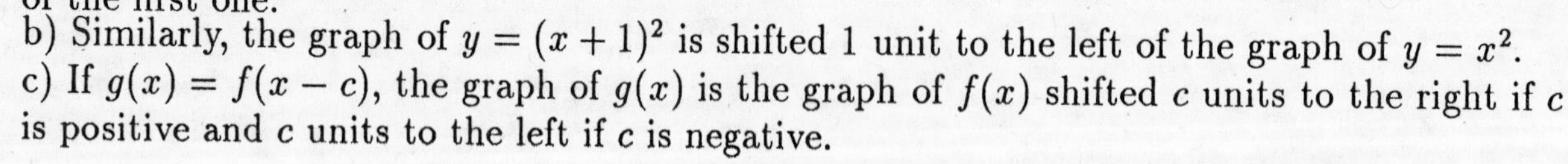

x	-3	-2	-1	0	1	2	3	4	5	6
$y = x^2$		4	1	0	1	4	9	16		
$y = (x-2)^2$				4	1	0	1	4	9	16
$y = (x+1)^2$	4	1	0	1	4	9	16			

57. The point $R(x_2, y_1)$ is at the intersection of the horizontal line $y = y_1$ and the vertical line $x = x_2$. Thus $PR = x_2 - x_1$ and $QR = y_2 - y_1$. By the Pythagorean theorem, $d = \sqrt{(x_2 - x_1)^2 + (y_2 - y_1)^2}$.

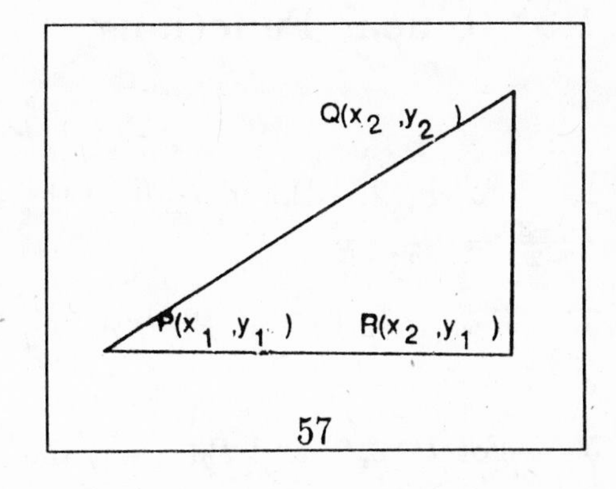

57

59. From exercise 57, $d = R = \sqrt{(x-a)^2 + (y-b)^2}$, where the point P is the center of the circle (a, b) and the point Q is any point (x, y) on the circle. Squaring both sides leads to $R^2 = (x-a)^2 + (y-b)^2$.

61. $f(x) = -x^2 - 2x + 15$. By the quadratic formula, $x^2 + 2x - 15 = 0$ has solutions $x = -1 \pm \sqrt{1+15}$, so $x_1 = -5$ and $x_2 = 3$. Since $f(0) = 15$, the intercepts are $(-5, 0)$, $(3, 0)$, and $(0, 15)$. From exercise 60, we see that the $x-$value of the vertex is at $x = -\frac{-2}{-2} = -1$ and $f(-1) = -1+2+15 = 16$. Thus the vertex is at $V(-1, 16)$.

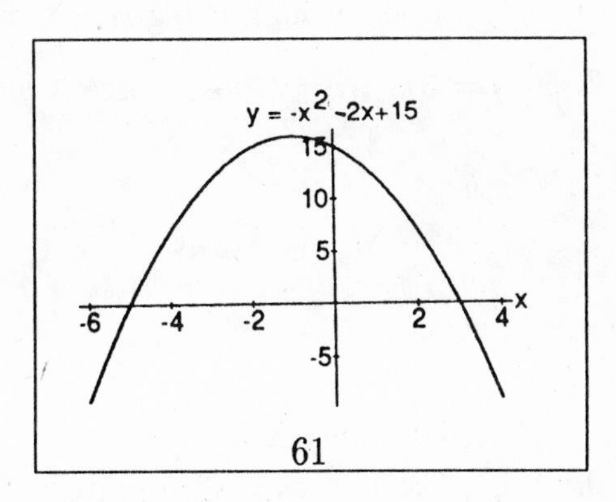

61

63. $f(x) = 6x^2 + 13x - 5$. By the quadratic formula, $6x^2 + 13x - 5 = 0$ has solutions $x = \frac{-13 \pm \sqrt{169+120}}{12}$ so $x_1 = -\frac{5}{2}$ and $x_2 = \frac{1}{3}$. Since $f(0) = -5$, the intercepts are $(-\frac{5}{2}, 0)$, $(\frac{1}{3}, 0)$, and $(0, -5)$. From exercise 60, we see that the $x-$value of the vertex is at $x = -\frac{13}{12}$ and $f(-\frac{13}{12}) = 6(\frac{169}{144}) + 13(-\frac{13}{12}) - 5 = -\frac{289}{24} \approx -12.0417$. Thus the vertex is at $V(-\frac{13}{12}, -\frac{289}{24})$.

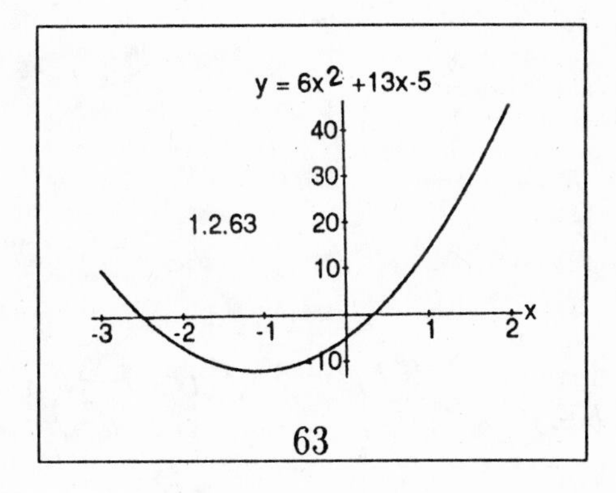

63

1.3 Linear Functions

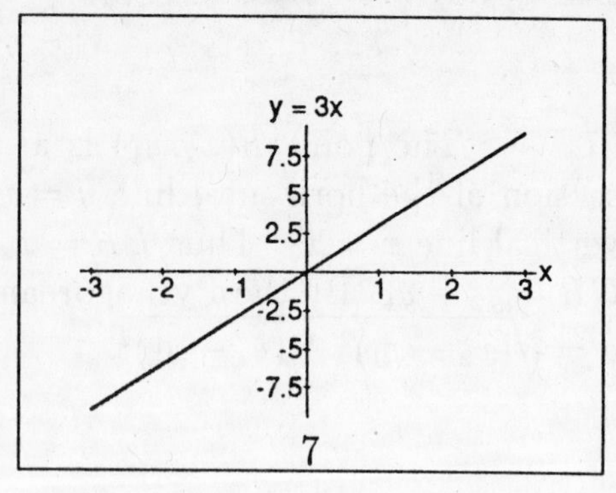

7

1. For $P_1(2,-3)$ and $P_2(0,4)$ the slope is $m = \frac{4-(-3)}{0-2} = -\frac{7}{2}$.

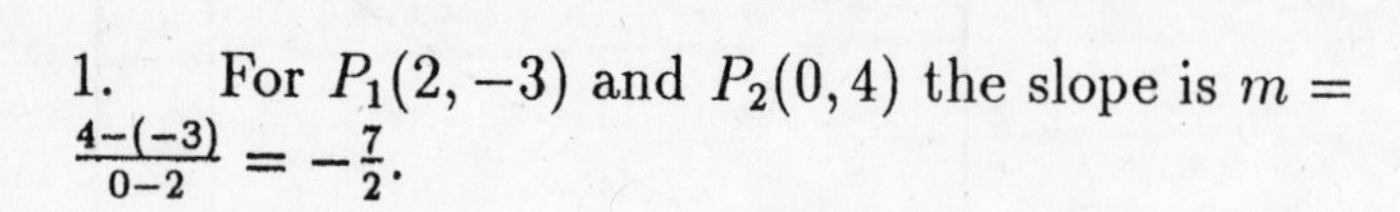

3. For $P_1(2,0)$ and $P_2(0,2)$ the slope is $m = \frac{2-0}{0-2} = -1$.

5. For $P_1(2,6)$ and $P_2(2,-4)$ the slope is $m = \frac{6-(-4)}{2-2}$, that is not defined, since the denominator is 0. The line through the given points is vertical.

7. $y = 3x$. $m = 3$, y–intercept $b = 0$.

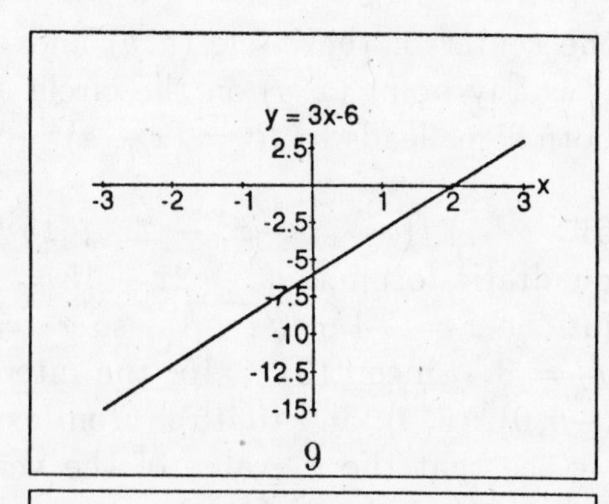

9

9. $y = 3x - 6$. $m = 3$, y–intercept $b = 6$.

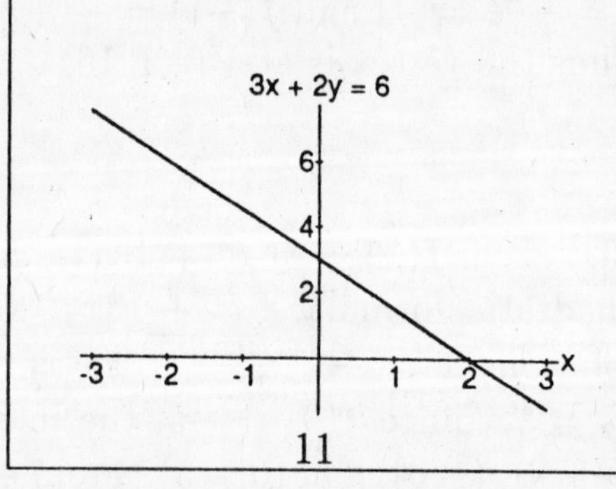

11

11. $3x + 2y = 6$ or $y = -\frac{3}{2}x + 3$. $m = -\frac{3}{2}$, $b = 3$.

13. $5y - 3x = 4$ or $y = \frac{3}{5}x + \frac{4}{5}$. $m = \frac{3}{5}$, $b = \frac{4}{5}$.

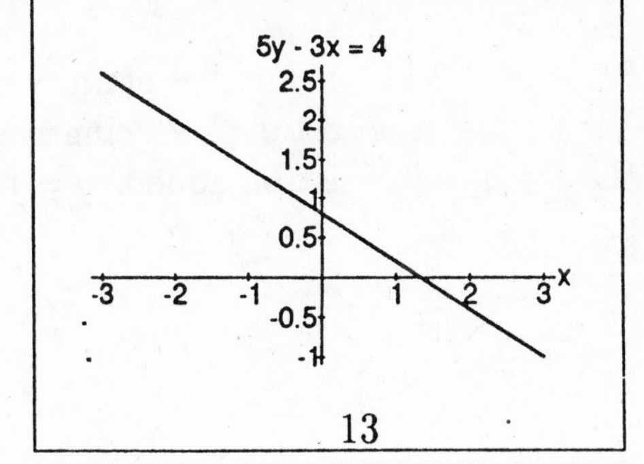

13

15. $\frac{x}{2} + \frac{y}{5} = 1$ or $y = -\frac{5}{2}x + 5$. $m = -\frac{5}{2}$, $b = 5$.

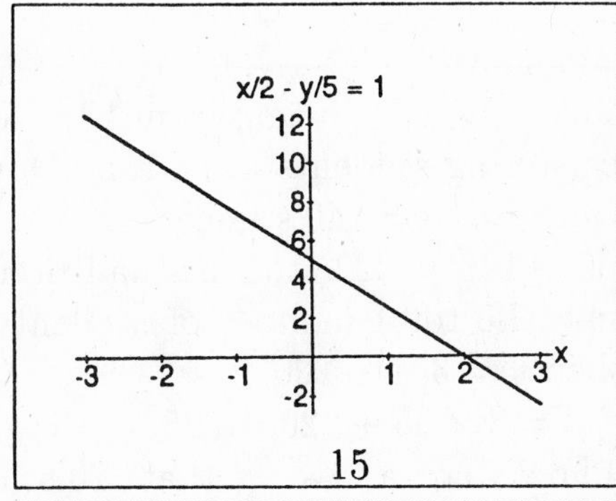

15

17. $x = -3$. The slope is not defined and there is no y intercept.

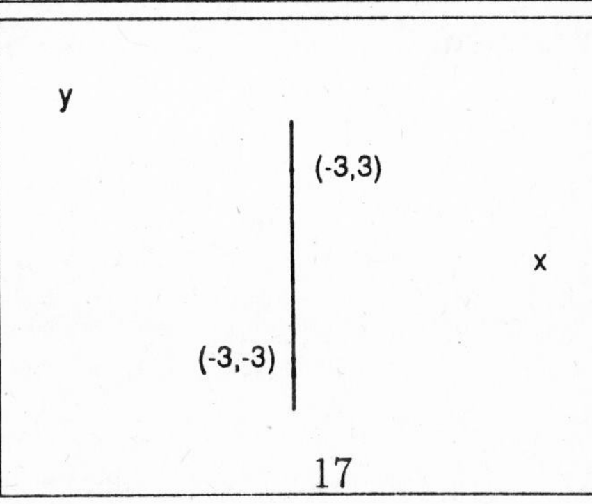

17

19. $m = 1$ and $P(2,0)$, so $y - 0 = (1)(x - 2)$ or $y = x - 2$.

21. $m = -\frac{1}{2}$ and $P(5,-2)$, so $y - (-2) = -(\frac{1}{2})(x - 5)$ or $y = -\frac{x}{2} + \frac{1}{2}$.

23. Since the line is parallel to the $x-$axis, it is horizontal and its slope is 0. For $P(2,5)$, the line is $y - 5 = 0(x - 2)$ or $y = 5$.

25. $m = \frac{1-0}{0-1}$ and for $P(1,0)$ the equation of the line is $y - 0 = -1(x - 1)$ or $y = -x + 1$. The equation would be the same if the point $(0,1)$ had been used.

27. $m = -\frac{1-(1/4)}{-(1/5)-(2/3)} = -\frac{45}{52}$ and $P(-\frac{1}{5},1)$, so $y - 1 = -(\frac{45}{52})(x + \frac{1}{5})$ or $45x + 52y - 43 = 0$.

29. The slope is 0 because the $y-$values are identical. Thus $y = 5$.

31. Let x be the number of units manufactured. Then $60x$ is the cost of producing x units, to which the fixed cost must be added. $y = 60x + 5,000$.

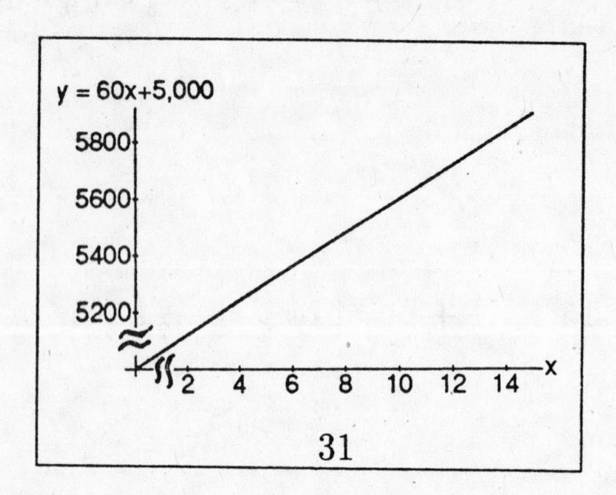

31

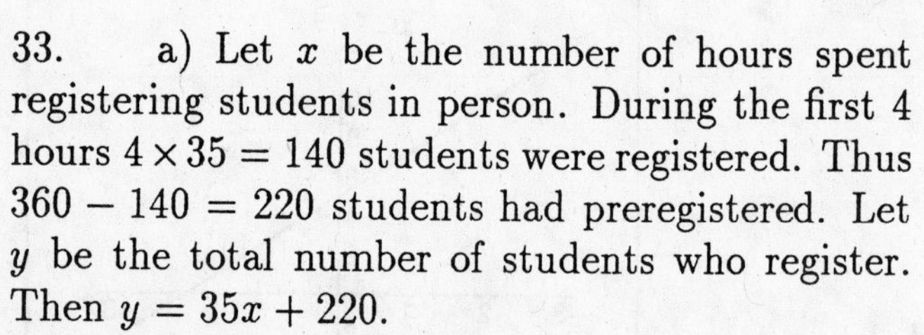
33. a) Let x be the number of hours spent registering students in person. During the first 4 hours $4 \times 35 = 140$ students were registered. Thus $360 - 140 = 220$ students had preregistered. Let y be the total number of students who register. Then $y = 35x + 220$.
b) $y = 3 \times 35 + 220 = 325$.
c) From part a), we see that 220 students had preregistered.

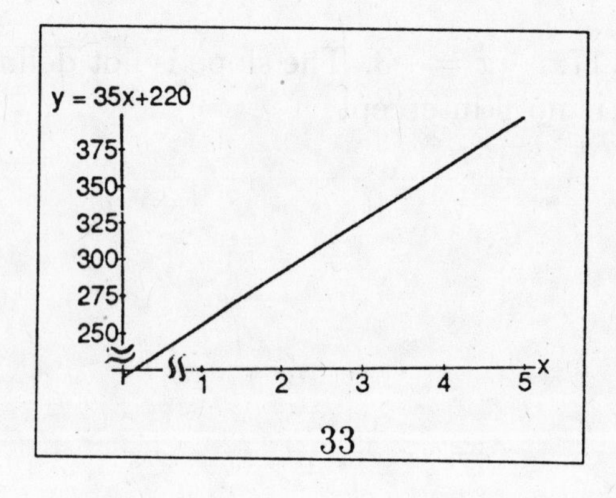

33

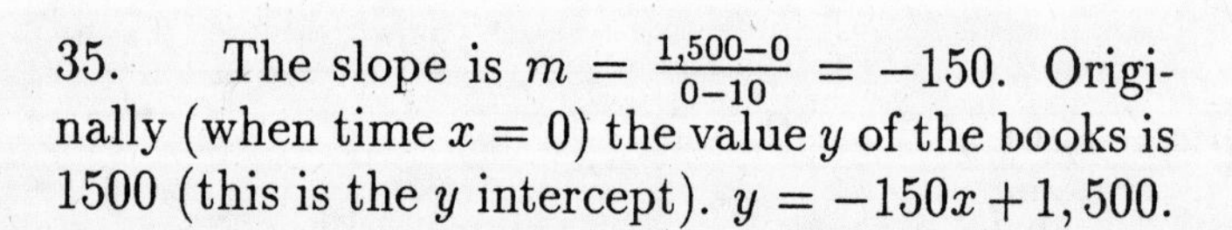
35. The slope is $m = \frac{1,500-0}{0-10} = -150$. Originally (when time $x = 0$) the value y of the books is 1500 (this is the y intercept). $y = -150x + 1,500$.

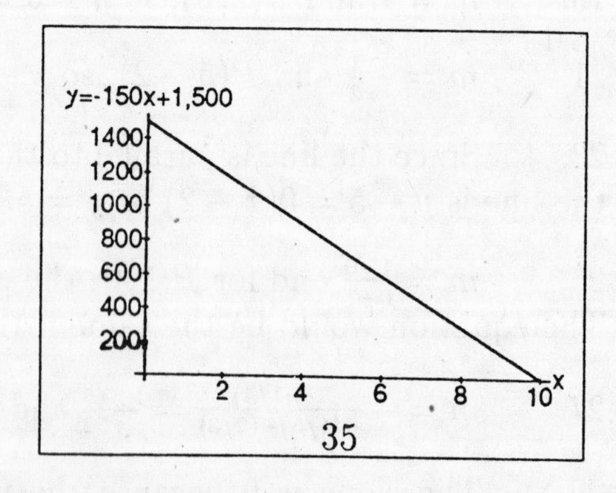

35

37. a) Let x be the number of days. The slope is $m = \frac{200-164}{12-21} = -4$. Thus $\frac{y-200}{x-12} = -4$ or $y = -4x + 248$.
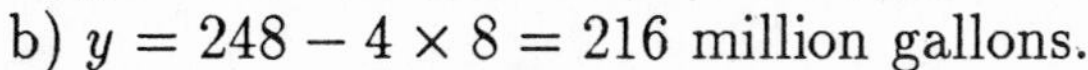
b) $y = 248 - 4 \times 8 = 216$ million gallons.

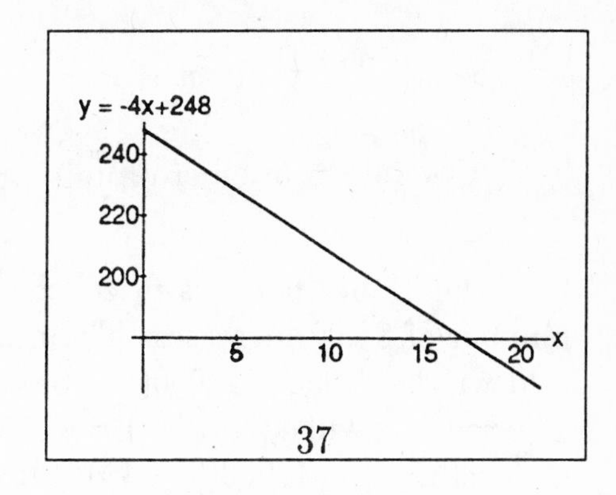

39. a) Let x be the temperature in degrees Celsius and y the temperature in degrees Fahrenheit.The slope is $m = \frac{212-32}{100-0} = \frac{9}{5}$. Thus $\frac{y-32}{x-0} = \frac{9}{5}$ or $y = \frac{9}{5}x + 32$.
b) $9 \times 15 + 160 = 5y$, $y = 59$ degrees F.
c) $9x = 5 \times 68 - 160$, $x = 20$ degrees C.

41. Since the number of radios sold per month decreases by $y = 400$ for each \$1.00 raise in sales price x, the slope is -400. The equation is then $y = -400x + 4,000$ as the line passes through the point $(5, 4000)$.

43. a) Two lines are $\|$ if their slopes are equal. All lines with equation $4x + 2y = C$ are $\|$. Since the point $(1, 3)$ must be on the line, $4 + 6 = C$ and $y = -2x + 5$ (after division by 2).
b) $2y - 3x = C$. $C = 4$ and $y = \frac{3}{2}x + 2$.
c) The desired slope is $\frac{2+1}{1-6} = -\frac{3}{5}$. The desired line has equation $\frac{y-5}{x+2} = -\frac{3}{5}$ or $y = -\frac{3}{5}x + \frac{19}{5}$.

1.4 Functional Models

1. This problem has two possible forms of the solution.
a) Assume the stream is along the length, say x. Then y is the width and $x + 2y = 1,000$ or $x = 1,000 - 2y$. The area is $A = xy = 2y(500 - y)$ square feet.
b) Now suppose the stream flows along the width, y, then $2x + y = 1,000$ or $x = \frac{1,000-y}{2}$. Then $A = xy = \frac{y(1,000-y)}{2}$ square feet.

3. Let x and y be the smaller and larger numbers, respectively. Then $x + y = 18$ or

$y = 18 - x$. The product is $P = xy = x(18 - x)$.

5. Revenue = number of units × price per unit = $x(35x + 15)$.

7. Let x be the sales price. $x - 6$ will be the number of \$1.00 increases. The number of lamps sold will be $3,000 - 1,000(x - 6) = 1,000(9 - x)$. The revenue will equal $R(x) = 1,000(9x - x^2)$ and the cost $C(x) = 1,000(36 - 4x)$. Since profit is revenue − cost, $P(x) = 1,000(9x - x^2 - 36 + 4x) = -1,000(x^2 - 13x + 36) = 1,000(9 - x)(x - 4)$. The optimal selling price appears to be \$6.50.

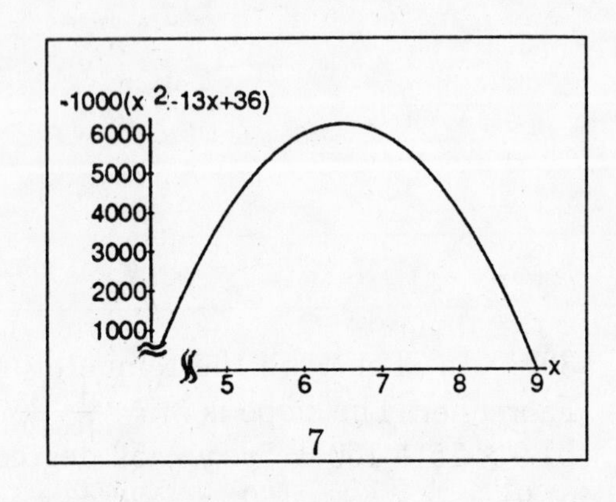

7

9. Let x be the number of additional trees planted per acre. The number of oranges per tree will be $400 - 4x$ and the number of trees per acre $60 + x$. The yield per acre is $Y(x) = \frac{\text{\# of oranges}}{\text{tree}} \frac{\text{\# of trees}}{\text{acre}} = (400 - 4x)(60 + x) = 4(100 - x)(60 + x)$. The number of trees for optimal yield appears to be $60 + 20 = 80$ trees.

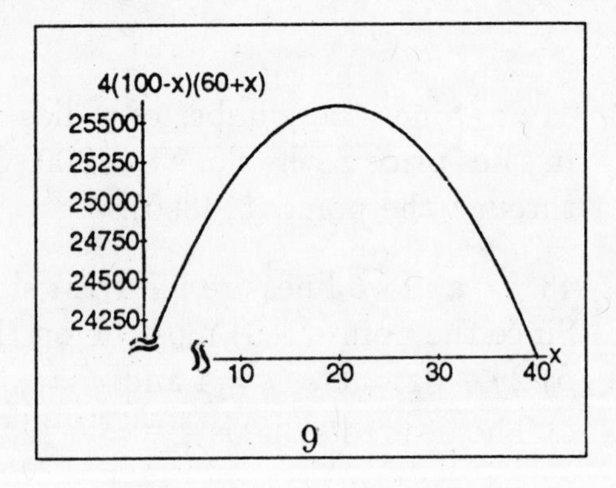

9

11. Let x be the number of additional days beyond 80 before the club takes all its glass to the recycling center. Let's assume that the same quantity of glass is collected daily, namely $\frac{24,000}{80} = 300$ lbs. The daily revenue for the first 80 days would be 300 cents. The reduction in daily revenue for x days beyond 80 is $3x$ or 1 cent per 100 lbs per day. The club's revenue on day $80 + x$ would be $(300 - 3x)(80 + x) = 3(80 + x)(100 - x)$. The key to this problem is understanding that all the glass is taken to the recycling center on day $80 + x$. x is estimated to be 10 from the graph.

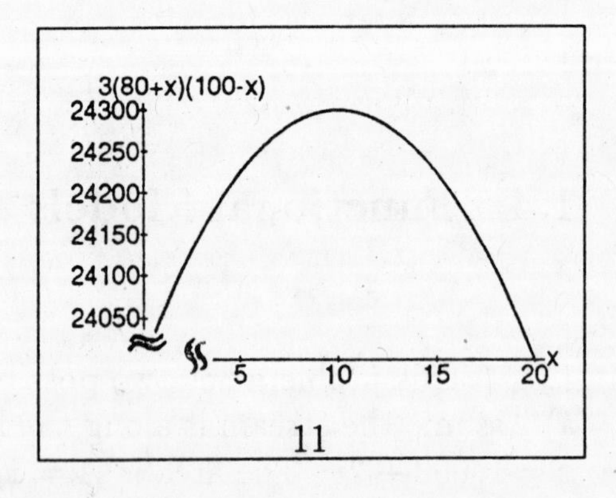

11

13. Let x be the length and y the width of the rectangle. Then $2x + 2y = 320$ or $y = 160 - x$. The area is length × width or $A(x) = x(160 - x)$. The length is estimated to be 80 meters from the graph below, which also happens to be the width. Thus the maximum area seems to correspond to that of a square.

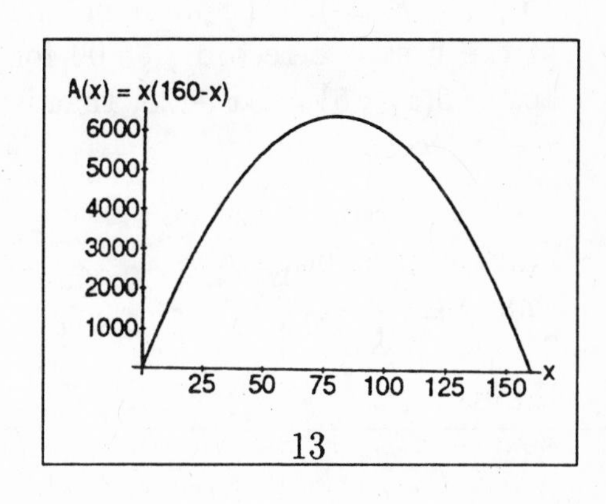

13

15. Let x be the length of the square base, then $2x$ is its depth. The volume is $V = x^2(2x) = 2x^3$

17. Let x be the side of the square base and y the height of the open box. The area of the base is x^2 square meters and that of each side is xy. The total cost is $4x^2 + 3(4xy) = 48$. Solving for y in terms of x we see that $12xy = 48 - 4x^2$, $3xy = 12 - x^2$ or $y = \frac{12-x^2}{3x}$. The volume of the box is $V = x^2y = \frac{x(12-x^2)}{3}$ cubic meters.

19. The volume of the soda can is $V = \pi r^2 h = 6.89\pi$. Thus $h = \frac{6.89}{r^2}$. The surface area consists of the top and bottom circles and the curved side. This curved side can be flattened out to a rectangle. The length is equal to the circumference of a circle of radius r and the width is the height of the can. $S = 2\pi r^2 + 2\pi rh = 2\pi r^2 + 2\pi r\,\frac{6.89}{r^2} = \frac{13.78\pi}{r} + 2\pi r^2$.

21. The surface area of the closed can is $S = 2\pi r^2 + 2\pi rh$ and its volume $V = \pi r^2 h$.
a) We can express r in terms of S and h. Thus the quadratic $2\pi r^2 + 2\pi rh - S = 0$ leads to $r = \frac{-2\pi h \pm \sqrt{4\pi^2h^2+8\pi S}}{4\pi} = \frac{-\pi h+\sqrt{\pi^2h^2+2\pi S}}{2\pi}$, since $r \geq 0$. So $V = \pi h(\frac{-\pi h+\sqrt{\pi^2h^2+2\pi S}}{2\pi})^2 = \frac{h}{4\pi}(-\pi h + \sqrt{\pi^2h^2 + 2\pi S})^2$.
We can also express h in terms of S and r. Now $h = \frac{S-2\pi r^2}{2\pi r}$. $V = \frac{r(S-2\pi r^2)}{2}$.
b) We can express r in terms of V and h. Thus $h = \frac{V}{\pi r^2}$ and $S = 2\pi r^2 + \frac{2V}{r}$.
We can also express h in terms of V and r. Now $r = \sqrt{\frac{V}{\pi h}}$ and $S = \frac{2V}{h} + 2\sqrt{\pi hV}$.

23. The surface area of the topless can is $S = \pi r^2 + 2\pi rh = 27\pi$. $r^2 + 2rh = 27$, $h = \frac{27}{2r} - \frac{r}{2}$. The volume is $V = \pi r^2 h = \frac{\pi r(27-r^2)}{2}$.

25. Let x be the number of records bought. The price per record is \$10.00 for each of the first five records, \$5.00 for record 6, 7, 8 or 9. The cost for $6 \leq x \leq 9$ records is $10x - 5(x - 5) = 5x + 25$, that is the regular price of \$10.00 minus a discount of \$5.00 for $x - 5$ records.

$$C(x) = \begin{cases} 10x & \text{if } 0 < x \leq 5 \\ 5x + 25 & \text{if } 5 < x \leq 9 \end{cases}$$

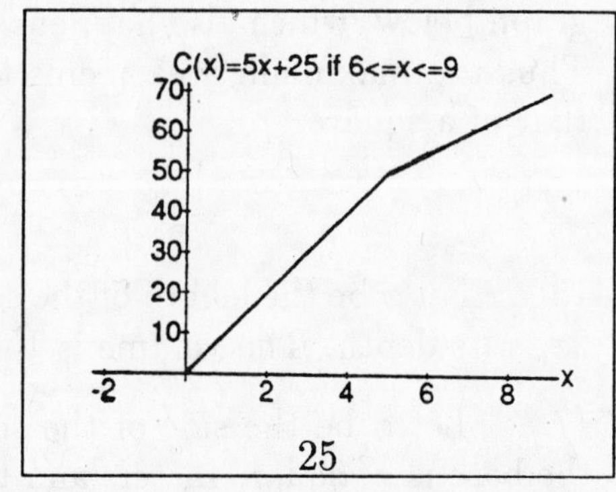

25

27. Let x be the number of passengers. There will be $x - 40$ passengers between $40 < x \leq 80$ (if the total number is below 80). The price for the second category is $60 - 0.5 \times (x - 40) = 80 - 0.5x$. The revenue generated in this category is $80x - 0.5x^2$.

$$R(x) = \begin{cases} 2,400 & \text{if } 0 < x \leq 40 \\ 80x - 0.5x^2 & \text{if } 40 < x < 80 \\ 40x & \text{if } x \geq 80 \end{cases}$$

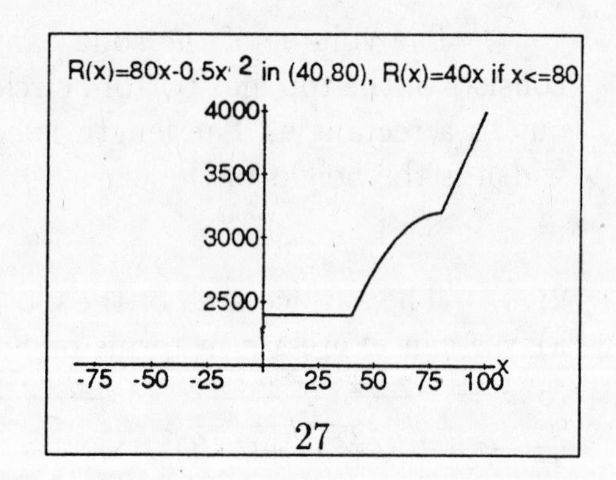

27

29. Let q be the amount of radium remaining and k a proportionality constant. Then $R(q) = kq$ (although $R(q) = -kq$ is usually used because decay means that less and less radium is left as time goes on, with $k > 0$.)

31. Let q be the number of people who have caught the disease. Then $n - q$ is the number of people who have not (yet) caught the disease out of a total population of n people. If k is the proportionality constant then $R(q) = kq(n - q)$.

33. Let x be the number of machines used and k_1, k_2 constants of proportionality. The setup cost is $k_1 x$, while the operating cost is $\frac{k_2}{x}$. The total cost is $C(x) = k_1 x + \frac{k_2}{x}$.

35. Let t be the time in hours, D_t the distance traveled by the truck, and D_c that of the car. $D_c = 80t$ and $D_t = 60t$ KPH. Since the vehicles travel at right angles to each other, the distance D between them is the hypothenuse of a right triangle. $D^2 = (80t)^2 + (60t)^2 = 10\sqrt{64t^2 + 36t^2} = 100t$ km.

37. The distance from the power plant to P is $3,000 - x$ and that across the water $\sqrt{810,000 + x^2}$. The cost overland is $4(3,000 - x)$ and that across the water $5\sqrt{810,000 + x^2}$. The total cost is $4(3,000 - x) + 5\sqrt{810,000 + x^2}$.

39. Let x be the number of machines used and t the number of hours of production. The number of kickboards produced per machine per hour is $30x$. It costs $20x$ to set up all the machines. The cost of supervision is $4.80t$. The number of kickboards produced by x machines in t hours is $30xt$ which must account for all $8,000$ kickboards. Solving $30xt = 8,000$ for t leads to $t = \frac{800}{3x}$. The cost of supervision can be rewritten as $\frac{4.80 \times 800}{3x} = \frac{1,280}{x}$. The total cost is $20x + \frac{1,280}{x}$.

41. a) The revenue is $R(x) = 175x$ and the cost is $C(x) = 25x + 600$, where x is the number of kayaks sold. For the break even point, $175x = 25x + 600$ or $x = 4$.
b) The profit is $P(x) = R(x) - C(x) = 175x - (25x + 600) = 150x - 600$. For $P(x) = 450$, $150x - 600 = 450$, $15x = 105$, $x = 7$.

43. With x representing the number of hours of court usage, the costs are $C_1(x) = 1000 + 3x$ and $C_2(x) = 800 + 4x$, for the two courts respectively. The cost at the clubs will be the same when $1000 + 3x = 800 + 4x$ or $x = 200$. Court 1 should be used if more than 200 hours of tennis are played, otherwise court 2 is cheaper.

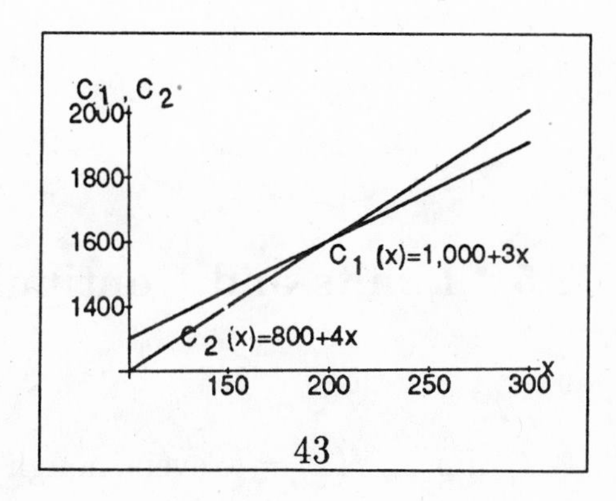

43

45. The equilibrium price is found when $S(p) = D(p)$, that is $4p + 200 = -3p + 480$ or $p = 40$. At this price the number of units supplied, as well as the number of unit demanded, is $D(40) = 480 - 3(40) = 360$ units.

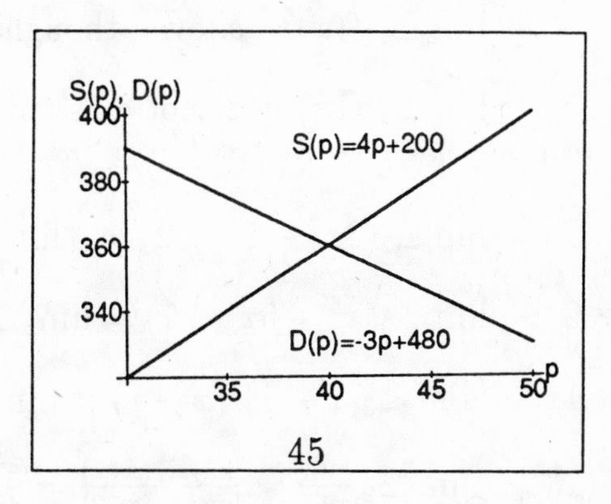

45

47. a) The equilibrium price is found when $S(p) = D(p)$, that is $p - 10 = \frac{5,600}{p}$, $p^2 - 10p - 5,600 = (p - 80)(p + 70) = 0$, or $p = 40$. Only the positive number $p = 80$ is meaningful in the present exercise. At this price the number of units supplied, as well as the number of unit demanded, is $D(40) = 360$ units.

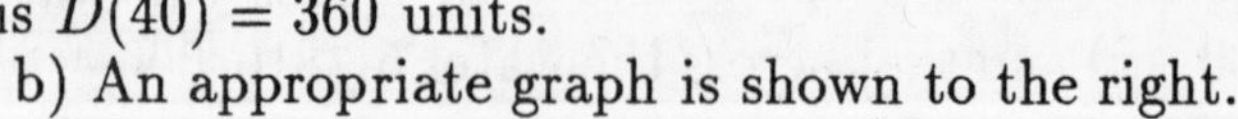
b) An appropriate graph is shown to the right.

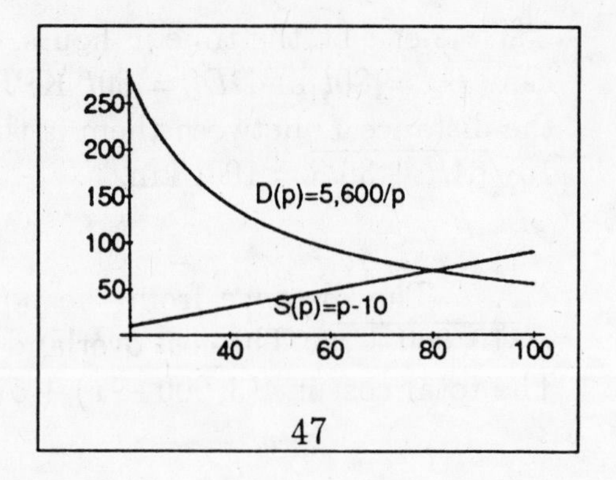

47

c) The supply curve crosses the p–axis when $S(p) = 0$, that is when $p = 10$. This means that the supplier is unwilling to provide the commodity unless the selling price exceeds \$10.

49. Let x be the number of hours since the second plane took off and y the distance traveled in flight. The first plane is ahead by $\frac{1}{2}$ hour or $0.5 \times 550 = 275$ miles. (Distance = velocity × time, just watch the units.) The equations for the distances are $y = 550x + 275$ and $y = 650x$ for the first and second planes, respectively. They will have covered the same distance when $650x = 550x + 275$, $100x = 275$, $x = 2.75$ hrs which is 2 hrs and 0.75×60 minutes or 2 hrs and 45 min. after the second plane took off.

1.5 Limits and Continuity

1. $\lim_{x \to a} f(x) = b$, even though $f(a)$ is not defined.

3. $\lim_{x \to a} f(x) = b$, even though $f(a) = c$.

5. $\lim_{x \to a} f(x)$ does not exist since as x approaches a from the left, the function becomes unbounded.

7. $\lim_{x \to 2}(3x^2 - 5x + 2) = 3\lim_{x \to 2} x^2 - 5\lim_{x \to 2} x + \lim_{x \to 2} 2 = 3(2)^2 - 5(2) + 2 = 4$

9. $\lim_{x \to 0}(x^5 - 6x^4 + 7) = \lim_{x \to 0} x^5 - 6\lim_{x \to 0} x^4 + \lim_{x \to 0} 7 = 0 - 6(0) + 7 = 7.$

11. $\lim_{x \to 3}(x - 1)^2(x + 1) = \lim_{x \to 3}(x - 1)^2 \lim_{x \to 3}(x + 1) = (3 - 1)^2(3 + 1) = 16.$

13. $\lim_{x \to 2} \frac{x+1}{x+2} = \frac{\lim_{x \to 2}(x+1)}{\lim_{x \to 2}(x+2)} = \frac{3}{4}.$

15. $\lim_{x\to 5} \frac{x+3}{5-x}$ does not exist since the limit of the denominator is zero while the limit of the numerator is not zero.

17. $\lim_{x\to 1} \frac{x^2-1}{x-1} = \lim_{x\to 1} \frac{(x+1)(x-1)}{x-1} = \lim_{x\to 1}(x+1) = 2.$

19. $\lim_{x\to 5} \frac{x^2-3x-10}{x-5} = \lim_{x\to 5} \frac{(x-5)(x+2)}{x-5} = \lim_{x\to 5}(x+2) = 7.$

21. $\lim_{x\to 4} \frac{(x+1)(x-4)}{(x-1)(x-4)} = \lim_{x\to 4} \frac{x+1}{x-1} = \lim_{x\to 4} \frac{\lim_{x\to 4}(x+1)}{\lim_{x\to 4}(x-1)} = \frac{5}{3}.$

23. $\lim_{x\to -2} \frac{x^2-x-6}{x^2+3x+2} = \lim_{x\to -2} \frac{(x-3)(x+2)}{(x+1)(x+2)} = \lim_{x\to -2} \frac{x-3}{x+1} = \frac{\lim_{x\to -2}(x-3)}{\lim_{x\to -2}(x+1)} = \frac{-5}{-1} = 5.$

25. $\lim_{x\to 4} \frac{\sqrt{x}-2}{x-4} = \lim_{x\to 4} \frac{\sqrt{x}-2}{x-4} \frac{\sqrt{x}+2}{\sqrt{x}+2} = \lim_{x\to 4} \frac{x-4}{(x-4)(\sqrt{x}+2)} = \lim_{x\to 4} \frac{1}{\sqrt{x}+2} = \frac{1}{4}.$

27. $\lim_{x\to 1} \frac{x-1}{\sqrt{x}-1} = \lim_{x\to 1} \frac{x-1}{\sqrt{x}-1} \frac{\sqrt{x}+1}{\sqrt{x}+1} = \lim_{x\to 1} \frac{(x-1)(\sqrt{x}+1)}{x-1} = \lim_{x\to 1}(\sqrt{x}+1) = 2.$

29. If $f(x) = 5x^2 - 6x + 1$, then $f(2) = 9$ and $\lim_{x\to 2} f(x) = 9$, and so f is continuous at $x = 2$.

31. $f(x) = \frac{x+2}{x+1}$, then $f(1) = \frac{3}{2}$ and $\lim_{x\to 1} f(x) = \lim_{x\to 1} \frac{x+2}{x+1} = \frac{\lim_{x\to 1}(x+2)}{\lim_{x\to 1}(x+1)} = \frac{3}{2}$. Hence f is continuous at $x = 1$.

33. If $f(x) = \frac{x+1}{x-1}$, $f(1)$ is undefined since the denominator is zero, and hence f is not continuous at $x = 1$.

35. If $f(x) = \frac{\sqrt{x}-2}{x-4}$, $f(4)$ is undefined since the denominator is zero, and hence f is not continuous at $x = 4$.

37. If $f(x) = \begin{cases} x+1 & \text{if } x \le 2 \\ 2 & \text{if } 2 \le x \end{cases}$ then $f(2) = 3$ and $\lim_{x\to 2} f(x)$ must be determined. As x approaches 2 from the right, $\lim_{x\to 2} f(x) = \lim_{x\to 2} 2 = 2$ and as x approaches 2 from the left, $\lim_{x\to 2} f(x) = \lim_{x\to 2}(x+1) = 3$. Hence the limit does not exists (since different limits are obtained from the left and the right), and so f is not continuous at $x = 2$.

39. If $f(x) = \begin{cases} x+1 & \text{if } x < 0 \\ x-1 & \text{if } 0 \le x \end{cases}$ then $f(0) = -1$ and $\lim_{x\to 0} f(x)$ must be determined. As x approaches 0 from the right, $\lim_{x\to 0} f(x) = \lim_{x\to 0}(x-1) = -1$ and as x approaches 0 from the left, $\lim_{x\to 0} f(x) = \lim_{x\to 0}(x+1) = 1$. Hence the limit does not exists (since different limits are obtained from the left and the right), and so f is not continuous at $x = 0$.

41. If $f(x) = \begin{cases} \frac{x^2-1}{x+1} & \text{if } x < -1 \\ x^2-3 & \text{if } -1 \le x \end{cases}$ then $f(-1) = -2$ and $\lim_{x\to -1} f(x)$ must be determined. As x approaches -1 from the right, $\lim_{x\to -1} f(x) = \lim_{x\to -1}(x^2-3) = -2$

and as x approaches 0 from the left, $\lim_{x\to -1} f(x) = \lim_{x\to -1} \frac{x^2-1}{x+1} = \lim_{x\to -1} \frac{(x-1)(x+1)}{x+1} = \lim_{x\to -1}(x-1) = -2$. Hence the limit exists and is equal to $f(-1)$, and so f is continuous at $x = -1$.

43. The polynomial $f(x) = x^5 - x^3$ is continuous for all values of x.

45. $f(x) = \frac{3x-1}{2x-6}$ is not continuous at $x = 3$, where the denominator is zero.

47. $f(x) = \frac{x^2-1}{x+1}$ is not continuous at $x = -1$, where the denominator is zero.

49. $f(x) = \frac{x}{(x+5)(x-1)}$ is not continuous at $x = -5$ or $x = 1$, where the denominator is zero.

51. $f(x) = \frac{x^2-2x+1}{x^2-x-2} = \frac{(x-1)^2}{(x-2)(x+1)}$ is not continuous at $x = -1$ or $x = 2$, where the denominator is zero.

53. The function $f(x) = \begin{cases} x^2 & \text{if } x \leq 2 \\ 9 & \text{if } 2 < x \end{cases}$ is possibly not continuous only at $x = 2$. As x approaches 2 from the right, $\lim_{x\to 2} f(x) = \lim_{x\to 2} 9 = 9$ and as x approaches 2 from the left, $\lim_{x\to 2} f(x) = \lim_{x\to 2} x^2 = 4$. Hence the limit does not exist and so f is not continuous at $x = 2$.

55. The function $f(x) = \begin{cases} Ax - 3 & \text{if } x < 2 \\ 3 - x + 2x^2 & \text{if } 2 \leq x \end{cases}$ then $f(x)$ is continuous everywhere except possibly at $x = 2$ since $Ax - 3$ and $3 - x + 2x^2$ are polynomials. Since $f(2) = 3 - 2 + 2(2)^2 = 9$, in order that $f(x)$ be continuous at $x = 2$, A must be chosen so that $\lim_{x\to 2} f(x) = 9$. As x approaches 2 from the right, $\lim_{x\to 2} f(x) = \lim_{x\to 2}(3 - x + 2x^2) = \lim_{x\to 2} 3 - \lim_{x\to 2} x + 2(\lim_{x\to 2} x)^2 = 3 - 2 + 2(2)^2 = 9$ and as x approaches from the left, $\lim_{x\to 2} f(x) = \lim_{x\to 2}(Ax - 3) = A \lim_{x\to 2} x - \lim_{x\to 2} 3 = 2A - 3$. The $\lim_{x\to 2} f(x) = 9$ exists when $2A - 3 = 9$ or $A = 6$. Thus $f(x)$ is continuous at $x = 2$ only when $A = 6$.

57. If $f(x) = \begin{cases} Ax^2 + 5x - 9 & \text{if } x < 1 \\ B & \text{if } x = 1 \\ (3-x)(A-2x) & \text{if } 1 < x \end{cases}$ then $f(x)$ is continuous everywhere except possibly at $x = 1$ since $Ax^2 + 5x - 9$ and $(3-x)(A-2x)$ are polynomials. Since $f(1) = B$, A and B need to be determined so that $\lim_{x\to 1} f(x) = B$. As x approaches 1 from the right, $\lim_{x\to 1}(3-x)(A-2x) = \lim_{x\to 1}(3-x)\lim_{x\to 1}(A-2x) = (3-1)(A-2) = 2(A-2)$ and as x approaches 1 from the left, $\lim_{x\to 1} f(x) = \lim_{x\to 1}(Ax^2+5x-9) = A(\lim_{x\to 1} x)^2 + 5\lim_{x\to 1} x - \lim_{x\to 1} 9 = A(1)^2 + 5(1) - 9 = A - 4$. The $\lim_{x\to 1} f(x)$ exists whenever $2(A-2) = A - 4$, $2A - 4 = A - 4$, or $A = 0$. When $A = 0$, $\lim_{x\to 1} f(x) = -4$ and thus, to have continuity at $x = 1$, $f(1) = B = -4$. The function is continuous for all x when $A = 0$ and $B = -4$.

59. On the open interval $0 < x < 1$, $f(x) = x(1 + \frac{1}{x}) = x + 1$ since $x \neq 0$. Thus, $f(x)$, a

polynomial on $0 < x < 1$, is continuous.
The function $f(x) = x(x + \frac{1}{x})$ is not continuous at $x = 0$ since $f(0)$ is not defined. The function $f(x)$ is defined at $x = 1$ since $f(1) = 1(1 + \frac{1}{1}) = 2$ and as x approaches 1 from the left $\lim_{x\to 1}(1 + \frac{1}{x}) = (\lim_{x\to 1} x)[\lim_{x\to 1}(1 + \frac{1}{x})] = 1(1 + \frac{1}{1}) = 2$. The function $f(x)$ is thus continuous on $0 < x \le 1$ (but not on $0 \le x \le 1$).

61. On the open interval $0 < x < 3$
$f(x) = \begin{cases} x(x-1) & \text{if } x < 3 \\ \frac{x^2-9}{x-3} & \text{if } 3 \le x \end{cases}$ can be rewritten as $f(x) = x(x-1) = x^2 - x$, a polynomialfunction and thus continuous on $0 < x < 3$.
On the closed interval, at the left hand end point $f(0) = 0(0-1) = 0$ and as x approaches 0 from the right $\lim_{x\to 0} f(x) = \lim_{x\to 0} x(x-1) = (\lim_{x\to 0} x)[\lim_{x\to 0}(x-1)] = 0(0-1) = 0$. Thus $f(x)$ is continuous at the left hand end point $x = 0$.
At the right hand end point $f(3) = \frac{3^2-9}{3-3} = \frac{0}{0}$ and $f(x)$ is not continuous at the right hand end point $x = 3$ since $f(x)$ is not defined there.

63. By hypothesis Tom's weight, $f(x)$, is continuous on $[a, b] = [0, 20]$. With $L = 100$, $f(0) \le 100 \le f(20)$ (or $f(0) \ge 100 \ge f(20)$) then there is at least one number c between 0 and 20 such that $f(c) = 100$. We'll assume that $f(0) \approx 20$ pounds at birth. $f(20) = 150$.

65. Let $f(x) = \sqrt[3]{x} - (x^2 + 2x - 1)$. $f(x)$ is continuous at all x and $f(0) = 1$, $f(1) = 1 - (1 + 2 - 1) = -1$. By problem 62, there is at least one number $0 \le c \le 1$ such that $f(c) = 0$, and $x = c$ is a solution.

Review Problems

1. a) The domain of the quadratic function $f(x) = x^2 - 2x + 6$ consists of all real numbers x.
b) Since division by zero is not possible, the domain of the rational function $f(x) = \frac{x-3}{x^2+x-2} = \frac{x-3}{(x+2)(x-1)}$ consists of all real numbers x except $x = -2$ and $x = 1$.
c) Since negative numbers do not have square roots, the domain of the function $f(x) = \sqrt{x^2 - 9} = \sqrt{(x+3)(x-3)}$ consists of all values of x for which the product $(x+3)(x-3) \ge 0$, that is for $x \le 3$ or $x \ge 3$. Another way of describing this domain is $|x| \ge 3$.

2. The price x months from now is $P(x) = 40 + \frac{30}{x+1}$ dollars. Hence:

a) The price 5 months from now is $P(5) = 40 + \frac{30}{5+1} = \$45$.
b) The change in price during the 5th month is $P(5) - P(4) = (40 + \frac{30}{5+1}) - (40 + \frac{30}{4+1}) = 45 - 46 = -1$. That is the price decreases by \$1.
c) The price will be \$43 when $43 = 40 + \frac{30}{x+1}$, $3 = \frac{30}{x+1}$, $3(x+1) = 30$, or $x = 9$, that is 9 months from now.
d) Since the term $\frac{30}{x+1}$ gets very small as x gets very large, the price $P(x) = 40 + \frac{30}{x+1}$ will approach \$40 in the long run.

3. a) If $g(u) = u^2+2u+1$ and $h(x) = 1-x$, then $g[h(x)] = g(1-x) = (1-x)^2+2(1-x)+1 = x^2 - 4x + 4$.
b) If $g(u) = \frac{1}{2u+1}$ and $h(x) = x + 2$, then $g[h(u)] = g(x+2) = \frac{1}{2(x+2)+1} = \frac{1}{2x+5}$.
c) If $g(u) = \sqrt{1-u}$ and $h(x) = 2x + 4$, then $g[h(x)] = g(2x+4) = \sqrt{1-(2x+4)} = \sqrt{-2x-3}$.

4. a) If $f(x) = x^2 - x + 4$, then $f(x-2) = (x-2)^2 - (x-2) + 4 = x^2 - 5x + 10$.
b) If $f(x) = \sqrt{x} + \frac{2}{x-1}$, then $f(x^2+1) = \sqrt{x^2+1} + \frac{2}{(x^2+1)-1} = \sqrt{x^2+1} + \frac{2}{x^2}$.
c) If $f(x) = x^2$, then $f(x+1) - f(x) = (x+1)^2 - x^2 = x^2 + 2x + 1 - x^2 = 2x + 1$.

5. a) $f(x) = (x^2+3x+4)^5$ can be written as $g[h(x)]$, where $g(u) = u^5$ and $h(x) = x^2+3x+4$.
b) $f(x) = (3x+1)^2 + \frac{5}{2(3x+2)^3}$ can be written as $g[h(x)]$, where $g(u) = u^2 + \frac{5}{2(u+1)^3}$ and $h(x) = 3x + 1$.

6. a) Since the smog level Q is related to the variable p by the equation $Q(p) = \sqrt{0.5p + 19.4}$ and the variable p is related to the variable t by the equation $p(t) = 8 + 0.2t^2$ it follows that the composite function $Q[p(t)] = \sqrt{0.5(8 + 0.2t^2) + 19.4} = \sqrt{4 + 0.1t^2 + 19.4} = \sqrt{23.4 + 0.1t^2}$ expresses the smog level as a function of the variable t.
b) The smog level 3 years from now will be $Q[p(3)] = \sqrt{23.4 + 0.1(3^2)} = \sqrt{24.3} \approx 4.93$ units.
c) Set $Q[p(t)]$ equal to 5 and solve for t to get $5 = \sqrt{23.4 + 0.1t^2}$, $25 = 23.4+0.1t^2$, $1.6 = 0.1t^2$, $t^2 = \frac{1.6}{0.1} = 16$ or $t = 4$ years from now.

7. If the graph of $y = 3x^2 - 2x + c$ passes through the point $(2, 4)$, then $y = 4$ and $x = 2$ so that $4 = 3(2)^2 - 2(2) + c$, $4 = 12 - 4 + c$, or $c = -4$.

8. a) Some points on the graph of $f(x) = x^2 + 2x - 8 = (x+4)(x-2)$ are shown below.

x	-6	-5	-4	-3	-1	0	1	2	3	4
$f(x)$	16	7	0	-5	-9	-8	-5	0	7	16

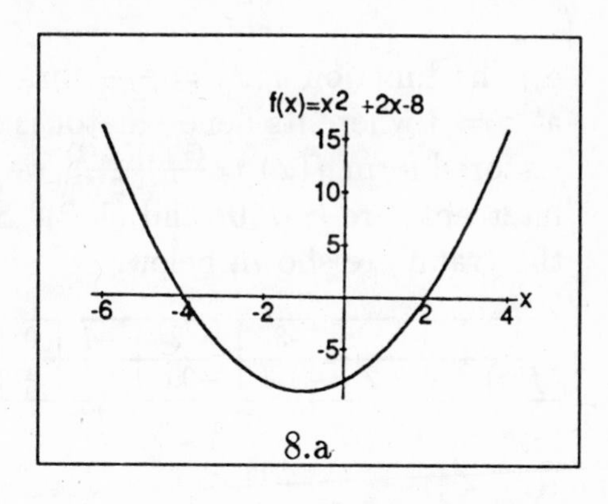

8.a

b) Some points on the graph of $f(x) = \begin{cases} \frac{1}{x^2} & \text{if } x < 0 \\ x & \text{if } 0 \le x < 3 \\ 4 & \text{if } 3 \le x \end{cases}$ are shown here.

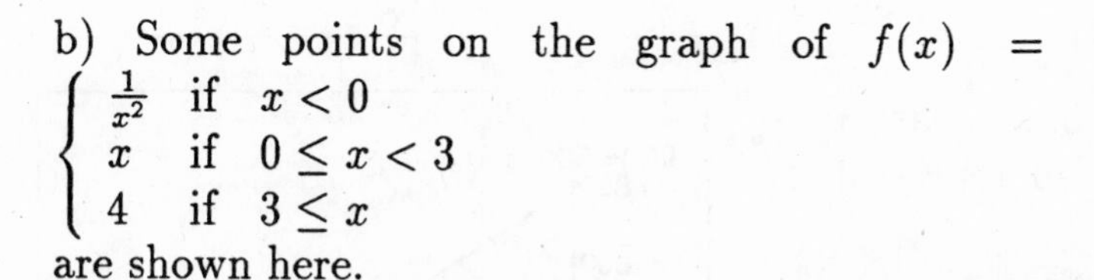

x	-3	-2	-1	-0.5	0	1	2	3	4	5	6
$f(x)$	$\frac{1}{9}$	$\frac{1}{4}$	1	4	0	1	2	4	4	4	4

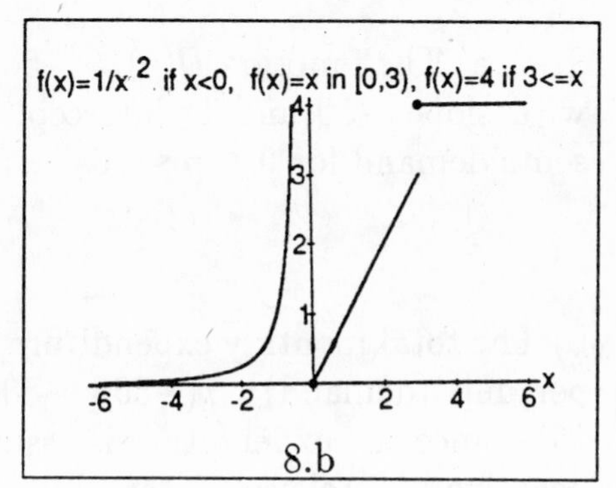

8.b

c) Since $x = 3$ is not in the domain of $f(x) = -\frac{2}{x-3}$, the graph breaks into two pieces at $x = 3$. Some points on the graph are shown below.

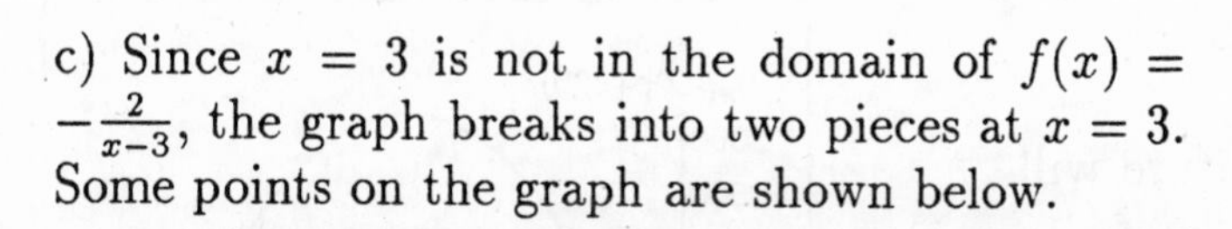

x	-97	1	2	2.99	3.01	4	5	103
$f(x)$	0.02	1	2	200	-200	-2	-1	-0.02

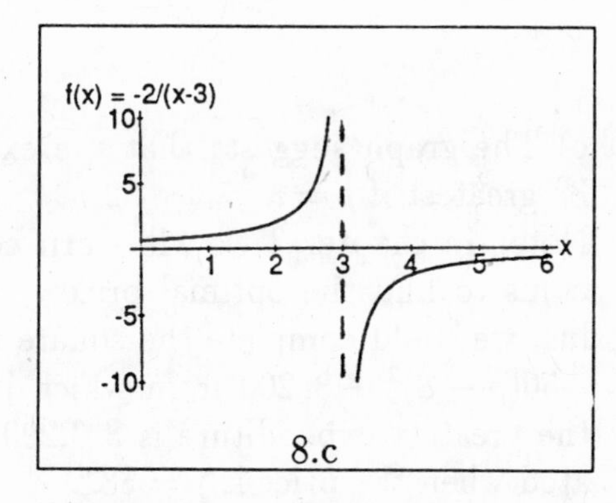

8.c

d) Since $x = 0$ is not in the domain of $f(x) = x + \frac{2}{x}$, the graph breaks into two pieces at $x = 0$. Some points on the graph are shown below.

x	± 0.01	± 1	± 2	± 4	± 100
$f(x)$	± 200.01	± 3	± 3	± 4.5	± 100.02

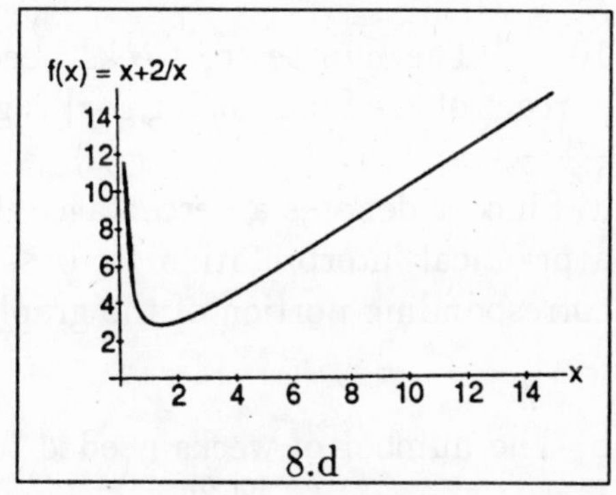

8.d

e) The function $f(x) = \frac{x^2-1}{x-3}$ has a discontinuity at $x = 3$ where its denominator is zero. From the factored form $f(x) = \frac{(x+1)(x-1)}{x-3}$ we see that the x-intercepts are $(-1, 0)$ and $(1, 0)$. Some points on the graph are shown below.

x	-50	-3	-2	-1	0	1	2	3	4	50
$f(x)$	-47.2	-1.3	-0.6	0	$\frac{1}{3}$	0	-3	∞	15	53.2

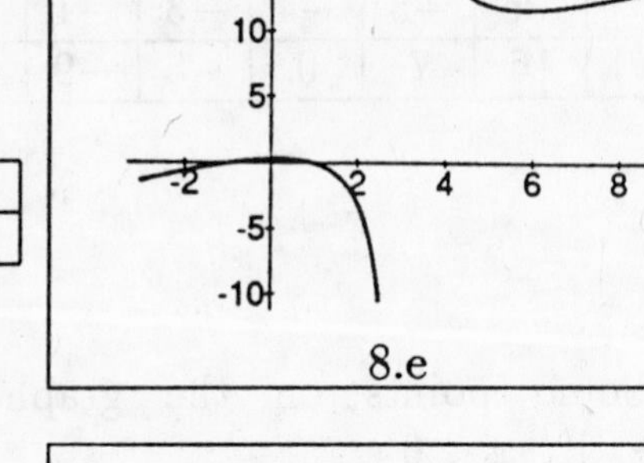

8.e

9. a) The function $D(p) = -50p + 800$ is linear with slope -50 and y–intercept 800 and represents demand for $0 \leq p \leq 16$.

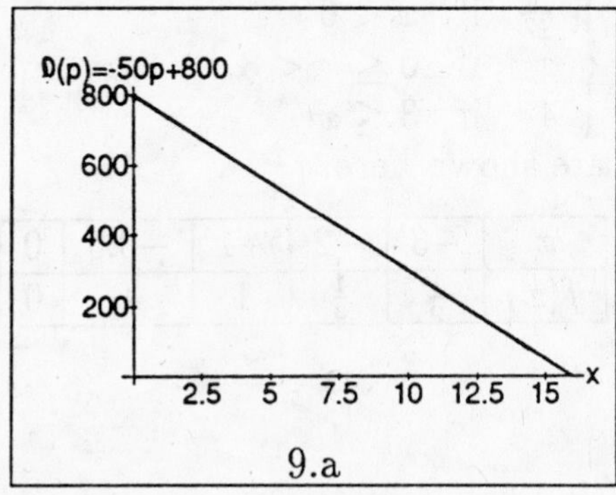

9.a

b) The total monthly expenditure is $E(p) =$(price per unit)(demand)$= p(-50p + 800) = -50p(p - 16)$. Since the expenditure is assumed to be non-negative, the relevant interval is $0 \leq p \leq 16$.

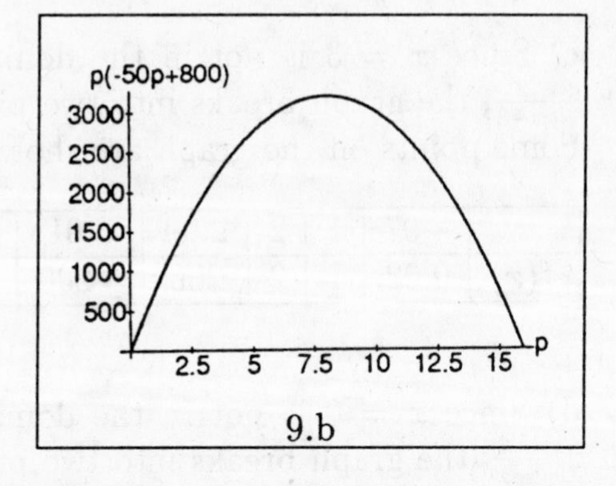

9.b

c) The graph suggests that the expenditure will be greatest if $p = 8$.
Note: In chapter 3 we will learn how to use calculus to find the optimal price. Without calculus, we could complete the square to get $E(p) = -50(p - 8)^2 + 3,200$ from which it is clear that the greatest expenditure is \$ 3,200 and is generated when the price is $p = \$8$.

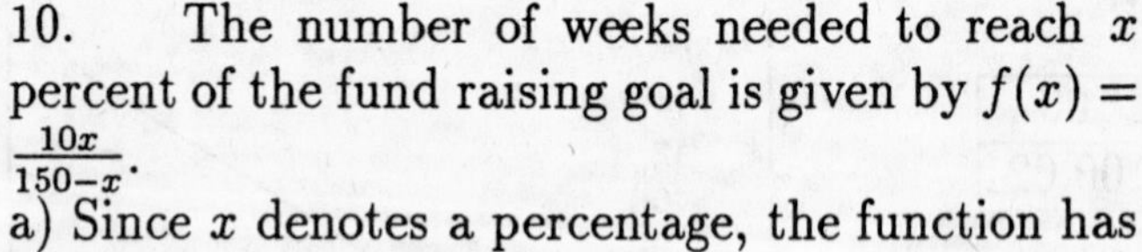

10. The number of weeks needed to reach x percent of the fund raising goal is given by $f(x) = \frac{10x}{150-x}$.
a) Since x denotes a percentage, the function has a practical interpretation for $0 \leq x \leq 100$. The corresponding portion of the graph is sketched.

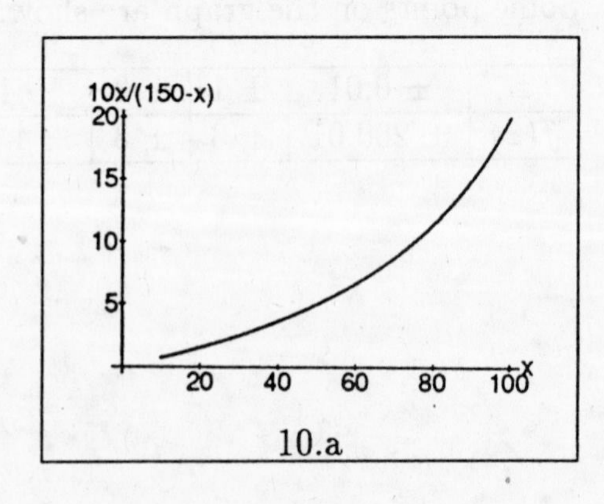

10.a

b) The number of weeks needed to reach 50 percent of the goal is $\frac{10(50)}{150-50} = 5$.

c) The number of weeks needed to reach 100 percent of the goal is $f(100) = \frac{10(100)}{150-100} = 20$.

11. a) If $y = 3x + 2$, $m = 3$ and $b = 2$.

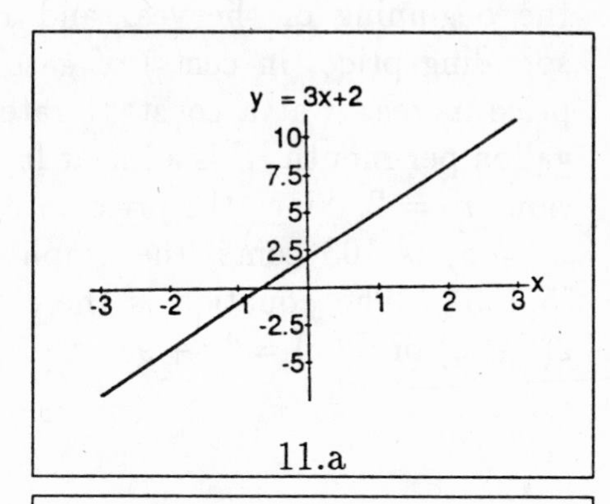

11.a

b) If $5x - 4y = 20$, then $y = \frac{5}{4}x - 5$, and so $m = \frac{5}{4}$ and $b = -5$.

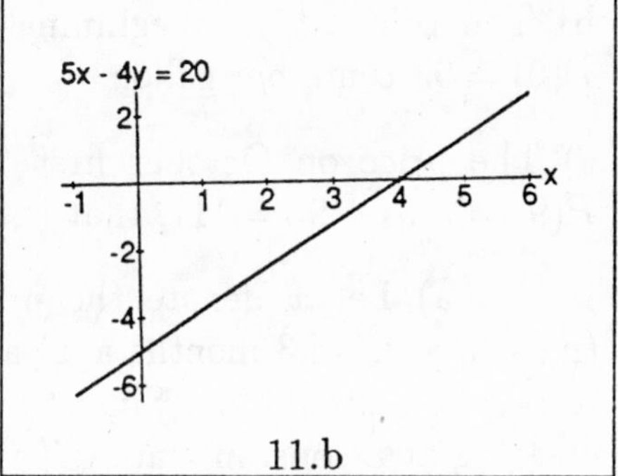

11.b

c) If $2y + 3x = 0$, then $y = -\frac{3}{2}x + 0$, and so $m = -\frac{3}{2}$ and $b = 0$.

d) If $\frac{x}{3} + \frac{y}{2} = 4$, then $\frac{y}{2} = -\frac{1}{3}x + 4$ or $y = -\frac{2}{3}x + 8$, and so $m = -\frac{2}{3}$ and $b = 8$.

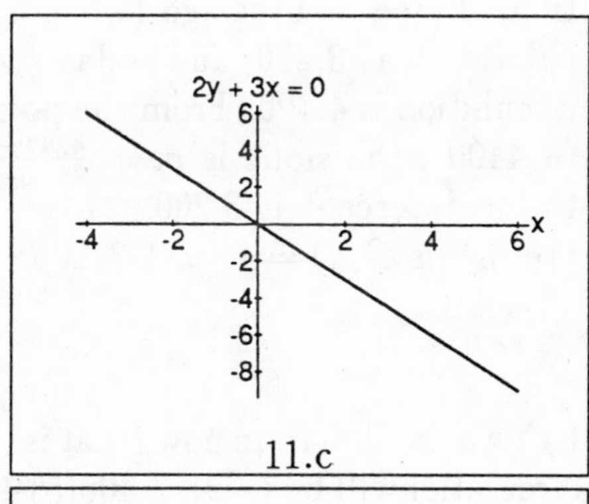

11.c

12. Since $m = 5$ and $b = -4$, the equation of the line is $y = 5x - 4$.

13. Since $m = -2$ and the point $(1, 3)$ is on the line, the equation of the line is $y - 3 = -2(x - 1)$ or $y = -2x + 5$.

14. From the points $(2, 4)$ and $(1, -3)$, the slope is $m = \frac{-3-4}{1-2} = 7$ and with $P = (2, 4)$, the equation of the line is $y - 4 = 7(x - 2)$ or $y = 7x - 10$. The same equation would result if $P(1, -3)$ were used.

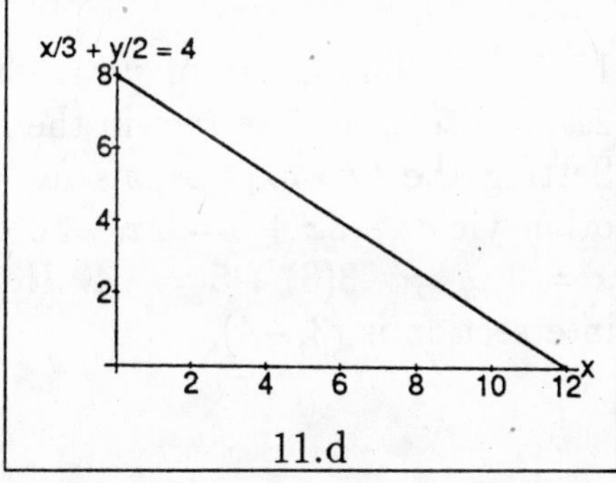

11.d

15. a) Let x denote the time in months since the beginning of the year and $P(x)$ the corresponding price (in cents) of gasoline. Since the price increases at a constant rate of 2 cents per gallon per month, P is a linear function of x with slope $m = 2$. Since the price on June first (when $x = 5$) is 103 cents, the graph passes through $(5, 103)$. The equation is therefore $P - 103 = 2(x - 5)$ or $P(x) = 2x + 93$.

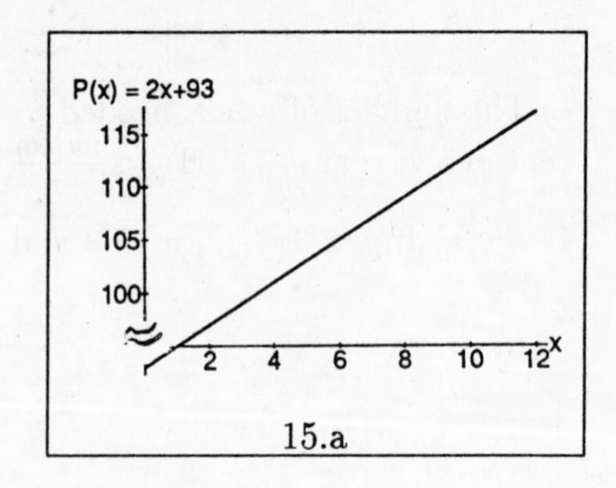

15.a

b) The price at the beginning of the year was $P(0) = 93$ cents per gallon.

c) The price on October first (when $x = 9$) is $P(9) = 2(9) + 93 = 111$, that is, \$1.11 per gallon.

16. a) Let x denote the number of months (measured from 3 months ago) and $C(x)$ the corresponding circulation. Since the circulation is increasing at a constant rate, $C(x)$ is a linear function. Three months ago (when x was 0), the circulation was 3,200, and today (when $x = 3$), the circulation is 4,400. From the points $(0, 3200)$ and $(3, 4400)$, the slope is $m = \frac{4{,}400-3{,}200}{3-0} = 400$ and the y–intercept is 3,200. Hence the equation of the line is $C(x) = 400x + 3,200$.

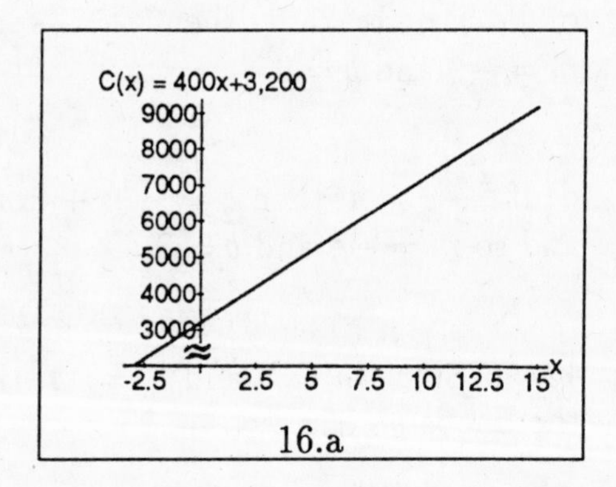

16.a

b) Two months from now (that is, when $x = 5$) the circulationwill be $C(x) = 400(5) + 3,200 = 5,200$.

17. a) The graph of $y = -3x + 5$ and $y = 2x - 10$ seem to intersect in the fourth quadrant. Setting the two expressions for y equal to each other yields $-3x + 5 = 2x - 10$ or $x = 3$. When $x = 3$, $y = -3(3) + 5 = -4$. Hence the point of intersection is $(3, -4)$.

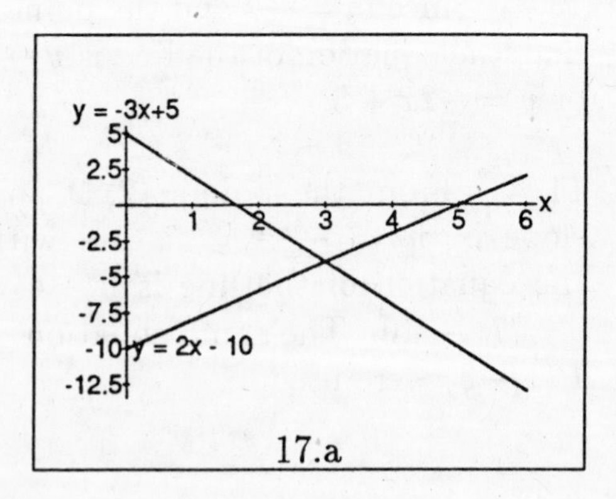

17.a

b) The graphs of $y = x + 7$ and $y = -2 + x$ are parallel lines with slope 1. Hence there are no points of intersection.

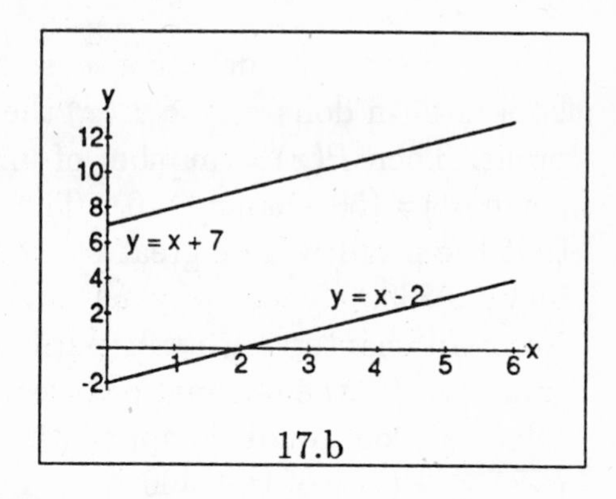

17.b

c) The graphs of $y = x^2 - 1$ and $y = 1 - x^2$ seem to intersect at $(-1, 0)$ and $(1, 0)$. Setting the two expressions for y equal to each other yields $x^2 - 1 = 1 - x^2$, $x^2 = 1$, or $x = \pm 1$. When $x = \pm 1$, $y = (\pm 1)^2 - 1 = 0$. Hence the points of intersection are $(-1, 0)$ and $(1, 0)$.

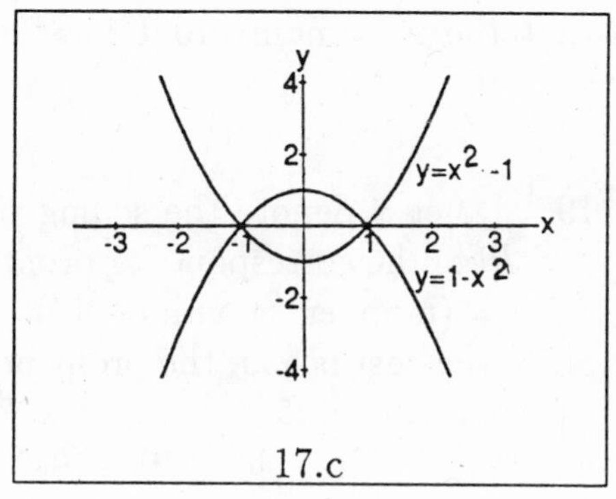

17.c

d) The graphs of $y = x^2$ and $y = 15 - 2x$ seem to intersect in the first and second quadrants. Setting the two expressions for y equal to each other yields $x^2 = 15 - 2x$, $x^2 + 2x - 15 = 0$, $(x + 5)(x - 3) = 0$, or $x = -5$ and $x = 3$. When $x = -5$, $y = 25$, and when $x = 3$, $y = 9$. Hence the points of intersection are $(-5, 25)$ and $(3, 9)$.

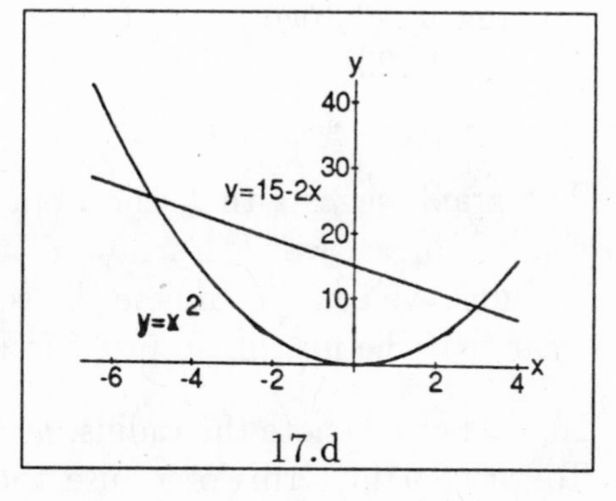

17.d

e) The graphs of $y = \frac{24}{x^2}$ and $y = 3x$ seem to intersect in the first quadrant. Setting the two expressions for y equal to each other yields $\frac{24}{x^2} = 3x$, $3x^3 = 24$, $x^3 = 8$ or $x = 2$. When $x = 2$, $y = 3(2) = 6$. Hence the point of intersection is $(2, 6)$.

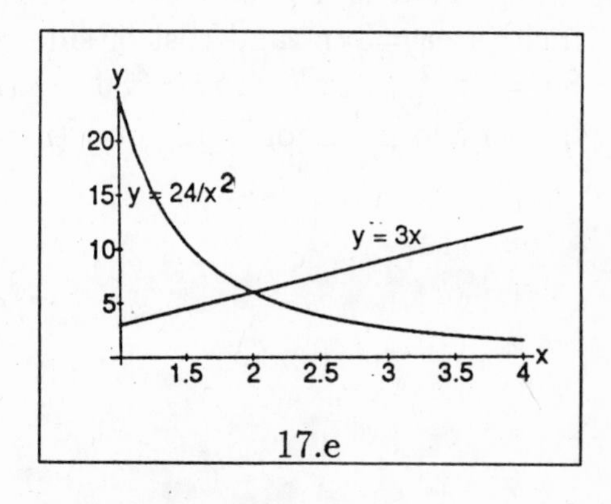

17.e

18. Let $P(x)$ denote the selling price of a bookcase (in dollars) and $P(x)$ the corresponding profit. Then $P(x)$ =(number of units sold)(profit per unit)= $(50 - x)(x - 10)$. The graph suggests that the profit will be greatest when x is approximately \$30.
Note: In chapter 3 we will learn how to use calculus to find the optimal price exactly. Without calculus, you could complete the square to get $P(x) = -(x - 30)^2 + 400$ from which it is clear that the maximum profit is \$400 and occurs at $x = 30$.

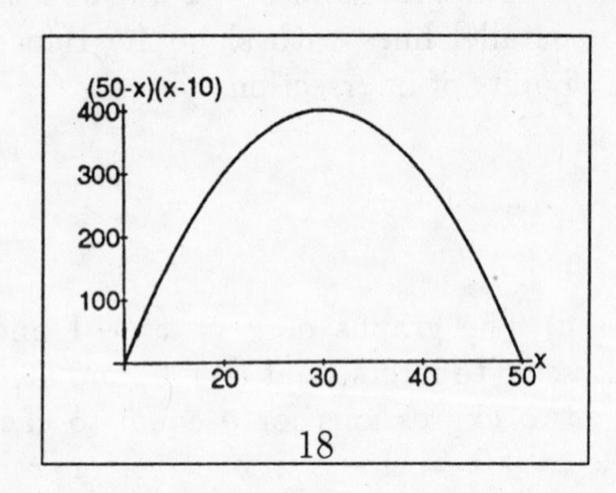

18

19. Let x denote the selling price (in dollars) and $P(x)$ the corresponding profit function. Then $P(x)$ =(number of units sold)(profit per unit). Since the cost is \$50, the profit per unit is $x - 50$. The number of \$5 reductions is $\frac{80-x}{5}$, and so the number of units sold is $40 + 10(\frac{80-x}{5}) = 200 - 2x$. Putting it all together, $P(x) = (200 - 2x)(x - 50) = -2(100 - x)(50 - x)$.

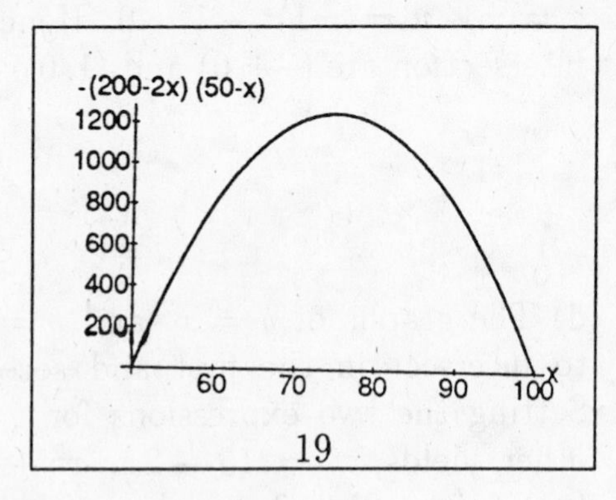

19

The graph sugests that the profit will be greatest when x is approximately \$75. Note: In chapter 3, we will learn how to use calculus to find the optimal price exactly. Without calculus, we could complete the square to get $P(x) = -2(x - 75)^2 + 1,250$ from which it is clear that the maximum profit is \$1,250 and occurs at $x = 75$.

20. Let r denote the radius, h the height, and V the volume of the can. Then, $V = \pi r^2 h$. To write h in terms of r, use the fact that the cost of constructing the can is to be 80 cents. That is, 80=cost of bottom + cost of side where cost of bottom = (cost per square inch)(area)=$3\pi r^2$ and cost of side = (cost per square inch)(area)= $2(2\pi rh) = 4\pi rh$. Hence $80 = 3\pi r^2 + 4\pi rh$, $4\pi rh = 80 - 3\pi r^2$ or $h = \frac{20}{\pi r} - \frac{3r}{4}$. Now substitute this expression for h into the formula for V to get $V(r) = \pi r^2(\frac{20}{\pi r} - \frac{3r}{4}) = 20r - \frac{3}{4}\pi r^3$.

21. Let x denote the number of machines used and $C(x)$ the corresponding cost function. Then, $C(x)$ =(set up cost)+(operating cost)= 80(number of machines)+5.76(number of hours) Since 400,000 medals are to be produced and each of the x machines can produce 200 medals per hour, number of hours $= \frac{400{,}000}{200x} = \frac{2{,}000}{x}$.

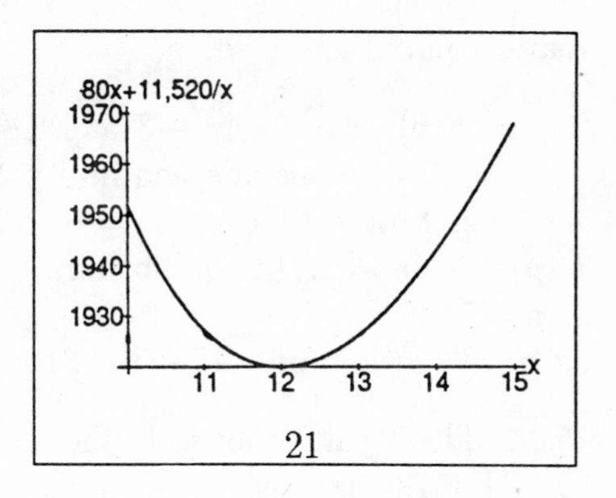

21

Putting it all together, $C(x) = 80x + 5.76(\frac{2{,}000}{x}) = 80x + \frac{11{,}520}{x}$. The graph suggests that the cost will be smallest when x is approximately 12. In chapter 3 you will learn how to use calculus to find the optimal number of machines exactly.

22. Let x denote time measured in half-hour units. The corresponding costs for the first and second plumbers, respectively, are $C_1(x) = 25 + 16x$ and $C_2(x) = 31 + 14x$. Setting the two cost functions equal to each other yields $31 + 14x = 25 + 16x$ or $x = 3$.

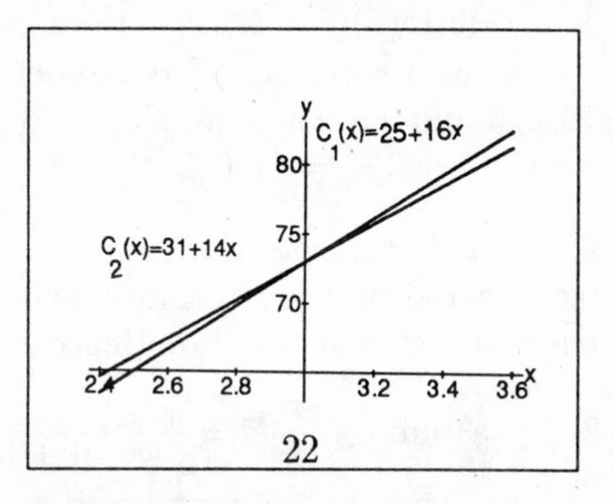

22

Thus, if a plumber is to be used for exactly 3 half-hours, $(1\frac{1}{2})$, the two plumbers charge the same. If a plumber is needed for fewer than $1\frac{1}{2}$ hours, the first plumber is the less expensive since the graph of C_1 is initially below that of C_2, and if a plumber is needed for more than $1\frac{1}{2}$ hours, the second plumber is the less expensive.

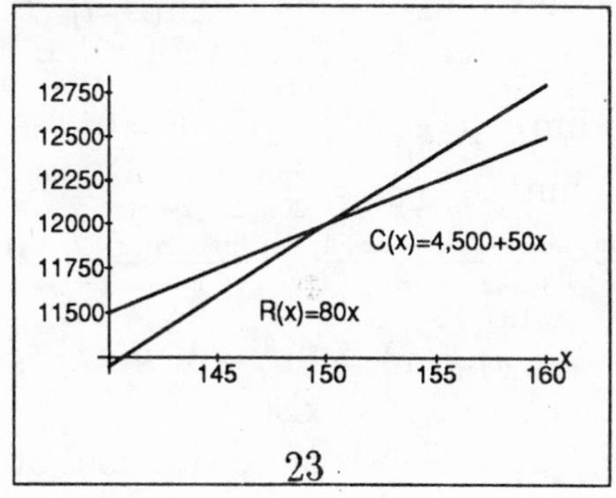

23

23. a) Let x denote the number of units manufactured and sold. $C(x)$ and $R(x)$ are the corresponding cost and revenue functions, respectively. Then $C(x) = 4,500 + 50x$ and $R(x) = 80x$. For the manufacturer to break even, the cost must be equal to the revenue. That is $4,500 + 50x = 80x$ or $x = 150$ units.

b) Let $P(x)$ denote the profit from the manufacture and sale of x units. Then, $P(x) = R(x) - C(x) = 80x - (4,500 + 50x) = 30x - 4,500$. When 200 units are sold, the profit is $P(200) = 30(200) - 4,500 = \$1,500$.

c) The profit will be \$900 when $900 = 30x - 4,500$ or $x = 180$, that is, when 180 units are manufactured and sold.

24. $S(p) = p^2 + Ap - 3$ and $D(p) = Cp + 32$ with $3 \le p \le 8$. For $p < 3$ we'll force $S(p) = 0$ (let's assume continuity and make $S(3) = 0$) so $9 + 3A - 3 = 0$ or $3A = -6$, $A = -2$. Similarly let $D(8) = 0$ so $8C + 32 = 0$ or $C = -4$. Thus $S(p) = p^2 - 2p - 3$ and $D(p) = -4p+32$. At equilibrium, $p^2-2p-3 = -4p+32$, $p^2+2p-35 = 0$, $(p+7)(p-5) = 0$, or $p = 5$.
At $p = 6$, $S(6) = 36 - 12 - 3 = 21$ and $D(6) = -24 + 32 = 8$. The difference is $21 - 8 = \$13$.

25. The royalties for publisher A are given by
$$P_a = \begin{cases} 0.12(5)x & \text{if } 0 < x \le 30,000 \\ 0.17(5)(x - 30,000) & \text{if } x \ge 30,000 \end{cases}$$
Similarly for publisher B,
$$P_b = \begin{cases} 0 & \text{if } 0 < x \le 4,000 \\ 0.15(6)(x - 4,000) & \text{if } x \ge 4,000 \end{cases}$$
Publisher A certainly offers the better deal if $x \le 4,000$. If $x = 30,000$, $P_a = 0.6(30,000) = 18,000$ and $P_b = 0.9(26,000) = 23,400$. For the break-even point $0.9(x - 4,000) = 0.6x$, $0.3x = 3,600$, $x = 12,000$.

26. Let x denote the number of relevant facts recalled, n the total number of relevant facts in the person's memory, and $R(x)$ the rate of recall. Then $n - x$ is the number of relevant facts not recalled. Hence $R(x) = k(n - x)$, where k is a constant of proportionality.

27. a) $\lim_{x\to2} \frac{x^2-3x}{x+1} = \frac{\lim_{x\to2}(x^2-3x)}{\lim_{x\to2}(x+1)} = \frac{4-6}{3} = -\frac{2}{3}$.
b) $\lim_{x\to1} \frac{x^2+x-2}{x^2-1} = \frac{\lim_{x\to1}(x^2+x-2)}{\lim_{x\to1}(x^2-1)} = \frac{0}{0}$. Simplifying before taking limits leads to $\lim_{x\to1} \frac{x^2+x-2}{x^2-1} = \frac{\lim_{x\to1}(x+2)(x-1)}{\lim_{x\to1}(x+1)(x-1)} = \frac{\lim_{x\to1}(x+2)}{\lim_{x\to1}(x+1)} = \frac{3}{2}$.
c) $\lim_{x\to1}(\frac{1}{x^2} - \frac{1}{x}) = 1 - 1 = 0$.
d) $\lim_{x\to2} \frac{x^3-8}{2-x} = \frac{\lim_{x\to2}(x^3-8)}{\lim_{x\to2}(2-x)} = \frac{0}{0}$. Simplifying before taking limits leads to $\lim_{x\to2} \frac{x^3-8}{2-x} = \frac{\lim_{x\to2}(x-2)(x^2+2x+4)}{\lim_{x\to2} -(x-2)} = \frac{\lim_{x\to2}(x^2+2x+4)}{\lim_{x\to2}(-1)} = -\lim_{x\to2}(x^2 + 2x + 4) = -12$.

28. a) $f(x) = 5x^3 - 3x + \sqrt{x}$ is not continuous for $x < 0$ since square roots of negative numbers do not exist.
b) $f(x) = \frac{x^2-1}{x+3}$ is not continuous at $x = -3$ since $f(-3) = \frac{10}{0}$ and division by 0 is undefined.
c) $g(x) = \frac{x^3+5x}{(x-2)(2x+3)}$ is not continuous at $x = 2$ and $x = -\frac{3}{2}$ since $g(2) = \frac{18}{0}$ and $g(-\frac{3}{2}) = \frac{-87/8}{0}$ and division by 0 is not defined.
d) $$h(x) = \begin{cases} x^3 + 2x - 33 & \text{if } x < 3 \\ \frac{x^2-6x+9}{x-3} & \text{if } 3 \le x \end{cases}$$
is not continuous at $x = 3$ since $h(3) = \frac{9-18+9}{3-3} = \frac{0}{0}$ and division by 0 is undefined.

29. a) $f(x) = \begin{cases} 2x+3 & \text{if } x < 1 \\ Ax-1 & \text{if } 1 \le x \end{cases}$

Then $f(x)$ is continuous everywhere except possibly at $x = 1$ since $2x + 3$ and $Ax - 1$ are polynomials. Since $f(1) = A - 1$, in order that $f(x)$ be continuous at $x = 1$, A must be chosen so that $\lim_{x\to1} f(x) = A - 1$. As x approaches 1 from the right, $\lim_{x\to1} f(x) = \lim_{x\to1}(Ax-1) = A-1$ and as x approaches 1 from the left, $\lim_{x\to1} f(x) = \lim_{x\to1}(2x+3) = 5$. $\lim_{x\to1} f(x)$ exists whenever $A-1 = 5$ or $A = 6$ and furthermore, for $A = 6$, $\lim_{x\to1} f(x) = 5$, $f(1) = 6 - 1 = 5$. Thus, $f(x)$ is continuous at $x = 1$ only when $A = 6$.

b) $f(x) = \begin{cases} \frac{x^2-1}{x+1} & \text{if } x < -1 \\ Ax^2+x-3 & \text{if } -1 \le x \end{cases}$

Then $f(x)$ is continuous everywhere except possibly at $x = -1$ since $\frac{x^2-1}{x+1}$ is a rational function and $Ax^2 + x - 3$ is a polynomial. Since $f(-1) = A - 4$, in order that $f(x)$ be continuous at $x = -1$, A must be chosen so that $\lim_{x\to-1} f(x) = A-4$. As x approaches -1 from the right, $\lim_{x\to-1} f(x) = \lim_{x\to-1}(Ax^2+x-3) = A-4$ and as x approaches -1 from the left, $\lim_{x\to-1} f(x) = \lim_{x\to-1} \frac{x^2-1}{x+1} = \lim_{x\to-1} \frac{(x+1)(x-1)}{x+1} = \lim_{x\to-1}(x-1) = -2$. $\lim_{x\to-1} f(x)$ exists whenever $A - 4 = -2$ or $A = 2$, and furthermore, for $A = 2$ $\lim_{x\to-1} f(x) = -2$ and $f(-1) = 2 - 4 = -2$. Thus $f(x)$ is continuous at $x = -1$ only when $A = 2$.

30. a) The population decreases when $5 \le t$. $-8t + 72 = 0$ when $t = 9$.

b) $f(2) = 5$ and $f(7) = -56 + 72 = 16$. Since $\lim_{x\to5} f(x) = 25 + 1 = 26$, and $5 < 10 < 26$, by the intermediate value property there exists a value $2 < c < 5 < 7$ such that $f(c) = 10$.

Chapter 2

Differentiation: Basic Concepts

2.1 The Derivative

1. If $f(x) = 5x - 3$, then $f(x + \Delta x) = 5(x + \Delta x) - 3$. The difference quotient DQ is $DQ = \frac{f(x+\Delta x)-f(x)}{\Delta x} = \frac{[5(x+\Delta x)-3]-[5x-3]}{\Delta x} = \frac{5x+5\Delta x-3-5x+3}{\Delta x} = \frac{5\Delta x}{\Delta x} = 5$. Now $\lim_{\Delta x\to 0} \frac{f(x+\Delta x)-f(x)}{\Delta x} = 5$. The slope is $m = f'(2) = 5$.

3. If $f(x) = 2x^2 - 3x + 5$, then $f(x + \Delta x) = 2(x + \Delta x)^2 - 3(x + \Delta x) + 5$. The difference quotient DQ is $DQ = \frac{f(x+\Delta x)-f(x)}{\Delta x} = \frac{[2(x+\Delta x)^2-3(x+\Delta x)+5]-[2x^2-3x+5]}{\Delta x} = \frac{2x^2+4x\Delta x+2(\Delta x)^2-3x-3\Delta x+5-2x^2+3x-5}{\Delta x} = \frac{4x\Delta x+2(\Delta x)^2-3\Delta x}{\Delta x} = 4x+2\Delta x-3$. $\lim_{\Delta x\to 0} \frac{f(x+\Delta x)-f(x)}{\Delta x} = 4x - 3$. The slope is $m = f'(0) = -3$.

5. If $f(x) = \frac{2}{x}$, then $f(x + \Delta x) = \frac{2}{x+\Delta x}$. The difference quotient DQ is $DQ = \frac{f(x+\Delta x)-f(x)}{\Delta x} = \frac{\frac{2}{x+\Delta x}-\frac{2}{x}}{\Delta x} = [\frac{\frac{2}{x+\Delta x}-\frac{2}{x}}{\Delta x}][\frac{x(x+\Delta x)}{x(x+\Delta x)}] = \frac{2x-2(x+\Delta x)}{\Delta x(x)(x+\Delta x)} = \frac{-2}{x(x+\Delta x)}$. $\lim_{\Delta x\to 0} \frac{f(x+\Delta x)-f(x)}{\Delta x} = -\frac{2}{x^2}$. The slope is $m = f'(\frac{1}{2}) = -8$.

7. If $f(x) = \sqrt{x}$, then $f(x + \Delta x) = \sqrt{x + \Delta x}$. The difference quotient DQ is $DQ = \frac{f(x+\Delta x)-f(x)}{\Delta x} = \frac{\sqrt{x+\Delta x}-\sqrt{x}}{\Delta x} = [\frac{\sqrt{x+\Delta x}-\sqrt{x}}{\Delta x}][\frac{\sqrt{x+\Delta x}+\sqrt{x}}{\sqrt{x+\Delta x}+\sqrt{x}}] = \frac{x+\Delta x-x}{\Delta x(\sqrt{x+\Delta x}+\sqrt{x})} = \frac{1}{\sqrt{x+\Delta x}+\sqrt{x}}$. $\lim_{\Delta x\to 0} \frac{f(x+\Delta x)-f(x)}{\Delta x} = \frac{1}{2\sqrt{x}}$. The slope is $m = f'(9) = \frac{1}{6}$.

9. $f(x) = x^2 + x + 1$. $f(x + \Delta x) = x^2 + 2x\Delta x + (\Delta x)^2 + x + \Delta x + 1$. The difference quotient DQ is $DQ = \frac{f(x+\Delta x)-f(x)}{\Delta x} = \frac{x^2+2x\Delta x+(\Delta x)^2+x+\Delta x+1-(x^2+x+1)}{\Delta x} = \frac{\Delta x(2x+\Delta x+1)}{\Delta x}$. Thus $f'(x) = \lim_{\Delta x\to 0}(2x + \Delta x + 1) = 2x + 1$. The slope is $m = f'(2) = 5$. $f(2) = 7$. $P(2, 7)$ is a point on the curve. The equation of the tangent line is $\frac{y-7}{x-2} = 5$ or $y = 5x - 3$.

11. $f(x) = \frac{3}{x^2}$. $f(x+\Delta x) = \frac{3}{(x+\Delta x)^2}$. $f(x+\Delta x) - f(x) = \frac{3}{(x+\Delta x)^2} - \frac{3}{x^2}$, which three times that of problem 6. Thus $f'(x) = -\frac{6}{x^3}$. The slope is $m = f'(\frac{1}{2}) = -6 \times 2^3 = -48$. $f(\frac{1}{2}) = 12$. $P(\frac{1}{2}, 12)$ is a point on the curve. The equation of the tangent line is $\frac{y-12}{x-1/2} = -48$ or $y = -48x + 36$.

13. $f(x) = x^2 - 3x = x(x-3)$. The lowest point occurs when the derivative (the slope of the tangent line) is zero. To find the derivative: $DQ = \frac{f(x+\Delta x) - f(x)}{\Delta x} = \frac{[(x+\Delta x)^2 - 3(x+\Delta x)] - [x^2 - 3x]}{\Delta x} = \frac{x^2 + 2x\Delta x + (\Delta x)^2 - 3x - 3\Delta x - x^2 + 3x}{\Delta x} = 2x + \Delta x - 3$. Thus $f'(x) = -\lim_{\Delta x \to 0} DQ = 2x - 3$. The derivative is 0 when $2x - 3 = 0$ or $x = \frac{3}{2}$. Since $f(\frac{3}{2}) = -\frac{9}{4}$, the lowest point occurs at $(\frac{3}{2}, -\frac{9}{4})$.

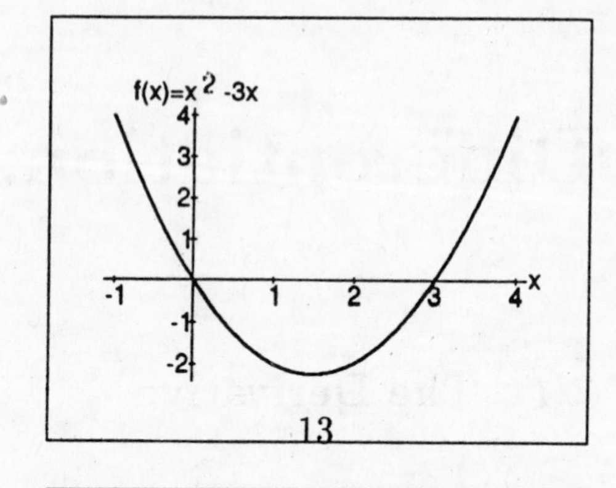

13

15. $f(x) = x^3 + 3x^2 = x^2(x+3)$. $DQ = \frac{f(x+\Delta x) - f(x)}{\Delta x} = \frac{[(x+\Delta x)^3 + 3(x+\Delta x)^2] - [x^3 + 3x^2]}{\Delta x} = \frac{x^3 + 3x^2\Delta x + 3x(\Delta x)^2 + (\Delta x)^3 + 3x^2 + 6x\Delta x + 3(\Delta x)^2 - x^3 - 3x^2}{\Delta x} = \frac{3x^2\Delta x + 3x(\Delta x)^2 + (\Delta x)^3 + 6x\Delta x + 3(\Delta x)^2}{\Delta x}$. Thus $f'(x) = -\lim_{\Delta x \to 0} DQ = 3x^2 + 6x = 3x(x+2)$. The derivative is 0 when $x = 0$ or $x = -2$. $f(0) = 0$ and $f(-2) = -8 + 12 = 4$. $(-2, 4)$ is a maximum and $(0, 0)$ is a minimum.

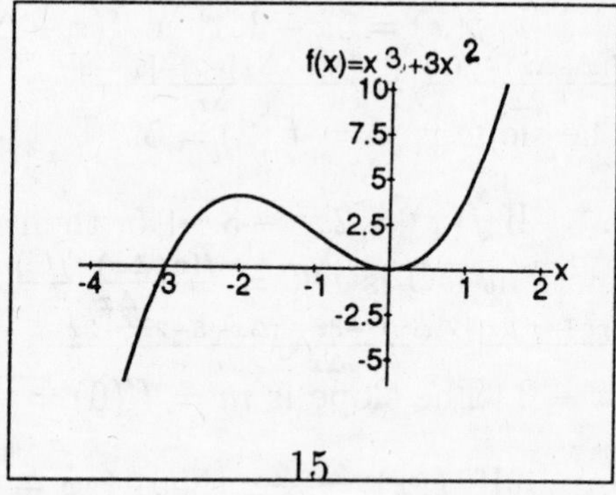

15

17. The number of tape recorders sold is $120 - x$ and the profit per tape recorder $x - 20$ (the sales price minus the cost). The total profit is the product of the profit per tape recorder times the number of tape recorders. $P(x) = (120 - x)(x - 20) = -(x^2 - 140x + 2,400)$. $P(x+\Delta x) - P(x) = -[x^2 + 2x\Delta x + (\Delta x)^2 - 140x - 140\Delta x + 2,400] + x^2 - 140x + 2,400 = -\Delta x(2x + \Delta x - 140)$. $P'(x) = \lim_{\Delta x \to 0}(2x + \Delta x - 140) = -2x + 140$. The slope of the profit curve is 0 when $x = \$70$.

19. a) $f(-2) = 4$, $f(-1.9) = 3.61$, $m_s = \frac{4 - 3.61}{-2 + 1.9} = -3.9$.
b) $f(x) = x^2$. $f(x+\Delta x) = x^2 + 2x\Delta x + (\Delta x)^2$. The difference quotient DQ is $DQ = \frac{f(x+\Delta x) - f(x)}{\Delta x} = \frac{x^2 + 2x\Delta x + (\Delta x)^2 - x^2}{\Delta x} = \frac{\Delta x(2x + \Delta x)}{\Delta x}$. Thus $f'(x) = \lim_{\Delta x \to 0}(2x + \Delta x) = 2x$. The slope is $m_t = f'(-2) = -4$.

21. a) $f(x) = 3x - 2$. $f(x+\Delta x) = 3x + 3\Delta x - 2$. The difference quotient DQ is

$DQ = \frac{f(x+\Delta x)-f(x)}{\Delta x} = \frac{3x+3\Delta x-2-3x+2}{\Delta x} = 3\Delta x$. Thus $f'(x) = 3$.
b) $m = f'(-1) = 3$. The equation of the tangent line is $\frac{y-(-5)}{x+1} = 3$ or $y = 3x - 2$.
c) The tangent to a line at a point on the line is the original line itself.

23. a) $y|_{x+\Delta x} - y|_x = x^2 + 2x\Delta x + (\Delta x)^2 + 3x + 3\Delta x - x^2 - 3x = \Delta x(2x + \Delta x + 3)$.
$y'_a = 2x + 3$.
b) If $y = x^2$ then $y'_{b1} = 2x$ (see problem 12.b). If $y = 3x$ then $y'_{b2} = 3$ (see problem 14.c).
c) $y'_a = y'_{b1} + y'_{b2}$.
d) A possible guess is that $f(x) = g(x) + h(x)$ could lead to $f'(x) = g'(x) + h'(x)$.

25. The difference quotient DQ is the change in f divided by the change in Δx. The derivative f' is the limit of DQ as $\Delta x \to 0$. With $\Delta f > 0$ (and $\Delta x > 0$, as a rule) this means that the function is rising. Similarly if $f'(x) < 0$ the graph of the function falls.

27. $f'(x) = x^2-x-6$ is such a function. $f'(x) = (x+2)(x-3) = x^2-x-6 = 0$ at $x = -2$ and $x = 3$ qualifies for the derivative function. Chapter 5 on antidifferentiation will show how to find a suitable function like $f(x) = \frac{1}{6}(2x^3 - 3x^2 - 36x) + 12$.

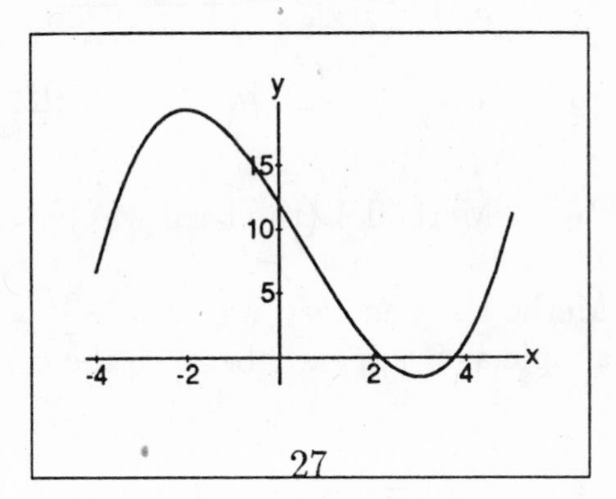

27

2.2 Techniques of Differentiation

1. $y = x^2 + 2x + 3$, $\frac{dy}{dx} = 2x + 2$.

3. $f(x) = x^9 - 5x^8 + x + 12$, $f'(x) = 9x^8 - 40x^7 + 1$.

5. $y = \frac{1}{x}+\frac{1}{x^2}-\frac{1}{\sqrt{x}} = x^{-1}+x^{-2}-x^{-1/2}$, $\frac{dy}{dx} = -x^{-2}+(-2x^{-3})-(-\frac{1}{2}x^{-3/2}) = -\frac{1}{x^2}-\frac{2}{x^3}+\frac{1}{2\sqrt{x^3}}$.

7. $f(x) = \sqrt{x^3} + \frac{1}{\sqrt{x^3}} = x^{3/2} + x^{-3/2}$, $f'(x) = \frac{3}{2}x^{1/2} - \frac{3}{2}x^{-5/2} = \frac{3}{2}\sqrt{x} - \frac{3}{2\sqrt{x^5}}$.

9. $y = -\frac{x^2}{16} + \frac{2}{x} - x^{3/2} + \frac{1}{3x^2} + \frac{x}{3} = -\frac{1}{16}x^2 + 2x^{-1} - x^{3/2} + \frac{1}{3}x^{-2} + \frac{1}{3}x$,

$\frac{dy}{dx} = -\frac{1}{16}(2x) + 2(-x^{-2}) - \frac{3}{2}x^{1/2} + \frac{1}{3}(-2x^{-3}) + \frac{1}{3} = -\frac{1}{8}x - \frac{2}{x^2} - \frac{3}{2}x^{1/2} - \frac{2}{3x^3} + \frac{1}{3}$.

11. $f(x) = (2x+1)(3x-2)$, $f'(x) = (2x+1)\frac{d}{dx}(3x-2) + (3x-2)\frac{d}{dx}(2x+1) = (2x+1)(3) + (3x-2)(2) = 12x - 1$.

13. $y = 10(3x+1)(1-5x)$, $\frac{dy}{dx} = 10\frac{d}{dx}[(3x+1)(1-5x)] = 10[(3x+1)\frac{d}{dx}(1-5x) + (1-5x)\frac{d}{dx}(3x+1)] = 10[(3x+1)(-5) + (1-5x)(3)] = -300x - 20$.

15. $f(x) = \frac{1}{3}(x^5 - 2x^3 + 1)$, $f'(x) = \frac{1}{3}[\frac{d}{dx}(x^5 - 2x^3 + 1)] = \frac{1}{3}(5x^4 - 6x^2)$.

17. $y = \frac{x+1}{x-2}$, $\frac{dy}{dx} = \frac{(x-2)\frac{d}{dx}(x+1) - (x+1)\frac{d}{dx}(x-2)}{(x-2)^2} = \frac{(x-2)(1)-(x+1)(1)}{(x-2)^2} = \frac{-3}{(x-2)^2}$

19. $f(x) = \frac{x}{x^2-2}$, $f'(x) = \frac{(x^2-2)\frac{d}{dx}(x) - x\frac{d}{dx}(x^2-2)}{(x^2-2)^2} = \frac{(x^2-2)(1)-(x)(2x)}{(x^2-2)^2} = \frac{-x^2-2}{(x^2-2)^2}$

21. Method 1 (the hard way):$y = \frac{3}{x+5}$, $\frac{dy}{dx} = \frac{(x+5)\frac{d}{dx}(3) - 3\frac{d}{dx}(x+5)}{(x+5)^2} = \frac{(x+5)(0)-3(1)}{(x+5)^2} = \frac{-3}{(x+5)^2}$.

Method 2 (the easy way):$y = \frac{3}{x+5} = 3(x+5)^{-1}$, $\frac{dy}{dx} = -(x+5)^{-2}$. Warning: This is not as simple as it appears, because the chain rule needs to be taken into account, to be introduced in section 5.

23. $f(x) = \frac{x^2-3x+2}{2x^2+5x-1}$, $f'(x) = \frac{(2x^2+5x-1)\frac{d}{dx}(x^2-3x+2) - (x^2-3x+2)\frac{d}{dx}(2x^2+5x-1)}{(2x^2+5x-1)^2} = \frac{(2x^2+5x-1)(2x-3)-(x^2-3x+2)(4x+5)}{(2x^2+5x-1)^2} = \frac{11x^2-10x-7}{(2x^2+5x-1)^2}$.

25. $y = x^5 - 3x^3 - 5x + 2$ and $P(1,-5)$. $y' = 5x^4 - 9x^2 - 5$. At $x = 1$, $y' = m = 5 - 9 - 5 = -9$. The equation of the tangent line is $\frac{y+5}{x-1} = -9$, $y + 5 = -9x + 9$ or $y = -9x + 4$.

27. $f(x) = \frac{x+1}{x-1}$ and $P(0,-1)$. $f'(x) = \frac{(x-1)(1)-(x+1)(1)}{(x-1)^2} = \frac{x-1-x-1}{(x-1)^2} = -\frac{2}{(x-1)^2}$. $f'(0) = \frac{-2}{1} = -2$. For the tangent line $\frac{y+1}{x-0} = -2$, $y + 1 = -2x$ or $y = -2x - 1$.

29. $f(x) = x^4 - 3x^3 + 2x^2 - 6$. $f(2) = 16 - 24 + 8 - 6 = -6$. $f'(x) = 4x^3 - 9x^2 + 4x$, the slope is $m = f'(2) = 4 \times 8 - 9 \times 4 + 4 \times 2 = 4$ and the equation of the tangent line is $\frac{y+6}{x-2} = 4$, $y + 6 = 4x - 8$ or $y = 4x - 14$.

31. $f(x) = \frac{x^2+2}{x^2-2}$. $f(-1) = \frac{1+2}{1-2} = -3$. $f'(x) = \frac{(x^2-2)(2x)-(x^2+2)(2x)}{(x^2-2)^2} = \frac{2x(-4)}{(x^2-2)^2}$. The slope is $m = f'(-1) = 8$ and the equation of the tangent line $\frac{y+3}{x+1} = 8$, $y + 3 = 8x + 8$ or $y = 8x + 5$.

33. a) $y = 2x^2 - 5x - 3$, $y' = 4x - 5$.
b) $y = (2x+1)(x-3)$, $y' = (2)(x-3) + (2x+1)(1) = 2x - 6 + 2x + 1 = 4x - 5$.

35. $\frac{d}{dx}(cf) = c\frac{d}{dx}f + (0)(f) = c\frac{df}{dx}$.

37. $f(x) = 3 - 2x - x^2$. $f'(x) = -2 - 2x$ which is 0 at $x = -1$, where the tangent line is horizontal, which could indicate a maximum or a minimum. $f(-1) = 3 + 2 - 1 = 4$.

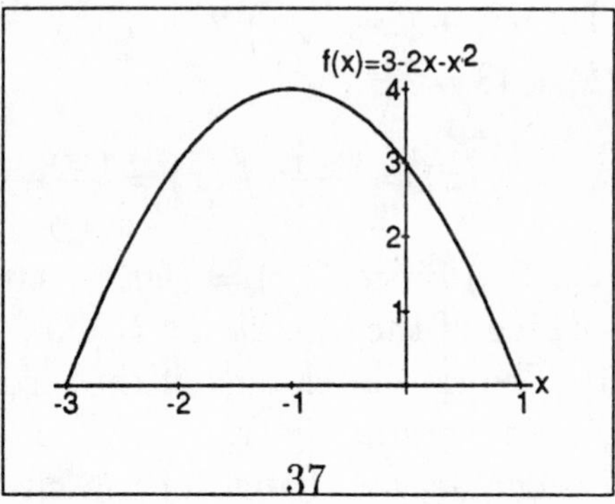

37

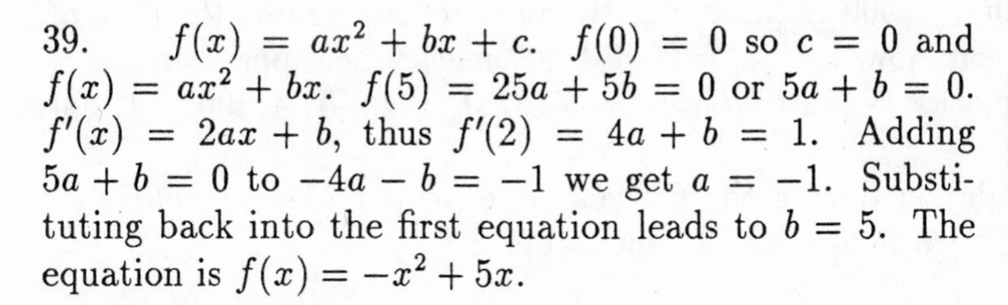

39. $f(x) = ax^2 + bx + c$. $f(0) = 0$ so $c = 0$ and $f(x) = ax^2 + bx$. $f(5) = 25a + 5b = 0$ or $5a + b = 0$. $f'(x) = 2ax + b$, thus $f'(2) = 4a + b = 1$. Adding $5a + b = 0$ to $-4a - b = -1$ we get $a = -1$. Substituting back into the first equation leads to $b = 5$. The equation is $f(x) = -x^2 + 5x$.

41. $y = 4x^2$ and $P(2,0)$. Note that P is not on the graph of the curve (its coordinates do not satisfy the equation of the curve.) $y' = 8x$. Let x_t be the abscissa (x-value) of the point of tangency. The slope is $m = 8x_t$. The equation of the tangent line is $\frac{y_t - 0}{x_t - 2} = 8x_t$, $y_t = 8x_t^2 - 16x_t$. The point of contact (tangency) is both on the curve and the tangent line. Thus $4x_t^2 = 8x_t^2 - 16x_t$ or $4x_t(x_t - 4) = 0$. This is satisfied for $x_t = x_1 = 0$ as well as $x_t = x_2 = 4$. The two points of contact have coordinates $(0,0)$ and $(4,64)$.

43. Let x be the length and y the width of the rectangular flower bed. $A = xy$. Now maximize this area, given a fixed perimeter $2p$, $x + y = p$ or $y = p - x$, and $A = x(p - x) = px - x^2$. $A' = p - 2x = 0$ when $x = \frac{p}{2}$, so $A_{max} = \frac{p^2}{4}$. Note that $0 \le x \le p$ and $A = 0$ at $x = 0$ and $x = p$. If the perimeter is $2p = 20$ ft, then $A_{max} = \frac{10^2}{4} = 25$ square feet.

45. a) $\frac{d}{dx}\left(\frac{fg}{h}\right) = \frac{h\frac{d}{dx}(fg) - (fg)\frac{d}{dx}h}{h^2} = \frac{(h)(fg' + f'g) - fgh'}{h^2}$.
b) $y = \frac{(2x+7)(x^2+3)}{3x+5}$. $\frac{dy}{dx} = \frac{(3x+5)[2x(2x+7)+2(x^2+3)] - (2x+7)(x^2+3)(3)}{(3x+5)^2} = \frac{(3x+5)[2x^2+4x^2+14x+6] - 3(2x^3+6x+7x^2+21)}{(3x+5)^2} = \frac{12x^3+51x^2+70x-33}{(3x+5)^2}$.

47. $y = cf(x)$, $y' = cf'(x) + c'f(x) = cf'(x)$ since $c' = 0$.

2.3 The Derivative as a Rate of Change

1. $f(x) = x^3 - 3x + 5$. $f'(x) = 3x^2 - 3$, $f'(2) = 3(2^2) - 3 = 9$.

3. $f(x) = (x^2+2)(x+\sqrt{x})$, $f'(x) = (x^2+2)\frac{d}{dx}(x+x^{1/2}) + (x+\sqrt{x})\frac{d}{dx}(x^2+2) = (x^2+2)(1+$

$\frac{1}{2}x^{-1/2}) + (x + \sqrt{x})(2x)$, $f'(4) = (4^2 + 2)[1 + \frac{1}{2}(4^{-1/2})][+(4 + \sqrt{4})(8) = (18)(\frac{5}{4}) + (6)(8) = \frac{45}{2} + 48 = \frac{141}{2}$.

5. $f(x) = \frac{2x-1}{3x+5}$, $f'(x) = \frac{(3x+5)(2)-(2x-1)(3)}{(3x+5)^2} = \frac{6x+10-(6x-3)}{(3x+5)^2} = \frac{13}{(3x+5)^2}$, $f'(1) = \frac{13}{8^2} = \frac{13}{64}$.

7. a) Since $C(t) = 100t^2 + 400t + 5,000$ is the circulation t years from now, the rate of change of the circulation t years from now is $C'(t) = 200t + 400$ newspapers per year.
b) The rate of change of the circulation 5 years from now is $C'(5) = 200(5) + 400 = 1,400$ newspapers per year.
c) The actual change in the circulation during the 6th year is $C(6) - C(5) = [100(6^2) + 400(6) + 5,000] - [100(5^2) + 400(5) + 5,000] = 1,500$ newspapers.

9. a) Since $f(x) = -x^3 + 6x^2 + 15x$ is the number of radios assembled x hours after 8:00 a.m., the rate at which the radios are being assembled x hours after 8:00 a.m. is $f'(x) = -3x^2 + 12x + 15$ radios per hour.
b) The rate of assembly at 9:00 a.m. (when $x = 1$) is $f'(1) = -3(1^2) + 12(1) + 15$=24 radios per hour.
c) The actual number of radios assembled between 9:00 a.m. (when $x = 1$) and 10:00 a.m. (when $x = 2$) is $f(2) - f(1) = [-2^3 + 6(2^2) + 15(2)] - [-1^3 + 6(1^2) + 15(1)] = 26$ radios.

11. a) Since $P(t) = 20 - \frac{6}{t+1} = 20 - 6(t+1)^{-1}$ is the population (in thousands) t years from now, the rate at which the population is changing t years from now is $P'(t) = 0 - (-1)(t+1)^{-2} = \frac{6}{(t+1)^2}$ thousand per year. [warning! watch out for the chain rule, to be introduced in section 5, which could involve additional steps in more complicated exercises.]
b) One year from now, the rate of change will be $P'(1) = \frac{6}{(1+1)^2} = 1.5$ thousand per year.
c) The actual population increase during the second year is $P(2) - P(1) = [20 - \frac{6}{2+1}] - [20 - \frac{6}{1+1}] = 1$ thousand.
d) Nine years from now the rate of change will be $P'(9) = \frac{6}{100}$ thousand or 60 people per year.
e) As t increases, $\frac{6}{(t+1)^2}$ approaches zero. Thus, the rate of population growth will approach 0 in the long run.

13. The speed of the first car C_1 is 60 kilometers per hour and the speed of the second car C_2 is 80 kilometers per hour. After t hours, C_1 has traveled $60t$ kilometers and C_2 has traveled $80t$ kilometers, as shown in the figure. Let $D(t)$ denote the distance between C_1 and C_2 at time t. By the pythagorean theorem, $[D(t)]^2 = (80t)^2 + (60t)^2 = 6,400t^2 + 3,600t^2 = 10,000t^2$ and so $D(t) = \sqrt{10,000t^2} = 100t$ kilometers. The rate of change of this distance is the derivative $D'(t) = 100$ kilometers per hour and is independent of the time t.

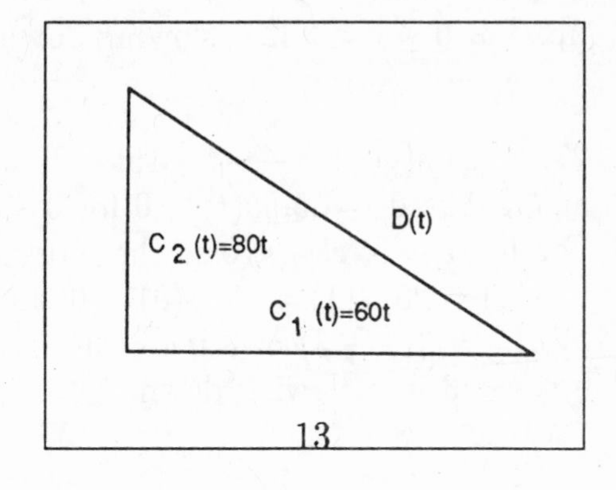

13

15. $f(x) = x(x+3)^2 = x^3 + 6x^2 + 9x$. $f'(x) = 3x^2 + 12x + 9$ and the percentage rate of change is $\frac{300(x^2+4x+3)}{x(x+3)^2}$. When $x = 3$ the rate is 66.67 %.

17. $A(t) = 0.1t^2 + 10t + 20$. a) $A'(t) = 0.2t + 10$, $A'(4) = 0.8 + 10$ or 10,800 people.
b) $A(4) = 0.1 \times 16 + 40 + 20 = 61.6$, so the percentage rate of change is $\frac{100 \times 10.8}{61.6} = 17.53$.

19. $P(t) = t^2 + 200t + 10,000 = (t+100)^2$. a) $P'(t) = 2t + 200 = 2(t+100)$. The percentage rate of change is $\frac{100 \times 2(t+100)}{(t+100)^2} = \frac{200}{t+100}$.
b) $\lim_{t \to \infty} P(t) = 0$.

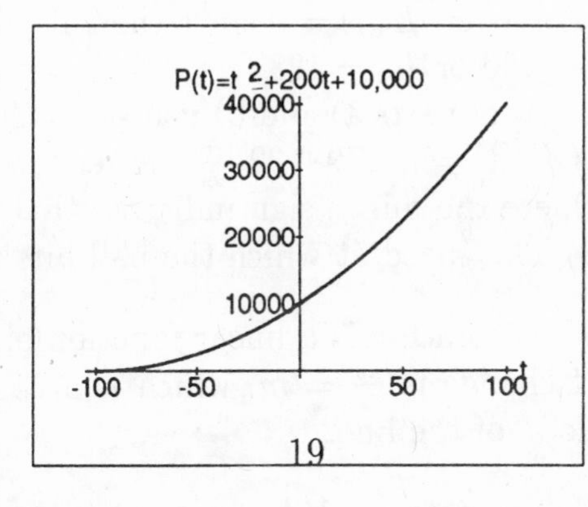

19

21. Let $t = 0$ for 1986 and $G(t)$ the GNP in billions of dollars. $G(0) = 125$, $G(2) = 155$, the slope of the line is $m = \frac{155-125}{2} = 15 = G'(t)$. $G(t) = 125 + 15t$, $G(5) = 125 + 75 = 200$. The percentage rate of change is $\frac{100 \times 15}{200} = 7.5$.

23. a) $s(t) = 3t^2 + 2t - 5$ for $0 \le t \le 1$. $v(t) = 6t + 2$. For $0 \le t \le 1$ $v(t) > 0$ so the object travels to the right.
b) $s(1) - s(0) = (3 + 2 - 5) - (-5) = 5$.
c) $a(t) = 6 > 0$ so the object is speeding up.

25. a) $s(t) = t^4 - 4t^3 + 8t$ for $0 \le t \le 4$. $v(t) = 4(t^3 - 3t^2 + 2)$. The positive roots are $t = 1$ and approximately $t = 2.73$. $v(t) > 0$ for $0 \le t < 1$ and $2.73 < t \le 4$, so the object travels to the right. For $1 < t < 2.73$, $v(t) < 0$ so the object travels to the left.

b) $s(0) = 0$, $s(1) = 1 - 4 + 8 = 5$, $s(2.73) = -4$, and $s(4) = 32$. Thus $\Delta s = 5 + 9 + 36 = 50$.
c) $a(t) = 12t(t-2) = 0$ if $t = 0$ and $t = 2$. $a(t) > 0$ for $2 < t \leq 4$ so the object is speeding up. For $0 \leq t < 2$ it is slowing down.

27. a) $s(t) = t^3 - 9t^2 + 15t + 25$ for $0 \leq t \leq 6$. $v(t) = 3t^2 - 18t + 15 = 3(t-1)(t-5) = 0$ at $t = 1$ and $t = 5$. $v(t) > 0$ for $0 \leq t < 1$ and $5 < t \leq 6$, so the object travels to the right. For $1 < t < 5$, $v(t) < 0$ so the object travels to the left.
b) $s(0) = 25$, $s(1) = 32$, $s(5) = 0$, and $s(6) = 7$. Thus $\Delta s = (32-25)+(32-0)+(7-0) = 46$.
c) $a(t) = 6(t-3) = 0$ if $t = 3$. $a(t) > 0$ for $3 < t \leq 6$ so the object is speeding up. For $0 \leq t < 3$ it is slowing down.

29. $s(t) = -16t^2 - 50t + H$. $s(3) = -16(3^2) - 50(3) + H = 0$, or $H = 294$ ft.
b) $v(t) = -32t - 50$. $v(3) = -32(3) - 50 = -146$ feet per second.

31. a) If after 2 seconds the ball passes you on the way down, then $H(2) = H_0$, where $H(t) = -16t^2 + S_0t + H_0$. Hence $-16(2^2) + (S_0)(2) + H_0 = H_0$, $-64 + 2S_0 = 0$, or $S_0 = 32$ $\frac{\text{ft}}{\text{sec}}$.
b) The height of the building is H_0 feet. From part a) you know that $H(t) = -16t^2 + 32t + H_0$. Moreover, $H(4) = 0$ since the ball hits the ground after 4 seconds. Thus, $-16(4^2) + 32(4) + H_0 = 0$ or $H_0 = 128$ feet.
c) From parts a) and b) you know that $H(t) = -16t^2 + 32t + 128$ and so the speed of the ball is $H'(t) = -32t + 32$ $\frac{\text{ft}}{\text{sec}}$. After 2 seconds, the speed will be $H'(2) = -32$ feet per second, where the minus sign indicates that the direction of motion is down.
d) The speed at which the ball hits the ground is $H'(4) = -96$ $\frac{\text{ft}}{\text{sec}}$.

33. Since y is a linear function of x, $y = mx + b$, where m and b are constants. The rate of change is $\frac{dy}{dx} = m$, which is a constant. Notice that this constant rate of change is the slope of the line.

35. $H(t) = -16t^2 + v_0t$. a) $t(-16t + v_0) = 0$, $t = \frac{v_0}{16}$.
b) $H'(t) = -32t + v_0$. $H'(\frac{v_0}{16}) = -2v_0 + v_0 = -v_0$.

2.4 Approximation by Differentials; Marginal Analysis

1. The rate of change in the function is approximately the derivative of the function times the change in its variable, that is, $f'(x)\Delta x =$. Since $f(x) = x^2 - 3x + 5$, then $f'(x) = 2x - 3$.

As x increases from 5 to 5.3, $\Delta x = 0.3$ and so $f'(3)\Delta x = [2(5)-3](0.3) = 2.1$ is an estimated change.

3. The percentage change in a function is $100\frac{\Delta f}{f(x)}$. Given $f(x) = x^2+2x-9$ and x increases from 4 to 4.3, then $\Delta x = 0.3$, $\Delta f = f(4.3)-f(4) = [(4.3)^2+2(4.3)-9]-[4^2+2(4)-9] = 3.09$ and $f(4) = 15$. Thus the percentage change in $f = 100\frac{\Delta f}{f(x)} = 100(\frac{3.09}{15}) = 20.6\%$.

5. Since the cost is $C(q) = 0.1q^3 - 0.5q^2 + 500q + 200$, the change in cost resulting from an increases in production from 4 units to 4.1 units ($\Delta q = 0.1$) is $\Delta C = C(4.1) - C(4) \approx [C'(4)](0.1)$. Since $C'(q) = 0.3q^2 - q + 500$ and $C'(4) = 500.80$, it follows that $\Delta C \approx (500.80)(0.1) = 50.08$. That is the cost will increase by approximately \$50.08.

7. Since the circulation will be $C(t) = 100t^2 + 400t + 5,000$ papers t years from now, the increase in circulation during the next 6 months ($\Delta t = 0.5$) will be $\Delta C = C(0.5) - C(0) \approx C'(0)(0.5)$. Since $C'(t) = 200t + 400$ and $C'(0) = 400$, it follows that $\Delta C \approx 400(0.5) = 200$ newspapers.

9. Since the average level of carbon monoxide in the air t years from now will be $Q(t) = 0.05t^2 + 0.1t + 3.4$ parts per million, the change in the carbon monoxide level during the next six months ($\Delta t = 0.5$) will be $\Delta Q = Q(0.5) - Q(0) \approx Q'(0)(0.5)$. Since $Q'(t) = 0.1t + 0.1$ and $Q'(0) = 0.1$, it follows that $\Delta Q \approx 0.1(0.5) = 0.05$ parts per million.

11. Since daily output is $Q(K) = 600K^{1/2}$ units when K thousand dollars is invested, the change in output due to an increase in capital investment from \$900,000 to \$900,800 ($\Delta K = 0.8$) is $\Delta Q = Q(900.8) - Q(900) \approx Q'(900)(0.8)$. Since $Q'(K) = 300K^{-1/2}$ and $Q'(900) = 10$, it follows that $\Delta Q \approx 10(0.8) = 8$ units.

13. $Q = 3,000K^{1/2}L^{1/3}$, $K = 400$, and $L = 1,331$. $Q' = 3,000 \times \sqrt[3]{1,331}(\frac{1}{2}K^{-1/2}) = 3,000 \times 11 \times \frac{1}{40} = 825$ units.

15. $A(r) = \pi r^2$, $A(12) = 144\pi\ ^2$. $\Delta r = 0.03 \times 12 = 0.36$ cm, $A'(r) = 2\pi r$, $A'(12) = 24\pi$, $\Delta A = 24\pi \times 0.36 = 27.14\ \text{cm}^2$.

17. $Q(L) = 300L^{2/3}$, $L = 512 = 2^9$, and $\Delta Q = 12.5$. $Q'(L) = 300 \times \frac{2}{3}L^{-1/3} = 200L^{-1/3}$, $Q'(512) = 200 \times 2^{-3} = 25$. $\Delta Q = 12.5 = 25\Delta L$, so $\Delta L = 0.5$ worker-hours.

19. At the end of 1991 (or the beginning of 1992), t will be 3 because it will be three years after 1988. $T(x) = 60x^{3/2}+40x+1,200$ and $T'(x) = 90x^{1/2}+40$. $T'(3) = 90\sqrt{3}+40 = 195.88$. $\Delta t = 0.5$ (for $\frac{1}{2}$ year) and $\Delta T = 195.88 \times 0.5 = 97.94$. Now $T(3) = 60 \times 3^{3/2} + 120 + 1,200 = 1631.77$. The percentage rate of change is $\frac{100 \times 97.94}{1631.77} \approx 6.0\ \%$.

21. $V = x^3$, $\Delta x = -0.02x$, $V' = 3x^2$. $\Delta V = (3x^2)(-0.02x) = -0.06x^3$. The percentage rate of change is $-\frac{6x^3}{x^3} = -6\ \%$. (As to the reference to example 4.2, note that 6% of 1,728

is 103.68.)

23. $V = \frac{4}{3}\pi r^3$, $\Delta r = 0.01r$, $V' = 4\pi r^2$. $\Delta V = (4\pi r^2)(0.01r) = 0.04\pi r^3$ and the percentage rate of change is $\frac{4\pi r^3}{\frac{4}{3}\pi r^3} = 3$ %.

25. $Q(K) = 400K^{1/2}$, $\Delta K = 0.01K$, $Q'(K) = 200K^{-1/2}$. The percentage rate of change is $\frac{100\times 200K^{-1/2}}{400K^{1/2}}(0.01K) = \frac{100\times 200\times 0.01}{400} = 0.5$ %.

27. $Q = 600K^{1/2}L^{1/3}$, $\Delta L = 0.02L$, $Q'(L) = 600K^{1/2}(\frac{1}{3}L^{-2/3}) = \frac{200K^{1/2}}{L^{2/3}}$. $\Delta Q = (\frac{200K^{1/2}}{L^{2/3}})(0.02L) = 4K^{1/2}L^{1/3}$. The percentage rate of change is $\frac{100\times 4K^{1/2}L^{1/3}}{600K^{1/2}L^{1/3}} = 0.67\%$.

29. $V = \frac{4}{3}\pi r^3$, $\Delta V = 0.08V$, $V' = 4\pi r^2$. $\Delta V = 4\pi r^2 \Delta r$. The percentage rate of change is $\frac{100\times 4\pi r^2 \Delta r}{\frac{4}{3}\pi r^3} = 300\frac{\Delta r}{r}$. $8 = (3)(100\frac{\Delta r}{r})$, $100\frac{\Delta r}{r} = \frac{8}{3} = 2.67$. The computations were performed with "+" in mind, although they are equally valid for "−". The % rate of change is then $\pm 2.67\%$.

31. $r = (\frac{1}{2})(8) = 4''$. $\Delta r = 0.2''$. $V(4) = \frac{256\pi}{3}$. $V(r) = \frac{4}{3}\pi r^3$, $V'(r) = 4\pi r^2$, $\Delta V = 4\pi r^2 \times 0.2 = 0.8\pi r^2$. $\Delta V|_{r=4} = 0.8\pi \times 16 = 12.8\pi$. The percentage rate of change is $\frac{100\times 12.8\pi}{256\pi/3} = 15$ %.

33. $C(q) = 0.1q^3 - 0.5q^2 + 500q + 200$. a) $C'(q) = 0.3q^2 - q + 500$. The cost of the 4th unit is $C'(3) = 0.3 \times 9 - 3 + 500 = \$499.70$.
b) $C(4) = 0.1 \times 64 - 0.5 \times 16 + 2,000 + 200$, $C(3) = 0.1 \times 27 - 0.5 \times 9 + 1,500 + 200$, and $C(4) - C(3) = \$500.20$.

35. a) $C'(x) = \frac{2}{5}x + 4$, $p(x) = 9 - \frac{x}{4}$, $R(x) = 9x - \frac{x^2}{4}$, $R'(x) = 9 - \frac{x}{2}$.
b) $C'(3) = \frac{6}{5} + 4 = \$5.2$.
c) $C(4) - C(3) = [\frac{1}{5}4^2 + 16 + 57] - [\frac{1}{5}3^2 + 12 + 57] = \$5.4$.
d) $R'(3) = 9 - \frac{3}{2} = 7.5$.
e) $R(4) - R(3) = 9(4) - 4 - 9(3) + \frac{9}{4} = 7.25$.

2.5 The Chain Rule

1. $y = u^2 + 1, u = 3x - 2$, $\frac{dy}{du} = 2u$, $\frac{du}{dx} = 3$, $\frac{dy}{dx} = \frac{dy}{du}\frac{du}{dx} = (2u)(3) = 6u = 6(3x - 2)$.

3. $y = \sqrt{u} = u^{1/2}$, $u = x^2 + 2x - 3$, $\frac{dy}{dx} = \frac{1}{2}u^{-1/2} = \frac{1}{2u^{1/2}}$, $\frac{du}{dx} = 2x + 2$, $\frac{dy}{dx} = \frac{dy}{du}\frac{du}{dx} =$

$(\frac{1}{2u^{1/2}})(2x+2) = \frac{x+1}{(x^2+2x-3)^{1/2}}$.

5. $y = \frac{1}{u^2} = u^{-2}$, $u = x^2+1$, $\frac{dy}{du} = -2u^{-3}$, $\frac{du}{dx} = 2x$, $\frac{dy}{dx} = \frac{dy}{du}\frac{du}{dx} = (\frac{-2}{u^3})(2x) = \frac{-4x}{(x^2+1)^3}$.

7. $y = \frac{1}{\sqrt{u}} = u^{-1/2}$, $u = x^2-9$, $\frac{dy}{du} = -\frac{1}{2}u^{-3/2} = \frac{-1}{2u^{3/2}}$, $\frac{du}{dx} = 2x$, $\frac{dy}{dx} = \frac{dy}{du}\frac{du}{dx} = (\frac{-1}{2u^{3/2}})(2x) = \frac{-x}{(x^2-9)^{3/2}}$.

9. $y = \frac{1}{u-1} = (u-1)^{-1}$, $u = x^2$, $\frac{dy}{du} = -(u-1)^{-2} = \frac{-1}{(u-1)^2}$, $\frac{du}{dx} = 2x$, $\frac{dy}{dx} = \frac{dy}{du}\frac{du}{dx} = (\frac{-1}{(u-1)^2})(2x) = \frac{-2x}{(x^2-1)^2}$.

11. $y = 3u^4-4u+5$, $u = x^3-2x-5$, $\frac{dy}{du} = 12u^3-4$, $\frac{du}{dx} = 3x^2-2$, $\frac{dy}{dx} = \frac{dy}{du}\frac{du}{dx} = (12u^3-4)(3x^2-2)$. When $x=2$, $u = 2^3-2(2)-5 = -1$ and so $\frac{dy}{dx} = [12(-1)^3-4][3(2^2)-2] = -160$.

13. $y = \sqrt{u} = u^{1/2}$, $u = x^2-2x+6$, $\frac{dy}{du} = \frac{1}{2}u^{-1/2} = \frac{1}{2u^{1/2}}$, $\frac{du}{dx} = 2x-2$, $\frac{dy}{dx} = \frac{dy}{du}\frac{du}{dx} = (\frac{1}{2u^{1/2}})(2x-2) = \frac{x-1}{u^{1/2}}$. When $x=2$, $u = 3^2-2(3)+6 = 9$, and so $\frac{dy}{dx} = \frac{3-1}{9^{1/2}} = \frac{2}{3}$.

15. $y = \frac{1}{u} = u^{-1}$, $u = 3-\frac{1}{x^2} = 3-x^{-2}$, $\frac{dy}{du} = -u^{-2} = \frac{-1}{u^2}$, $\frac{du}{dx} = -(-2x^{-3}) = \frac{2}{x^3}$, $\frac{dy}{dx} = \frac{dy}{du}\frac{du}{dx} = (\frac{-1}{u^2})(\frac{2}{x^3})$. When $x = \frac{1}{2}$, $u = 3-(\frac{1}{(1/2)^2}) = 3-4 = -1$, $\frac{dy}{dx} = (\frac{-1}{(-1)^2})(\frac{2}{(1/2)^3}) = \frac{-2}{1/8} = -16$.

17. $f(x) = (2x+1)^4$, $f'(x) = 4(2x+1)^3\frac{d}{dx}(2x+1) = 4(2x+1)^3(2) = 8(2x+1)^3$.

19. $f(x) = (x^5-4x^3-7)^8$, $f'(x) = 8(x^5-4x^3-7)^7\frac{d}{dx}(x^5-4x^3-7) = 8(x^5-4x^3-7)^7(5x^4-12x^2) = 8x^2(x^5-4x^3-7)^7(5x^2-12)$.

21. $f(x) = \frac{1}{5x^2-6x+2} = (5x^2-6x+2)^{-1}$, $f'(x) = -(5x^2-6x+2)^{-2}\frac{d}{dx}(5x^2-6x+2) = -\frac{10x-6}{(5x^2-6x+2)^2}$.

23. $f(x) = \frac{1}{\sqrt{4x^2+1}} = (4x^2+1)^{-1/2}$, $f'(x) = \frac{-1}{2}(4x^2+1)^{-3/2}\frac{d}{dx}(4x^2+1) = -\frac{1}{2}(4x^2+1)^{-3/2}(8x) = \frac{-4x}{(4x^2+1)^{3/2}}$.

25. $f(x) = \frac{3}{(1-x^2)^4} = 3(1-x^2)^{-4}$, $f'(x) = -12(1-x^2)^{-5}\frac{d}{dx}(1-x^2) = -12(1-x^2)^{-5}(-2x) = \frac{24x}{(1-x^2)^5}$.

27. $f(x) = (1+\sqrt{3x})^5$, $f'(x) = 5(1+\sqrt{3x})^4\frac{d}{dx}(1+\sqrt{3x}) = 5(1+\sqrt{3x})^4\frac{d}{dx}(1+\sqrt{3}\sqrt{x}) = 5(1+\sqrt{3x})^4\frac{d}{dx}(1+\sqrt{3}x^{1/2}) = 5(1+\sqrt{3x})^4(\sqrt{3}\frac{1}{2}x^{-1/2}) = 5\sqrt{3}(1+\sqrt{3x})^4(\frac{1}{2})x^{-1/2} = \frac{15(1+\sqrt{3x})^4}{2\sqrt{3x}}$.

29. $f(x) = (x+2)^3(2x-1)^5$, $f'(x) = (x+2)^3\frac{d}{dx}(2x-1)^5 + (2x-1)^5\frac{d}{dx}(x+2)^3 = (x+2)^3[5(2x-1)^4(2)] + (2x-1)^5[3(x+2)^2(1)] = (x+2)^2(2x-1)^4[10(x+2)+3(2x-1)] = (x+2)^2(2x-1)^4(16x+17)$.

31. $f(x) = \sqrt{\frac{3x+1}{2x-1}} = [\frac{3x+1}{2x-1}]^{1/2}$, $f'(x) = \frac{1}{2}[\frac{3x+1}{2x-1}]^{-1/2}\frac{d}{dx}[\frac{3x+1}{2x-1}] = \frac{1}{2}[\frac{3x+1}{2x-1}]^{-1/2}[\frac{(2x-1)(3)-(3x+1)(2)}{(2x-1)^2}] =$ $\frac{1}{2}[\frac{(3x+1)^{-1/2}}{(2x-1)^{-1/2}}][\frac{-5}{(2x-1)^2}] = \frac{-5}{2(3x+1)^{1/2}(2x-1)^{3/2}}$.

33. $f(x) = \frac{(x+1)^5}{(1-x)^4}$, $f'(x) = \frac{(1-x)^4\frac{d}{dx}(x+1)^5-(x+1)^5\frac{d}{dx}(1-x)^4}{(1-x)^8} = \frac{(1-x)^4(5)(x+1)^4(1)-(x+1)^5(4)(1-x)^3(-1)}{(1-x)^8} =$ $\frac{(1-x)^3(x+1)^4[5(1-x)+4(x+1)]}{(1-x)^8} = \frac{(x+1)^4(9-x)}{(1-x)^5}$.

35. $f(x) = \frac{3x+1}{\sqrt{1-4x}}$, $f'(x) = \frac{\sqrt{1-4x}\frac{d}{dx}(3x+1)-(3x+1)\frac{d}{dx}[(1-4x)^{1/2}]}{(\sqrt{1-4x})^2} =$ $\frac{\sqrt{1-4x}(3)-(3x+1)\frac{1}{2}(1-4x)^{-1/2}\frac{d}{dx}(1-4x)}{1-4x} = \frac{3\sqrt{1-4x}-\frac{1}{2}(3x+1)(1-4x)^{-1/2}(-4)}{1-4x} = \frac{3\sqrt{1-4x}+2(3x+1)(1-4x)^{-1/2}}{1-4x} =$ $\frac{(1-4x)^{-1/2}[3(1-4x)+2(3x+1)]}{1-4x} = \frac{3-12x+6x+2}{(1-4x)^{3/2}} = \frac{5-6x}{(1-4x)^{3/2}}$.

37. $f(x) = (3x^2+1)^2$. $f'(x) = 2(3x^2+1)(6x)$, $m = f'(-1) = -48$. $f(-1) = 16$. For the tangent line $\frac{y-16}{x+1} = -48$, $y - 16 = -48x - 48$, $y = -48x - 32$.

39. $f(x) = (2x-1)^{-6}$. $f'(x) = -6(2x-1)^{-7}(2) = -\frac{12}{(2x-1)^7}$, $m = f'(1) = -12$. $f(1) = 1$. For the tangent line $\frac{y-1}{x-1} = -12$, $y - 1 = -12x + 12$, $y = -12x + 13$.

41. $f(x) = (3x+5)^2$. a) $f'(x) = 2(3x+5)(3) = 6(3x+5)$.
b) $f'(x) = (3x+5)(3) + (3)(3x+5) = 6(3x+5)$.

43. a) Since $f(t) = \sqrt{10t^2+t+236} = (10t^2+t+236)^{1/2}$ is the factory's gross annual earnings (in thousand-dollar units) t years after its formation in 1988, the rate at which the earnings are growing at that time is $f'(t) = \frac{1}{2}(10t^2+t+236)^{-1/2}(20t+1) = \frac{20t+1}{2(10t^2+t+236)^{1/2}}$ thousand dollars per year. The rate of growth in 1992 (when $t = 4$) is $f'(4) = \frac{20(4)+1}{2(10(4)^2+4+236)^{1/2}} =$ $\frac{81}{2(400)^{1/2}} = \frac{81}{40} = 2.025$. That is, in 1992 the gross annual earnings were increasing at the rate of \$2,025 per year.
b) The percentage rate of the earnings increases in 1992 was $100[\frac{f'(4)}{f(4)}] = \frac{100(2.025)}{\sqrt{10(4^2)+4+236}} =$ 10.125 percent per year.

45. Since the population t years from now is $p(t) = 20 - \frac{6}{t+1} = 20 - 6(t+1)^{-1}$ thousand and the average daily level of carbon monoxide in the air is $c(p) = 0.5\sqrt{p^2+p+58} = 0.5(p^2+p+58)^{1/2}$ parts per million when the population is p thousand, the rate of change of carbon monoxide with respect to time is $\frac{dc}{dt} = \frac{dc}{dp}\frac{dp}{dt} = [(0.5)(\frac{1}{2})(p^2+p+58)^{-1/2}(2p+1)][\frac{6}{(t+1)^2}] =$ $[\frac{2p+1}{4(p^2+p+58)^{1/2}}][\frac{6}{(t+1)^2}]$ parts per million per year. Two years from now, $t = 2$, $p(2) = 18$, and the rate of change will be $\frac{dc}{dt} = [\frac{2(18)+1}{4(18^2+18+58)^{1/2}}][\frac{6}{(2+1)^2}] \approx 0.31$ parts per million per year.

47. Since $D(p) = \frac{4,374}{p^2} = 4,374p^{-2}$ pounds of coffee are sold per week if the price is

p dollars per pound and the price is $p(t) = 0.02t^2 + 0.1t + 6$ dollars per pound, $\frac{dD}{dt} = \frac{dD}{dp}\frac{dp}{dt} = 4,374(-2p^{-3})(0.04t + 0.1) = [-\frac{8,748}{p^3}](0.04t + 0.1t)$ pounds per week. 10 weeks from now, $t = 10$, $p(10) = 9$, and the rate of change of the demand with respect to time is $\frac{dD}{dt} = [-\frac{8,748}{9^3}][0.04(10) + 0.1] = -6$ pounds per week which signifies a decrease.

49. The population is $p(t) = 12 - \frac{6}{t+1} = 12 - 6(t+1)^{-1}$ thousand, and the carbon monoxide level is $c(p) = 0.6\sqrt{p^2 + 2p + 24} = 0.6(p^2 + 2p + 24)^{1/2}$ units. The corresponding derivatives are $\frac{dp}{dt} = 6(t+1)^{-2} = \frac{6}{(t+1)^2}$ and $\frac{dc}{dp} = 0.3(p^2 + 2p + 24)^{-1/2}(2p + 2) = \frac{0.6(p+1)}{\sqrt{p^2+2p+24}}$. The rate of change of the carbon monoxide level with respect to time is $\frac{dc}{dt} = \frac{dc}{dp}\frac{dp}{dt} = \frac{0.6(p+1)}{\sqrt{p^2+2p+24}}\frac{6}{(t+1)^2}$ and the percentage rate of change is $100\frac{dc/dt}{c}$. When $t = 2$, $p = p(2) = 12 - \frac{6}{3} = 10$ and $c = c(10) = 0.6\sqrt{10^2 + 2(10) + 24} = 0.6\sqrt{144} = 7.2$. Hence, when $t = 2$, the percentage rate of change is $100[\frac{0.6(10+1)}{\sqrt{144}}\frac{6}{(2+1)^2}][\frac{1}{7.2}] \approx 5.09$ percent per year.

51. To prove that $\frac{d}{dx}[h(x)]^2 = 2h(x)h'(x)$, use the product rule to get $\frac{d}{dx}[h(x)]^2 = \frac{d}{dx}[h(x)h(x)] = h(x)h'(x) + h'(x)h(x) = 2h(x)h'(x)$.

2.6 Implicit Differentiation

1. $x^2 + y^2 = 25$, $2x + 2y\frac{dy}{dx} = 0$, $\frac{dy}{dx} = -\frac{x}{y}$.

3. $x^3 + y^3 = xy$, $3x^2 + 3y^2\frac{dy}{dx} = x\frac{dy}{dx} + y$, $(3y^2 - x)\frac{dy}{dx} = y - 3x^2$, $\frac{dy}{dx} = \frac{y-3x^2}{3y^2-x}$.

5. $y^2 + 2xy^2 - 3x + 1 = 0$, $2y\frac{dy}{dx} + [2x(2y\frac{dy}{dx}) + y^2(2)] - 3 = 0$, $(2y + 4xy)\frac{dy}{dx} = 3 - 2y^2$, $\frac{dy}{dx} = \frac{3-2y^2}{2y+4xy} = \frac{3-2y^2}{2y(1+2x)}$.

7. $(2x + y)^3 = x$, $3(2x + y)^2[2 + \frac{dy}{dx}] = 1$, $2 + \frac{dy}{dx} = \frac{1}{3(2x+y)^2}$, $\frac{dy}{dx} = \frac{1}{3(2x+y)^2} - 2$.

9. $(x^2+3y^2)^5 = 2xy$, $5(x^2+3y^2)^4\frac{d}{dx}(x^2+3y^2) = 2x\frac{dy}{dx}+2y$, $5(x^2+3y^2)^4(2x+y\frac{dy}{dx}) = 2x\frac{dy}{dx}+2y$, $10x(x^2+3y^2)^4+30y(x^2+3y^2)^4\frac{dy}{dx} = 2x\frac{dy}{dx}+2y$, $[30y(x^2+3y^2)^4-2x]\frac{dy}{dx} = 2y-10x(x^2+3y^2)^4$, $\frac{dy}{dx} = \frac{2y-10x(x^2+3y^2)^4}{30y(x^2+3y^2)^4-2x}$

11. $x^2 = y^3$, $2x = 3y^2\frac{dy}{dx}$, $\frac{dy}{dx} = \frac{2x}{3y^2}$. When $x = 8$, the original equation gives $8^2 = y^3$, $y^3 = 64$, or $y = 4$. Substituting in the equation for $\frac{dy}{dx}$ yields $\frac{dy}{dx} = \frac{2(8)}{3(4^2)} = \frac{1}{3}$, that is, the

slope of the tangent line at the point $(8,4)$ is $\frac{1}{3}$.

13. $xy = 2$, $x\frac{dy}{dx} + y = 0$, $\frac{dy}{dx} = -\frac{y}{x}$. When $x = 2$, the original equation gives $2y = 2$ or $y = 1$. Substituting in the equation for $\frac{dy}{dx}$ yields $\frac{dy}{dx} = -\frac{1}{2}$. That is, the slope of the tangent line at the point $(2,1)$ is $-\frac{1}{2}$.

15. $(1-x+y)^3 = x+7$, $3(1-x+y)^2[-1+\frac{dy}{dx}] = 1$, $-1+\frac{dy}{dx} = \frac{1}{3(1-x+y)^2}$, or $\frac{dy}{dx} = \frac{1}{3(1-x+y)^2}+1$. When $x = 1$, the original equation gives $(1-1+y)^3 = 1+7$, $y^3 = 8$, or $y = 2$. Substituting in the equation for $\frac{dy}{dx}$ yields $\frac{dy}{dx} = \frac{1}{3(1-1+2)^2}+1 = \frac{1}{12}+1 = \frac{13}{12}$. Thus the slope of the tangent line at $(1,2)$ is $\frac{13}{12}$.

17. $(2xy^3+1)^3 = 2x-y^3$, $3(2xy^3+1)^2\frac{d}{dx}(2xy^3+1) = 2-3y^2\frac{dy}{dx}$, $3(2xy^3+1)^2(6xy^2\frac{dy}{dx}+2y^3) = 2-3y^2\frac{dy}{dx}$, $[18xy^2(2xy^3+1)^2+3y^2]\frac{dy}{dx} = 2-6y^3(2xy^3+1)^2$, $\frac{dy}{dx} = \frac{2-6y^3(2xy^3+1)^2}{18xy^2(2xy^3+1)^2+3y^2}$. When $x = 0$, the original equation gives $1^3 = -y^3$ or $y = -1$. Hence $\frac{dy}{dx} = \frac{2-6(-1)^3(0+1)^2}{0(0+1)^2+3(-1)^2} = \frac{8}{3}$.

19. For explicit differentiation, $xy+2y = x^2$ is solved for y in terms of x. Thus $y = \frac{x^2}{x+2}$. Differentiate using the quotient rule. $\frac{dy}{dx} = \frac{(x+2)(2x)-x^2(1)}{(x+2)^2} = \frac{x^2+4x}{(x+2)^2} = \frac{x(x+4)}{(x+2)^2}$.
For implicit differentiation, $xy+2y = x^2$, $[x\frac{dy}{dx}+y]+2\frac{dy}{dx} = 2x$, $(x+2)\frac{dy}{dx} = 2x-y$, $\frac{dy}{dx} = \frac{2x-y}{x+2}$.
But since $y = \frac{x^2}{x+2}$, $\frac{dy}{dx} = \frac{2x-[\frac{x^2}{x+2}]}{x+2} = \frac{2x^2+4x-x^2}{(x+2)^2} = \frac{x(x+4)}{(x+2)^2}$.

21. For explicit differentiation, $xy-x = y+2$ is solved for y in terms of x. Thus $y = \frac{x+2}{x-1}$. Differentiate using the quotient rule. $\frac{dy}{dx} = \frac{(x-1)(1)-(x+2)(1)}{(x-1)^2} = \frac{-3}{(x-1)^2}$.
For implicit differentiation, $xy-x = y+2$, $[x\frac{dy}{dx}+y]-1 = \frac{dy}{dx}$, $(x-1)\frac{dy}{dx} = 1-y$, or $\frac{dy}{dx} = \frac{1-y}{x-1}$.
But since $y = \frac{x+2}{x-1}$, $\frac{dy}{dx} = \frac{1-[\frac{x+2}{x-1}]}{x-1} = \frac{(x-1)-(x+2)}{(x-1)^2} = \frac{-3}{(x-1)^2}$.

23. If the output is to remain unchanged, the equation relating inputs x and y can be written as $Q = 2x^3+3x^2y^2+(1+y)^3$ where Q is a constant representing the current output. By implicit differentiation of this equation, $0 = 6x^2+6x^2y\frac{dy}{dx}+6xy^2+3(1+y)^2\frac{dy}{dx}$, $-[6x^2y+3(1+y)^2]\frac{dy}{dx} = 6x^2+6xy^2$, $\frac{dy}{dx} = -\frac{6x^2+6xy^2}{6x^2y+3(1+y)^2}$. Use the approximation formula $\Delta y \approx \frac{dy}{dx}\Delta x$ from section 4, with $x = 30$, $y = 20$, and $\Delta x = -0.8$ to get $\Delta y \approx [-\frac{6(30^2)+6(30)(20^2)}{6(30^2)(20)+3(21^2)}](-0.8) \approx 0.57$. That is, to maintain the current level of output, input y should be decreased by approximately 0.57 unit to offset a decrease in input of 0.8 unit.

25. If the output is to remain unchanged, the equation relating inputs x and y can be written as $Q = 0.06x^2+0.14xy+0.05y^2$ where Q is a constant representing the current output. By implicit differentiation of this equation, $0 = 0.12x+0.14(x\frac{dy}{dx}+y)+0.1y\frac{dy}{dx}$, $(0.14x+0.1y)\frac{dy}{dx} = -0.12x-0.14y$, $\frac{dy}{dx} = \frac{-0.12x-0.14y}{0.14x+0.1y} = -\frac{0.06x+0.07y}{0.07x+0.05y}$. Use the approximation

formula $\Delta y \approx \frac{dy}{dx}\Delta x$ from section 4, with $x = 60$, $y = 300$, and $\Delta x = -1$ to get $\Delta y \approx [-\frac{(0.06)(60)+(0.07)(300)}{(0.07)(60)+(0.05)(300)}](-1) \approx 1.28125$ hours of unskilled labor. That is, to maintain the current level of output, input y should be increased by 1.28125 hours to offset a decrease in input of 1 hour.

27. $\frac{x^2}{a^2} - \frac{y^2}{b^2} = 1$, $\frac{2x}{a^2} - \frac{2y\frac{dy}{dx}}{b^2} = 0$, $\frac{dy}{dx} = \frac{b^2x}{a^2y}$. Thus the slope at (x_0, y_0) is $m = \frac{b^2x_0}{a^2y_0}$ and the equation of the line becomes $y - y_0 = \frac{b^2x_0}{a^2y_0}(x - x_0)$, $\frac{y_0y}{b^2} - \frac{y_0^2}{b^2} = \frac{x_0x}{a^2} - \frac{x_0^2}{a^2}$, $\frac{x_0x}{a^2} - \frac{y_0y}{b^2} = \frac{x_0^2}{a^2} - \frac{y_0^2}{b^2} = 1$, because the point (x_0, y_0) is on the curve.

29. a) $x^2+y^2 = 6y-10$, $x^2+y^2-6y+9 = -10+9$, $x^2+(y-3)^2 = -1$ which is impossible since the sum of two squares is not negative (in the real number system).
b) $x^2 + y^2 = 6y - 10$, $2x + 2y\frac{dy}{dx} = 6\frac{dy}{dx}$, $\frac{dy}{dx}(2y - 6) = -2x$, $\frac{dy}{dx} = \frac{-2x}{2(y-3)} = \frac{-x}{y-3}$.

31. $x^2+y^2 = \sqrt{x^2 + y^2}+x$ (expression 1). We can rewrite this as $[\sqrt{x^2 + y^2}]^2 - \sqrt{x^2 + y^2} - x = 0$ and use the quadratic formula. $\sqrt{x^2 + y^2} = \frac{1\pm\sqrt{1+4x}}{2}$ (expression 2). Differentiating expression 1 yields $2x + 2y\frac{dy}{dx} = \frac{1}{2}(x^2 + y^2)^{-1/2}(2x + 2y\frac{dy}{dx}) + 1$, $(2x + 2y\frac{dy}{dx})[1 - \frac{1}{2\sqrt{x^2+y^2}}] = 1$, $(x + y\frac{dy}{dx})\frac{2\sqrt{x^2+y^2}-1}{\sqrt{x^2+y^2}} = 1$, $y\frac{dy}{dx} = \frac{\sqrt{x^2+y^2}-x(2\sqrt{x^2+y^2}-1)}{2\sqrt{x^2+y^2}-1}$, $\frac{dy}{dx} = \frac{\sqrt{x^2+y^2}(1-2x)+x}{2y\sqrt{x^2+y^2}-y}$.
For a horizontal tangent line, $\frac{dy}{dx} = 0$. The numerator is 0 if $\sqrt{x^2 + y^2}(1-2x) = -x$, or using expression 2, $(1 \pm \sqrt{1 + 4x})(1 - 2x) = -2x$, $1 + (1 - 2x)\sqrt{1 + 4x} = 0$, $1 = (1 - 2x)^2(1 + 4x)$, $1 = (1 - 4x + 4x^2)(1 + 4x) = 1 - 4x + 4x^2 + 4x - 16x^2 + 16x^3$, $4x^2(4x - 3) = 0$, so $x = \frac{3}{4}$ or $x = 0$.
For $x = 0.75$ and using expression 2 with +, $\sqrt{0.5625 + y^2} = \frac{1\pm2}{2} = 1.5$, $y^2 = \frac{27}{16}$, or $y = \pm1.3$. Thus the tangent line is horizontal at $(0.75, \pm1.3)$. These are two of six desired points. Is there a better way to solve this problem?
For $x = 0$ (and $y = 0$) $\frac{dy}{dx} = \frac{0}{0}$ which presents a slight difficulty. This is undetermined. Perhaps a point near $(0, 0)$ will help. Using expression 2, it can be shown that the point $(-0.001, 0.000045)$ is on the curve. $\frac{dy}{dx} = -0.02$ at this point. This suggests that $\frac{dy}{dx} = 0$ at the origin and leads us to believe that the tangent line is horizontal at the origin. And, yes indeed, there is a simpler way of solving this problem.
For vertical tangent lines, the slope should be undefined, that is $y(2\sqrt{x^2 + y^2} - 1) = 0$. This is 0 if $y = 0$. From expression 1, $x^2 = x + x$, so $x = 0$ (and we already discussed the origin) or $x = 2$. The curve has a vertical tangent line at $(2, 0)$.
Now, for $\sqrt{x^2 + y^2} = \frac{1}{2}$, or by using expression 2, $\frac{1\pm\sqrt{1+4x}}{2} = \frac{1}{2}$, $\sqrt{1 + 4x} = 0$ or $x = -\frac{1}{4}$. To find y, $\sqrt{\frac{1}{16} + y^2} = \frac{1\pm0}{2}$, $y = \pm0.433$. The curve has a vertical tangent line at $(-0.25, \pm0.433)$. This problem is relatively simple when the equation is converted from cartesian coordinates to polar coordinates, but this topic is beyond the scope of this course. The name for this curve is "cardioid".

2.7 Higher Order Derivatives

1. $f(x) = 5x^{10} - 6x^5 - 27x + 4$, $f'(x) = 50x^9 - 30x^4 - 27$, $f''(x) = 450x^8 - 120x^3$,

3. $y = 5\sqrt{x} + \frac{3}{x^2} + \frac{1}{3\sqrt{x}} + \frac{1}{2} = 5x^{1/2} + 3x^{-2} + \frac{1}{3}x^{-1/2} + \frac{1}{2}$. $\frac{dy}{dx} = \frac{5}{2}x^{-1/2} - 6x^{-3} - \frac{1}{6}x^{-3/2}$, $\frac{d^2y}{dx^2} = -\frac{5}{4}x^{-3/2} + 18x^{-4} + \frac{1}{4}x^{-5/2} = -\frac{5}{4x^{3/2}} + \frac{18}{x^4} + \frac{1}{4x^{5/2}}$.

5. $f(x) = (3x+1)^5$. $f'(x) = 5(3x+1)^4(3) = 15(3x+1)^4$, $f''(x) = 60(3x+1)^3(3) = 180(3x+1)^3$.

7. $y = (x^2+5)^8$. $\frac{dy}{dx} = 8(x^2+5)^7(2x) = 16x(x^2+5)^7$, $\frac{d^2y}{dx^2} = 16x[7(x^2+5)^6(2x)] + (x^2+5)^7(16) = 16(x^2+5)^6[14x^2 + (x^2+5)] = 16(x^2+5)^6(15x+5) = 80(x^2+5)^6(3x^2+1)$.

9. $f(x) = \sqrt{1+x^2} = (1+x^2)^{1/2}$. $f'(x) = \frac{1}{2}(1+x^2)^{-1/2}(2x) = \frac{x}{\sqrt{1+x^2}}$, $f''(x) = \frac{(1+x^2)^{1/2}(1) - x[\frac{1}{2}(1+x^2)^{-1/2}(2x)]}{1+x^2} = \frac{(1+x^2)^{-1/2}[(1+x^2)-x^2]}{1+x^2} = \frac{1}{(1+x^2)^{3/2}}$.

11. $y = \frac{2}{1+x^2} = 2(1+x^2)^{-1}$. $\frac{dy}{dx} = -2(1+x^2)^{-2}(2x) = -\frac{4x}{(1+x^2)^2}$, $\frac{d^2y}{dx^2} = -\frac{(1+x^2)^2(4) - 4x[2(1+x^2)(2x)]}{(1+x^2)^4} = -\frac{4(1+x^2)[(1+x^2)-4x^2]}{(1+x^2)^4} - \frac{4(1-3x^2)}{(1+x^2)^3} = \frac{4(3x^2-1)}{(1+x^2)^3}$.

13. $f(x) = x(2x+1)^4$. $f'(x) = x(4)(2x+1)^3(2) + (2x+1)^4(1) = (2x+1)^3[8x + (2x+1)] = (2x+1)^3(10x+1)$, $f''(x) = (2x+1)^3(10) + (10x+1)(3)(2x+1)^2(2) = 2(2x+1)^2[5(2x+1) + 3(10x+1)] = 2(2x+1)^2(10x+5+30x+3) = 2(2x+1)^2(40x+8) = 16(2x+1)^2(5x+1)$.

15. $y = (\frac{x}{x+1})^2$. $\frac{dy}{dx} = 2(\frac{x}{x+1})\frac{(x+1)(1)-x(1)}{(x+1)^2} = \frac{2x}{(x+1)^3}$, $\frac{d^2y}{dx^2} = \frac{(x+1)^3(2) - 2x(3)(x+1)^2(1)}{(x+1)^6} = \frac{(x+1)^2(2x+2-6x)}{(x+1)^6} = \frac{2(1-2x)}{(x+1)^4}$.

17. $2y^2 - 5x^2 = 3$. $4y\frac{dy}{dx} - 10x = 0$, or $\frac{dy}{dx} = \frac{10x}{4y} = \frac{5x}{2y}$, $\frac{d^2y}{dx^2} = \frac{2y(5) - 5x[2\frac{dy}{dx}]}{(2y)^2} = \frac{10y - 10x\frac{dy}{dx}}{4y^2} = \frac{5y - 5x\frac{dy}{dx}}{2y^2}$. Since $\frac{dy}{dx} = \frac{5x}{2y}$, $\frac{d^2y}{dx^2} = \frac{5y - 5x(\frac{5x}{2y})}{2y^2} = \frac{10y^2 - 25x^2}{4y^3} = \frac{5(2y^2-5x^2)}{4y^3}$. From the original equation $2y^2 - 5x^2 = 3$ so $\frac{d^2y}{dx^2} = \frac{5(3)}{4y^3} = \frac{15}{4y^3}$.

19. $ax^2 + by^2 = 1$. $2ax + 2by\frac{dy}{dx} = 0$ or $\frac{dy}{dx} = -\frac{2ax}{2by} = -\frac{ax}{by}$. $\frac{d^2y}{dx^2} = -\frac{by(a) - ax(b\frac{dy}{dx})}{(by)^2} = -\frac{aby - abx\frac{dy}{dx}}{b^2y^2}$. Since $\frac{dy}{dx} = -\frac{ax}{by}$, $\frac{d^2y}{dx^2} = -\frac{aby - abx(-\frac{ax}{by})}{b^2y^2} = -\frac{ab^2y^2 + a^2bx^2}{b^2y^3} = -\frac{ab(by^2+ax^2)}{b^2y^3} = -\frac{a(by^2+ax^2)}{b^2y^3}$. From the original equation $by^2 + ax^2 = 1$ so $\frac{d^2y}{dx^2} = -\frac{a}{b^2y^3}$.

21. $Q(t) = -t^3 + 8t^2 + 15t$, 8 a.m. corresponds to $t = 0$. a) $Q'(t) = -3t^2 + 16t + 15$ and 9 a.m. means $t = 1$. $Q'(1) = -3 + 16 + 15 = 28$ units per hour.
b) $\frac{d}{dt}Q'(t) = -6t + 16$, $\frac{d}{dt}Q'(1) = 10$ units per hour square.

c) The approximation is $10 \times 0.25 = 2.5$ because $0.25 = 15$ minutes.
d) $Q'(1.25) = -4.6875 + 20 + 15 = 30.3125$, $Q'(1) = 28$, $Q'(1.25) - Q'(1) = 2.3125$.

23. $P(t) = -t^3 + 9t^2 + 48t + 200$, t years from now. a) $P'(t) = -3t^2 + 18t + 48$ t years from now, $P'(3) = -27 + 54 + 48 = 75$ 3 years from now.
b) $P''(t) = -6t + 18$, $P''(3) = 0$ 3 years from now.
c) 4 years from now, that is at the beginning of the fourth year, $P''(4) = -6$, $\Delta P = (-6)(\frac{1}{12}) = -0.5$ which corresponds to a decrease of 500 people.
d) one month $= \frac{1}{12} \approx 0.8333333$. $P'(4.08333333) = 71.479$, $P'(4) = 72$, $P'(4.08333333) - P'(4) = -0.521$ or a decrease of 521 people.

25. $H(t) = -16t^2 + S_0 t + H_0$. a) $H'(t) = -32t + S_0$ and the acceleration is $H''(t) = -32$.
b) The acceleration is not a function of time.
c) The force (of gravity $F = mg$, $g = -32$ ft square per sec) points in the negative direction. The object is pulled down.

27. $f(x) = x^5 - 2x^4 + x^3 - 3x^2 + 5x - 6$. $f'(x) = 5x^4 - 8x^3 + 3x^2 - 6x + 5$. $f''(x) = 20x^3 - 24x^2 + 6x - 6$. $f'''(x) = 60x^2 - 48x + 6$. $f^{(4)}(x) = 120x - 48$. Note that $f^{(4)}$ means 4th derivative while f^4 means 4th power.

29. $f(x) = \frac{1}{\sqrt{3}}x^{-1/2} - 2x^{-2} + \sqrt{2}$. $f'(x) = -\frac{1}{2\sqrt{3}}x^{-3/2} + 4x^{-3}$. $f''(x) = \frac{3}{4\sqrt{3}}x^{-5/2} - 12x^{-4}$. $f'''(x) = -\frac{15}{8\sqrt{3}}x^{-7/2} + 48x^{-5}$.

Review Problems

1. a) $f(x) = x^2 - 3x + 1$. $\frac{f(x+\Delta x)-f(x)}{\Delta x} = \frac{[(x+\Delta x)^2-3(x+\Delta x)+1]-[x^2-3x+1]}{\Delta x} = \frac{x^2+2x\Delta x+(\Delta x)^2-3x-3\Delta x+1-x^2+3x-1}{\Delta x} = \frac{2x\Delta x+(\Delta x)^2-3\Delta x}{\Delta x} = 2x + \Delta x - 3$. As $\Delta x \to 0$, this difference quotient approaches $2x - 3$, so $f'(x) = 2x - 3$.
b) $f(x) = \frac{1}{x-2}$. $\frac{f(x+\Delta x)-f(x)}{\Delta x} = \frac{[\frac{1}{(x+\Delta x)-2}]-[\frac{1}{x-2}]}{\Delta x} = [\frac{\frac{1}{x+\Delta x-2}-\frac{1}{x-2}}{\Delta x}][\frac{(x+\Delta x-2)(x-2)}{(x+\Delta x-2)(x-2)}] = \frac{(x-2)-(x+\Delta x-2)}{\Delta x(x+\Delta x-2)(x-2)} = \frac{-1}{(x+\Delta x-2)(x-2)}$. As $\Delta x \to 0$ this difference quotient approaches $\frac{-1}{(x-2)^2}$, so $f'(x) = \frac{-1}{(x-2)^2}$.

2. a) $f(x) = 6x^4 - 7x^3 + 2x + \sqrt{2}$. $f'(x) = 24x^3 - 21x^2 + 2$.
b) $f(x) = x^3 - \frac{1}{3x^5} + 2\sqrt{x} - \frac{3}{x} + \frac{1-2x}{x^3} = x^3 - \frac{1}{3}x^{-5} + 2x^{1/2} - 3x^{-1} + x^{-3} - 2x^{-2}$. $f'(x) = 3x^2 + \frac{5}{3}x^{-6} + x^{-1/2} + 3x^{-2} - 3x^{-4} + 4x^{-3} = 3x^2 + \frac{5}{3x^6} + \frac{1}{\sqrt{x}} + \frac{3}{x^2} - \frac{3}{x^4} + \frac{4}{x^3}$.
c) $y = \frac{2-x^2}{3x^2+1}$. $\frac{dy}{dx} = \frac{(3x^2+1)(-2x)-(2-x^2)(6x)}{(3x^2+1)^2} = \frac{2x(-3x^2-1-6+3x^2)}{(3x^2+1)^2} = \frac{-14x}{(3x^2+1)^2}$.

d) $y = (2x+5)^3(x+1)^2$. $\frac{dy}{dx} = (2x+5)^3(2)(x+1)(1) + (x+1)^2(3)(2x+5)^2(2) = 2(2x+5)^2(x+1)[(2x+5)+3(x+1)] = 2(2x+5)^2(x+1)(5x+8)$.
e) $f(x) = (5x^4 - 3x^2 + 2x + 1)^{10}$. $f'(x) = 10(5x^4 - 3x^2 + 2x + 1)^9(20x^3 - 6x + 2) = 20(5x^4 - 3x^2 + 2x + 1)^9(10x^3 - 3x + 1)$.
f) $f(x) = \sqrt{x^2+1} = (x^2+1)^{1/2}$. $f'(x) = \frac{1}{2}(x^2+1)^{-1/2}(2x) = \frac{x}{(x^2+1)^{1/2}}$.
g) $y = (x+\frac{1}{x})^2 - \frac{5}{\sqrt{3x}} = (x+x^{-1})^2 - \frac{5}{\sqrt{3}}x^{-1/2}$. $\frac{dy}{dx} = 2(x+x^{-1})\frac{d}{dx}(x+x^{-1}) + \frac{5}{2\sqrt{3}}x^{-3/2}\frac{d}{dx}(x) = 2(x+x^{-1})(1-x^{-2}) + \frac{5}{2\sqrt{3}}x^{-3/2} = 2(x+\frac{1}{x})(1-\frac{1}{x^2}) + \frac{5}{2\sqrt{3}x^{3/2}}$.
h) $y = (\frac{x+1}{1-x})^2$. $\frac{dy}{dx} = 2(\frac{x+1}{1-x})\frac{d}{dx}(\frac{x+1}{1-x}) = 2(\frac{x+1}{1-x})\frac{(1-x)(1)-(x+1)(-1)}{(1-x)^2} = 2(\frac{x+1}{1-x})\frac{2}{(1-x)^2} = \frac{4(x+1)}{(1-x)^3}$.
i) $f(x) = (3x+1)\sqrt{6x+5} = (3x+1)(6x+5)^{1/2}$. $f'(x) = (3x+1)(\frac{1}{2})(6x+5)^{-1/2}(6) + (6x+5)^{1/2}(3) = \frac{3(3x+1)}{(6x+5)^{1/2}} + 3(6x+5)^{1/2} = \frac{9(3x+2)}{\sqrt{6x+5}}$.
j) $f(x) = \frac{(3x+1)^3}{(1-3x)^4}$. $f'(x) = \frac{(1-3x)^4\frac{d}{dx}(3x+1)^3 - (3x+1)^3\frac{d}{dx}(1-3x)^4}{(1-3x)^8} = \frac{(1-3x)^4[3(3x+1)^2(3)] - (3x+1)^3[4(1-3x)^3(-3)]}{(1-3x)^8} = \frac{3(1-3x)^3(3x+1)^2[3(1-3x)+4(3x+1)]}{(1-3x)^8} = \frac{3(3x+1)^2(3-9x+12x+4)}{(1-3x)^5} = \frac{3(3x+1)^2(3x+7)}{(1-3x)^5}$.
k) $y = \sqrt{\frac{1-2x}{3x+2}} = (\frac{1-2x}{3x+2})^{1/2}$. $\frac{dy}{dx} = \frac{1}{2}(\frac{1-2x}{3x+2})^{-1/2}\frac{d}{dx}(\frac{1-2x}{3x+2}) = \frac{1}{2}(\frac{1-2x}{3x+2})^{-1/2}\frac{(3x+2)(-2)-(1-2x)(3)}{(3x+2)^2} = (\frac{1}{2})(\frac{(1-2x)^{-1/2}}{(3x+2)^{-1/2}})(\frac{-6x-4-3+6x}{(3x+2)^2}) = \frac{-7}{2(1-2x)^{1/2}(3x+2)^{3/2}}$.

3. a) $f(x) = x^2 - 3x + 2$. $f'(x) = 2x - 3$. $f(1) = 0$. The slope of the tangent line at $(1,0)$ is $m = f'(1) = -1$. The equation of the tangent line is thus $y - 0 = -(x-1)$ or $y = -x + 1$.
b) $f(x) = \frac{4}{x-3}$. $f'(x) = \frac{-4}{(x-3)^2}$. $f(1) = -2$. The slope of the tangent line at $(1,-2)$ is $m = f'(1) = -1$. The equation of the tangent line is thus $y - (-2) = -(x-1)$ or $y = -x - 1$.
c) $f(x) = \frac{x}{x^2+1}$. $f'(x) = \frac{(x^2+1)(1)-x(2x)}{(x^2+1)^2} = \frac{1-x^2}{(x^2+1)^2}$. $f(0) = 0$. The slope of the tangent line at $(0,0)$ is $m = f'(0) = 1$. The equation of the tangent line is thus $y - 0 = x - 0$ or $y = x$.
d) $f(x) = \sqrt{x^2+5} = (x^2+5)^{1/2}$. $f'(x) = \frac{1}{2}(x^2+5)^{-1/2}(2x) = \frac{x}{\sqrt{x^2+5}}$. $f(-2) = 3$. The slope of the tangent line at $(-2,3)$ is $m = f'(-2) = -\frac{2}{3}$. The equation of the tangent line is thus $y - 3 = -\frac{2}{3}(x+2)$ or $y = -\frac{2}{3}x + \frac{5}{3}$.

4. a) The rate of change of $f(t) = t^3 - 4t^2 + 5t\sqrt{t} - 5 = t^3 - 4t^2 + 5t^{3/2} - 5$ is $f'(t) = 3t^2 - 8t + \frac{15}{2}t^{1/2}$ at any value of $t \geq 0$ and when $t = 4$, $f'(4) = 48 - 32 + \frac{15}{2}(2) = 31$.
b) The rate of change of $f(t) = t^3(t^2-1) = t^5 - t^3$ is $f'(t) = 5t^4 - 3t^2$ at any value of t and when $t = 0$, $f'(0) = 0$.

5. a) $f(t) = t^2 - 3t + \sqrt{t} = t^2 - 3t + t^{1/2}$. The percentage rate of change is $100(\frac{f'(t)}{f(t)}) = 100(\frac{2t-3+\frac{1}{2}t^{-1/2}}{t^2-3t+t^{1/2}}) = 100(\frac{4t^{3/2}-6t^{1/2}+1}{2t^{5/2}-6t^{3/2}+2t})$.
b) $f(t) = t^2(3-2t)^3$. The percentage rate of change is $100(\frac{f'(t)}{f(t)}) =$

$100(\frac{(t^2)(3)(3-2t)^2(-2)+(3-2t)^3(2t)}{t^2(3-2t)^3}) = 100(3-2t)^2(\frac{-6t^2+6t-4t^2}{t^2(3-2t)^3}) = 100(\frac{-10t^2+6t}{t^2(3-2t)}) = 100(\frac{2t(3-5t)}{t^2(3-2t)}) = 100(\frac{(3-5t)}{t(3-2t)})$.
c) $f(t) = \frac{1}{t+1} = (t+1)^{-1}$. The percentage rate of change is $100(\frac{f'(t)}{f(t)}) = 100(\frac{-(t+1)^{-2}}{(t+1)^{-1}}) = -100(\frac{1}{t+1})$.

6. a) Since $N(x) = 6x^3 + 500x + 8,000$ is the number of people using the system after x weeks, the rate at which use of the system is changing after x weeks is $N'(x) = 18x^2 + 500$ people per week and the rate after 8 weeks is $N'(8) = 1,652$ people per week.
b) The actual increase in the use of the system during the 8th week is $N(8) - N(7) = 1,514$ people.

7. a) Since $Q(x) = 50x^2 + 9,000x$ is the weekly output when x workers are employed, the marginal output is $Q'(x) = 100x + 9,000$. The change in output due to an increase from 30 to 31 workers is approximately $Q'(30) = 12,000$ units.
b) The actual increase in output is $Q(31) - Q(30) = 12,050$ units.

8. Since the population t months from now will be $P(t) = 3t + 5t^{3/2} + 6,000$, the rate of change of the population will be $P'(t) = 3 + \frac{15}{2}t^{1/2}$, and the percentage rate of change 4 months from now will be $100[\frac{P'(4)}{P(4)}] = 100[\frac{18}{6,052}] \approx 0.30$ percent per month.

9. Since daily output is $Q(L) = 20,000L^{1/2}$ units when L worker-hours are used, a change in the work force from 900 worker-hours to 885 worker-hours ($\Delta L = -15$) results in a change in the output of $\Delta Q = Q(900) - Q(885) \approx Q'(900)(-15)$. Since $Q'(L) = 10,000L^{-1/2}$ and $Q'(900) = 10,000(900)^{-1/2} = \frac{1,000}{3}$ it follows that $\Delta Q \approx \frac{1,000}{3}(-15) = -5,000$, that is, a decrease in output of 5,000 units.

10. The gross national product t years after 1990 is $N(t) = t^2 + 6t + 300$ billion dollars. The derivative is $N'(t) = 2t + 6$. At the beginning of the second quarter of 1994, $t = 4.25$. The change in t during this quarter is $\Delta t = 0.25$. Hence the percentage change in N is $100\frac{N'(4.25)\Delta t}{N(4.25)} = 100\frac{[2(4.25)+6](0.25)}{4.25^2+6(4.25)+300} \approx 1.055$ percent. {If the year had been 1998 instead of 1994, the percentage change would have been $100\frac{N'(8.25)\Delta t}{N(8.25)} = 100\frac{[2(8.25)+6](0.25)}{8.25^2+6(8.25)+300} \approx 1.347$ percent. }

11. Let A denote the level of air pollution and p the population. Then $A = kp^2$, where k is a positive constant of proportionality. If the population increases by 5 percent, the change in population is $\Delta p = 0.05p$. The corresponding increase in the level of air pollution is $\Delta A = A(p + 0.05p) - A(p) \approx A'(p)(0.05p) = 2kp(0.05p) = 0.1kp^2 = 0.1A$. That is, an increase of 5 percent in the population causes an increase of 10% in the level of pollution.

12. The output is $Q(L) = 600L^{2/3}$. The derivative is $Q'(L) = 400L^{-1/3}$. We are

given that the percentage change in Q is 1 percent, and that the goal is to find the percentage change in L, which can be represented as $100\frac{\Delta L}{L}$. Apply the formula for the percentage change in $Q = 100\frac{Q'(L)\Delta L}{Q(L)}$ with 1 on the left-hand side and solve for $\frac{\Delta L}{L}$ as follows: $1 \approx 100\frac{400L^{-1/3}\Delta L}{600L^{2/3}} = 100(\frac{2}{3})(\frac{\Delta L}{L})$, $100\frac{\Delta L}{L} \approx \frac{3}{2} = 1.5$ percent. That is, an increase in labor of approximately 1.5 percent is required to increase output by 1 percent.

13. a) $y = 5u^2 + u - 1$, $u = 3x + 1$. $\frac{dy}{du} = 10u + 1$, $\frac{du}{dx} = 3$, $\frac{dy}{dx} = \frac{dy}{du}\frac{du}{dx} = (10u + 1)(3) = 3[10(3x + 1) + 1] = 3(30x + 11)$.
b) $y = \frac{1}{u^2}$, $u = 2x + 3$, $\frac{dy}{du} = \frac{-2}{u^3}$, $\frac{du}{dx} = 2$, $\frac{dy}{dx} = \frac{dy}{du}\frac{du}{dx} = (\frac{-2}{u^3})(2) = \frac{-4}{(2x+3)^3}$.

14. a) $y = u^3 - 4u^2 + 5u + 2$, $u = x^2 + 1$. $\frac{dy}{du} = 3u^2 - 8u + 5$, $\frac{du}{dx} = 2x$, $\frac{dy}{dx} = \frac{dy}{du}\frac{du}{dx} = (3u^2 - 8u + 5)(2x)$. When $x = 1$, $u = 2$, and so $\frac{dy}{dx} = [3(2^2) - 8(2) + 5][2(1)] = 2$.
b) $y = \sqrt{u} = u^{1/2}$, $u = x^2 + 2x - 4$, $\frac{dy}{du} = \frac{1}{2u^{1/2}}$, $\frac{du}{dx} = 2x + 2$, $\frac{dy}{dx} = \frac{dy}{du}\frac{du}{dx} = (\frac{1}{2u^{1/2}})(2x + 2) = \frac{x+1}{u^{1/2}}$. When $x = 2$, $u = 4$, and so $\frac{dy}{dx} = \frac{2+1}{4^{1/2}} = \frac{3}{2}$

15. Since $C(q) = 0.1q^2 + 10q + 400$ is the total cost of producing q units and $q(t) = t^2 + 50t$ is the number of units produced during the first t hours, then $\frac{dC}{dq} = 0.2q + 10$ (dollars per unit) and $\frac{dq}{dt} = 2t + 50$ (units per hour). The rate of change of cost with respect to time is $\frac{dC}{dt} = \frac{dC}{dq}\frac{dq}{dt} = (0.2q + 10)(2t + 50)$ dollars per hour. After 2 hours, $t = 2$, $q(2) = 104$, and so $\frac{dC}{dt} = [0.2(104) + 10][2(2) + 50] = \$1,663.20$ per hour.

16. a) $s(t) = 2t^3 - 21t^2 + 60t - 25$ for $1 \le t \le 6$. $v(t) = 6(t^2 - 7t + 10) = 6(t - 2)(t - 5)$. The positive roots are $t = 2$ and $t = 5$. $v(t) > 0$ for $1 \le t \le 2$ and $5 \le t \le 6$, so the object travels to the right. For $2 \le t \le 5$, $v(t) < 0$ so the object travels to the left. $a(t) = 6(2t - 7) = 0$ if $t = \frac{7}{2}$. $a(t) > 0$ for $\frac{7}{2} \le t \le 6$ so the object is speeding up. For $1 \le t \le \frac{7}{2}$ it is slowing down.
b) $s(1) = 2 - 21 + 60 - 25 = 16$, $s(2) = 16 - 84 + 120 - 25 = 27$, $s(5) = 250 - 21(25) + 300 - 25 = 0$, and $s(6) = 432 - 21(36) + 360 - 25 = 11$. Thus $\Delta s = (27 - 16) + (27 - 0) + (11 - 0) = 49$.

17. a) $s(t) = \frac{2t+1}{t^2+12}$ for $0 \le t \le 3$. $v(t) = \frac{(t^2+12)(2)-(2t+1)(2t)}{(t^2+12)^2} = \frac{2t^2+24-4t^2-2t}{(t^2+12)^2} = \frac{-2t^2-2t+24}{(t^2+12)^2} = 0$ if $t^2 + t - 12 = (t + 4)(t - 3) = 0$. $v(t) < 0$ for $t < 3$, so the object travels to the left. $a(t) = \frac{2}{(t^2+12)^4}[(t^2 + 12)^2(-2t - 1) - (-t^2 - t + 12)(2)(t^2 + 12)(2t)] = \frac{2}{(t^2+12)^3}(-2t^3 - 24t - t^2 - 12 + 4t^3 + 4t^2 - 48t) = \frac{2}{(t^2+12)^3}(2t^3 + 3t^2 - 72t - 12)$. $a(t) > 0$ for $0 < t < 3$ the object is slowing down.
b) $s(0) = \frac{1}{12}$, $s(3) = \frac{1}{3}$, Thus $\Delta s = \frac{1}{3} - \frac{1}{12} = \frac{1}{4}$. Note: $a(t) = 0$ when $t = 5.388$, $t = -6.722$, and $t = -0.166$, all of which are outside of the relevant interval.

18. $s(t) = 88t - 8t^2$, $v(t) = 88 - 16t = 0$ at the instant t_1 the car stops. Thus $11 - 2t_1 = 0$ or $t_1 = 5.5$ seconds. The distance required to stop is then $s(\frac{11}{2}) = 8[(11)(\frac{11}{2}) - (\frac{11}{2})^2] = 22(11) = 242$ feet.

19. The population is $p(t) = 10 - \frac{20}{(t+1)^2} = 10 - (t+1)^{-2}$ and the carbon monoxide level is $c(p) = 0.8\sqrt{p^2+p+139} = 0.8(p^2+p+139)^{1/2}$ units. By the chain rule, the rate of change of the carbon monoxide level with respect to time is $\frac{dc}{dt} = \frac{dc}{dp}\frac{dp}{dt} = [0.4(p^2+p+139)^{-1/2}][40(t+1)^{-3}] = \frac{0.4(2p+1)}{\sqrt{p^2+p+139}}\frac{40}{(t+1)^3}$. At $t = 1$ $p = p(1) = 10 - \frac{20}{4} = 5$ and $c = c(5) = 0.8\sqrt{169} = 10.4$. The percentage rate of change is $100\frac{dc/dt}{c} = [\frac{0.4(10+1)}{\sqrt{169}}\frac{40}{(1+1)^3}][\frac{1}{10.4}] \approx 16.27$ percent per year.

20. a) $5x + 3y = 12$, $5(1) + 3\frac{dy}{dx} = 0$ or $\frac{dy}{dx} = -\frac{5}{3}$.
b) $x^2y = 1$, $x^2\frac{dy}{dx} + y(2x) = 0$ or $\frac{dy}{dx} = -\frac{2xy}{x^2} = -\frac{2y}{x}$.
c) $(2x+3y)^5 = x+1$, $5(2x+3y)^4(2+3\frac{dy}{dx}) = 1$, $10(2x+3y)^4 + 15(2x+3y)^4\frac{dy}{dx} = 1$, $15(2x+3y)^4\frac{dy}{dx} = 1 - 10(2x+3y)^4$, $\frac{dy}{dx} = \frac{1-10(2x+3y)^4}{15(2x+3y)^4}$.
d) $(1-2xy^3)^5 = x+4y$, $5(1-2xy^3)^4\frac{d}{dx}(1-2xy^3) = 1+4\frac{dy}{dx}$, $5(1-2xy^3)^4(-6xy^2\frac{dy}{dx} - 2y^3) = 1+4\frac{dy}{dx}$, $-30xy^2(1-2xy^3)^4\frac{dy}{dx} - 10y^3(1-2xy^3)^4 = 1+4\frac{dy}{dx}$, $[-4-30xy^2(1-2xy^3)^4]\frac{dy}{dx} = 1+10y^3(1-2xy^3)^4$, $\frac{dy}{dx} = \frac{1+10y^3(1-2xy^3)^4}{-4-30xy^2(1-2xy^3)^4}$.

21. a) $xy^3 = 8$, $x(3y^2\frac{dy}{dx}) + y^3 = 0$ or $\frac{dy}{dx} = \frac{-y^3}{3xy^2} = -\frac{y}{3x}$. At $x = 1$, the original equation gives $y^3 = 8$ or $y = 2$. To find the slope of the tangent at the point $(1,2)$, substitute into the equation for $\frac{dy}{dx}$ to get the slope $m = \frac{dy}{dx} = -\frac{2}{3}$.
b) $x^2y - 2xy^3 + 6 = 2x + 2y$, $x^2\frac{dy}{dx} + y(2x) - 2[x(3y^2\frac{dy}{dx}) + y^3(1)] = 2 + 2\frac{dy}{dx}$, $x^2\frac{dy}{dx} + 2xy - 6xy^2\frac{dy}{dx} - 2y^3 = 2 + 2\frac{dy}{dx}$. At $x = 0$, the original equation gives $0^2y - 2(0)y^3 + 6 = 2(0) + 2y$, $6 = 2y$, or $y = 3$. To find the slope of the tangent line at $(0,3)$, substitute into the derivative equation and solve for $\frac{dy}{dx}$ to get $0^2\frac{dy}{dx} + 2(0)y - 6(0)y^2\frac{dy}{dx} - 2y^3 = 2 + 2\frac{dy}{dx}$, $-54 = 2 + 2\frac{dy}{dx}$ or the slope is $m = \frac{dy}{dx} = -28$.

22. By the approximation formula from section 4, $\Delta y \approx \frac{dy}{dx}\Delta x$. To find $\frac{dy}{dx}$, differentiate the equation $Q = x^3 + 2xy^2 = 2y^3$ implicitly WRT x, where Q is a constant representing the current level of output. You get $0 = 3x^2 + 4xy\frac{dy}{dx} + 2y^2 + 6y^2\frac{dy}{dx}$ or $\frac{dy}{dx} = -\frac{3x^2+2y^2}{4xy+6y^2}$. At $x = 10$ and $y = 20$, $\frac{dy}{dx} = -\frac{3(10)^2+2(20)^2}{4(10)(20)+6(20)^2} \approx -0.344$. Let's use the approximation formula with $\frac{dy}{dx} \approx -0.344$ and $\Delta x = 0.5$ to get $\Delta y \approx -0.344(0.5) = -0.172$ unit. That is, to maintain the current level of output, input y should be decreased by approximately 0.172 unit to offset a 0.5 unit increase in input x.

23. a) $f(x) = 6x^5 - 4x^3 + 5x^2 - 2x + \frac{1}{x}$, $f'(x) = 30x^4 - 12x^3 + 10x - 2 - \frac{1}{x^2}$ and $f''(x) = 120x^3 - 24x^2 + 10 + \frac{2}{x^3}$.
b) $y = (3x^2+2)^4$, $\frac{dy}{dx} = 4(3x^2+2)^3(6x) = 24x(3x^2+2)^3$, $\frac{d^2y}{dx^2} = 24x[3(3x^2+2)^2(6x)] + (3x^2+2)^3(24) = 24(3x^2+2)^2[18x^2+3x^2+2] = 24(3x^2+2)^2[21x^2+2]$.
c) $f(x) = \frac{x-1}{(x+1)^2}$, then $f'(x) = \frac{(x+1)^2(1)-(x-1)(2)(x+1)(1)}{(x+1)^4} = \frac{(x+1)[(x+1)-2(x-1)]}{(x+1)^4} = \frac{x+1-2x+2}{(x+1)^3} = \frac{3-x}{(x+1)^3}$ and $f''(x) = \frac{(x+1)^3(-1)-(3-x)(3)(x+1)^2(1)}{(x+1)^6} = \frac{(x+1)^2[-(x+1)-3(3-x)]}{(x+1)^6} = \frac{-x-1-9+3x}{(x+1)^4} = \frac{2x-10}{(x+1)^4} =$

$\frac{2(x-5)}{(x+1)^4}$.

24. $3x^2-2y^2=6$, $6x-4y\frac{dy}{dx}=0$ or $\frac{dy}{dx}=\frac{6x}{4y}=\frac{3x}{2y}$. Thus $\frac{d^2y}{dx^2}=\frac{2y(3)-3x[2\frac{dy}{dx}]}{(2y)^2}=\frac{3y-3x\frac{dy}{dx}}{2y^2}$. Since $\frac{dy}{dx}=\frac{3x}{2y}$ $\frac{d^2y}{dx^2}=\frac{3y-3x(\frac{3x}{2y})}{2y^2}=\frac{6y^2-9x^2}{4y^3}$. From the original equation $6y^2-9x^2=3(2y^2-3x^2)=-3(3x^2-2y^2)=-3(6)=-18$ and so $\frac{d^2y}{dx^2}=\frac{-18}{4y^3}=\frac{-9}{2y^3}$.

25. The production function is $Q(t)=-t^3+9t^2+12t$. a) The rate of production is the derivative $Q'(t)=-3t^2+18t+12$. At 9:00 a.m. $t=1$ and the rate of production is $Q'(1)=27$ units per hour.
b) The rate of change of the rate of production is the second derivative $Q''(t)=-6t+18$. At 9:00 a.m., this rate is $Q''(1)=12$ units per hour per hour.
c) The change in the rate of production between 9:00 a.m. and 9:06 a.m. is given by the approximation formula $\Delta Q'\approx Q''(t)\Delta t$. At $t=1$ and $\Delta t=0.1$ hour, this gives $Q''(1)\Delta t=12(0.1)=1.2$ units per hour.
d) The actual change in the worker's rate of production between 9:00 a.m. and 9:06 a.m. is $Q'(1.1)-Q'(1)\approx 1.17$ units per hour.

26. a) $y=2x^5+5x^4-2x+\frac{1}{x}$, $\frac{dy}{dx}=10x^4+20x^3-2-\frac{1}{x^2}$, $\frac{d^2y}{dx^2}=40x^3+60x^2+\frac{2}{x^3}$, $\frac{d^3y}{dx^3}=120x^2+120x-\frac{6}{x^4}$, $\frac{d^4y}{dx^4}=240x+120+\frac{24}{x^5}$.
b) $f(x)=\sqrt{3x}+\frac{3}{2x^2}=\sqrt{3}x^{1/2}+\frac{3}{2}x^{-2}$, $f'(x)=\frac{1}{2}\sqrt{3}x^{-1/2}-3x^{-3}$, $f''(x)=-\frac{1}{4}\sqrt{3}x^{-3/2}+9x^{-4}$, $f'''(x)=\frac{3}{8}\sqrt{3}x^{-5/2}-36x^{-5}$, $f^{(4)}(x)=-\frac{15}{16}\sqrt{3}x^{-7/2}+180x^{-6}=-\frac{15\sqrt{3}}{16x^{7/2}}+\frac{180}{x^6}$. Note in this last derivative that the superscript (4) is in parentheses. f^4, without parentheses, would have meant the 4th power of f.

Chapter 3

Additional Applications of the Derivative

3.1 Relative Extrema; Curve Sketching With the First Derivative

1.

x	$-\infty$		-2		2		∞
$f'(x)$		$-$		$+$		$-$	
		Falling		Rising		Falling	

3.

x	$-\infty$		-4		-2		0		2		∞
$f'(x)$		$+$		$-$		$-$		$+$		$-$	
		Rising		Falling		Falling		Rising		Falling	

5. $f(x) = x^2 - 4x + 5$. $f'(x) = 2x - 4 = 2(x - 2) = 0$ when $x = 2$. $f(2) = 1$. If $x > 2$ $f'(x) > 0$ else $f'(x) < 0$ which indicates that $m(2, 1)$ is a relative minimum.

↘ 2 ↗
min

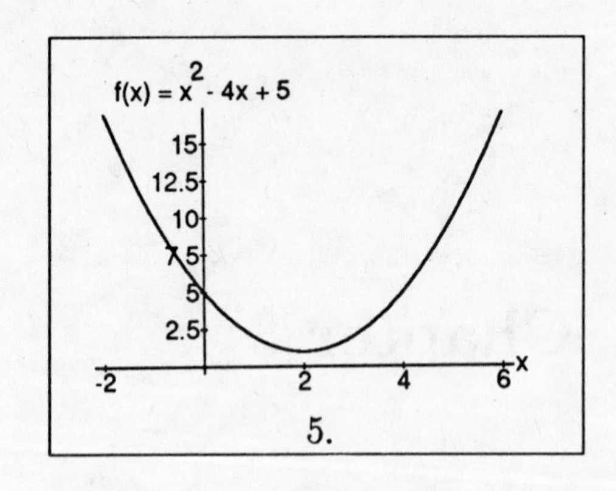

5.

7. $f(x) = x^3 - 3x - 4$. $f'(x) = 3x^2 - 3 = 3(x+1)(x-1) = 0$ when $x = -1, 1$. $f(-1) = -2$, $f(1) = -6$. If $x < -1$ and $x > 1$ $f'(x) > 0$ else $f'(x) < 0$ which indicates that $M(-1, -2)$ is a relative maximum, $m(1, -6)$ is a relative minimum.

↗ −1 ↘ 1 ↗
max min

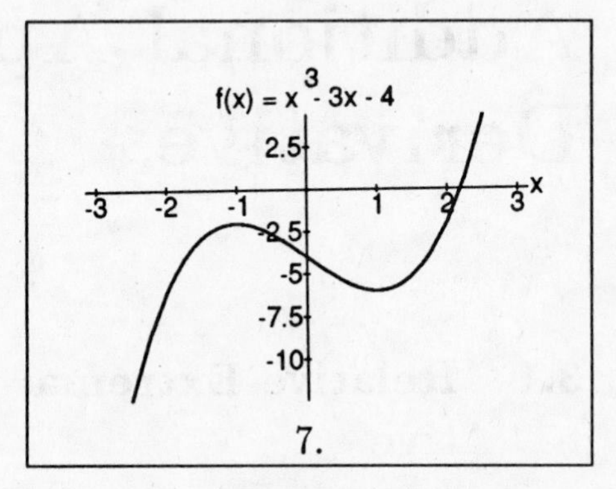

7.

9. $f(x) = x^5 - 5x^4 + 100$. $f'(x) = 5x^4 - 20x^3 = 5x^3(x - 4) = 0$ when $x = 0, 4$. $f(0) = 100$, $f(4) = -156$. $M(0, 100)$ is a relative maximum, $m(4, -156)$ is a relative minimum.

↗ 0 ↘ 4 ↗
max min

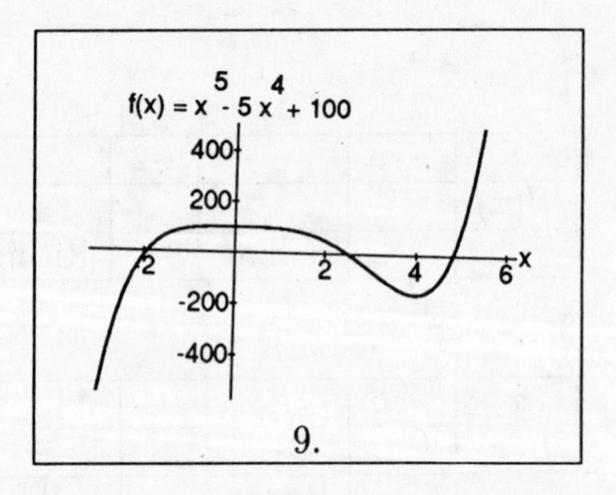

9.

11. $f(x) = 3x^4 - 8x^3 + 6x^2 + 2$. $f'(x) = 12x^3 - 24x^2 + 12x = 12x(x^2 - 2x + 1) = 12(x-1)^2 = 0$ when $x = 0, 1$. $f(0) = 2$, $f(1) = 3$. $m(0, 2)$ is a relative minimum.

↘ 0 ↗ 1 ↗
min none

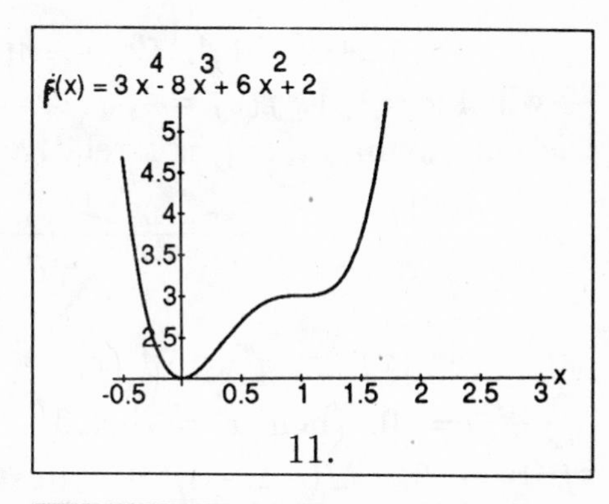

11.

13. $f(x) = 2x^3 + 6x^2 + 6x + 5$. $f'(x) = 6x^2 + 12x + 6 = 6(x^2 + 2x + 1) = 6(x+1)^2 \geq 0$. There are no relative extrema. Note that $f'(-1) = 0$ but $(-1, 3)$ is not an extremum.

↗

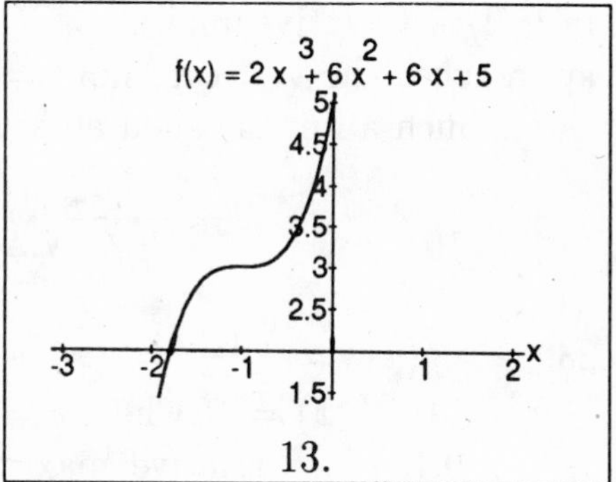

13.

15. $f(x) = (x-1)^5$. $f'(x) = 5(x-1)^4 = 0$ when $x = 1$. $f(1) = 0$ but $(1, 0)$ is not a relative extremum.

↗ 1 ↗
none

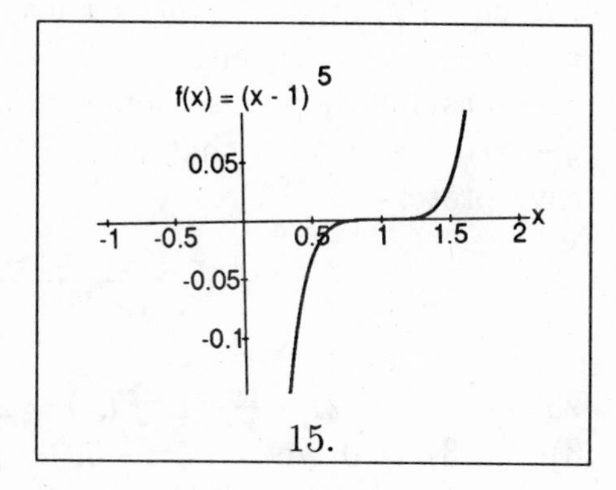

15.

17. $f(x) = (x^2-1)^5$. $f'(x) = 5(x^2-1)^4(2x) = 0$ when $x = -1, 0, 1$. $f(-1) = 0$, $f(0) = -1$, $f(1) = 0$. $m(0, -1)$ is a relative minimum. $(-1, 0)$ and $(1, 0)$ are not extrema. Note the symmetry WRT the $y-$axis since x can be replaced by $-x$ without changing the original equation.

↘ −1 ↘ 0 ↗ 1 ↗
none min none

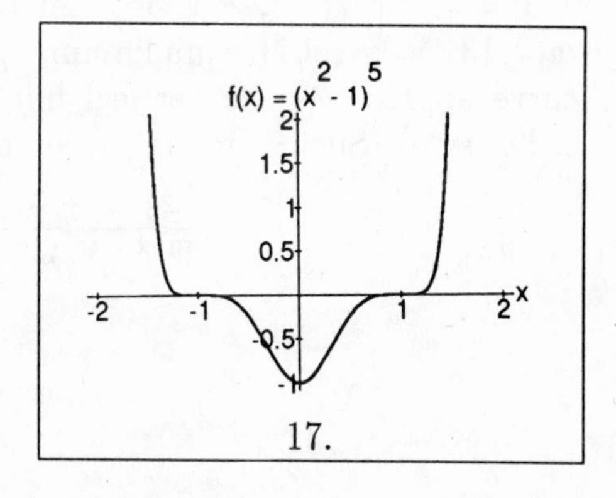

17.

19. $f(x) = (x^3-1)^4$. $f'(x) = 4(x^3-1)^3(3x^2) = 0$ when $x = 0, 1$. $f(0) = 1$, $f(1) = 0$. $(0, 1)$ is not an extremum, $m(1, 0)$ is a relative minimum.

$\searrow$ 0 $\nearrow$ 1 $\nearrow$
none min

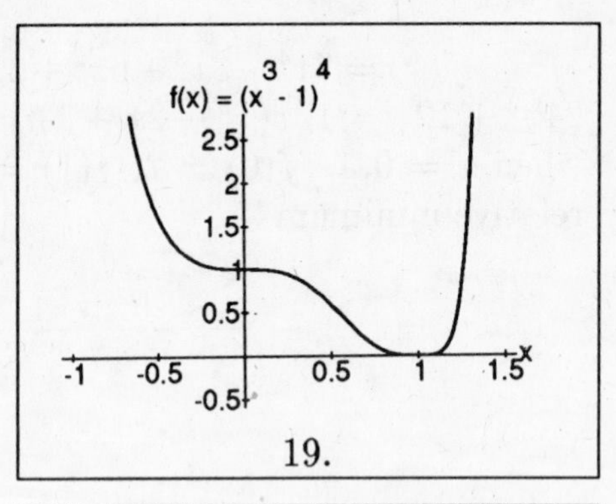

19.

21. $f(x) = \frac{x^2}{x+2}$. $f'(x) = \frac{(x+2)(2x)-x^2}{(x+2)^2} = \frac{x^2+4x}{(x+2)^2} = 0$ when $x = -4, 0$. $f(-4) = -8$, $f(0) = 0$. $M(-4, -8)$ is a relative maximum, $m(0, 0)$ is a relative minimum. Note that the curve approaches the vertical line $x = -2$ (from either side). Such a line is called an asymptote.

$\nearrow$ -4 $\searrow$ -2 $\searrow$ 0 $\nearrow$
max V.A. min

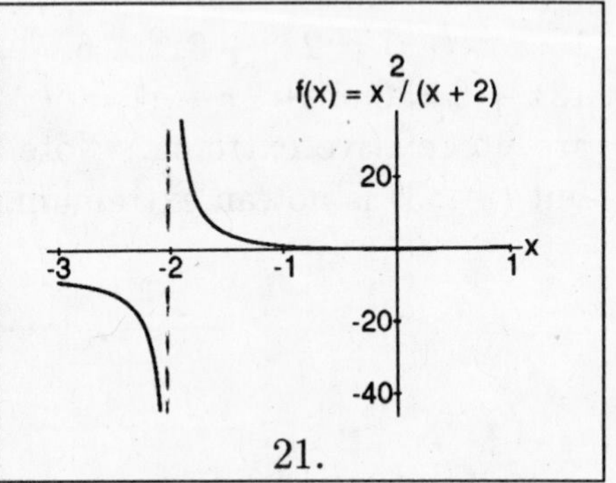

21.

23. $f(x) = \frac{1}{x^2-9} = (x^2-9)^{-1}$. $f'(x) = -(x^2-9)^{-2}(2x) = 0$ when $x = 0$. $f(0) = -\frac{1}{9}$. $M(0, -0.11)$ is a relative maximum. Note that the curve approaches the vertical lines $x = -3$ and $x = 3$ (from either side). Such lines are called asymptotes. Also note symmetry WRT the $y-$axis. Note also that $x = 0$ is a horizontal asymptote.

$\nearrow$ -3 $\nearrow$ 0 $\searrow$ 3 $\searrow$
V.A. max V.A.

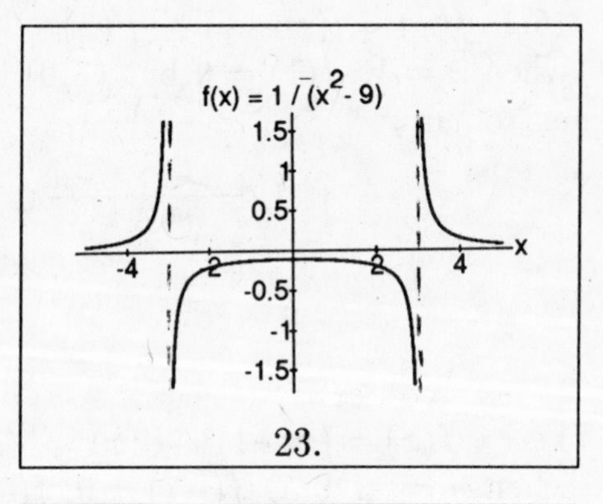

23.

25. $f(x) = 2x + \frac{18}{x} + 1$. $f'(x) = 2 - \frac{18}{x^2} = \frac{2}{x^2}(x+3)(x-3) = 0$ when $x = -3, 3$. $f(-3) = -11$, $f(3) = 13$. $M(-3, -11)$ is a relative maximum, $m(3, 13)$ is a relative minimum. Note that the curve approaches the vertical line $x = 0$ (from either side). Such a line is called an asymptote.

$\nearrow$ -3 $\searrow$ 0 $\searrow$ 3 $\nearrow$
max V.A. min

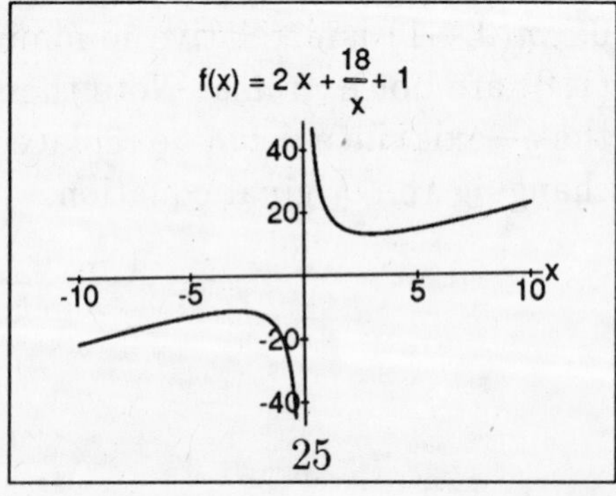

25

27. $f(x) = 1 + x^{1/3}$. $f'(x) = \frac{1}{3}x^{-2/3} > 0$. There are no relative extrema. Note that $f'(0)$ is not defined, which signifies that the tangent line at $(0, 1)$ is vertical.

↗

27.

29. $f(x) = 2 + (x-1)^{2/3}$. $f'(x) = \frac{2}{3}(x-1)^{-1/3}$ is not defined when $x = 1$. There are no relative extrema for which the tangent line is horizontal, but $(1, 2)$ is a relative minimum because there are no lower values of y in the vicinity of $(1, 2)$. Note also that the tangent line at $(1, 2)$ is vertical. We say that there is a cusp at $m(1, 2)$.

↘ 1 ↗
min

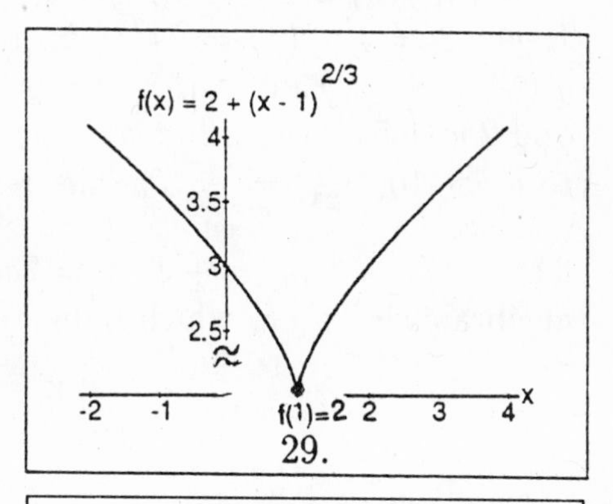

29.

31. $f(x) = 3x^5 - 5x^3 + 4$. $f'(x) = 15x^4 - 15x^2 = 15x^2(x^2-1) = 15x^2(x-1)(x+1) = 0$ at $x = -1$, $x = 0$ and $x = 1$. There is a relative maximum at $(-1, 6)$ and a relative minimum at $(1, 2)$. The point $(0, 4)$ is not an extremum, even though the tangent line is horizontal for this point.

↗ −1 ↘ 0 ↘ 1 ↗
max no ext. min

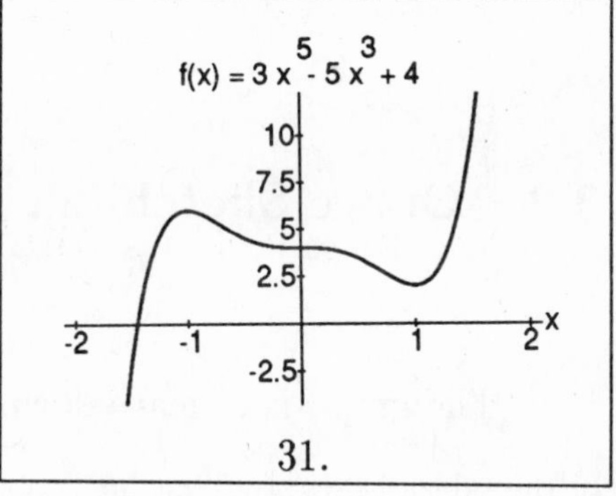

31.

33. a) Since $f'(x) > 0$ when $x < -5$ and $x > 1$, $f(x)$ is increasing on these intervals.
b) Since $f'(x) < 0$ when $-5 < x < 1$, $f(x)$ is decreasing on this interval.
c) Since $f(-5) = 4$ and $f(1) = -1$, $f(x)$ has a relative maximum at $(-5, 4)$ and a relative minimum at $(1, -1)$.

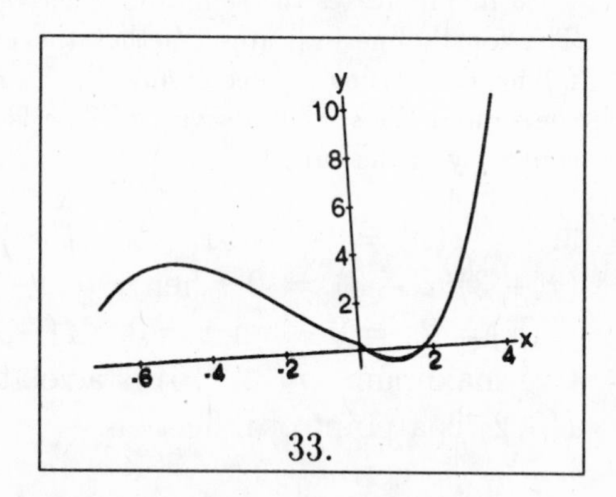
33.

35. a) Since $f'(x) > 0$ when $x > 2$, $f(x)$ is increasing on this interval.
b) Since $f'(x) < 0$ when $x < 0$ and $0 < x < 2$, $f(x)$ is decreasing on these intervals. Thus $f(x)$ has a relative minimum at $x = 2$.
c) Since $f(x)$ is undefined at $x = 0$, $f(x)$ has a break at $x = 0$.

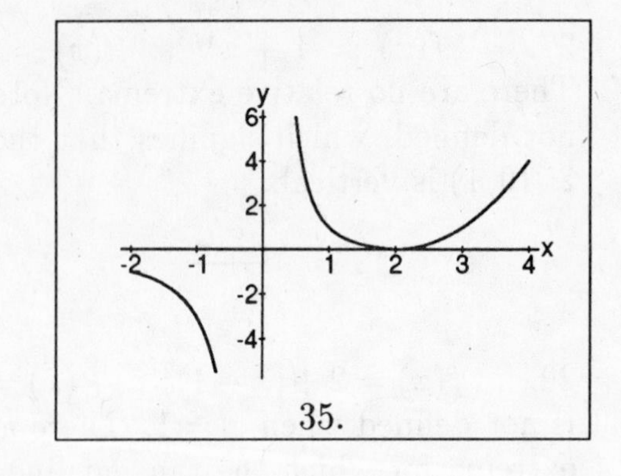

35.

37. Since $f(x) = ax^2+bx+c$ crosses the $y-$axis at (0,3), $f(0) = 3$ and $a(0^2) + b(0) + c = 3$.
Hence $c = 3$. $f(x) = ax^2 + bx + 3$ has a relative maximum at (5,12) so $f'(5) = 0$. Thus $f'(x) = 2ax+b$ and $2a(5)+b = 0$ or $b = -10a$. Note that (5,12) is on the curve, so $f(5) = 12$ and $25a + 5b + 3 = 12$, $25a + 5b = 25a + 5(-10a) = 9$, or $a = -\frac{9}{25}$. Substituting for b leads to $b = -10(-\frac{9}{25}) = \frac{18}{5}$. The desired function is $f(x) = -\frac{9}{25}x^2 + \frac{18}{5} + 3$.

39. $y = ax^2 + bx + c$, $y' = 2ax + b = 0$ at $x_1 = -\frac{b}{2a}$. The first derivative changes its algebraic sign at x_1 which indicates an extremum.

3.2 Curve Sketching: Concavity and the Second Derivative Test

1. The graph is concave downward ($f'' < 0$) for $x < 2$, upward ($f'' > 0$) for $x > 2$.

In the tables below, the first line exhibits values of $-\infty < x < \infty$. The second line shows values of y, The third line indicates the sign of first derivative.
The fourth line indicates whether the curve is rising or falling (increasing or decreasing). The fifth line reflects the sign of the second derivative. The last line shows the concavity of the curve.

3. $f(x) = \frac{x^3}{3} - 9x + 2$. $f'(x) = x^2 - 9 = (x+3)(x-3) = 0$ when $x = -3$ and $x = 3$. $f''(x) = 2x = 0$ when $x = 0$. $M(-3, 20)$ is a relative maximum, $m(3, -16)$ is a relative minimum. $I(0, 2)$ is a point of inflection.

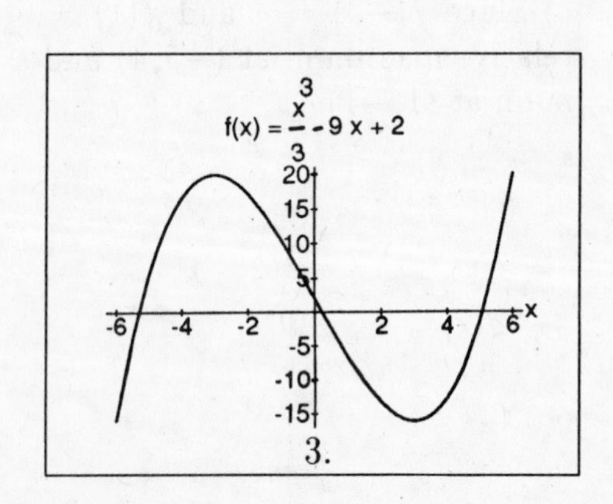

3.

x	$-\infty$		-3		0		3		∞
$f(x)$	$-\infty$		20		2		-16		∞
$f'(x)$	+	+	0	−	−	−	0	+	+
		inc		dec	dec	dec	0	inc	
$f''(x)$	−	−	−	−	0	+	+	+	+
		down	down	down		up	up	up	

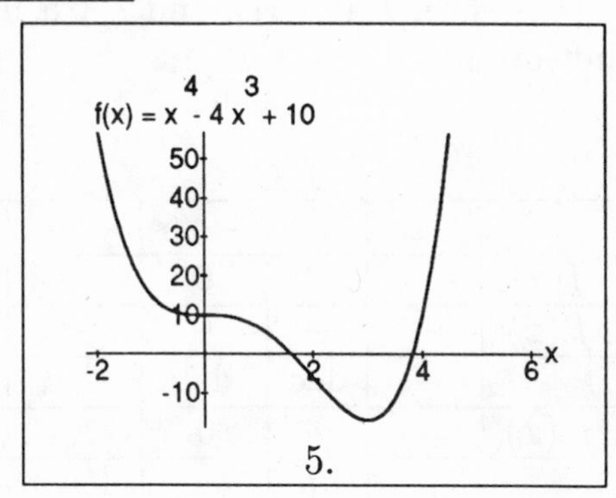

5.

5. $f(x) = x^4 - 4x^3 + 10$. $f'(x) = 4x^3 - 12x^2 = 4x^2(x-3) = 0$ when $x = 0, 3$. $f''(x) = 12x^2 - 24x = 12x(x-2) = 0$ when $x = 0$ and $x = 2$. $m(3,-17)$ is a relative minimum, $(0,10)$ is not an extremum, and $I(0,10)$ as well as $I(2,-6)$ are points of inflection.

x	$-\infty$		0		2		3		∞
$f(x)$	$-\infty$		10		-6		-17		∞
$f'(x)$	−	−	0	−	−	−	0	+	+
		dec	0	dec	−	dec	0	inc	
$f''(x)$	+	+	0	−	0	+	+	+	+
		up	0	down	0	up	up	up	

7. $f(x) = (x-2)^3$. $f'(x) = 3(x-2)^2 = 0$ when $x = 2$. $f''(x) = 6(x-2) = 0$ when $x = 2$. $I(2,0)$ is a point of inflection.

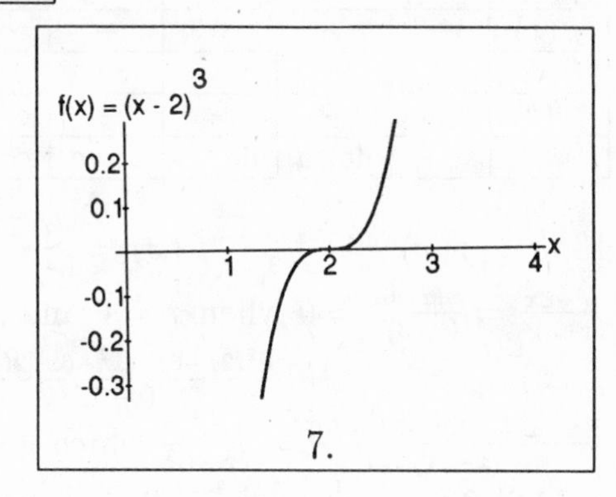

7.

x	$-\infty$		2		∞
$f(x)$	$-\infty$		0		∞
$f'(x)$	+	+	0	+	+
		inc	0	inc	
$f''(x)$	−	−	0	+	+
		down	0	up	

9. $f(x) = (x^2 - 5)^3$. $f'(x) = 3(x^2 - 5)(2x) = 6x(x^2 - 5)^2 = 0$ when $x = -\sqrt{5}$, $x = 0$, and $x = \sqrt{5}$. $f''(x) = 6x[2(x^2-5)(2x)]+(x^2-5)^2(6) = 6(x^2 - 5)(5x^2 - 5) = 30(x^2 - 5)(x+1)(x-1) = 0$ when $x = -\sqrt{5}$, $x = -1$, $x = 1$, and $x = \sqrt{5}$. $m(0, -125)$ is a relative minimum. $I(-\sqrt{5}, 0)$, $I(-1, -64)$, $I(1, -64)$, and $I(\sqrt{5}, 0)$ are points of inflection.

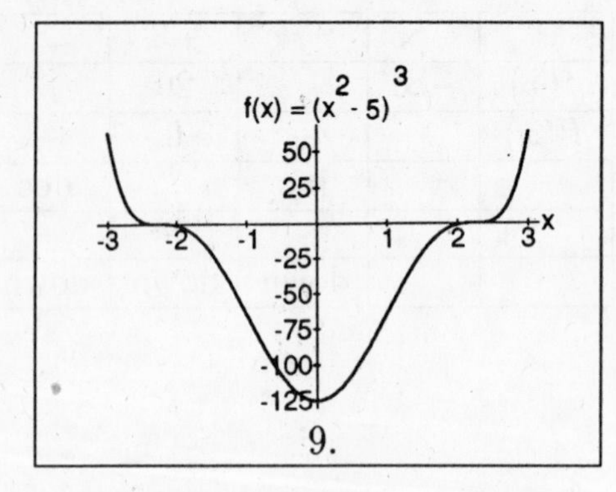

9.

x	$-\infty$		$-\sqrt{5}$		-1		0		1		$\sqrt{5}$		∞
$f(x)$	$-\infty$		0		-64		-125		-64		0		∞
$f'(x)$	$-$	$-$	$-$	$-$	$-$	$-$	0	$+$	$+$	$+$	$+$	$+$	$+$
		dec	dec	dec	dec	dec	0	inc	inc	inc	inc	inc	$+$
$f''(x)$	$+$	$+$	0	$-$	0	$+$	$+$	$+$	0	$-$	0	$+$	$+$
		up		down		up	up	up		down	0	up	

11 $f(x) = x + \frac{1}{x}$. $f'(x) = 1 - \frac{1}{x^2} = \frac{x^2-1}{x^2} = \frac{(x-1)(x+1)}{x^2} = 0$ when $x = -1$ and $x = 1$. $f''(x) = \frac{2}{x^3}$is undefined at $x = 0$. $(-1, -2)$ is a relative maximum, $(1, 2)$ is a relative minimum.

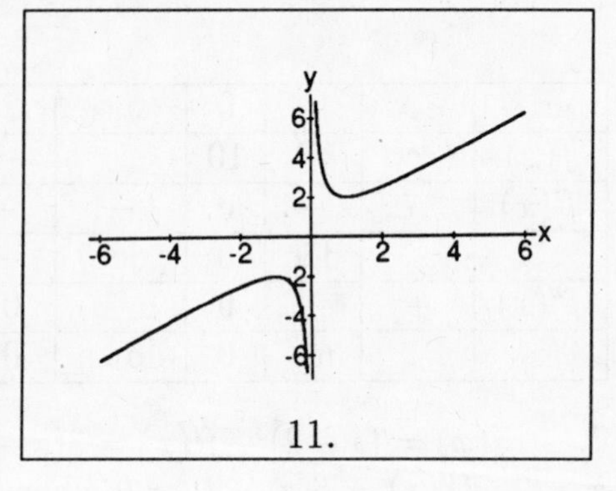

11.

x	$-\infty$		-1		0		1		∞
$f(x)$	$-\infty$		-2		∞		2		∞
$f'(x)$	$+$	$+$	0	$-$	∞	$-$	0	$+$	$+$
		inc		dec	∞	dec	0	inc	
$f''(x)$	$-$	$-$	$-$	$-$	∞	$+$	$+$	$+$	$+$
		down	down	down	∞	up	up	up	

13. $f(x) = \frac{x^2}{x-3}$. $f'(x) = \frac{(x-3)(2x)-x^2(1)}{(x-3)^2} = \frac{x^2-6x}{(x-3)^2} = \frac{x(x-6)}{(x-3)^2} = 0$ when $x = 0$ and $x = 6$.
$f''(x) = \frac{(x-3)^2(2x-6)-(x^2-6x)[2(x-3)(1)]}{(x-3)^4} = \frac{2(x-3)[(x-3)^2-x(x-6)]}{(x-3)^4} = \frac{18}{(x-3)^3}$ which is never 0. $x = 3$ is not a critical point because it is not in the domain of the function. $(0, 0)$ is a relative maximum, $(6, 12)$ is a relative minimum.

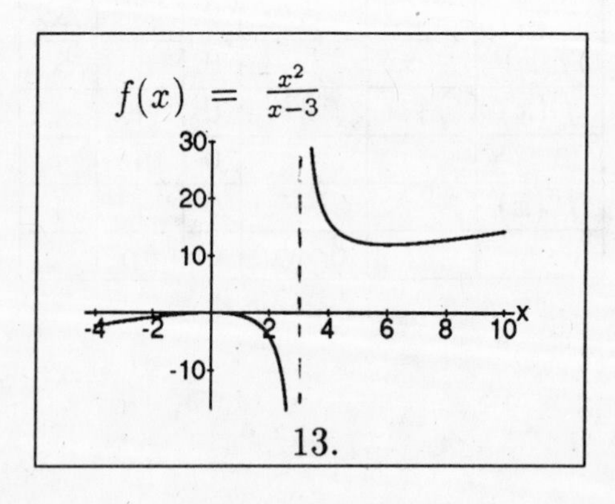

13.

x	$-\infty$		0		3		6		∞
$f(x)$	$-\infty$		0		∞		12		∞
$f'(x)$	+	+	0	−	∞	−	0	+	+
		inc	0	dec	∞	dec	0	inc	
$f''(x)$	−	−	−	−	∞	+	+	+	+
		down	down	down	∞	up	up	up	

15. $f(x) = (x+1)^{1/3}$. $f'(x) = \frac{(x+1)^{-2/3}}{3} = \frac{1}{3(x+1)^{2/3}}$ is never 0. It is undefined when $x = -1$. $f''(x) = -\frac{2(x+1)^{-5/3}}{9} = \frac{-2}{9(x+1)^{5/3}}$ which is never 0 but undefined at $x = -1$. $I(-1, 0)$ is an inflection point.

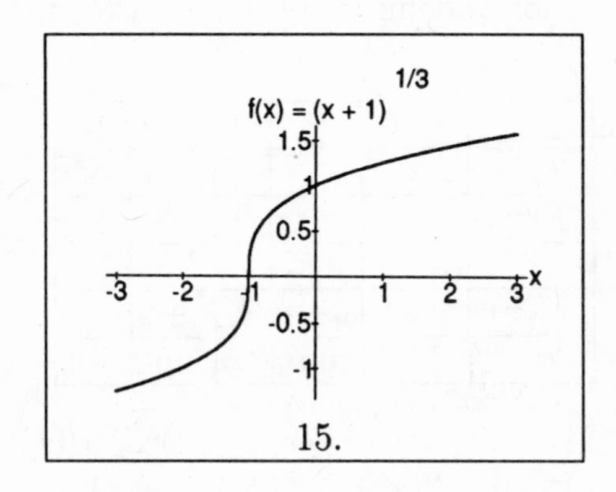

15.

x	$-\infty$		-1		∞
$f(x)$	$-\infty$		0		∞
$f'(x)$	+	+	∞	+	+
		inc		inc	
$f''(x)$	+	+	∞	−	−
		up		down	

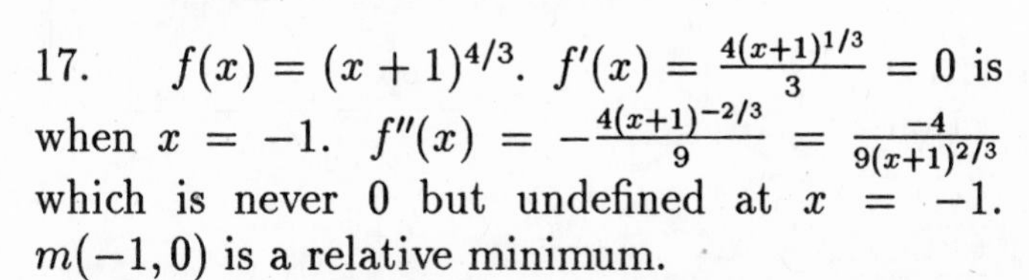

17. $f(x) = (x+1)^{4/3}$. $f'(x) = \frac{4(x+1)^{1/3}}{3} = 0$ is when $x = -1$. $f''(x) = -\frac{4(x+1)^{-2/3}}{9} = \frac{-4}{9(x+1)^{2/3}}$ which is never 0 but undefined at $x = -1$. $m(-1, 0)$ is a relative minimum.

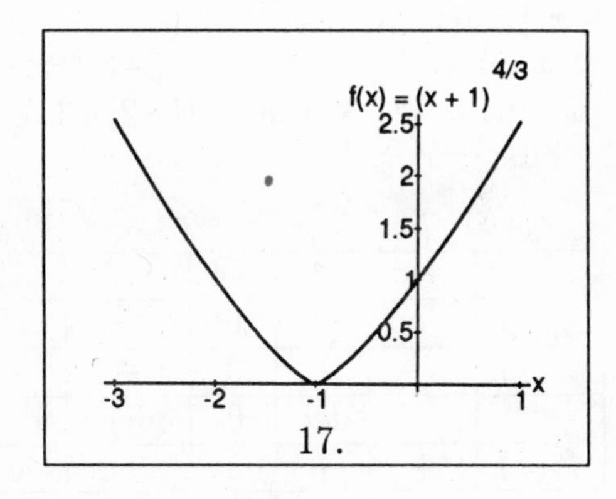

17.

x	$-\infty$		-1		∞
$f(x)$	$-\infty$		0		∞
$f'(x)$	−	−	∞	+	+
		dec		inc	
$f''(x)$	+	+	∞	+	+
		up		up	

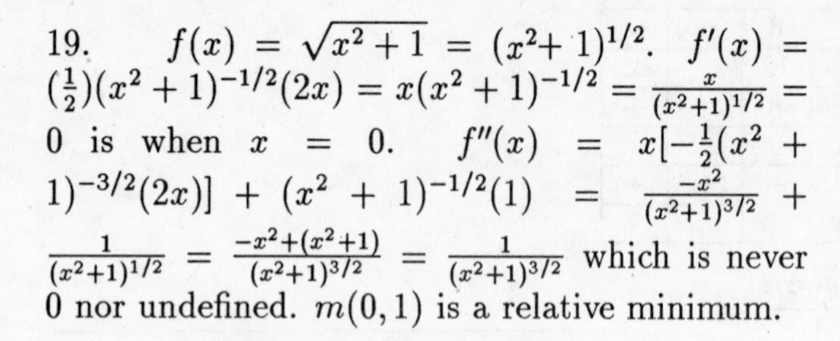
19. $f(x) = \sqrt{x^2+1} = (x^2+1)^{1/2}$. $f'(x) = (\frac{1}{2})(x^2+1)^{-1/2}(2x) = x(x^2+1)^{-1/2} = \frac{x}{(x^2+1)^{1/2}} = 0$ is when $x = 0$. $f''(x) = x[-\frac{1}{2}(x^2+1)^{-3/2}(2x)] + (x^2+1)^{-1/2}(1) = \frac{-x^2}{(x^2+1)^{3/2}} + \frac{1}{(x^2+1)^{1/2}} = \frac{-x^2+(x^2+1)}{(x^2+1)^{3/2}} = \frac{1}{(x^2+1)^{3/2}}$ which is never 0 nor undefined. $m(0,1)$ is a relative minimum.

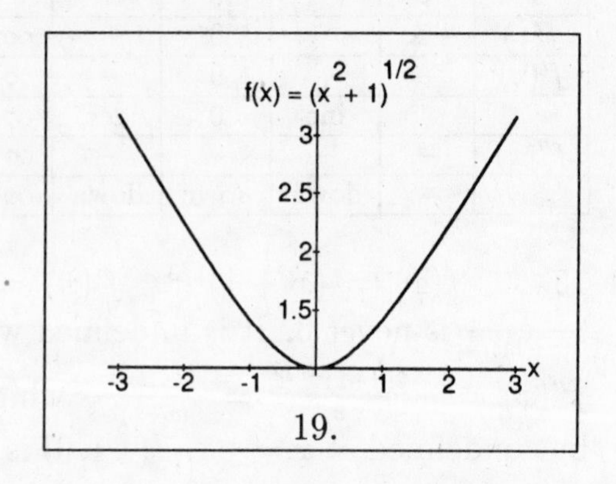

19.

x	$-\infty$		0		∞
$f(x)$	∞		1		∞
$f'(x)$	$-$	$-$	0	$+$	$+$
		dec		inc	
$f''(x)$	$+$	$+$	1	$+$	$+$
		up		up	

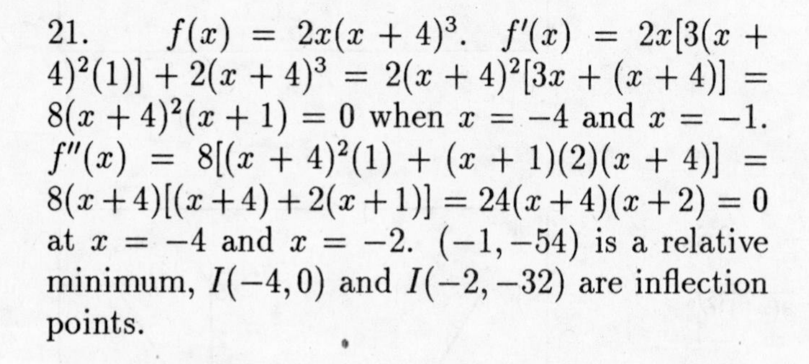
21. $f(x) = 2x(x+4)^3$. $f'(x) = 2x[3(x+4)^2(1)] + 2(x+4)^3 = 2(x+4)^2[3x+(x+4)] = 8(x+4)^2(x+1) = 0$ when $x = -4$ and $x = -1$. $f''(x) = 8[(x+4)^2(1) + (x+1)(2)(x+4)] = 8(x+4)[(x+4)+2(x+1)] = 24(x+4)(x+2) = 0$ at $x = -4$ and $x = -2$. $(-1,-54)$ is a relative minimum, $I(-4,0)$ and $I(-2,-32)$ are inflection points.

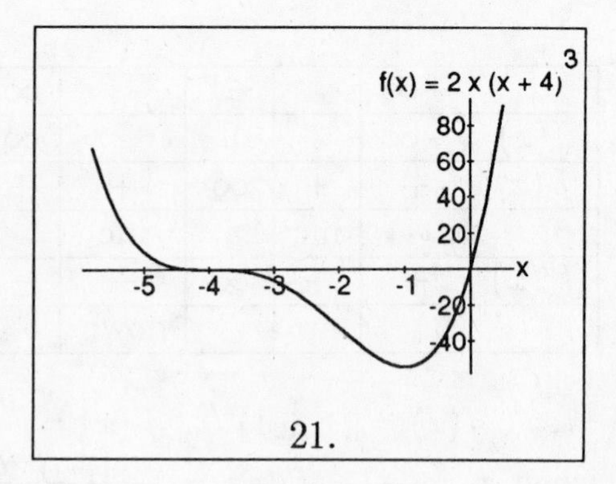

21.

x	$-\infty$		-4		-2		-1		∞
$f(x)$	$-\infty$		0		-32		-54		∞
$f'(x)$	$-$	$-$	0	$-$	$-$	$-$	0	$+$	$+$
		dec	0	dec	dec	dec	0	inc	
$f''(x)$	$+$	$+$	0	$-$	0	$+$	$+$	$+$	$+$
		up		down		up	up	up	up

23. $f(x) = \frac{2}{1+x^2} = 2(1+x^2)^{-1}$. $f'(x) = -2(1+x^2)^{-2}(2x) = \frac{-4x}{(1+x^2)^2} = 0$ when $x = 0$. $f''(x) = \frac{(1+x^2)^2(-4)-(-4x)(2)(1+x^2)(2x)]}{(1+x^2)^4} = \frac{4(1+x^2)[-(1+x^2)+4x^2]}{(1+x^2)^4} = \frac{4(3x^2-1)}{(1+x^2)^3} = 0$ at $I_1 = (-\frac{1}{\sqrt{3}}, \frac{3}{2})$ and $I_2 = (\frac{1}{\sqrt{3}}, \frac{3}{2})$. $(0,2)$ is a relative maximum, I_1 and I_2 are inflection points.

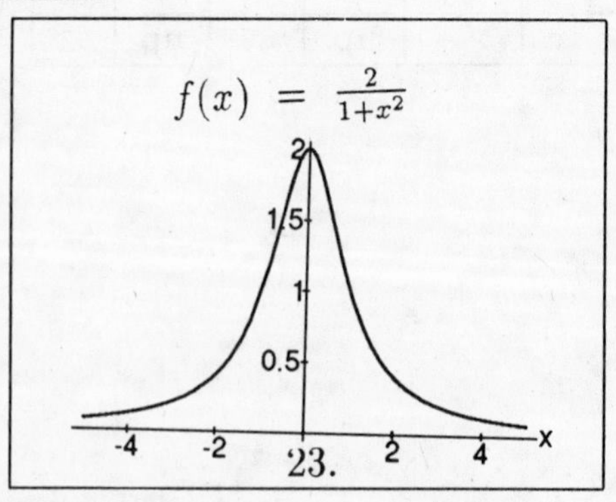

23.

x	$-\infty$		$-\frac{1}{\sqrt{3}}$		0		$\frac{1}{\sqrt{3}}$		∞
$f(x)$	$-\infty$		$\frac{3}{2}$		2		$\frac{3}{2}$		∞
$f'(x)$	+	+	+	+	0	−	−	−	−
		inc	inc	inc		dec	dec	dec	dec
$f''(x)$	+	+	0	−	−	−		+	+
	up	up		down	down	down		up	up

25. $f(x) = \frac{(x-2)^3}{x^2}$. $f'(x) = \frac{x^2[3(x-2)^2(1)]-(x-2)^3(2x)}{x^4} = \frac{x(x-2)^2[3x-2(x-2)]}{x^4} = \frac{(x-2)^2(x+4)}{x^3} = 0$ when $x = -4$ and $x = 2$. $f''(x) = \frac{x^3[(x-2)^2(1)+(x+4)(2)(x-2)]-(x-2)^2(x+4)(3x^2)}{x^6} = \frac{x^2(x-2)[x(x-2)+2x(x+4)-3(x-2)(x+4)]}{x^6} = \frac{(x-2)[x^2-2x+2x^2+8x-3x^2-6x+24]}{x^4} = \frac{24(x-2)}{x^4} = 0$ at $x = 2$. $(-4, -\frac{27}{2})$ is a relative maximum and $I(2,0)$ is an inflection points.

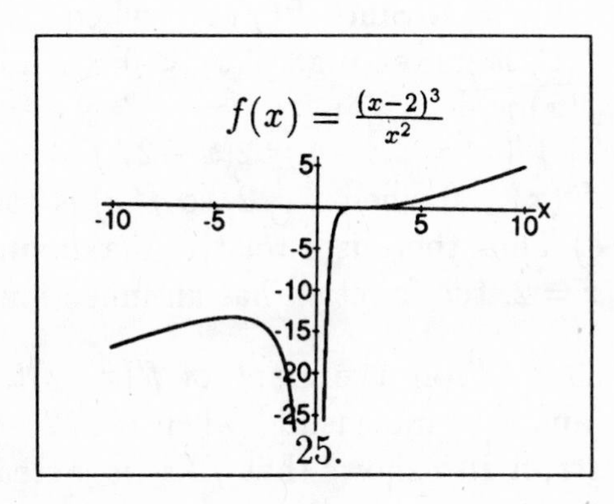

25.

x	$-\infty$		-4		0		2		∞
$f(x)$	$-\infty$		$-\frac{27}{2}$		∞		0		∞
$f'(x)$	+	+	0	−	∞	+	0	+	+
	inc	inc		dec		inc		inc	
$f''(x)$	−	−	−	−	∞	+	0	+	+
		down	down	down		up		up	up

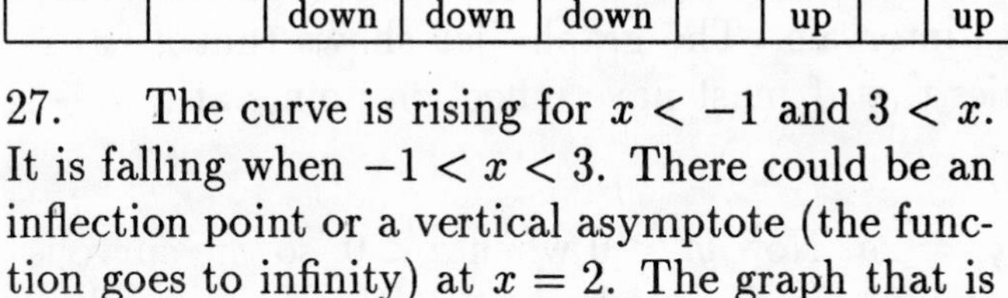
27. The curve is rising for $x < -1$ and $3 < x$. It is falling when $-1 < x < 3$. There could be an inflection point or a vertical asymptote (the function goes to infinity) at $x = 2$. The graph that is shown is not unique.

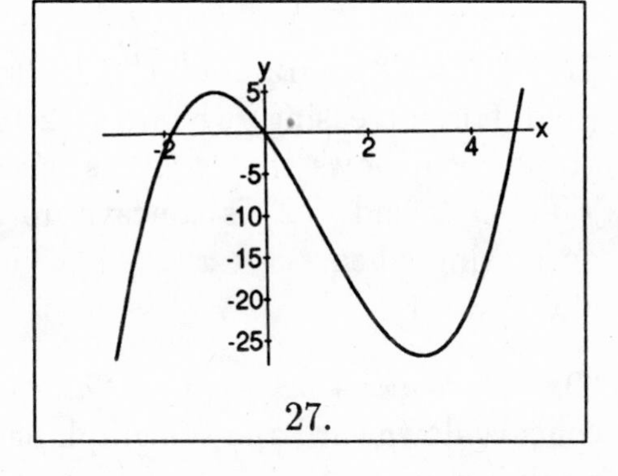

27.

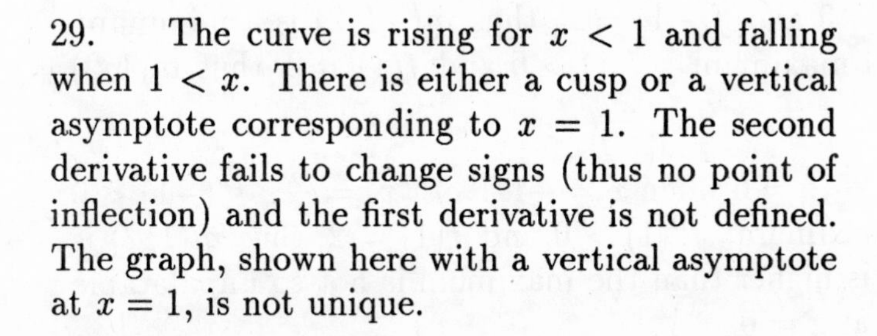
29. The curve is rising for $x < 1$ and falling when $1 < x$. There is either a cusp or a vertical asymptote corresponding to $x = 1$. The second derivative fails to change signs (thus no point of inflection) and the first derivative is not defined. The graph, shown here with a vertical asymptote at $x = 1$, is not unique.

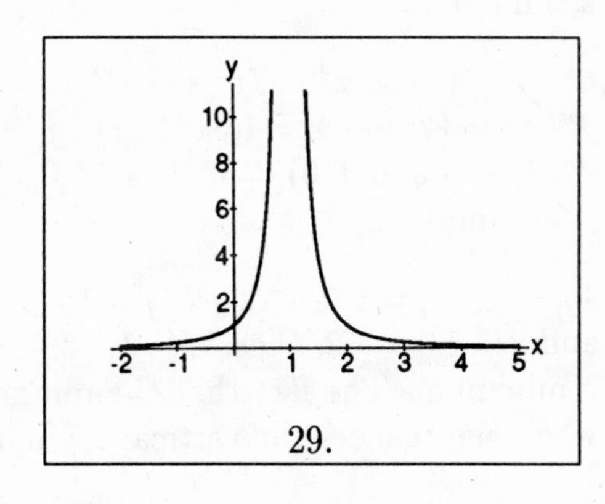

29.

31. The curve is rising for $x > 0$ ($x \neq 2$) and falling when $x < 0$. The function is not defined at $x = 2$. The curve is concave upward when $x < 2$ and concave downward when $x > 2$. The curve shown here is not unique

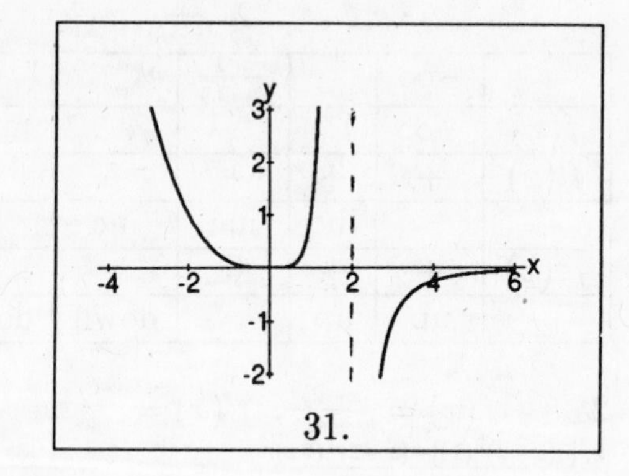

31.

33. a) $f'(x) = x^2 - 4x = x(x-4) = 0$ at $x = 0$ and $x = 4$. Since $f'(x) > 0$ when $x < 0$ and $x > 4$, $f(x)$ is increasing. $f'(x) < 0$ when $0 < x < 4$, so $f(x)$ is decreasing.
b) $f''(x) = 2x - 4 = 2(x-2) = 0$ at $x = 2$. $f''(x) < 0$ when $x < 2$, so $f(x)$ is concave down. $f''(x) > 0$ when $x > 2$, so $f(x)$ is concave up.
c) Thus there is a relative maximum when $x = 0$ and a relative minimum at $x = 4$. When $x = 2$, the function has an inflection point.

35. From the graph of $f'(x)$, $f'(x) < 0$ (and f is decreasing) when $x < 2$ and $f'(x) > 0$ (and f is increasing) when $x > 2$. Hence f must have a relative minimum at $x = 2$. The graph also shows that $f'(x)$ is increasing for all x, which implies that $f''(x) > 0$ and $f(x)$ is concave up for all x.

37. From the graph of $f'(x)$, $f'(x) \leq 0$ (and f is decreasing) when $x < 2$ and $f'(x) > 0$ (and f is increasing) when $x > 2$. Hence f must have a relative minimum at $x = 2$. The graph also shows that $f'(x)$ is increasing when $x < -3$ and $x > -1$, which implies that $f''(x) > 0$ and $f(x)$ is concave up on these intervals. The graph also shows that $f'(x)$ is decreasing when $-3 < x < -1$, which implies that f must have inflection points at $x = -3$ and $x = -1$.

39. $y = ax^2 + bx + c$, $y' = 2ax + b$, and $y'' = 2a$. Now $y'' < 0$ when $a < 0$, so the curve is concave downward and the graph has a maximum. Similarly, when $a > 0$, the graph exhibits a minimum.

41. $f(x) = x^4 - 2x^2 + 3$, $f'(x) = 4x^3 - 4x = 4x(x+1)(x-1) = 0$ when $x = -1, 0, 1$. $f''(x) = 12x^2 - 4 = 4(3x^2 - 1)$. $f''(-1) > 0$ and $f(-1) = 2$, thus $m(-1,2)$ is a minimum. $f''(0) < 0$ and $f(0) = 3$, thus $M(0,3)$ is a maximum. $f''(1) > 0$ and $f(1) = 2$, thus $m(1,2)$ is a minimum.

43. $f(x) = x + x^{-1}$, $f'(x) = 1 - x^{-2} = \frac{x^2-1}{x^2} = 0$ when $x = -1, 1$. $f''(x) = \frac{2}{x^3}$. $f''(-1) < 0$ and $f(-1) = -2$, thus $M(-1,-2)$ is a maximum. $f''(1) > 0$ and $f(1) = 2$, thus $m(1,2)$ is a minimum. The fact that the minimum is higher than the maximum is not so unpalatable when one realizes the vertical asymptote at $x = 0$.

45. $f(x) = \frac{x^2}{x-2}$, $f'(x) = \frac{(x-2)(2x)-x^2}{(x-2)^2} = \frac{x(x-4)}{(x-2)^2} = 0$ when $x = 0, 4$.

$f''(x) = \frac{(x-2)^2(2x-4)-(x^2-4x)(2)(x-2)}{(x-2)^4} = \frac{8}{(x-2)^3}$. $f''(0) < 0$ and $f(0) = 0$, thus $M(0,0)$ is a maximum. $f''(4) > 0$ and $f(4) = 8$, thus $m(4,8)$ is a minimum. The fact that the minimum is higher than the maximum is not so unpallatable when one realizes the vertical asymptote at $x = 2$.

3.3 Absolute Maxima and Minima

1. $f(x) = x^2 + 4x + 5$, $-3 \le x \le 1$. $f'(x) = 2x + 4 = 2(x+2) = 0$ at $x = -2$, which is in the interval. $f(-2) = 1$, $f(-3) = 2$, and $f(1) = 10$. Thus $f(1) = 10$ is the absolute maximum while $f(-2) = 1$ is the absolute minimum.

3. $f(x) = \frac{x^3}{3} - 9x + 2$, $0 \le x \le 2$. $f'(x) = x^2 - 9 = (x+3)(x-3) = 0$ at $x = -3$ and $x = 3$, which are not in the interval. $f(0) = 2$ and $f(2) = -\frac{40}{3}$. Thus $f(0) = 2$ is the absolute maximum while $f(2) = -\frac{40}{3}$ is the absolute minimum.

5. $f(x) = 3x^5 - 5x^3$, $-2 \le x \le 0$. $f'(x) = 15x^4 - 15x^2 = 15x^2(x^2-1) = 15x^2(x+1)(x-1) = 0$ at $x = -1$, $x = 0$, and $x = 1$, of which only $x = -1$ and $x = 0$ are in the interval. $f(0) = 0$, $f(-1) = 2$, and $f(-2) = -56$. Thus $f(-1) = 2$ is the absolute maximum while $f(-2) = -56$ is the absolute minimum.

7. $f(x) = (x^2-4)^5$, $-3 \le x \le 2$. $f'(x) = 5(x^2-4)^4(2x) = 10x(x-2)^4(x+2)^4 = 0$ at $x = -2$, $x = 0$, and $x = 2$, all of which are in the interval. $f(0) = -1{,}024$, $f(-2) = 0$, $f(2) = 0$, and $f(-3) = 3{,}125$. Thus $f(-3) = 3{,}125$ is the absolute maximum while $f(0) = -1{,}024$ is the absolute minimum.

9. $f(x) = x + \frac{1}{x}$, $\frac{1}{2} \le x \le 3$. $f'(x) = 1 - \frac{1}{x^2} = \frac{x^2-1}{x^2} = \frac{(x-1)(x+1)}{x^2} = 0$ at $x = -1$ and $x = 1$, of which only $x = 1$ is in the interval. Note that $f'(x)$ is undefined when $x = 0$, which is not in the domain of f and hence not a critical point. $f(1) = 2$, $f(\frac{1}{2}) = \frac{5}{2}$, and $f(3) = \frac{10}{3}$. Thus $f(3) = \frac{10}{3}$ is the absolute maximum while $f(1) = 2$ is the absolute minimum.

11. $f(x) = x + \frac{1}{x}$, $0 < x$. $f'(x) = 1 - \frac{1}{x^2} = \frac{x^2-1}{x^2} = \frac{(x-1)(x+1)}{x^2} = 0$ at $x = -1$ and $x = 1$, of which only $x = 1$ is in the interval. Note that $f'(x)$ is undefined when $x = 0$, which is not in the domain of f and hence not a critical point. $f(1) = 2$ and there are no other endpoints. Since $f'(x) < 0$ when $0 < x < 1$ (i.e. f is decreasing) and $f'(x) > 0$ when $x > 1$ (i.e. f is increasing), it follows that there is no absolute maximum and $f(1) = 2$ is the absolute minimum.

13. $f(x) = \frac{1}{x}$, $0 < x$. $f'(x) = -\frac{1}{x^2} \neq 0$. There are no endpoints and no critical points.

Hence there is no absolute maximum and no absolute minimum.

15. $f(x) = \frac{1}{x+1} = (x+1)^{-1}, 0 \leq x$. $f'(x) = -(x+1)^{-2} = -\frac{1}{(x+1)^2} < 0$ for all x. Hence the graph of f begins at $f(0) = 1$ and decreases for all $x > 0$. Hence $f(0) = 1$ is the absolute maximum and there is no absolute minimum.

17. a) The membership of the association x years after 1980 is given by the function $f(x) = 100(2x^3 - 45x^2 + 264x)$. The period of time between 1980 and 1994 corresponds to the interval $0 \leq x \leq 14$. $f'(x) = 100(6x^2 - 90x + 264) = 600(x^2 - 15x + 44) = 600(x-4)(x-11) = 0$ when $x = 4$ and $x = 11$. $f(11) = 12,100$, $f(4) = 46,400$, $f(0) = 0$, and $f(14) = 36,400$. $f(4) = 46,400$ is the absolute maximum. Hence the membership was greatest in 1984, four years after the founding of the association, when there were 46,400 members.
b) The period of time between 1980 and 1994 corresponds to the interval $1 \leq x \leq 14$. $f(1) = 22,100$, $f(11) = 12,100$, $f(4) = 46,400$, $f(14) = 36,400$. Thus the membership was smallest in 1991, eleven years after the founding of the association, when there were 12,100 members.

19. Let $P(x)$ denote the profit if the price is x dollars per radio. Then $P(x) =$(number of radios sold)(profit per radio)$= (20 - x)(x - 5) = -100 + 25x - x^2$ for $0 \leq x \leq 20$. $P'(x) = 25 - 2x = 0$, when $x = 12.5$. $P(12.5) = 56.25$, $P(0) = 0$, and $P(20) = 0$. Thus the profit is greatest when $x = 12.5$, that is, when the price is \$12.50 per radio.

21. Let S denote the speed of the blood, R the radius of the artery, and r the distance from the central axis. Poiseuille's law states that $S(r) = c(R^2 - r^2)$, where c is a positive constant. The relevant interval is $0 \leq r \leq R$. $S'(r) = -2cr = 0$ at $r = 0$ (the left-hand endpoint of the interval). With $S(0) = cR^2$ and $S(R) = 0$, we see that the speed of the blood is greatest when $r = 0$, that is, at the central axis.

23. Let R denote the rate at which the population changes, P the current population size, and B the upper bound for the population. Then $R = k(P)(B - P) = kPB - kP^2$ where k is a positive constant of proportionality. $\frac{dR}{dP} = kB - 2kP = 0$ when $P = \frac{B}{2}$. Moreover the derivative is positive (and the function is increasing) when $P < \frac{B}{2}$ and the derivative is negative (and the function is decreasing) when $P > \frac{B}{2}$. Hence R has an absolute maximum at $P = \frac{B}{2}$. That is the rate of change is greatest when the population size P is $\frac{B}{2}$ or 50% of its upper bound.

25. Let v be the speed at which the truck is driven and k_1 as well as k_2 positive constants of proportionality. The driver's wages added to the cost of fuel lead to a cost function $C = \frac{k_1}{v} + k_2 v = k_1 v^{-1} + k_2 v$. $C' = -k_1 v^{-2} + k_2 = 0$ when $\frac{k_1}{v^2} = k_2$ or $\frac{k_1}{v} = k_2 v$ (driver's wages equal cost of fuel). With $C' = -k_1 v^{-2} + k_2$ falling to the left of the critical point and rising to the right, the critical point is a minimum for the cost function.

27. The average cost $A = \frac{C(q)}{q}$ where $C(q)$ is the cost of q units. $A' = \frac{qC'-C}{q^2} = 0$ when $C' = \frac{C}{q} = A$.

29. $P = \frac{w^2}{2\rho Sv} + \frac{\rho Av^3}{2}$. $P' = -\frac{w^2}{2\rho Sv^2} + \frac{3Av^2\rho}{2} = \frac{-w^2+3A\rho^2Sv^4}{2\rho Sv^2} = 0$ when $v^4 = \frac{w^2}{3A\rho^2S}$ or $v = (\frac{w^2}{3A\rho^2S})^{1/4}$. $P'' = \frac{w^2}{\rho Sv^3} + 3Av\rho > 0$ so P is a relative minimum.

3.4 Practical Optimization Problems

1. Let x denote the number that exceeds its square, x^2, by the largest amount. Then $f(x) = x - x^2$ is the function to be maximized. There is no restriction on the domain of $f(x)$. $f'(x) = 1 - 2x = 0$ when $x = \frac{1}{2}$. $f''(x) = -2 < 0$ so that $(\frac{1}{2}, f(\frac{1}{2})) = (\frac{1}{2}, \frac{1}{4})$ is a relative maximum. Since the graph of $f(x)$ is a parabola opening down. the relative maximum is an absolute maximum. Thus, $x = \frac{1}{2}$ is the number that maximizes $f(x) = x - x^2$.

3. Let x and y denote two positive numbers. Then xy is their product and $x + y$ is their sum. The goal is to maximize $f = xy$, which is a function of two variables. Since $x + y = 50$ or $y = 50 - x$, substituting for y in the formula for f yields $f(x) = x(50 - x) = 50x - x^2$, a function of one variable. $f'(x) = 50 - 2x = 0$ at $x = 25$, and $f''(x) = -2 < 0$ so that $(25, f(25)) = (25, 625)$ is a relative maximum. Since the graph of $f(x)$ is a parabola opening down, the relative maximum is the absolute maximum. Thus $x = 25$ and $y = 25$ have the properties that their sum is 50 and their product, 625, is as large as possible.

5. Let x denote the number of \$1 increases and $P(x)$ the corresponding profit function. Then $P(x)$ =(number of skateboards sold)(profit per skateboard). For each \$1 increase, 3 fewer skateboards than the current 50 will be sold, and so the total number sold will be $50 - 3x$. Each skateboard will sell for $40 + x$ dollars (the current price plus the number of \$1 increases), and the cost of each skateboard is \$25. Hence the profit per skateboard is $(40+x)-25 = 15+x$ dollars. Putting it all together, $P(x) = (50-3x)(15+x) = 750+5x-3x^2$. The relevant interval is $0 \leq x \leq 16$. If there are more than 16 reductions the total number sold will be a negative number. $P'(x) = 5 - 6x = 0$ when $x = \frac{5}{6}$, $P''(x) = -6 < 0$ and so a relative maximum is achieved when $x = \frac{5}{6}$. Since the graph of $P(x)$ is a parabola opening down, the relative maximum is the absolute maximum. Since x needs to be a non-negative integer, the profit is computed at $x = 0$ and $x = 1$; $P(0) = 750$ and $P(1) = 752$. At the right-hand endpoint, $P(16) = 750 + 5(16) - 3(16)^2 = 62$. Thus the greatest profit is \$752 which is generated when there is only one \$1 increase, that is, the skateboard sells for \$41.

7. Let x be the number of \$0.50 reductions. Then $5x$ more cards will be sold. The sales price per card will be $10 - 0.50x$, the number of cards sold $50 + 5x$, and the total cost $C = 5(50 + 5x)$. The revenue is $R = (50 + 5x)(10 - 0.50x)$. Since profit $P(x) = R - C = (50 + 5x)(10 - 0.50x - 5) = 25(10 - 0.1x^2)$, $P'(x) = 25(-0.2x) = 0$ at $x = 0$. A reasonable interval for x is $0 \leq x \leq 10$ because the sales price $10 - 0.50x$ is not less than the cost of \$5.00. $P(0) = 250$, $P(10) = 0$. Profit is maximized for $x = 0$ at a sales price per card of \$10.00. $P(x)$ is a parabola pointing downward, so the relative maximum is also the absolute maximum.

9. Let x denote the number of additional trees to be planted and $N(x)$ the corresponding yield. Then $N(x)$ =(number of oranges per tree)(number of trees). Since there are 60 trees to begin with and x additional trees are planted, the total number is $60 + x$. For each additional tree, the average yield of 400 oranges per tree is decreased by 4. Thus, for x additional trees, the average yield per tree is $400 - 4x$. Thus $N(x) = (400 - 4x)(60 + x) = 4(100 - x)(60 + x) = 4(6,000 + 40x - x^2)$. The relevant interval is $0 \leq x \leq 100$. To find the critical points, set the derivative equal 0. $N'(x) = 4(40 - 2x) = 8(20 - x) = 0$ or $x = 20$. $N(20) = 25,600$, $N(0) = 2,400$, and $N(100) = 0$. Then the greatest possible yield is 25,600 oranges, which is generated by planting 20 additional trees, that is 80 trees are planted.

11. If x is the number of additional days and $R(x)$ the corresponding revenue, $R(x) =$ (number of pounds collected)(price per pound). Over the period of 80 days, 24,000 pounds of glass have been collected at a rate of 300 pounds per day, and so for each day over 80, an additional 300 pounds will be collected. Thus, the total number of pounds collected and sold is $24,000 + 300x$. Currently, the recycling center pays 1 cent per pound. For each additional day, it reduces the price it pays by 1 cent per 100 pounds, that is, by $\frac{1}{100}$ cents per pound. Hence, after x additional days, the price per pound will be $1 - \frac{1}{100}$ cents. $R(x) = (24,000 + 300x)(1 - \frac{x}{100}) = 24,000 + 60x - 3x^2$. The relevant interval is $0 \leq x \leq 100$. To find the critical points, $R'(x) = 60 - 6x = 0$ or $x = 10$. Now $R(10) = 24,300$, $R(0) = 24,000$, and $R(100) = 0$ shows that the most profitable time to conclude the project will be 10 days from now.

13. Let x be the length of the field and y the width. $2x + 2y = 320$ or $y = 160 - x$. The area is $A = xy = x(160 - x) = 160x - x^2$. $A' = 2(80 - x) = 0$ when $x = 80$. $A'' < 0$ which means that $x = y = 80$ maximizes the enclosed area. The field is a square.

15. Let x and y denote the dimensions of the rectangle, A the (fixed) area, and P the perimeter. The goal is to minimize the perimeter $P = 2x + 2y$, which is a function of two variables. The fact that the area is to be A gives $A = xy$ or $y = \frac{A}{x}$. Thus $P(x) = 2x + \frac{2A}{x}$ (where A is a positive constant). The relevant interval is $x > 0$. $P'(x) = 2 - \frac{2A}{x^2} = 0$, $x^2 = A$ or $x = \pm\sqrt{A}$. For $x > 0$ and with $P''(x) = \frac{4A}{x^3} > 0$, the minimal perimeter is attained when $x = \sqrt{A}$ and $y = \frac{A}{x} = \frac{A}{\sqrt{A}} = \sqrt{A}$, making the rectangle a square.

17. Consider the circle with radius $r = 4$ units and center at the origin of the coordinate system. Let $P(x, y)$ be a point on this circle and $A(-4, 0)$, $B(4, 0)$. Then $x^2 + y^2 = 16$. We want to maximize the area of the triangle ABP. Now $d_1 = AP = \sqrt{(x+4)^2 + y^2} = \sqrt{16 + 8x + x^2 + y^2} = \sqrt{16 + 8x + 16} = 2\sqrt{2x + 8}$ and similarly $d_2 = BP = 2\sqrt{8 - 2x}$. Area $A(x) = \frac{d_1 d_2}{2} = 4\sqrt{(4+x)(4-x)} = 4\sqrt{16 - x^2}$. $A'(x) = \frac{-4x}{\sqrt{16-x^2}} = 0$ at $x = 0$. $A''(x) = \frac{-4}{16-x^2}(\sqrt{16 - x^2} + \frac{x2}{\sqrt{16-x^2}})$, $A''(0) = -\frac{1}{4}(4 + 0) < 0$ which shows that $x = 0$ gives a relative maximum. With $-4 \leq x \leq 4$ and $A(-4) = A(4) = 0$, $x = 0$ and $y = 4$ is the absolute maximum for (right) triangle ABP. The dimensions of the triangle are 8, $4\sqrt{2}$, and $4\sqrt{2}$. Note: From geometry, it is known that a triangle with one vertex on a circle and one side the diameter of the circle is a right triangle. With this piece of knowledge, the given exercise is much simplified.

19. Let x be the dimension of one side of the square base and y the dimension of the height. The cost of both the top and bottom is $2x^2 + 2x^2 = 4x^2$. The cost of the four sides is $4xy$. Since the volume $V = x^2y = 250$, $y = \frac{250}{x^2}$. The total cost is $C = 4(x^2 + \frac{250}{x})$. $C' = \frac{8}{x^2}(x^3 - 125) = 0$ when $x = 5$ and $y = 10$. $C''(5) > 0$ which means that $x = 5$ minimizes the cost. The box can be constructed for not less than \$ 300.00.

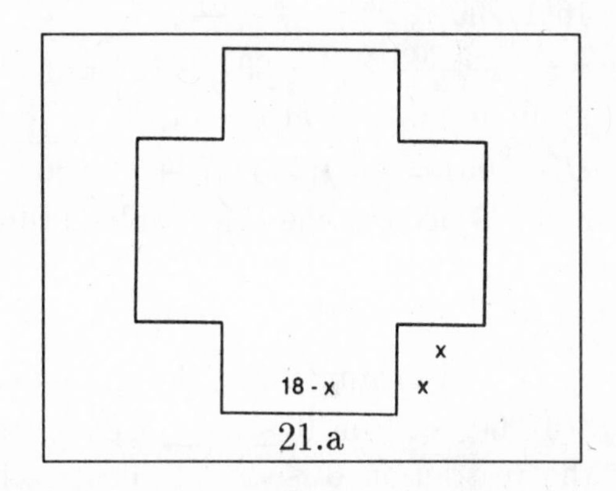

21.a

21. Label the square piece of cardboard as indicated. in the figure. Let V denote the volume of the box. Then $V(x) = (18 - 2x)^2(x) = 324x - 72x^2 + 4x^3$. The relevant interval is $0 \leq x \leq 9$. $V'(x) = 324 - 144x + 12x^2 = 12(27 - 12x + x^2) = 12(x - 9)(x - 3) = 0$ when $x = 9$ and $x = 3$. Now $V(0) = 0$, $V(3) = 432$, and $V(9) = 0$ so the greatest volume occurs at $x = 3$. The sides of the box are then$18 - 2(3) = 12$ inches long. Hence the dimensions of the greatest box are 12 inches by 12 inches by 3 inches.

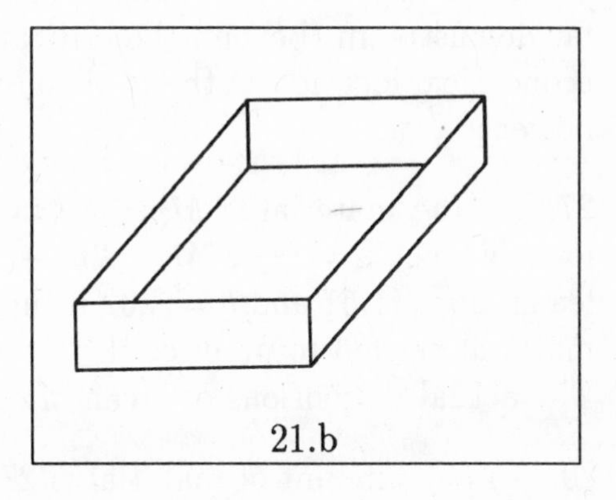
21.b

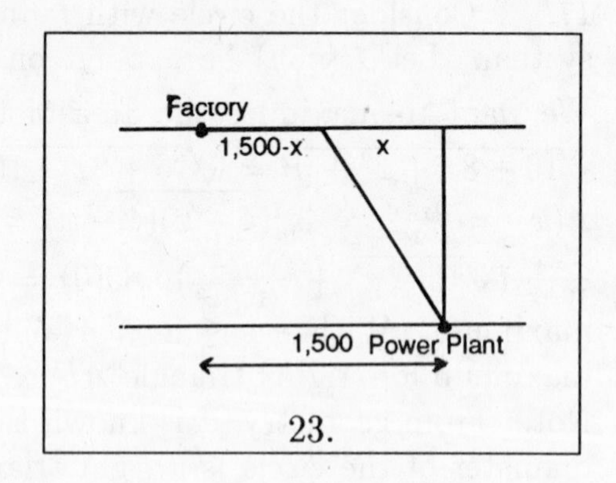

23. Let x be the distance indicated in the picture. We need to minimize the cost $C(x)$ =cost under water + cost over land.
Since the distance over land is $1,500 - x$ and the distance under water is $\sqrt{x^2 + (1,200)^2}$ (by the pythagorean theorem), the total cost is $C(x) = 25\sqrt{x^2 + (1,200)^2} + 20(1,500 - x)$. The relevant interval is $0 \le x \le 1,500$.
$C'(x) = 25(\frac{1}{2})[x^2 + (1,200)^2]^{-1/2}(2x) - 20 = \frac{25x}{\sqrt{x^2+(1,200)^2}} - 20 = 0$ when $25x = 20\sqrt{x^2 + (1,200)^2}$ or $5x = 4\sqrt{x^2 + (1,200)^2}$, $25x^2 = 16[x^2 + (1,200)^2]$, $9x^2 = 16(1,200)^2$, $x^2 = \frac{16(1,200)^2}{9}$, $x = \pm\frac{4(1,200)}{3} = \pm 1,600$, neither of which is in the interval $0 \le x \le 1,500$. Thus the absolute minimum occurs at one of the endpoints. Now $C(0) = 25(1,200) + 20(1,500) = 60,000$ and $C(1,500) = 25\sqrt{(1,500)^2 + (1,200)^2} + 0 \approx 48,023$, which shows that the minimum cost of \$48,023 occurs when the cable is run entirely under water.

25. In example 4.3, the minimum value of the cost function $C(x) = 5\sqrt{(900)^2 + x^2} + 4(3,000 - x)$ on $0 \le x \le 3,000$ was shown to be $C(1,200) = 14,700$. Suppose now that the restriction $0 \le x \le 3,000$ is changed to $0 < x$. Then, as in example 4.3, $C'(x) = \frac{5x}{\sqrt{(900)^2+x^2}} - 4 = 0$ when $x = 1,200$. Since $C'(x) < 0$ (and hence C is decreasing) when $0 < x < 1,200$ and $C'(x) > 0$ (and hence C is increasing) when $1,200 < x$, it follows that C has an absolute minimum on the interval $0 < x$ when $x = 1,200$. Thus, no matter how far downstream (beyond the critical point at 1,200 meters) the factory is located, the most economical location of the cable on the opposite bank is 1,200 meters downstream from the power plant.

27. The material is $M = (2)(\pi r^2) + 2\pi rh$ and the volume $V = \pi r^2 h = 6.89\pi$. Solving for h we get $h = \frac{6.89}{r^2}$. $M = 2r^2 + \frac{13.78}{r}$. $M' = 4r - \frac{13.78}{r^2} = 0$ at $r^3 = \frac{13.78}{4} = 3.445$. This leads to $r = 1.51$ and $h = 3.02$ inches. $M'' > 0$ indicates that these dimensions minimize the material needed to produce the can.
The actual dimensions of a can of cola differ because of ease of handling and packaging.

29. The amount of material is $27\pi = \pi r^2 + 2\pi rh$. Dividing by π and solving for h leads to $h = \frac{27}{2}r^{-1} - \frac{r}{2}$. The volume is $V = \pi r^2 h = \frac{\pi}{2}(27r - r^3) = V(r)$. $V' = \frac{\pi}{2}(27 - 3r^2) = 0$ when $r = 3$. $V''(3) < 0$, from which we deduce that $r = 3$ indicates a maximum value for the volume. $V(3) = \frac{\pi}{2}(81 - 27) = 27\pi$ cubic inches.

31. The material will cost $C = (3)(\pi r^2) + (2)(2\pi rh)$. The volume is $V = \pi r^2 h$ which leads to $h = \frac{V}{\pi r^2}$. Thus $C = 3\pi r^2 + \frac{4V}{r}$. $C' = 6\pi r - \frac{4V}{r^2} = 0$ if $3\pi r^3 = 2V$. $r^3 = \frac{2V}{3\pi} = \frac{2\pi r^2 h}{3\pi}$ or $r = \frac{2h}{3}$.

33. a) Let x denote the number of machines used and $C(x)$ the corresponding total cost. Then $C(x)$ = set-up cost + operating cost = 20(number of machines) +4.80(number of hours). Since each machine produces 30 kickboards per hour, x machines produce $30x$ kickboards per hour and the number of hours required to produce 8,000 kickboards is $\frac{8{,}000}{30x}$. Hence, $C(x) = 20x + 4.8(\frac{8{,}000}{30x}) = 20x + \frac{1{,}280}{x}$. Since the firm owns only 10 machines, the relevant interval is $0 \leq x \leq 10$. $C'(x) = 20 - \frac{1{,}280}{x^2} = \frac{20(x^2-64)}{x^2} = 0$ or $x = 8$. $C''(x) = \frac{2{,}560}{x^3} > 0$, hence the total cost C has an absolute minimum when $x = 8$ (the graph of $C(x)$ is concave upward), that is, when 8 machines are used.
b) When $x = 8$, the supervisor earns $\frac{1{,}280}{8} = \$160$.
c) The cost of setting up 8 machines is 20(8)=\$160

35. We want to maximize the rate of output on the interval $0 \leq t \leq 4$. Since $Q(t) = -t^3 + 6t^2 + 15t$ is the output, the rate of output is $R(t) = Q'(t) = -3t^2 + 12t + 15$. $R'(t) = Q''(t) = -6t + 12 = -6(t-2) = 0$ or $t = 2$. $R(0) = 15$, $R(2) = 27$, and $R(4) = 15$. Thus an absolute maximum occurs at $t = 2$ and an absolute minimum when $t = 0$ and $t = 4$. That is, the worker is performing most efficiently at 10:00 a.m. and least efficiently at 8:00 a.m. as well as 12:00 noon.

37. The slope of the tangent line is given by the derivative. Hence, to find the point on the curve $y = 2x^3 - 3x^2 + 6x$ at which the tangent has smallest slope, minimize the derivative function $m(x) = y'(x) = 6x^2 - 6x + 6$. $m'(x) = y''(x) = 12x - 6 = 0$ when $x = \frac{1}{2}$. From the original equation $y = 2(\frac{1}{2})^3 - 3(\frac{1}{2})^2 + 6(\frac{1}{2}) = \frac{5}{2}$. With $m''(x) = 12 > 0$, the curve is concave upward, and the slope is minimal at $(\frac{1}{2}, \frac{5}{2})$. The value of the slope at this point is $m(\frac{1}{2}) = \frac{9}{2}$.

39. The rate of population growth is the derivative $R(t) = P'(t) = -3t^2 + 18t + 48$. We need to optimize $R(t)$ on the interval $0 \leq t \leq 5$. $R'(t) = P''(t) = -6t + 18 = 0$ when $t = 3$. Now $R(0) = 48$, $R(3) = 75$, and $R(5) = 63$. Thus the rate of growth is greatest 3 years from now, when it is 75 thousand people per year. It is smallest now (at $t = 0$), when it is 48 thousand people per year.

41. Let n denote the number of floors and $A(n)$ the corresponding average cost. Since the total cost is $C(n) = 2n^2 + 500n + 600$ (thousand dollars), $A(n) = \frac{C(n)}{n} = 2n + 500 + \frac{600}{n}$. The relevant interval is $n > 0$. $A'(n) = 2 - \frac{600}{n^2} = \frac{2(n^2-300)}{n^2} = 0$ when $n = \sqrt{300} \approx 17.32$. $A''(n) = \frac{1{,}200}{n^3} > 0$ when $n > 0$, $A(n)$ has its absolute minimum when $n = \sqrt{300}$. In the context of this practical problem, the optimal value of n (which denotes the number of floors) must be an integer. $A(17) = 569{,}294$ and $A(18) = 569{,}333$. Thus 17 floors should be built to minimize the average cost per floor.

43. Let x denote the number of hours after 8:00 a.m. at which the 15 minute coffee break begins. Then $f(x)$ is the number of radios assembled before the break. After the break, $4-x$ hours remain until lunchtime at 12:15 (4 hours 15 minutes minus x hours before the break minus 15 minutes for the break). Thus the number of radios assembled during the $4-x$ hours after the break is $g(4-x)$, and the total number of radios assembled between 8:00 a.m. and 12:15 p.m. is $N(x) = f(x)+g(x-4) = -x^3+6x^2+15x-(\frac{1}{3})(4-x)^3+(4-x)^2+23(4-x)$. The relevant interval is $0 \le x \le 4$. $N'(x) = -3x^2+12x+15-(4-x)^2(-1)+2(4-x)(-1)-23 = -3x^2+12x+15+16-8x+x^2-8+2x-23 = -2x^2+6x = -2x(x-3) = 0$ at $x=0$ and $x=3$. $N(0) = 86.67$, $N(3) = 95.67$, and $N(4) = 92$. Thus the worker will assemble the maximum number of radios by lunch time if the coffee break begins when $x = 3$, that is, at 11:00 a.m.

3.5 Applications to Business and Economics

1. Let x denote the number of transistors in each shipment and $C(x)$ the corresponding (variable) cost. Then $C(x)$ =(storage cost) + (ordering cost). The storage cost is $\frac{0.9x}{2} = 0.45x$. (See example 5.1 in the text for an explanation of the storage cost.) Since 600 transistors are used each year, $\frac{600}{x}$ is the number of shipments, and so the ordering cost is $30(\frac{600}{x}) = \frac{18{,}000}{x}$. The relevant interval is $0 < x \le 600$. $C'(x) = 0.45 - \frac{18{,}000}{x^2} = \frac{0.45x^2-18{,}000}{x^2} = \frac{0.45(x^2-40{,}000)}{x^2} = 0$ when $x^2 = 40,000$ or $x = 200$. Since this is the only critical point in the interval, and since $C''(x) = \frac{36{,}000}{x^3}$ (signifying a minimum for the critical point), C has an absolute minimum when $x = 200$, that is, when there are 200 cases per shipment and the transistors are ordered $\frac{600}{200} = 3$ times a year.

3. Let x denote the number of maps per batch and $C(x)$ the corresponding cost. Then, $C(x)$ = (storage cost) + (production cost) + (set-up cost). The storage cost is $(\frac{x}{2})(0.20) = 0.1x$. (See example 5.1 for an explanation of the storage cost.) The production cost is $0.06(16,000) = 960$. Since x maps are produced per batch and 16,000 maps are needed, the number of batches is $\frac{16{,}000}{x}$. The set-up cost is $100(\frac{16{,}000}{x}) = \frac{1{,}600{,}000}{x}$. Thus $C(x) = 0.1x + 960 + \frac{1{,}600{,}000}{x}$. The relevant interval is $0 < x \le 16,000$. $C'(x) = 0.1 - \frac{1{,}600{,}000}{x^2} = \frac{0.1x^2-1{,}600{,}000}{x^2} = \frac{0.1(x^2-16{,}000{,}000)}{x^2} = 0.1(\frac{(x-4{,}000)(x+4{,}000)}{x^2}) = 0$ when $x = 4,000$. Since this is the only critical point in the relevant interval, $C''(x) = \frac{3{,}200{,}000}{x^3}$, which indicates that C has its absolute minimum when $x = 4,000$ maps per batch.

5. a) If the total cost is $C(q) = 3q^2 + 5q + 75$, the average cost is $A(q) = \frac{C(q)}{q} = \frac{3q^2+5q+75}{q} = 3q + 5 + \frac{75}{q}$. The goal is to find the absolute minimum of $A(q)$ on the interval $q > 0$. $A'(q) = 3 - \frac{75}{q^2} = \frac{3q^2-75}{q^2} = \frac{3(q-5)(q+5)}{q^2} = 0$ on the interval $q > 0$ only when $q = 5$. Since $A''(q) = \frac{150}{q^3} > 0$ when $q = 5$, it follows that the average cost is minimal on $q > 0$ when $q = 5$, that is, when 5 units are produced.

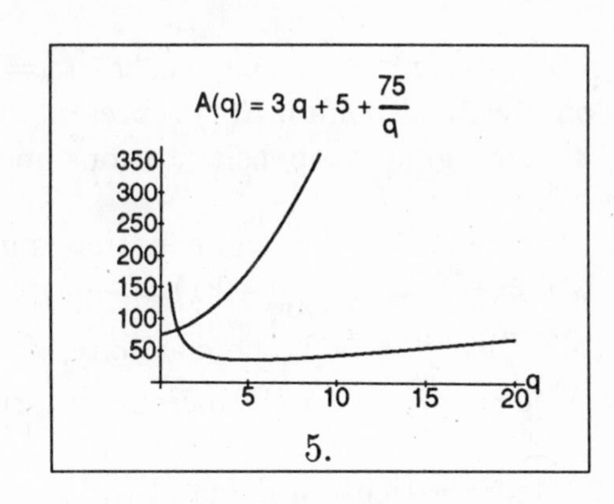

5.

b) The marginal cost is the derivative $C'(q) = 6q + 5$ of the total cost function and equals average cost when $6q + 5 = 3q + 5 + \frac{75}{q}$, $3q^2 = 75$, $q^2 = 25$, or $q = 5$ which is the same level of production as that in part a) for which average cost is minimal.
c) The marginal cost $C'(q) = 6q + 5$ is a linear function and its graph is a straight line with slope 6 and vertical intercept 5. To graph the average cost function $A(q)$, observe from part a) that $A''(q) = \frac{5(q+5)(q-5)}{q^2} < 0$ for $0 < q < 5$ and $A''(q) > 0$ for $q > 5$. Hence $A(q)$ is decreasing for $0 < q < 5$, increasing for $5 < q$, and has a relative minimum at $q = 5$. See the graph.

7. a) If the total revenue is $R(q) = -2q^2 + 68q - 128$, the average revenue per unit is $A(q) = \frac{R(q)}{q} = -2q + 68 - \frac{128}{q}$.
The marginal revenue is $R'(q) = -4q + 68$ and is equal to the average revenue when $-4q + 68 = -2q + 68 - \frac{128}{q}$, $2q = \frac{128}{q}$, $q^2 = 64$, or $q = 8$.

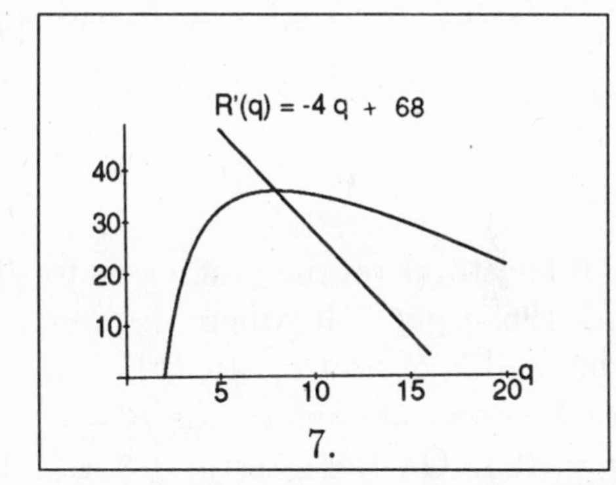

7.

b) $A'(q) = -2 + \frac{128}{q^2} = \frac{-2q^2+128}{q^2} = \frac{-2(q+8)(q-8)}{q^2}$. If $0 < q < 8$, $A'(q) > 0$, and $A(q)$ is increasing. If $8 < q$, $A'(q) < 0$, and $A(q)$ is decreasing.
c) The average and marginal revenue functions are sketched on the same graph. Notice that the average revenue has a maximum at $q = 8$ and q–intercepts are $q = 2$ and $q = 32$.

9. Let x be the number of units and $C(x)$ the cost. Total cost = wages + production cost + maintenance cost = $1,200 + 1.2x + \frac{10^5}{x^2}$. $C'(x) = 1.2 - \frac{200,000}{x^3} = 0$ if $x = \frac{100}{\sqrt[3]{6}} = 55.03$. $C(55) = 1299.06$, $C(0)$ is not defined, and $C(100,000) = 121,200$. Thus, the minimum cost occurs when 55 units are produced.

11. Let x be the total national income and $C(x)$ the total national consumption. The national savings is $S(x) = x - C(x) = x - 8 - 0.8x - 0.8\sqrt{x}$. The marginal propensity to

save is $S'(x) = 1 - 0.8 - 0.4x^{-1/2} = 0$ or $x = 4$. Since $S''(x) = 0.2x^{-3/2} > 0$ for $x = 4$ (the only critical point), $S(4)$ represents the absolute minimum in national savings.
The marginal propensity to consume is $C'(x) = 0.8 + \frac{0.4}{\sqrt{x}}$.

13. a) If the demand equation is $q = 60 - 0.1p$ (for $0 \le p \le 600$), the elasticity of demand is $\eta = \frac{p}{q}\frac{dq}{dp} = \frac{p}{60-0.1p}(-0.1) = -\frac{0.1p}{60-0.1p}$.
b) When $p = 200$, the elasticity of demand is $\eta = -\frac{0.1(200)}{60-0.1(200)} = -0.5$. That is, when the price is $p = 200$, a 1% increase in price will produce a decrease in demand of approximately 0.5 %.
c) The elasticity of demand will be -1 when $-1 = -\frac{0.1p}{60-0.1p}$, $60 - 0.1p = 0.1p$, $0.2p = 60$, or $p = \frac{60}{0.2} = 300$.

15. a) If the demand equation is $q = 500 - 2p$ (for $0 \le p \le 250$), the elasticity of demand is $\eta = \frac{p}{q}\frac{dq}{dp} = \frac{p}{500-2p}(-2) = -\frac{p}{250-p}$. The demand is of unit elasticity when $|\eta| = 1$, that is, when $\frac{p}{250-p} = 1$, $p = 250 - p$, or $p = 125$. If $0 \le p < 125$, $|\eta| = \frac{p}{250-p} < \frac{125}{250-125} = 1$ and so the demand is inelastic. If $125 < p \le 250$, $|\eta| = \frac{p}{250-p} > \frac{125}{250-125} = 1$ and so the demand is elastic.

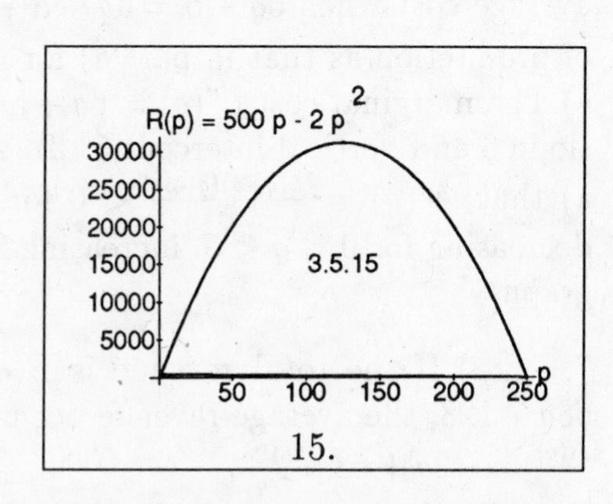

15.

b) The total revenue increases for $0 \le p < 125$ (where the demand is inelastic), decreases for $125 < p \le 250$ (where the demand is elastic), and has a maximum at $p = 125$ (where the demand is of unit elasticity).
c) The revenue function is $R = pq = (500 - 2p)p = 500p - 2p^2$. $R'(p) = 500 - 4p = 0$ when $p = 125$. On the interval $0 \le p < 125$, $R'(p)$ is positive and so $R(p)$ is increasing. On the interval $125 < p \le 250$, $R'(p)$ is negative and so $R(p)$ is decreasing. $R(p)$ has a relative maximum at the critical point $p = 125$.
d) The graphs of the demand and revenue functions are drawn in the pictures. Notice that for this linear demand function, the price of unit elasticity and maximum revenue is the midpoint $p = 125$ of the relevant interval $0 \le p \le 250$.

17. If the demand is $q = \frac{a}{p^m} = ap^{-m}$, the elasticity of demand is $\eta = \frac{p}{q}\frac{dq}{dp} = \frac{p}{a/p^m}(-amp^{-m-1}) = \frac{p^{m+1}}{a}(-\frac{am}{p^{m+1}}) = -m$, a constant which is independent of the price p. Thus, at any price, a 1% increase in price will produce a decrease in demand of approximately m percent.

19. a) Let x be the number of units produced, $p(x)$ the price per unit, t the tax per unit, and $C(x)$ the total cost. With $C(x) = \frac{7x^2}{8} + 5x + 100$, $p(x) = 15 - \frac{3x}{8}$, the revenue

is $R(x) = xp(x) = 15x - \frac{3x^2}{8}$. Now profit $P(x)$ = revenue − taxation − cost. Thus $P(x) = 15x - \frac{3x^2}{8} - tx - \frac{7x^2}{8} - 5x - 100$. $P'(x) = 15 - \frac{3x}{4} - t - \frac{7x}{4} - 5 = 0$ when $\frac{5x}{2} = 10 - t$ or $x = \frac{2(10-t)}{5}$.
b) The government share is $G(x) = tx = (\frac{2}{5})(10t - t^2)$. $G'(x) = (\frac{2}{5})(10 - 2t) = 0$ if $t = 5$. This represents an absolute maximum since $G''(x) < 0$ and there is only one critical point.
c) From part a), with $t = 0$, $x = \frac{2(10-0)}{5} = 4$, and with $t = 5$, $x = \frac{2(10-5)}{5} = 2$. The price per unit for the two quantities produced is, respectively, $p(4) = 15 - \frac{3(4)}{8} = \$13.50$ and $p(2) = 15 - \frac{3(2)}{8} = \$14.25$. The difference between the two unit prices is 14.25 − 13.50 or 75 cents.

21. $R = pq$, $\frac{dR}{dq} = (p)(1) + q\frac{dp}{dq} = p + \frac{p}{\frac{p}{q}\frac{dq}{dp}} = (1 + \frac{1}{\eta})p$.

23. $p = 600 - 2q^2$. a) $\eta = \frac{p}{q} / \frac{dp}{dq} = \frac{600-2q^2}{q}\frac{1}{-4q} = -\frac{300-q^2}{2q^2}$. The demand is of unit elasticity when $|\eta| = 1$ or $2q^2 = 300 - q^2$, $q = 10$. If $0 \le q < 10$ then $|\eta| < 1$, the demand is inelastic. If $10 < q \le \sqrt{300}$ then $|\eta| > 1$, hence the demand is elastic.
b) The total revenue increases when the demand is inelastic, that is when $0 \le q < 10$. The total revenue decreases when the demand is elastic, that is when $10 < q \le \sqrt{300}$. When $q = 10$, the revenue is maximized.
c) $R(q) = pq = 600q - 2q^3$. $R'(q) = 600 - 6q^2 = 0$ when $q = 10$, for which the revenue is maximized, since $R''(q) < 0$.

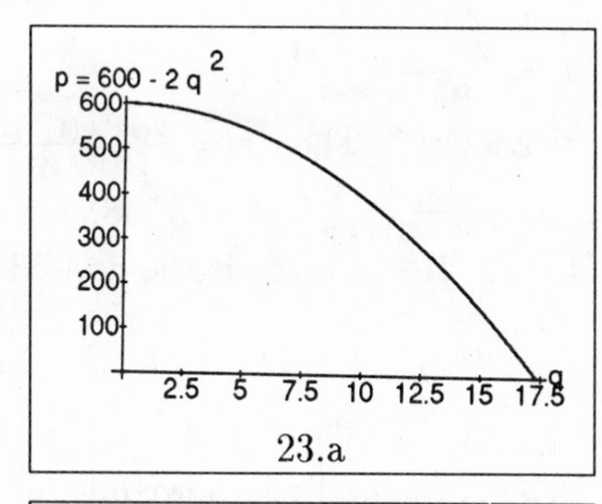

23.a

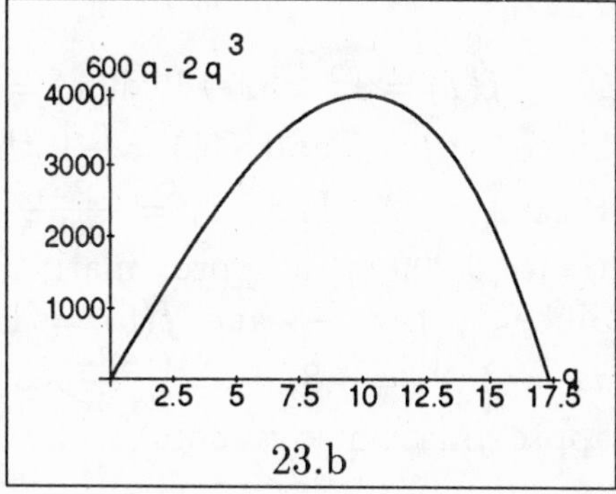

23.b

3.6 Newton's Method for Approximating Roots

$$\boxed{x_n = x_{n-1} - \frac{f(x_{n-1})}{f'(x_{n-1})}}$$

1. $f(x) = x^2 - 12$, $f'(x) = 2x$, $x - \frac{f(x)}{f'(x)} = x - \frac{x^2-12}{2x} = \frac{x^2+12}{2x}$. Thus $x_n = \frac{x_{n-1}^2+12}{2x_{n-1}}$.
Pick $x_0 = 3$. Then $x_1 = \frac{3^2+12}{2(3)} = 3.5$, $x_2 = \frac{(3.5)^2+12}{2(3.5)} = 3.4643$, $x_3 = \frac{(3.4643)^2+12}{2(3.4643)} = 3.4641$.
Hence 3.464 is the desired approximation to the root of $x^2 - 12 = 0$.

3. $f(x) = x^3 + x^2 - 1$, $f'(x) = 3x^2 + 2x$, $x - \frac{f(x)}{f'(x)} = x - \frac{x^3+x^2-1}{3x^2+2x} = \frac{2x^3+x^2+1}{3x^2+2x}$. Thus $x_n = \frac{2x_{n-1}^3+x_{n-1}^2+1}{3x_{n-1}^2+2x_{n-1}}$.
Pick $x_0 = 1$. Then $x_1 = \frac{2(1)^3+(1)^2+1}{3(1)^2+2(1)} = 0.8$,
$x_2 = \frac{2(0.8)^3+(0.8)^2+1}{3(0.8)^2+2(0.8)} = 0.7568$, $x_3 = \frac{2(0.7568)^3+(0.7568)^2+1}{3(0.7568)^2+2(0.7568)} = 0.7549$, $x_4 = \frac{2(0.7549)^3+(0.7549)^2+1}{3(0.7549)^2+2(0.7549)} =$ 0.7549. Hence 0.755 is the desired approximation to the root of $x^3 + x^2 - 1 = 0$.

5. $f(x) = x^3 - 2x - 5$, $f'(x) = 3x^2 - 2$, $x - \frac{f(x)}{f'(x)} = x - \frac{x^3-2x-5}{3x^2-2} = \frac{2x^3+5}{3x^2-2}$. Thus $x_n = \frac{2x_{n-1}^3+5}{3x_{n-1}^2-2}$.
Pick $x_0 = 2$. Then $x_1 = \frac{2(2)^3+5}{3(2)^2-2} = 2.1$, $x_2 = \frac{2(2.1)^3+5}{3(2.1)^2-2} = 2.095$, $x_3 = \frac{2(2.095)^3+5}{3(2.095)^2-2} = 2.095$. Hence 2.095 is the desired approximation to the root of $x^3 - 2x - 5 = 0$.

7. $f(x) = x^2 - 5x + 1$, $f'(x) = 2x - 5$, $x - \frac{f(x)}{f'(x)} = x - \frac{x^2-5x+1}{2x-5} = \frac{x^2-1}{2x-5}$. Thus $x_n = \frac{x_{n-1}^2-1}{2x_{n-1}-5}$.
Since $f(0) = 1$ and $f(1) = -3$, there is a root between 0 and 1.
Pick $x_0 = 0$. Then $x_1 = \frac{0^2-1}{2(0)-5} = 0.2$, $x_2 = \frac{(0.2)^2-1}{2(0.2)-5} = 0.2087$, $x_3 = \frac{(0.2087)^2-1}{2(0.2087)-5} = 0.2087$.
Hence 0.209 is an approximation to one of the roots.
Since $f(4) = -3$ and $f(5) = 1$, there is a root between 4 and 5. Pick $x_0 = 5$. Then $x_1 = \frac{5^2-1}{2(5)-5} = 4.8$, $x_2 = \frac{(4.8)^2-1}{2(4.8)-5} = 4.7913$, $x_3 = \frac{(4.7913)^2-1}{2(4.7913)-5} = 4.7913$. Hence 4.791 is the other approximation to a root.

9. Observe that $\sqrt{2}$ is a root of the equation $f(x) = 0$, where $f(x) = x^2 - 2$. $f'(x) = 2x$, $x - \frac{f(x)}{f'(x)} = x - \frac{x^2-2}{2x} = \frac{x^2+2}{2x}$. Thus $x_n = \frac{x_{n-1}^2+2}{2x_{n-1}}$.
Pick $x_0 = 1$. Then $x_1 = \frac{1^2+2}{2(1)} = 1.5$, $x_2 = \frac{(1.5)^2+2}{2(1.5)} = 1.41667$, $x_3 = \frac{(1.41667)^2+2}{2(1.41667)} = 1.414216$, $x_4 = \frac{(1.414216)^2+2}{2(1.414216)} = 1.414214$, $x_4 = \frac{(1.414214)^2+2}{2(1.414214)} = 1.414214$. Hence $\sqrt{2} \approx 1.41421$.

11. Observe that $\sqrt[3]{9}$ is a root of the equation $f(x) = 0$, where $f(x) = x^3 - 9$. $f'(x) = 3x^2$, $x - \frac{f(x)}{f'(x)} = x - \frac{x^3-9}{3x^2} = \frac{2x^3+9}{3x^2}$. Thus $x_n = \frac{2x_{n-1}^3}{3x_{n-1}^2}$.

Pick $x_0 = 2$. Then $x_1 = \frac{2(2)^3+9}{3(2)^2} = 2.083333$, $x_2 = \frac{2(2.083333)^3+9}{3(2.083333)^2} = 2.080089$, $x_3 = \frac{2(2.080089)^3+9}{3(2.080089)^2} = 2.080084$, $x_4 = \frac{2(2.080084)^3+9}{3(2.080084)^2} = 2.080084$. Hence $\sqrt[3]{9} \approx 2.08008$.

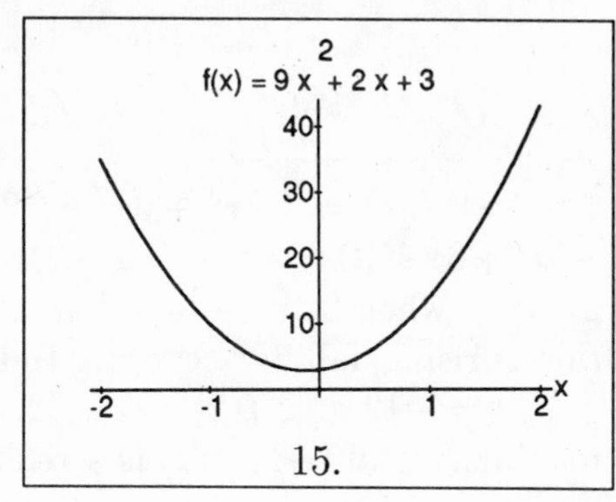

15.

13. Observe that $\sqrt{\frac{1}{26}}$ is a root of the equation $f(x) = 0$, where $f(x) = 26x^2 - 1$. $f'(x) = 52x$, $x - \frac{f(x)}{f'(x)} = x - \frac{26x^2-1}{52x} = \frac{26x^2+1}{52x}$. Thus $x_n = \frac{26x_{n-1}^2+1}{52x_{n-1}}$.

Pick $x_0 = 0.2$. Then $x_1 = \frac{26(0.2)^2+1}{52(0.2)} = 0.196$, $x_2 = \frac{26(0.196)^2+1}{52(0.196)} = 0.19613$, $x_2 = \frac{26(0.19613)^2+1}{52(0.19613)} = 0.19612$, $x_2 = \frac{26(0.19612)^2+1}{52(0.19612)} = 0.19612$, Hence $\sqrt{\frac{1}{26}} \approx 0.19612$.

15. $f(x) = 9x^2+2x+3 = 9(x^2+\frac{2x}{9}+\frac{1}{81})+3-\frac{1}{9} = 9(x+\frac{1}{9})^2+\frac{26}{9}$. The graph of this curve is a parabola opening upward, with vertex at $(-\frac{1}{9}, \frac{26}{9})$, which never crosses the x-axis.

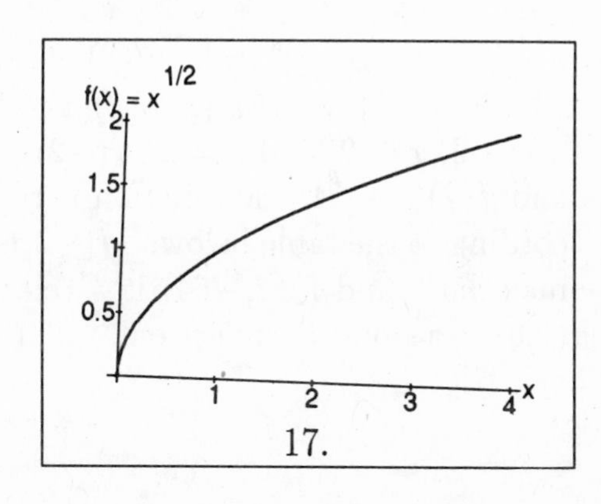

17.

17. $f(x) = \sqrt{x} = y$ or $y^2 = x$. The graph of this function is a parabola pointing to the right with vertex at the origin. We are interested in only the upper branch of this parabola.

Pick any point with $x > 0$. The tangent line at this point intersects the x-axis at $x < 0$. Now $x < 0$ is not in the domain of the function and Newton's method fails.

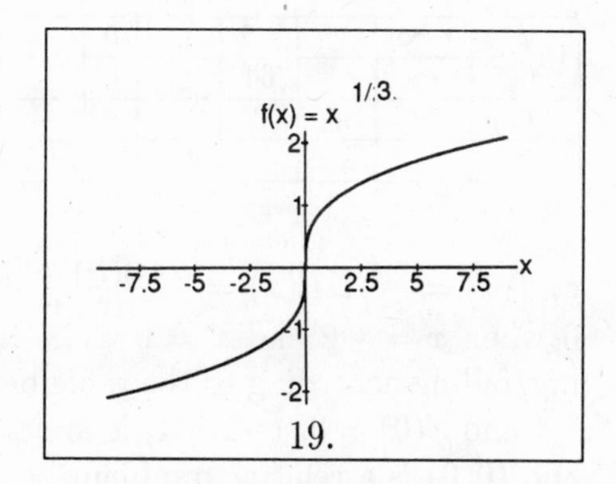

19.

19. a) $f(x) = \sqrt[3]{x}$, $f'(x) = \frac{1}{3x^{2/3}}$, $x - \frac{x^{1/3}}{(3x^{2/3})^{-1}} = x - 3x = -2x$. Thus $x_1 = -2x_0$, $x_2 = -2(-2x_0) = 4x_0$, $x_3 = -2x_2 = -8x_0, \ldots$.

Review Problems

1. a) $f(x) = -2x^3 + 3x^2 + 12x - 5$. $f'(x) = -6x^2 + 6x + 12 = -6(x^2 - x - 2) = -6(x+1)(x-2) = 0$ when $x = -1$ and $x = 2$. The function is rising/falling according to the table below. $f(-1) = -12$ and $f(2) = 15$. $(2,15)$ is a relative maximum and $(-1,-12)$ is a relative minimum.

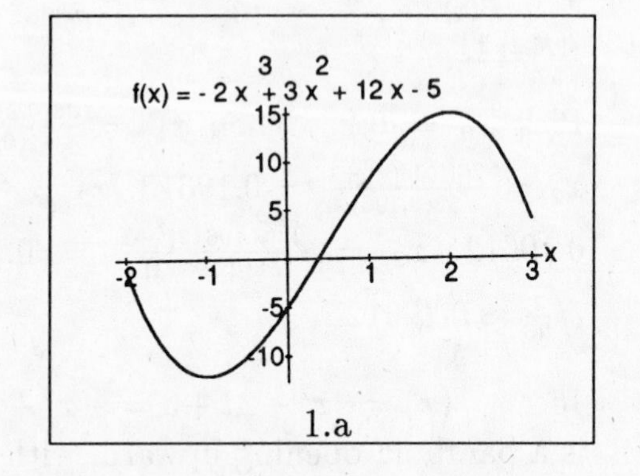

1.a

x	$-\infty$		-1		2		∞
$f(x)$	∞		-12		15		$-\infty$
$f'(x)$	$-\infty$	$-$	0	$+$	0	$-$	$-\infty$
		dec		inc		dec	

b) $f(x) = 3x^5 - 20x^3$. $f'(x) = 15x^4 - 60x^2 = 15x^2(x^2 - 4) = 15x^2(x-2)(x+2) = 0$ when $x = -2$, $x = 0$, and $x = 2$. $f(-2) = 64$, $f(0) = 0$, and $f(2) = -64$.The function is rising/falling according to the table below. $M(-2,64)$ is a relative maximum and $m(2,-64)$ is a relative minimum. (0,0) is a point of inflection.

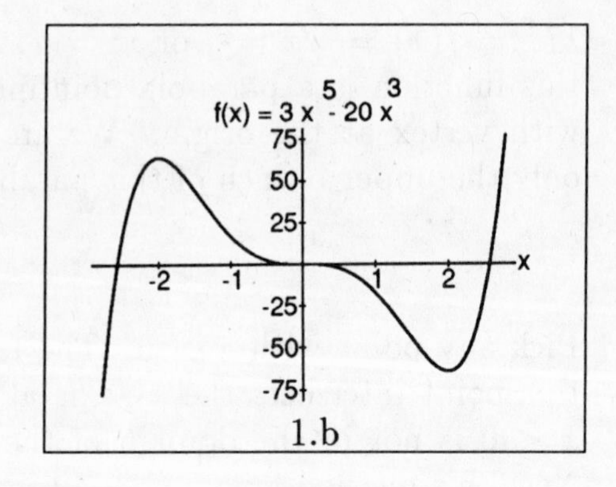

1.b

x	$-\infty$		-2		0		2		∞
$f(x)$	$-\infty$		64		0		-64		∞
$f'(x)$	$+$	$+$	0	$-$	0	$-$	0	$+$	$+$
		inc		dec		dec		inc	

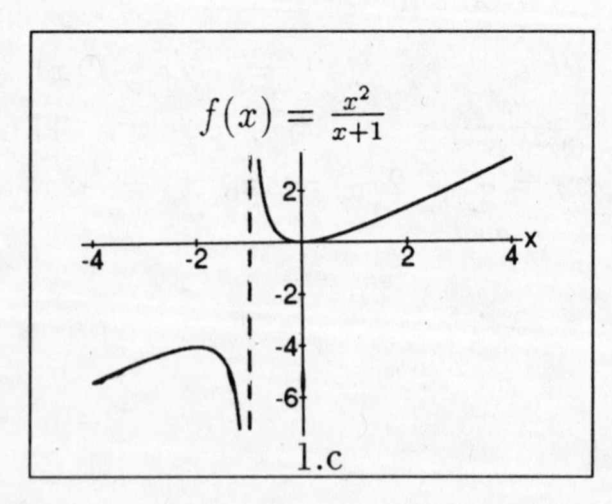

1.c

c) $f(x) = \frac{x^2}{x+1}$. $f'(x) = \frac{(x+1)(2x)-(x^2)(1)}{(x+1)^2} = \frac{x(x+2)}{(x+1)^2} = 0$ when $x = -2$ and $x = 0$. The function is rising/falling according to the table below. $f(-2) = -4$ and $f(0) = 0$. $(-2,-4)$ is arelative maximum and $(0,0)$ is a relative minimum.

x	$-\infty$		-2		-1		0		∞
$f(x)$	$-\infty$		-4		∞		0		∞
$f'(x)$	1	+	0	−		−	0	+	1
		inc		dec		dec		inc	

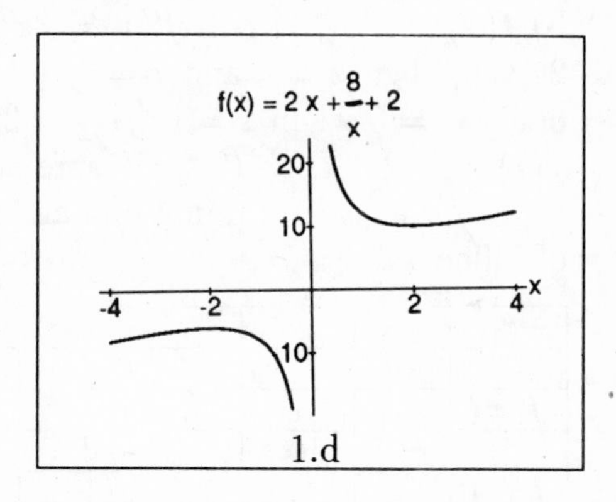

1.d

d) $f(x) = 2x + \frac{8}{x} + 2$. $f'(x) = 2 - \frac{8}{x^2} = \frac{2x^2-8}{x^2} = \frac{2(x-2)(x+2)}{x^2} = 0$ when $x = -2$ and $x = 2$. The function is rising/falling according to the table below. $f(-2) = -6$ and $f(2) = 10$. $(-2, -6)$ is arelative maximum and $(2, 10)$ is a relative minimum.

x	$-\infty$		-2		0		2		∞
$f(x)$	$-\infty$		-6		∞		10		∞
$f'(x)$	1	+	0	−		−	0	+	1
		inc		dec		dec		inc	

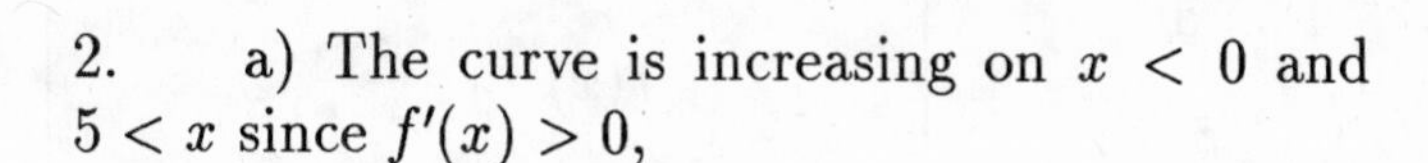

2. a) The curve is increasing on $x < 0$ and $5 < x$ since $f'(x) > 0$,

b) the curve is decreasing on $0 < x < 5$ since $f'(x) < 0$,

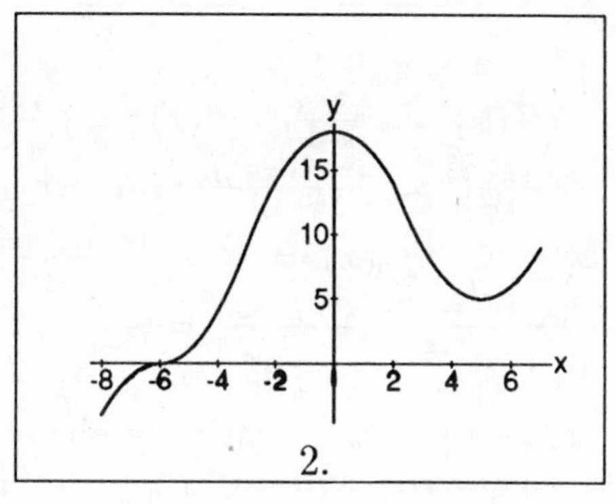

2.

c) the curve is concave upward on $-6 < x < -3$ and $2 < x$ since $f''(x) > 0$,
d) the curve is concave downward on $x < -6$ and $-3 < x < 2$ since $f''(x) < 0$.

One such curve could be constructed piecemeal as follows: $f(x) = -(x+6)^2$ on $x < -6$, $f(x) = (x+6)^2$ on $-6 < x < -3$, $f(x) = -x^2 + 18$ on $-3 < x < 2$, $f(x) = (x-5)^2 + 5$ on $2 < x$.

3. a) $f(x) = x^2 - 6x + 1$. $f'(x) = 2x - 6 = 2(x-3) = 0$ when $x = 3$. $f''(x) = 2 > 0$, $f(3) = -8$. Thus $(3, -8)$ is a relative minimum (actually the absolute minimum).

x	$-\infty$		3		∞
$f(x)$	∞		-8		∞
$f'(x)$	$-\infty$	−	0	+	∞
		dec		inc	
$f''(x)$	2	+	+	+	2
		up	up	up	

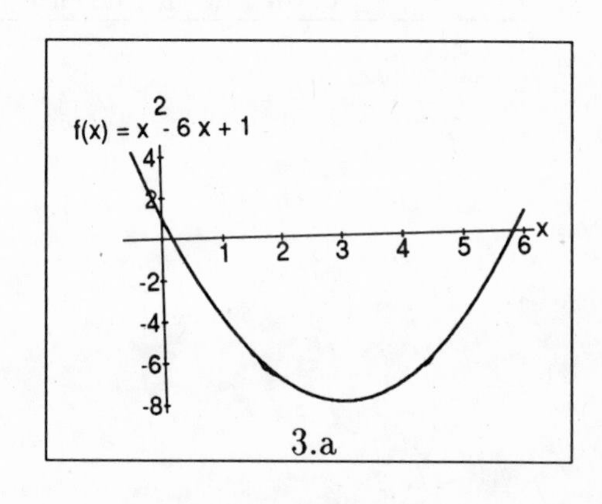

3.a

b) $f(x) = x^3 - 3x^2 + 2$. $f'(x) = 3x^2 - 6x = 3x(x-2) = 0$ when $x = 0$ and $x = 2$. $f''(x) = 6x - 6 = 6(x-1) = 0$ when $x = 1$. $f(0) = 2$, $f(1) = 0$, and $f(2) = -2$. Thus $(0, 2)$ is a relative maximum, $(2, -2)$ is a relative minimum, and $(1, 0)$ is a point of inflection.

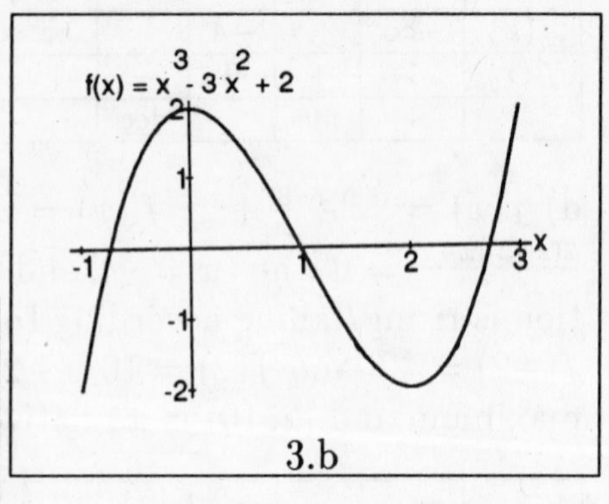

3.b

x	$-\infty$		0		1		2		∞
$f(x)$	$-\infty$		2		0		-2		∞
$f'(x)$		+	0	−	−	−	0	+	
		inc		dec	dec	dec		inc	
$f''(x)$		−	−	−	0	+	+	+	
		down	down	down		up	up	up	

c) $f(x) = \frac{x^2+3}{x-1}$. $f'(x) = \frac{(x-1)(2x)-(x^2+3)(1)}{(x-1)^2} = \frac{x^2-2x-3}{(x-1)^2} = \frac{(x-3)(x+1)}{(x-1)^2} = 0$ when $x = -1$ and $x = 3$. $f''(x) = \frac{(x-1)^2(2x-2)-(x^2-2x-3)(2)(x-1)}{(x-1)^4} = \frac{2(x-1)[(x-1)^2-(x^2-2x-3)]}{(x-1)^4} = \frac{8}{(x-1)^3} \neq 0$. $f(-1) = -2$ and $f(3) = 6$. Thus $(-1, -2)$ is a relative maximum, $(3, 6)$ is a relative minimum, and the graph changes concavity at $x = 1$ even though this is not in the domain of the function.

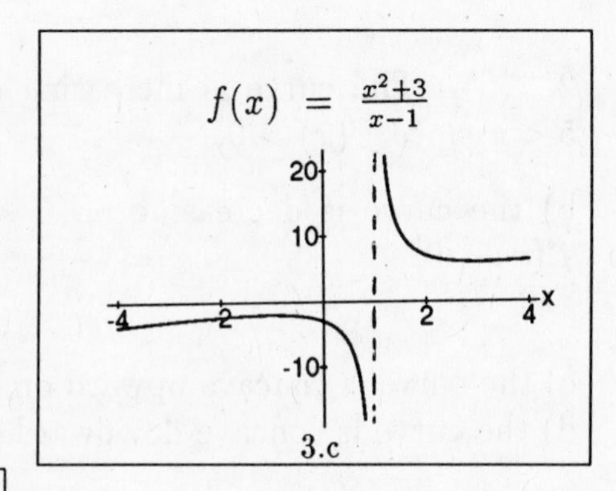

3.c

x	$-\infty$		-1		1		3		∞
$f(x)$	$-\infty$		-2		∞		6		∞
$f'(x)$	1	+	0	−		−	0	+	1
		inc		dec		dec		inc	
$f''(x)$		−	−	−		+	+	+	
		down	down	down		up	up	up	

d) $f(x) = \frac{x-1}{(x+1)^2}$. $f'(x) = \frac{(x+1)^2(1)-(x-1)[2(x+1)(1)]}{(x+1)^4} = \frac{(x+1)[(x+1)-2(x-1)]}{(x+1)^4} = \frac{3-x}{(x+1)^3} = 0$ when $x = 3$. $f''(x) = \frac{(x+1)^3(-1)-(3-x)[3(x+1)^2(1)]}{(x+1)^6} = \frac{(x+1)^2[-(x+1)-3(3-x)]}{(x+1)^6} = \frac{-x-1-9+3x}{(x+1)^4} = \frac{2(x-5)}{(x+1)^4} = 0$ when $x = 5$. $f(3) = \frac{1}{8}$ and $f(5) = \frac{1}{9}$. Thus $(3, \frac{1}{8})$ is a relative maximum, $(5, \frac{1}{9})$ is a point of inflection.

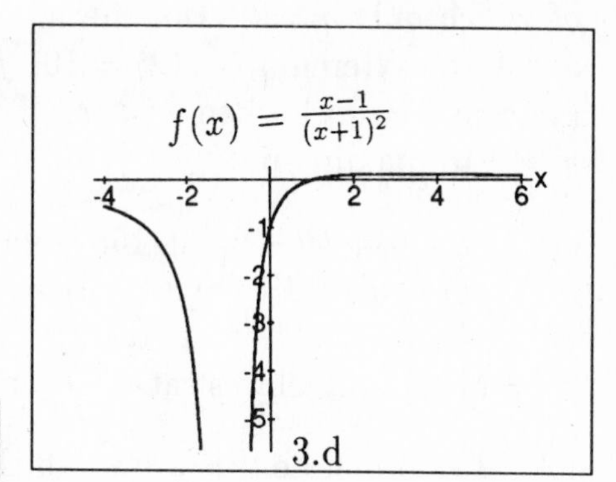

3.d

x	$-\infty$		-1		3		5		∞
$f(x)$	0		∞		$\frac{1}{8}$		$\frac{1}{9}$		0
$f'(x)$		$-$			0	$-$	$-$	$-$	
		dec				dec	dec	dec	
$f''(x)$		$-$		$-$		$-$	0	$+$	
		down		down		down		up	

4. a) $f(x) = -2x^3 + 3x^2 + 12x - 5$, $f'(x) = -6x^2 + 6x + 12 = -6(x+1)(x-2) = 0$ when $x = -1$ and $x = 2$. $f''(x) = -12x + 6 = -6(2x-1)$. $f(-1) = -12$, $f(2) = 15$. $f''(-1) = 18 > 0$, so f has a relative minimum at $(-1, -12)$. Similarly $f''(2) = -18 < 0$, so f has a relative maximum at $(2, 15)$.
b) $f(x) = \frac{x^2}{x+1}$, $f'(x) = \frac{(x+1)(2x)-(x^2)(1)}{(x+1)^2} = \frac{x^2+2x}{(x+1)^2} = \frac{x(x+2)}{(x+1)^2} = 0$ when $x = -2$ and $x = 0$. $f''(x) = \frac{(x+1)^2(2x+2)-(x^2+2x)(2)(x+1)(1)}{(x+1)^4} = \frac{2}{(x+1)^3} = 0$. $f(0) = 0$, $f(-2) = -4$. $f''(0) = 2 > 0$, so f has a relative minimum at $(0,0)$. Similarly $f''(-2) = -4 < 0$, so f has a relative maximum at $(-2, -4)$.
c) $f(x) = 2x + \frac{8}{x} + 2$, $f'(x) = 2 - \frac{8}{x^2} = \frac{2(x+2)(x-2)}{x^2} = 0$ when $x = -2$ and $x = 2$. $f''(x) = \frac{16}{x^3} = 0$. $f(-2) = -6$, $f(2) = 10$. $f''(-2) = -2 < 0$, so f has a relative maximum at $(-2, -6)$. Similarly $f''(2) = 2 > 0$, so f has a relative minimum at $(2, 10)$.

5. a) $f(x) = -2x^3 + 3x^2 + 12x - 5$, $f'(x) = -6x^2 + 6x + 12 = -6(x+1)(x-2) = 0$ when $x = -1$ and $x = 2$, both of which are in the interval $-3 \le x \le 3$. $f(-1) = -12$, $f(2) = 15$, $f(-3) = 40$, $f(3) = 4$. Thus $f(-3) = 40$ is the absolute maximum and $f(-1) = -12$ the absolute minimum.
b) $f(x) = -3x^4 + 8x^3 - 10$, $f'(x) = -12x^3 + 24x^2 = -12x^2(x-2) = 0$ when $x = 0$ and $x = 2$, both of which are in the interval $0 \le x \le 3$. $f(0) = -10$, $f(2) = 6$, and $f(3) = -37$. Thus $f(2) = 6$ is the absolute maximum and $f(3) = -37$ the absolute minimum.
c) $f(x) = \frac{x^2}{x+1}$, $f'(x) = \frac{(x+1)(2x)-(x^2)(1)}{(x+1)^2} = \frac{x(x+2)}{(x+1)^2} = 0$, when $x = 0$ and $x = -2$, of which only $x = 0$ is in the interval $-\frac{1}{2} \le x \le 1$. Note that the derivative is undefined at $x = -1$, which is not in the domain of f. $f(-\frac{1}{2}) = \frac{1}{2}$, $f(0) = 0$, and $f(1) = \frac{1}{2}$. Thus $f(-\frac{1}{2}) = f(1) = \frac{1}{2}$ is the absolute maximum and $f(0) = 0$ the absolute minimum.

d) $f(x) = 2x + \frac{8}{x} + 2$, $f'(x) = 2 - \frac{8}{x^2} = \frac{2x^2-8}{x^2} = \frac{2(x+2)(x-2)}{x^2} = 0$ when $x = -2$ and $x = 2$, of which only $x = 2$ is in the interval $0 < x$. There are no endpoints, so the only possible absolute extremum is $f(2) = 10$. $f'(x) < 0$ (f is decreasing) when $0 < x < 2$ and $f'(x) > 0$ (f is increasing) when $2 < x$. Thus $f(2) = 10$ is the absolute minimum and there is no absolute maximum.

6. $S(t) = t^3 - 9t^2 + 15t + 45$ on the interval $0 \leq t \leq 7$. $S'(t) = 3t^2 - 18t + 15 = 3(t^2 - 6t + 5) = 3(t-1)(t-5) = 0$ when $t = 1$ and $t = 5$. $S(0) = 45$, $S(1) = 52$, $S(5) = 20$, and $S(7) = 52$. Thus the traffic is moving fastest at 1:00 p.m. and 7:00 p.m. when its speed is 52 MPH and slowest at 5:00 p.m. when its speed is 20 MPH.

7. Let R denote the rate at which the rumor is spreading, Q the number of people who have heard the rumor, and P the total population of the community. Then $R(Q) = kQ(P - Q)$ where k is a positive constant of proportionality. $R'(Q) = kQ(-1) + (P-Q)(k) = Pk - 2kQ = 0$ when $2kQ = Pk$ or $Q = \frac{P}{2}$, which is in the interval $0 \leq Q \leq P$. $R''(Q) = -2k < 0$ which means that the absolute maximum is reached when half the people have heard the rumor.

8. Let x be the number of \$1.00 increases. The sales price per lamp is $6 + x$ and the corresponding number of lamps $3,000 - 1,000x = 1,000(3 - x)$. The total cost $C = (3,000 - 1,000x) \times 4 = 4,000(3-x)$ and the total revenue $R = 1,000(3-x)(6+x)$. The profit is $P = R - C = 1,000(3-x)(6+x) - 4,000(3-x) = 1,000(6+x-x^2)$. $P' = 1,000(1-2x) = 0$ when $x = \frac{1}{2}$. $P''(0.50) < 0$, so $x = 0.50$ maximizes the profit. The corresponding sales price is \$6.50. (Note that the graph of p is a parabola pointing downward.)

9. Label the sides of the rectangular plot as indicated in the figure and let A denote the area. Then $A = xy + xy = 2xy$, a function of two variables. The fact that 300 meters of fencing are to be used implies $4x + 3y = 300$ or $y = \frac{300-4x}{3} = 100 - \frac{4x}{3}$
Thus $A(x) = 2x(100 - \frac{4x}{3}) = 200x - \frac{8x^2}{3}$. $A'(x) = 200 - \frac{16x}{3} = 0$ or $x = 37.5$. $A''(x) = -\frac{16}{3} < 0$, $x = 37.5$ indicates the absolute maximum. The corresponding value of y is $y = 100 - \frac{4(37.5)}{3} = 50$. Each plot of land should be 37.5 by 50 meters.

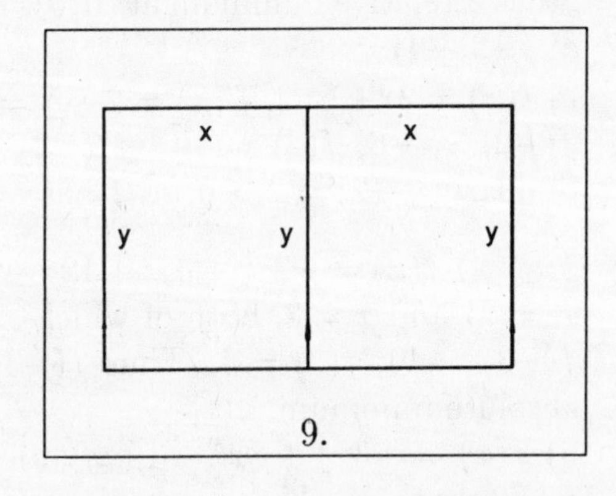

9.

10. a) Let x be the length and y the width of rectangular pasture, and let A denote the area. Then $A = xy$, a function of two variables. The fact that 320 ft of fencing are to be used implies $2x + 2y = 320$ or $y = 160 - x$. Thus $A(x) = x(160 - x) = 160x - x^2$. $A'(x) = 160 - 2x = 0$ or $x = 80$. $A''(x) = -2 < 0$, $x = 80$ indicates the absolute maximum. The corresponding value of y is $y = 160 - 80 = 80$. The pasture should be a square with 80

ft per side.
b) The fact that 320 ft of fencing are to be used implies $x+2y=320$ or $x=2(160-y)$. Thus $A(y)=2y(160-y)=2(160y-y^2)$. $A'(y)=2(160-2y)=0$ or $y=80$. $A''(y)=-4<0$, $y=80$ indicates the absolute maximum. The corresponding value of x is $x=2(160-80)=160$. The pasture measures 160 ft (on the unfenced side) by 80 ft (each for both fenced sides).

11. Let r denote the radius, h the height, C the (fixed) cost (in cents), and V the volume. $V=\pi r^2 h$, a function of two variables. To write h in terms of r, use the fact that the cost is to be C cents. That is C =cost of bottom + cost of side = 3(area of bottom) + 2(area of side) or $C=3\pi r^2+2(2\pi rh)=3\pi r^2+4\pi rh$. To solve for h, $4\pi rh=C-3\pi r^2$ or $h=\frac{C-3\pi r^2}{4\pi r}$ where C is a constant. Thus $V(r)=\pi r^2(\frac{C-3\pi r^2}{4\pi r})=\frac{r(C-3\pi r^2)}{4}=\frac{rC}{4}-\frac{3\pi r^3}{4}$. $V'(r)=\frac{C}{4}-\frac{9\pi r^2}{4}=0$ when $\frac{C}{4}=\frac{9\pi r^2}{4}$ or $C=9\pi r^2$. Substituting in the formula for cost leads to $3\pi r^2+4\pi rh=9\pi r^2$, $4\pi rh=6\pi r^2$ or $h=\frac{3r}{2}$. $V''(r)=-\frac{9\pi r}{2}<0$, so the volume is maximized when the height is 1.5 times the radius of the cylindrical container.

12. $f(t)=-t^3+7t^2+200t$ is the number of letters the clerk can sort in t hours. The clerk's rate of output is $R(t)=f'(t)=-3t^2+14t+200$ letters per hour. The relevant interval is $0\le t\le 4$. $R'(t)=f''(t)=-6t+14=0$ when $t=\frac{7}{3}$. $R(\frac{7}{3})=216.33$, $R(0)=200$, and $R(4)=208$. Thus the rate of output is greatest when $t=\frac{7}{3}$ hours, that is, after 2 hours and 20 minutes.

13. Let x denote the number of additional trees planted and $f(x)$ the corresponding total yield. The total number of trees is $60+x$. The average yield per tree is 475 lemons when 60 trees are planted will decrease by 5 lemons for each additional tree. Thus, the average yield per tree is $475-5x$. Putting it all together, $f(x)$ =(number of trees)(yield per tree)= $(60+x)(475-5x)=5(60+x)(95-x)=5(5,700+35x-x^2)$. $f'(x)=5(35-2x)=0$ or $x=17.5$. For $x<17.5$, $f'(x)>0$ and $f(x)$ is increasing. For $x>17.5$, $f'(x)<0$ and $f(x)$ is decreasing. Hence $f(x)$ has its absolute maximum when $x=17.5$. Since the number of additional trees planted must be a whole number, compare $f(17)=30,030$ and $f(18)=30,030$ to conclude that planting either 17 or 18 additional trees (for a total of either 77 or 78 trees) will maximize the total yield.

14. Let x denote the number of machines used and $C(x)$ the corresponding cost of producing the 400,000 medals. Then $C(x)$ =set-up cost + operating cost = 80(number of machines) + 5.76(number of hours). Each machine can produce 200 medals per hour, so x machines can produce $200x$ medals per hour, and it will take $\frac{400{,}000}{200x}$ hours to produce the 400,000 medals. Hence $C(x)=80x+5.76(\frac{400{,}000}{200x})=80x+\frac{11{,}520}{x}$. $C'(x)=80-\frac{11{,}520}{x^2}=\frac{80(x^2-144)}{x^2}=\frac{80(x-12)(x+12)}{x^2}=0$ when $x=12>0$. For $0<x<12$, $C'(x)<0$ and $C(x)$ is decreasing. For $12<x$, $C'(x)>0$ and $C(x)$ is increasing. Thus $C(x)$ has its absolute minimum when $x=12$ machines are used.

15. Let Q be the point on the opposite bank straight across from the starting point. With

$PQ = x$, the distance along the bank is $1 - x$. The distance across the water is given by the pythagorean theorem to be $\sqrt{1+x^2}$. The time T is T =time in the water + time on the land = $\frac{\text{distance in the water}}{\text{speed in the water}} + \frac{\text{distance on the land}}{\text{speed on the land}} = \frac{\sqrt{1+x^2}}{4} + \frac{1-x}{5} = (\frac{1}{4})(1+x^2)^{1/2} + (\frac{1}{5})(1-x)$. The relevant interval is $0 \le x \le 1$ and $T'(x) = (\frac{1}{4})(\frac{1}{2})(1+x^2)^{-1/2}(2x) - \frac{1}{5} = \frac{x}{4\sqrt{1+x^2}} - \frac{1}{5} = 0$ if $\frac{x}{4\sqrt{1+x^2}} = \frac{1}{5}$, $5x = 4\sqrt{1+x^2}$, $25x^2 = 16(1+x^2) = 16 + 16x^2$, $9x^2 = 16$, $x^2 = \frac{16}{9}$, or $x = \pm\frac{4}{3}$. Neither of these critical points lies in the interval $0 \le x \le 1$. Hence, the absolute minimum must occur at an endpoint. $T(0) = \frac{1}{4} + \frac{1}{5} = 0.45$ hour and $T(1) = \frac{\sqrt{2}}{4} = 0.354$ hour. The minimum time thus occurs if $x = 1$, that is, if you row all the way to the town.

16. $f(x) = Ax^3 + Bx^2 + C$, $f'(x) = 3Ax^2 + 2Bx$, and $f''(x) = 6Ax + 2B$. $E(2, 11)$ is an extremum, so $f'(2) = 12A + 4B = 0$ or $B = -3A$. $I(1, 5)$ is a point of inflection, so $f''(1) = 6A + 2B = 0$ or $B = -3A$. Since the points E and I are on the curve, their coordinates must satisfy its equation. Thus $f(2) = 8A + 4B + C = 8A - 12A + C = 11$ and $f(1) = A + B + C = A - 3A + C = 5$. Solving simultaneously leads to $A = -3$, $B = 9$, and $C = -1$ and $f(x) = -3x^3 + 9x^2 - 1$.

17. a) $C(x) = x^3 - 24x^2 + 350x + 400$. $C'(x) = 3x^2 - 48x + 350 = 0$ and $[C'(x)]' = 6x - 48 = 0$ if $x = 8$. This leads to a relative minimum since $[C'(x)]'' = 6 > 0$.
b) $A(x) = x^2 - 24x + 350 + \frac{400}{x}$, $A'(x) = 2x - 24 - \frac{400}{x^2} = (\frac{1}{x^2})(2x^3 - 24x^2 - 400) = 0$ when $f(x) = 2x^3 - 24x^2 - 400 = 0$. $x = 13.156$ is the only solution (by trial and error or from a computer program).

18. Let $A(x, y)$ and $B(-x, y)$, $(x > 0)$, be points on the parabola $y = (\frac{1}{14})(48 - x^2)$ to form the isosceles triangle OAB. Since AB is horizontal, the area is $A = (\frac{1}{2})(2xy) = \frac{48x - x^3}{14}$. $A'(x) = \frac{48 - 3x^2}{14} = 0$ when $x^2 = 16$ or $x = 4 > 0$. $A''(4) < 0$ reveals a relative maximum for $A(x)$. Thus $y = \frac{48-16}{14} = \frac{16}{7}$ and the largest area is $A = \frac{192 - 64}{14} = \frac{64}{7}$ square units.

19. The relationship between the number of Moppsy dolls and Floppsy dolls is given by $y = \frac{82 - 10x}{10 - x}$ with the relevant interval $0 \le x \le 8$. Suppose Moppsy sells for \$2 and Floppsy sells for \$1 each. The revenue is then $R(x) = x + \frac{2(82-10x)}{10-x} = \frac{164 - 20x + 10x - x^2}{10 - x} = \frac{x^2 + 10x - 164}{x - 10}$. $R'(x) = \frac{(x-10)(2x+10) - (x^2+10x-164)(1)}{(x-10)^2} = \frac{x^2 - 20x + 64}{(x-10)^2} = 0$ when $x = 10 \pm 6$. Pick $x = 4 > 0$ to be in the relevant interval. $R(0) = 16.4$, $R(4) = \frac{16 + 40 - 164}{4 - 10} = 18$, and $R(8) = \frac{64 + 80 - 164}{8 - 10} = 10$. Now $y = \frac{82 - 40}{10 - 4} = 7$ which means that revenue is maximized when 4 hundred Floppsies and 7 hundred Moppsies are sold.

20. Let Q be the point on the shore straight across from the oil rig and P the point on the shore where the pipe starts on land. With $PQ = x$, the distance along the bank is $8 - x$. The distance across the water is given by the pythagorean theorem to be $\sqrt{9 + x^2}$. The cost C is C =cost in the water + cost on the shore = $(1.5)\sqrt{9 + x^2} + (1)(8 - x)$, using 1 unit of price. $C'(x) = \frac{(1.5)(2x)}{2\sqrt{9+x^2}} - 1 = 0$ when $2.25x^2 = 9 + x^2$, $1.25x^2 = 9$, or $x = 2.6833 > 0$, which is in the

relevant interval $0 \leq x \leq 8$. $C(0) = (1.5)(3) + 8 = 12.45$, $C(2.6833) = 6.04 + 5.31 = 11.35$, and $C(8) = (1.5)\sqrt{9+64} = 12.81$. The minimum occurs thus occurs if $x = 2.6833$, that is, if $8 - x = 5.3167$ miles or 28,072 ft of pipe is laid on the shore.

21. Let x be the dimension of one side of the equilateral triangle, which is also the dimension of the base of the rectangle. y is the height of the rectangle. The perimeter of the window is $3x+2y = 20$ giving $y = \frac{20-3x}{2}$. We need to find the height h of the isosceles triangle. With x as the hypotenuse of one-half of the isosceles triangle, $h = x\sin(\frac{\pi}{3}) = \frac{\sqrt{3}x}{2}$. The base of the right triangle is $\frac{x}{2}$. Thus the area of the (whole) isosceles triangle is $A_t = \frac{\sqrt{3}x^2}{4}$. The area of the rectangle is $A_r = xy = \frac{x(20-3x)}{2}$. Since twice as much light passes through the rectangle as through the stained glass isosceles triangle, the light is $L = k[\frac{\sqrt{3}x^2}{4} + (2)\frac{x(20-3x)}{2}]$. k is a proportionality constant which does not affect the outcome of the problem. With $L = k[(\frac{\sqrt{3}}{4} - 3)x^2 + 20x]$, $L' = k[(\frac{\sqrt{3}}{2} - 6)x + 20] = 0$ when $x = \frac{20}{6-\sqrt{3}/2} = 3.8956$ feet. Then $y = (\frac{1}{2})(20 - 11.6868) = 4.1566$ feet. $L'' < 0$ which signifies that $x = 3.8956$ is the absolute maximum of the curve, whose graph is a parabola pointing downward.

22. Let $f(x) = x^2 - 55$. Then $f'(x) = 2x$ and $x - \frac{f(x)}{f'(x)} = x - \frac{x^2-55}{2x} = \frac{x^2+55}{2x}$. Hence $x_n = \frac{x_{n-1}^2+55}{2x_{n-1}}$. Since $f(7) < 0$ and $f(8) > 0$, the root must lie between 7 and 8. With $x_0 = 7$, $x_1 = \frac{7^2+55}{2(7)} = 7.428571$, $x_2 = \frac{(7.428571)^2+55}{2(7.428571)} = 7.416209$, $x_3 = \frac{(7.416209)^2+55}{2(7.416209)} = 7.416199$, $x_3 = \frac{(7.416199)^2+55}{2(7.416199)} = 7.416199$. Hence the root is 7.41620, rounded to five decimal places.

23. Let $f(x) = x^3+3x^2+1$. Then $f'(x) = 3x^2+6x$ and $x-\frac{f(x)}{f'(x)} = x-\frac{x^3+3x^2+1}{3x^2+6x} = \frac{2x^3+3x^2-1}{3x(x+2)}$. Hence $x_n = \frac{2x_{n-1}^3+3x_{n-1}^2-1}{3x_{n-1}(x_{n-1}+2)}$. Since $f(-3) > 0$ and $f(-4) = -15 < 0$, the root must lie between -3 and -4. With $x_0 = -3$, $x_1 = \frac{2(-3)^3+3(-3)^2-1}{3(-3)(-3+2)} = -3.1111$, $x_2 = \frac{2(-3.1111)^3+3(-3.1111)^2-1}{3(-3.1111)(-3.1111+2)} = -3.1038$, $x_3 = \frac{2(-3.1038)^3+3(-3.1038)^2-1}{3(-3.1038)(-3.1038+2)} = -3.1038$, Hence the root is -3.104.

24. Let $f(x) = 2x^4 - 8x^3 + 8x^2 - 1$. Then $f'(x) = 8x^3 - 24x^2 + 16x$ and $x - \frac{f(x)}{f'(x)} = x - \frac{2x^4-8x^3+8x^2-1}{8x^3-24x^2+16x} = \frac{6x^4-16x^3+8x^2+1}{8(x^3-3x^2+2x)}$. Hence $x_n = \frac{6x_{n-1}^4-16x_{n-1}^3+8x_{n-1}^2+1}{8(x_{n-1}^3-3x_{n-1}^2+2x_{n-1})}$ With $x_0 = -0.5$, $x_1 = \frac{5.375}{-15} = -0.358$, $x_2 = \frac{2.858}{-9.171} = -.312$, $x_3 = \frac{2.322}{-7.571} = -.307$, $x_4 = \frac{2.270}{-7.405} = -.307$. Hence one of the roots is -0.307.
Synthetic division leads to the numbers and the resulting third degree equation below:

2	−8	8	0	1	−0.307
	−0.614	2.644	−3.268	1.003	
2	−8.614	10.644	−3.268	0.003	

Now we need to find the roots of $g(x) = 2x^3 - 8.614x^2 + 10.644x - 3.268$. $g'(x) = 6x^2 -$

$17.228x + 10.644$, $x - \frac{g(x)}{g'(x)} = x - \frac{2x^3-8.614x^2+10.644x-3.268}{6x^2-17.228x+10.644} = \frac{4x^3-8.614x^2+3.268}{6x^2-17.228x+10.644}$. Thus $x_n = \frac{4x_{n-1}^3-8.614x_{n-1}^2+3.268}{6x_{n-1}^2-17.228x_{n-1}+10.644}$. With $x_0 = 0$, $x_1 = \frac{4x(0)^3-8.614(0)^2+3.268}{6(0)^2-17.228(0)+10.644} = \frac{3.268}{10.644} = .307$, $x_2 = \frac{2.5718}{5.92} = 0.434$, $x_3 = \frac{1.972}{4.297} = 0.459$, $x_4 = \frac{1.84}{4.01} = 0.459$.
Another synthetic division leads to

2	−8.614	10.644	−3.268	0.459
	0.918	−3.532	3.265	
2	−7.696	7.112	−0.003	

We are now faced with solving a quadratic equation, namely $h(x) = 2x^2 - 7.696x + 7.112 = 0$. From the quadratic formula, $x = \frac{-(-7.696)\pm\sqrt{(-7.696)^2-4(2)(7.112)}}{2(2)} = \frac{-7.696\pm1.527}{4}$, so $x = -1.542$ or -2.306.

Chapter 4

Exponential and Logarithmic Functions

4.1 Exponential Functions

1. The following calculations are for a calculator with an Algebraic Operating System (AOS) and an e^x key.
For e^2, punch 2, then e^x, to get 7.389.
For e^{-2}, punch 2, then $\pm$ followed by e^x, to get 0.135.
For $e^{0.05}$, punch 0.05, then e^x, to get 1.051.
For $e^{-0.05}$, punch 0.05, then $\pm$ followed by e^x, to get 0.951.
For e^0, punch 0, then e^x, to get 1.
For e, punch 1, then e^x, to get 2.718
For $\sqrt{e} = e^{1/2}$, punch 0.5, then e^x, to get 1.649.
For $\frac{1}{\sqrt{e}} = e^{1/2}$, punch 0.5, followed by $\pm$, then e^x, to get 0.607.

3. $y = 3^x$ and $y = 4^x$. The x−axis is a horizontal asymptote.

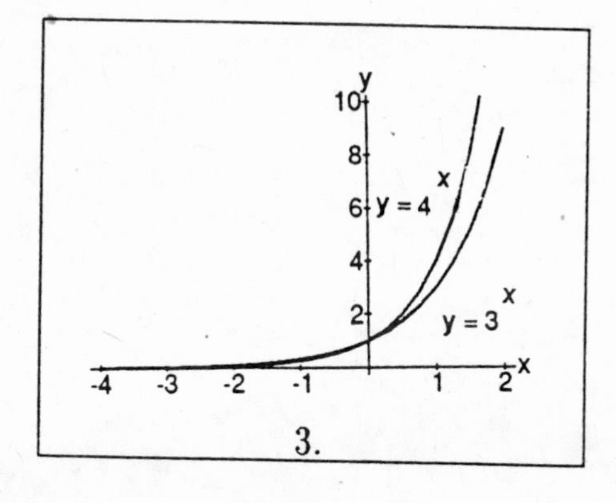

3.

5. Since $e \approx 2.718$ $y = e^x$ is "close" to $y = 3^x$. The $x-$axis is a horizontal asymptote.

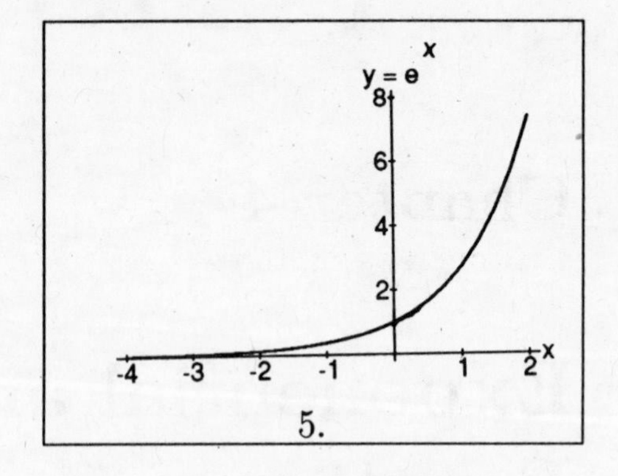

5.

7. The graph of this $y = 2 + e^x$ is identical to the one in problem 5 raised by 2 units. The line $y = 2$ is a horizontal asymptote.

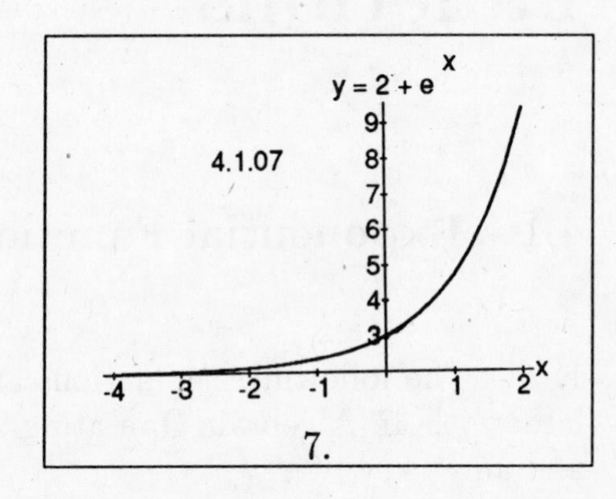

7.

9. $y = 2 - 3e^x$. The line $y = 2$ is a horizontal asymptote.

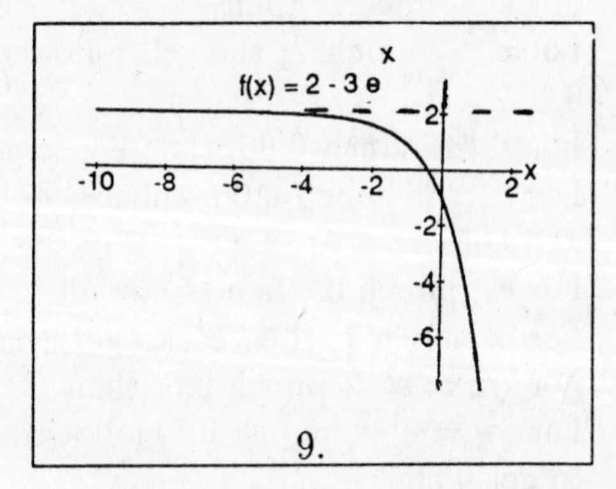

9.

11. $y = 5 - 3e^{-x}$. The line $y = 5$ is a horizontal asymptote.

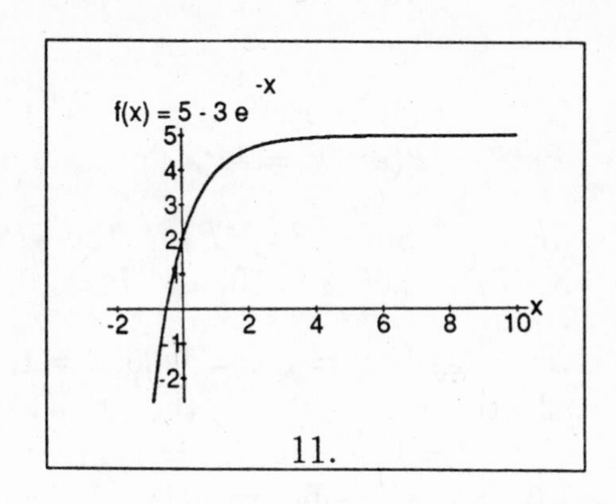

11.

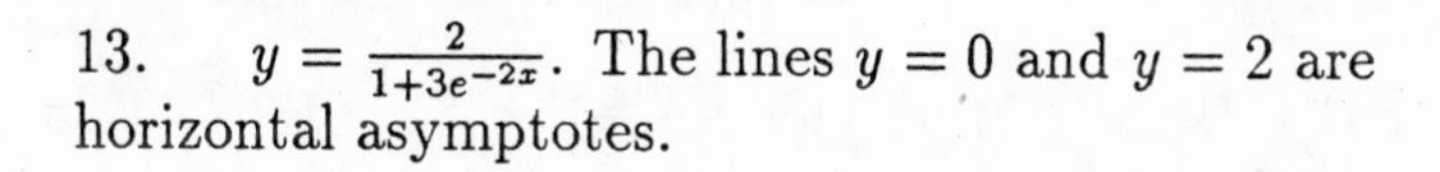

13. $y = \frac{2}{1+3e^{-2x}}$. The lines $y = 0$ and $y = 2$ are horizontal asymptotes.

15. $y = 1 - \frac{3}{1+2e^{-3x}}$. The lines $y = 1$ and $y = -2$ are horizontal asymptotes.

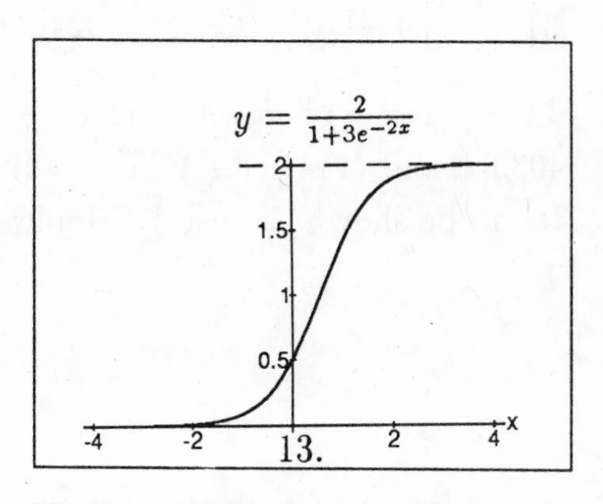

13.

17. $f(x) = e^{kx}$, $f(1) = e^k = 20$, $f(2) = e^{2k} = (e^k)^2 = (20)^2 = 400$.

19. $f(x) = Ae^{kx}$, $f(0) = A = 20$, $f(2) = 20e^{2k} = 40$, so $e^{2k} = 2$. $f(8) = 20e^{8k} = 20(e^{2k})^4 = 20(2^4) = 320$.

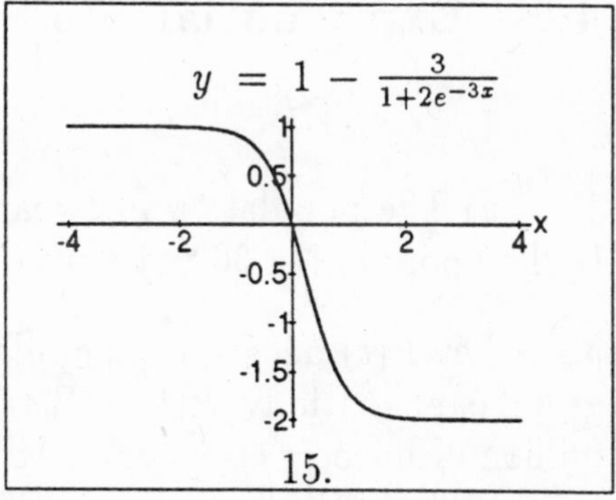

15.

21. $f(x) = 50 - Ae^{kx}$, $f(0) = 50 - A = 30$, so $A = 20$. $f(4) = 50 - 20e^{4k} = 5$, so $e^{4k} = \frac{9}{4}$. $f(2) = 50 - 20e^{2k} = 50 - 20(e^{4k})^{1/2} = 50 - 20(\frac{9}{4})^{1/2} = 20$.

23. If P dollars is invested at an annual interest rate r and interest is compounded k times per year, the balance after t years will be $B(t) = (P)(1 + \frac{r}{k})^{kt}$ dollars and if interest is compounded continuously, the balance will be $B(t) = Pe^{rt}$ dollars.

a) If $P = 1,000$, $r = 0.07$, $t = 10$, and $k = 1$, then $B(10) = 1,000(1 + \frac{0.07}{1})^{10} = 1,000(1.07)^{10} = \$1,967.15$.

b) If $P = 1,000$, $r = 0.07$, $t = 10$, and $k = 4$, then $B(10) = 1,000(1 + \frac{0.07}{4})^{40} = \$2,001.60$.

c) If $P = 1,000$, $r = 0.07$, $t = 10$, and $k = 12$, then $B(10) = 1,000(1 + \frac{0.07}{12})^{120} = \$2,009.66$.

d) If $P = 1,000$, $r = 0.07$, $t = 10$, and interest is compounded continuously, then $B(10) = 1,000e^{0.7} = \$2,013.75$.

25. $B(t) = Pe^{rt}$. $t = 10$ and $B(10) = 2P$. Substituting leads to $e^{10r} = 2$. $B(20) = Pe^{20r} = P(e^{10r})^2 = 4P$. Thus the original investment will quadruple.

27. a) Since $B = (P)e^{rt}$, then $P = Be^{-rt}$.
b) If $r = 0.06$, $t = 10$, and $B = 10,000$, then from part a), $P = 10,000e^{-0.6} = \$5,488.12$.

29. a) If $P = 1$, $r = 0.06$, $t = 1$, and $k = 4$, then $B(1) = (1)(1 + \frac{0.06}{4})^4 = \$1.061364$. The effective interst rate is then $100(1.0614 - 1) = 6.14$ %.
b) If $P = 1$, $r = 0.06$, $t = 1$, and interest is compounded continuously, then $B(1) = 1e^{0.06} = 1.06184$. The effective interst rate is then $100(1.0618 - 1) = 6.18$ %.

31. $(1+\frac{1}{1,000})^{1,000} = 2.716924$, $(1+\frac{1}{2,000})^{2,000} = 2.717603, \ldots, (1+\frac{1}{50,000})^{50,000} = 2.7182546$.

33. $(1 + \frac{3}{10})^{20} = 190.5$, $(1 + \frac{3}{100})^{200} = 369.36$, $(1 + \frac{3}{1,000})^{2,000} = 399.82$, $(1 + \frac{3}{10,000})^{20,000} = 403.07$, and $(1 + \frac{3}{100,000})^{200,000} = 403.39$suggests that $\lim_{n\to\infty}(1 + \frac{3}{n})^{2n}$ tends to 403 or 403.5. It can be shown, by using l'Hôpital's rule that $\lim_{n\to\infty}(1 + \frac{3}{n})^{2n} = e^6 = 403.43$

4.2 Exponential Models

1. a) The population in t years will be $P(t) = 50e^{0.02t}$ million. Thus $P(0) = 50$ million.
b) The population 30 years from now will be $P(30) = 50e^{0.02(30)} = 50e^{0.6} = 91.11$ million.

3. Let $P(t)$ denote the population (in millions) t years after 1988. Since the population grows exponentially, $P(t) = P_0e^{kt}$, where P_0 is the initial population (in 1988), which was 60 million. Hence $P(t) = 60e^{kt}$. Moreover, since the population was 90 million in 1993 (when $t = 5$), $90 = P(5) = 60e^{5k}$ or $e^{5k} = \frac{3}{2}$. The population in 2003 will be $P(15) = 60e^{15k} = 60(e^{5k})^3 = 60(\frac{3}{2})^3 = 202.5$ million.

5. Let $G(t)$ denote the gross national product (GNP) in billions t years after 1980. Since the GNP grows exponentially and was 100 billion in 1980 (when $t = 0$), $G(t) = 100e^{kt}$. Moreover, since the GNP was 180 billion in 1990 (when $t = 10$), $180 = G(10) = 100e^{10k} =$ or $e^{10k} = \frac{180}{100} = \frac{9}{5}$. The GNP in 2000, (when $t = 20$) is $G(20) = 100e^{20k} = 100(e^{10k})^2 = 100(\frac{9}{5})^2 = 324$ billion dollars.

7. The population density x miles from the center of the city is $D(x) = 12e^{-0.07x}$ thousand

people per square mile.
a) At the center of the city, the density is $D(0) = 12$ thousand people per square mile.
b) Ten miles from the center, the density is $D(10) = 12e^{-0.07(10)} = 12e^{-0.7} = 5.959$ that is 5,959 people per square mile.

9. Let $Q(t)$ denote the amount of the radioactive substance present after t years. Since the decay is exponential and 500 grams were present initially, $Q(t) = 500e^{-kt}$. Moreover, since 400 grams are present 50 years later, $400 = Q(50) = 500e^{-50k}$ or $e^{-50k} = \frac{4}{5}$. The amount present after 200 years will be $Q(200) = 500e^{-200k} = 500(e^{-50k})^4 = 500(\frac{4}{5})^4 = 204.8$ grams.

11. a) The reliability function is $f(t) = 1 - e^{-0.03t}$. As t increases without bound, $e^{-0.03t}$ approaches 0 and so $f(t)$ approaches 1. Furthermore, $f(0) = 0$. The graph is like that of a learning curve.

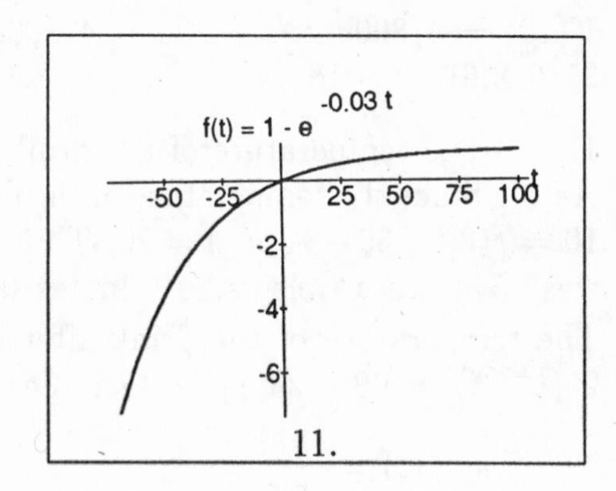

11.

b) The fraction of tankers that sink in fewer than 10 days is $f(10) = 1 - e^{-0.03(10)} = 1 - e^{-0.3}$. The fraction of tankers that remain afloat for at least 10 days is therefore $1 - f(10) = 1 - (1 - e^{-0.3}) = e^{-0.3} = 0.7408$.

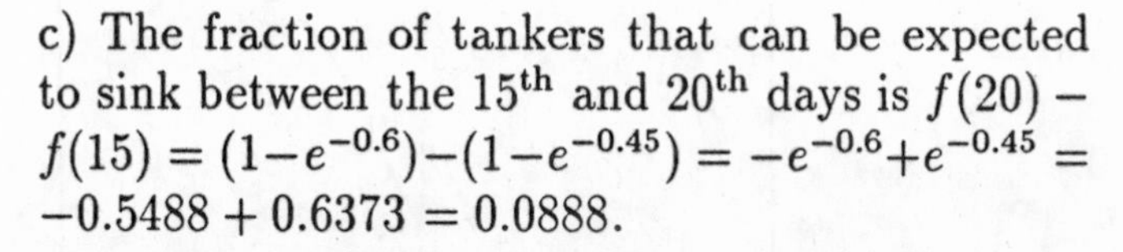
c) The fraction of tankers that can be expected to sink between the 15th and 20th days is $f(20) - f(15) = (1-e^{-0.6})-(1-e^{-0.45}) = -e^{-0.6}+e^{-0.45} = -0.5488 + 0.6373 = 0.0888$.

13. a) The number of facts recalled after t minutes is $Q(t) = (A)(1 - e^{-kt})$, where k is a positive constant. As t increases without bound, e^{-kt} approaches 0 and so $Q(t)$ approaches A. Moreover, $Q(0) = 0$.

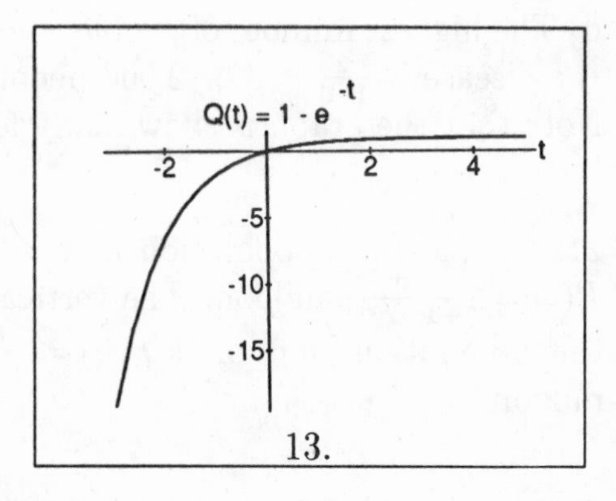

13.

b) As t increases without bound, $Q(t)$ approaches A, the total number of relevant facts in the person's memory.

15. a) The resale value of the machine when it is t years old is $V(t) = 4,800e^{-t/5} + 400$ dollars. As t increases without bound, $e^{-t/5}$ approaches 0 and $V(t)$ approaches 400 dollars. Moreover, $V(0) = 5,200$ dollars.

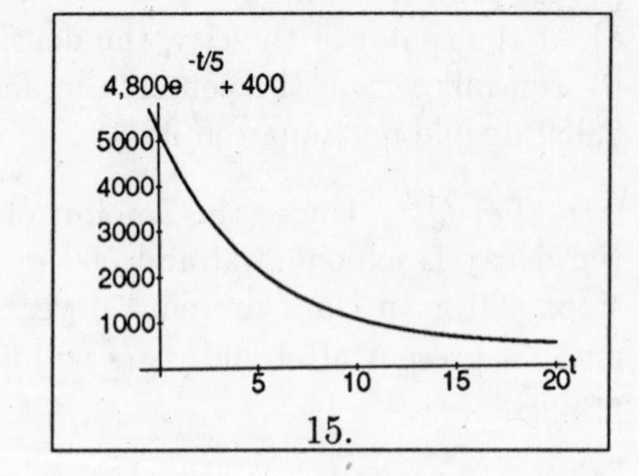

15.

b) When the machine was new, its value was $V(0) = \$5,200$.

c) after 10 years, the value of the machine is $V(10) = 4,800e^{-10/5} + 400 = 4,800e^{-2} + 400 = \$1,049.61$

17. The temperature of the drink t minutes after leaving the refrigerator is $f(t) = 30 - Ae^{-kt}$. Since the temperature of the drink when it left the refrigerator was 10 degrees Celsius, $10 = f(0) = 30 - A$ or $A = 20$. Thus $f(t) = 30 - 20e^{-kt}$. Since the temperature of the drink was 15 degrees Celsius 20 minutes later, $15 = f(20) = 30 - 20e^{-20k}$ or $e^{-20k} = \frac{-15}{-20} = \frac{3}{4}$. The temperature of the drink after 40 minutes is therefore $f(40) = 30 - 20e^{-40k} = 30 - 20(e^{-20k})^2 = 30 - 20(\frac{3}{4})^2 = 18.75$ degrees.

19. a) $f(t) = \frac{2}{1+3e^{-0.8t}}$.
b) $f(0) = 0.5$ thousand people (500 people).
c) $f(3) = \frac{2}{1+3\times 0.0907} = 1.572$, so 1,572 people have caught the disease.
d) The highest number of people who can contract the disease is $\frac{2}{1+0} = 2$ or 2,000 people.
Note that the graph is shown only for $t \geq 0$.

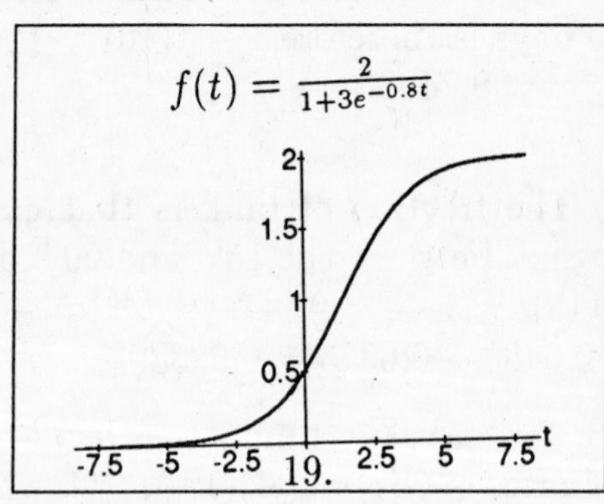

19.

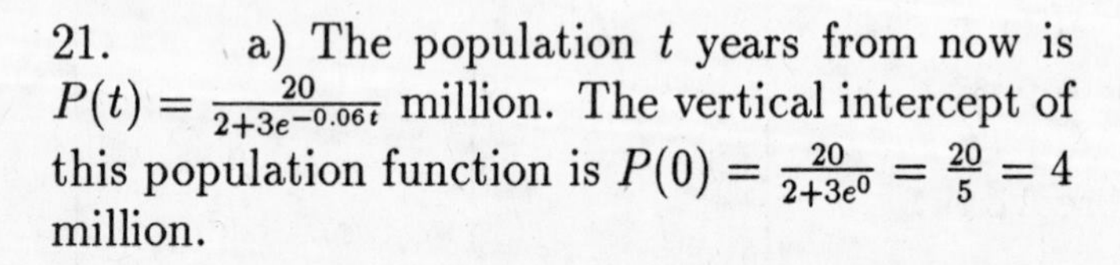

21. a) The population t years from now is $P(t) = \frac{20}{2+3e^{-0.06t}}$ million. The vertical intercept of this population function is $P(0) = \frac{20}{2+3e^{0}} = \frac{20}{5} = 4$ million.

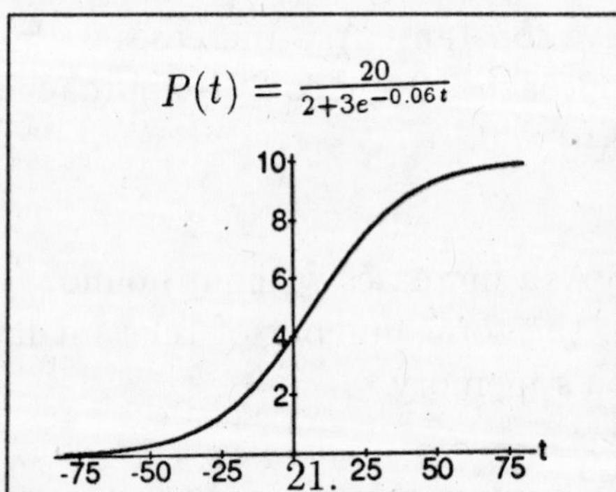

21.

As t increases without bound, $e^{-0.06t}$ approaches 0. Hence $\lim_{t\to\infty} P(t) = \frac{20}{2+0} = 10$ million. As t decreases without bound, $e^{-0.06t}$ increases without bound. Hence, the denominator $2 + e^{-0.06t}$ increases without bound and so $\lim_{t\to-\infty} P(t) = 0$.
b) The current population is $P(0) = 4$ million.
c) Fifty years from now, the population will be $P(50) = \frac{20}{2+3e^{-3}} = 9.76$ million.

d) In the long run, (as t increases without bound), $e^{-0.06t}$ approaches 0, and so $P(t)$ approaches $\frac{20}{2} = 10$ million.

23. a) Let $N(t)$ be the number of years from now. $N(0) = 500(0.03)^{(0.4)^0} = 500(0.03)^1 =$ 15 employees.
b) $N(5) = 500(0.03)^{(0.4)^5} = 500(0.03)^{0.01024} = (500)(0.96473) = 482.4$ or 482 employees in five years.

4.3 The Natural Logarithm

1. We'll use a calculator with an Algebraic Oeprating System (AOS) to find $\ln a$. First enter the number a, then press the ln key. This gives $\ln 1 = 0$, $\ln 2 = 0.6931472$, $\ln 5 = 1.6094379$, $\ln \frac{1}{5} = \ln 0.2 = -1.6094379$.
Now, for $\ln e^n$, enter n, then press the e^x key, followed by the ln key. Thus $\ln e^1 = 1$ and $\ln e^2 = 2$. These results should not be surprising since the natural logarithm and the exponential functions are inverses. $\ln e^n = n$.
For $\ln 0$ and $\ln(-2)$ the calculator displays an error signal (such as the letter E or a flashing light). This is due to the domain of the logarithmic function which is $x > 0$. (Ask yourself "what exponent of e will give -2 ?" Answer: $e^x = -2$ is impossible since $e^x > 0$ for all real x. Remember that "log" means "exponent".)

3. $\ln e^3 = 3 \ln e = 3 \times 1 = 3$ since $\ln u^v = v \ln u$ and $\ln e = 1$.

5. Let's give the expression a name, say A, so that we can handle it. $A = e^{\ln 5}$ which can be written as $\ln A = \ln 5$. A solution (the only solution) is A=5. Thus $e^{\ln 5} = 5$.
Actually this result is immediate from the inverse relationship between e^x and $\ln x$.

7. Let's call the given expression A. $A = e^{3\ln 2 - 2\ln 5} = e^{\ln 2^3 - \ln 5^2}$ (since $v \ln u = \ln u^v$). $A = e^{\ln 8 - \ln 25} = e^{\ln \frac{8}{25}}$ (since $\ln \frac{u}{v} = \ln u - \ln v$). $A = \frac{8}{25}$ because of the inverse relationship between e^x and $\ln x$.

9. $2 = e^{0.06x}$ is equivalent to $\ln 2 = 0.06x$, from which $x = \frac{\ln 2}{0.06} = 11.55$.

11. $3 = 2 + 5e^{-4x}$, $\frac{1}{5} = e^{-4x}$ (using arithmetic), $-4x = \ln \frac{1}{5} = -\ln 5$ (since $\ln \frac{u}{v} = \ln u - \ln v$ and $\ln 1 = 0$), from which $x = \frac{\ln 5}{4} = 0.402$.

13. $-\ln x = \frac{t}{50} + C$ or $\ln x = -\frac{t}{50} - C$. Thus $x = e^{-C-t/50} = (e^{-C})(e^{-t/50})$ because $a^{r+s} = a^r a^s$ (which) certainly applies when $a = e$.

15. $\ln x = \frac{1}{3}(\ln 16 + 2\ln 2) = \frac{1}{3}(\ln 16 + \ln 2^2) = \frac{1}{3}\ln[(16)(2^2)] = \frac{1}{3}\ln 2^6 = \ln(2^6)^{1/3} = \ln 2^2$. Thus $\ln x = \ln 4$ which is valid when $x = 4$.

17. $3^x = e^2$ is valid if the logarithms of both members are taken. $\ln 3^x = \ln e^2 = 2\ln e = 2$, $x \ln 3 = 2$, or $x = \frac{2}{\ln 3} = 1.82$.

19. $a^{x+1} = b$ if $\ln a^{x+1} = \ln b$, $(x+1)\ln a = \ln b$, $x = \frac{\ln b}{\ln a} - 1$.

21. $\ln \frac{1}{\sqrt{ab^3}} = 0 - \ln(ab^3)^{1/2} = -\frac{1}{2}\ln(ab^3) = -\frac{1}{2}(\ln a + \ln b^3) = -\frac{1}{2}(\ln a + 3\ln b) = -\frac{1}{2}(2+9) = -5.5$.

23. $B(t) = Pe^{rt}$. After a certain time the investment will have grown to $B(t) = 2P$ at the interest rate of 0.06. Thus $2P = Pe^{0.06t}$, $2 = e^{0.06t}$, $\ln 2 = 0.06t$, and $t = \frac{\ln 2}{0.06} = 11.55$ years.

25. The balance after t years is $B(t) = Pe^{rt}$, where P is the initial investment and r is the interest rate compounded continuously. Since money doubles in 13 years, $2P = B(13) = Pe^{13r}$, $2 = e^{13r}$, $\ln 2 = 13r$, or $r = \frac{\ln 2}{13} = 0.0533$. Thus the annual interest rate is 5.33%.

27. $Q(t) = Q_0 e^{-0.003t}$. $\frac{1}{2} = e^{-0.003t}$, $-0.003t = 0 - \ln 2$, and $t = \frac{\ln 2}{0.003} = 231.05$ years.

29. $Q(t) = Q_0 e^{-kt}$. $\frac{1}{2}Q_0 = Q_0 e^{-k\lambda}$, $\frac{1}{2} = e^{-k\lambda}$, $0 - \ln 2 = -k\lambda$, $k = \frac{\ln 2}{\lambda}$. Thus $Q(t) = Q_0 e^{-(\ln 2/\lambda)t}$.

31. $f(x) = 20 - 15e^{-0.2x}$. $12 = 20 - 15e^{-0.2x}$, $\frac{8}{15} = e^{-0.2x}$, $-0.2x = \ln \frac{8}{15}$, $x = \frac{-\ln 8 + \ln 15}{0.2} = 3.143$ or 3,143 books.

33. $t = 0$ in 1980, $t = 10$ in 1990, and $t = 20$ in 2000. a) The points $(t, G) = (0, 100)$ and $(10, 180)$ are on the line. The y–intercept is $b = 100$ and the slope $m = \frac{180-100}{10-0} = 8$. Thus $G = 8t + 100$ and at $t = 20$ $G = 8 \times 20 + 100 = 260$ billion dollars.
b) $G = G_0 e^{kt}$, $G_0 = 100$, $G = 100e^{kt}$. $180 = 100e^{10k}$, $k = \frac{\ln 1.8}{10}$, and $G = 100e^{(\ln 1.8/10)t}$. At $t = 20$ $G = 100e^{2\ln 1.8} = 324$ billion dollars. [Alternately, $G(20) = 100(e^{10k})^2 = 100(1.8)^2$].

35. $t = 0$ in 1960 and $t = 15$ in 1975. $P(t) = \frac{40}{1+Ce^{-kt}}$ so that $P(0) = \frac{40}{1+C} = 3$ from which $C = \frac{37}{3}$. $P(15) = \frac{40}{1+(\frac{37}{3})e^{-15k}} = 4$, $9 = \frac{37}{3}e^{-15k}$, $k = -\ln \frac{27}{37}/15 = 0.021$. Thus $P(t) = \frac{40}{1+(\frac{37}{3})e^{-0.021t}}$.
In the year 2,000, $t = 40$ and $P(40) = \frac{40}{1+(\frac{37}{3})e^{-0.021(40)}} = \frac{40}{1+(12.3333)0.4317} = \frac{40}{6.3243} = 6.325$ billion people.

37. $R(t) = R_0 e^{-kt}$, $\frac{1}{2}R_0 = R_0 e^{-kt}$ and $k = \frac{\ln 2}{5{,}730}$ from example 3.9. Thus $\ln \frac{1}{2} = -\frac{\ln 2}{5{,}730}t$ or $t = 5{,}730$ years.

39. For carbon dating we use the formula $R(t) = R_0 e^{-(\ln 2)t/5{,}730}$. Here $t = 3.8(10^6)$ years

and $e^{-(\ln 2)(3.8)(10^6)/5,730} = e^{-459.7} \approx 0$ by using a nine-digit decimal calculator. With such a calculator, $e^{-20} = 0.000000002$.

41. a) The air pressure is $f(s) = e^{-1.25(10^{-4})s}$ atmospheres at s meters above sea level. When $f(s) = 0.25$, $0.25 = e^{-1.25(10^{-4})s}$, $\ln 4 = 1.25(10^{-4})s$, $s = \frac{(\ln 4)(10,000)}{1.25} = 11,090$ meters.
b) At $s = 7,000$, $f(7,000) = e^{-1.25(10^{-4})(7)(10^3)} = e^{-0.875} = 0.417$ atmospheres.

43. $e^{v \ln u} = (e^{\ln u})^v = u^v = e^{\ln u^v}$, so, since the bases are the same, $v \ln u = \ln u^v$.

45. We are given the line $L{:}y = x$ whose slope is 1. Consider the points $A(a, b)$ and $B(b, a)$. The midpoint $M(\frac{a+b}{2}, \frac{a+b}{2})$ is on the line L. The distance $MB = \sqrt{(\frac{2b-a-b}{2})^2 + (\frac{2a-a-b}{2})^2} =$ $(\frac{1}{2})\sqrt{(b-a)^2 + (a-b)^2} = \frac{|a-b|}{\sqrt{2}}$. Similarly $MA = \sqrt{(\frac{a+b-2a}{2})^2 + (\frac{a+b-2b}{2})^2} =$ $(\frac{1}{2})\sqrt{(b-a)^2 + (a-b)^2} = \frac{|a-b|}{\sqrt{2}}$. Thus $MA = MB$.
The slope of the line AB is $m = \frac{a-b}{b-a} = -1$, so AB is perpendicular to L with slope 1.

4.4 Differentiation of Logarithmic and Exponential Functions

1. $f(x) = e^{5x}$, $f'(x) = e^{5x}\frac{d}{dx}(5x) = 5e^{5x}$.

3. $f(x) = e^{x^2+2x-1}$, $f'(x) = e^{x^2+2x-1}\frac{d}{dx}(x^2 + 2x - 1) = (2x + 2)e^{x^2+2x-1}$.

5. $f(x) = 30 + 10e^{-0.05x}$, $f'(x) = 0 + 10e^{-0.05x}\frac{d}{dx}(-0.05x) = -0.5e^{-0.05x}$.

7. $f(x) = (x^2 + 3x + 5)e^{6x}$, $f'(x) = (x^2 + 3x + 5)\frac{d}{dx}e^{6x} + e^{6x}\frac{d}{dx}(x^2 + 3x + 5) = (x^2 + 3x +$ $5)e^{6x}(6) + e^{6x}(2x + 3) = (6x^2 + 20x + 33)e^{6x}$.

9. $f(x) = \frac{x}{e^x} = xe^{-x}$, $f'(x) = x\frac{d}{dx}e^{-x} + e^{-x}\frac{d}{dx}(x) = -xe^{-x} + e^{-x} = \frac{1-x}{e^x}$.

11. $f(x) = (1 - 3e^x)^2$, $f'(x) = 2(1 - 3e^x)\frac{d}{dx}(1 - 3e^x) = 2(1 - 3e^x)(-3e^x) = -6e^x(1 - 3e^x)$.

13. $f(x) = e^{\sqrt{3x}} = e^{(3x)^{1/2}}$, $f'(x) = e^{\sqrt{3x}}\frac{d}{dx}(3^{1/2}x^{1/2}) = e^{\sqrt{3x}}[3^{1/2}(\frac{1}{2})x^{-1/2}) = \frac{3}{2\sqrt{3x}}e^{\sqrt{3x}}$.

15. $f(x) = \ln x^3 = 3 \ln x$, $f'(x) = 3(\frac{1}{x})\frac{d}{dx}x = \frac{3}{x}$.

17. $f(x) = \ln(x^2 + 5x - 2)$, $f'(x) = \frac{1}{x^2+5x-2}\frac{d}{dx}(x^2 + 5x - 2) = \frac{2x+5}{x^2+5x-2}$.

19. $f(x) = x^2 \ln x$, $f'(x) = x^2\frac{d}{dx}(\ln x) + \ln x\frac{d}{dx}(x^2) = x^2(\frac{1}{x}) + (\ln x)(2x) = x + 2x \ln x =$

$x(1+2\ln x)$.

21. $f(x)=\frac{\ln x}{x}$, $f'(x)=\frac{x\frac{d}{dx}(\ln x)-(\ln x)\frac{d}{dx}x}{x^2}=\frac{x(\frac{1}{x})-(\ln x)(1)}{x^2}=\frac{1-\ln x}{x^2}$.

23. $f(x)=\ln(\frac{x+1}{x-1})$, $f'(x)=\frac{x-1}{x+1}\frac{d}{dx}(\frac{x+1}{x-1})=\frac{x-1}{x+1}\frac{(x-1)(1)-(x+1)(1)}{(x-1)^2}=\frac{x-1}{x+1}\frac{-2}{(x-1)^2}=\frac{-2}{(x-1)(x+1)}=\frac{-2}{x^2-1}$.

25. $f(x)=\ln e^{2x}=2x$, $f'(x)=\frac{1}{e^{2x}}\frac{d}{dx}e^{2x}=\frac{1}{e^{2x}}(2e^{2x})=2$ (this was obviously the hard way of getting the answer).

27. $e^{xy}=xy^3$. Using implicit differentiation leads to $e^{xy}\frac{d}{dx}(xy)=y^3+3xy^2\frac{dy}{dx}$, $e^{xy}(x\frac{dy}{dx}+y)=y^3+3xy^2\frac{dy}{dx}$, $xe^{xy}\frac{dy}{dx}+ye^{xy}=y^3+3xy^2\frac{dy}{dx}$, $(xe^{xy}-3xy^2)\frac{dy}{dx}=y^3-ye^{xy}$, $\frac{dy}{dx}=\frac{y^3-ye^{xy}}{xe^{xy}-3xy^2}=\frac{y(y^2-e^{xy})}{x(e^{xy}-3y^2)}$. From the original form of the exercise we see that $\frac{dy}{dx}=\frac{y(y^2-xy^3)}{x(xy^3-3y^2)}=\frac{y(1-xy)}{x(xy-3)}$

Alternate method: Take the logarithm of both members of $e^{xy}=xy^3$ and simplify. $\ln e^{xy}=\ln xy^3$ or $xy=\ln x+3\ln y$. Using implicit differentiation leads to $x\frac{dy}{dx}+y=\frac{1}{x}+\frac{3\frac{dy}{dx}}{y}$, $(x-\frac{3}{y})\frac{dy}{dx}=\frac{1}{x}-y$, $(\frac{xy-3}{y})\frac{dy}{dx}=\frac{1-xy}{x}$, $\frac{dy}{dx}=\frac{y(1-xy)}{x(xy-3)}$.

29. $\ln\frac{y}{x}=x^2y^3$ can be rewritten $\ln y-\ln x=x^2y^3$. Differentiating implicitly leads to $\frac{1}{y}\frac{dy}{dx}-\frac{1}{x}=2xy^3+3x^2y^2\frac{dy}{dx}$, $x\frac{dy}{dx}-y=2x^2y^4+3x^3y^3\frac{dy}{dx}$, $(x-3x^3y^3)\frac{dy}{dx}=2x^2y^4+y$, $\frac{dy}{dx}=\frac{y(2x^2y^3+1)}{x(1-3x^2y^3)}$,

31. a) The population t years from now will be $P(t)=50e^{0.02t}$ million. Hence the rate of change of the population t years from now will be $P'(t)=50e^{0.02t}(0.02)=e^{0.02t}$ and the rate of change 10 years from now will be $P'(t)=e^{0.2}=1.22$ million per year.
b) The percentage rate of change t years from now will be $100[\frac{P'(t)}{P(t)}]=100(\frac{e^{0.02t}}{50e^{0.02t}})=\frac{100}{50}=2$ % per year, which is a constant, independent of time.

33. a) The value of the machine after t years is $Q(t)=20,000e^{-0.4t}$ dollars. Hence the rate of depreciation after t years is $Q'(t)=20,000e^{-0.4t}(-0.4)=-8,000e^{-0.4t}$ and the rate after 5 years is $Q'(5)=-8,000e^{-2}=\$1,082.68$ per year.
b) The percentage rate of change t years from now will be $100[\frac{Q'(t)}{Q(t)}]=100(\frac{-8,000e^{-0.4t}}{20,000e^{-0.4t}})=\frac{100(-8,000)}{20,000}=-40$ % per year, which is a constant, independent of time.

35. Q decreases exponentially, so $Q(t)=Ae^{-kt}$, where A and k are constants and $k>0$. $Q'(t)=-kAe^{-kt}$, so the percentage rate of change is $100[\frac{Q'(t)}{Q(t)}]=100(\frac{-kAe^{-kt}}{Ae^{-kt}}=-100k$ which is a constant, independent of time.

37. a) The world's population in billions t years after 1960 is $P(t)=\frac{40}{1+12e^{-0.08t}}=40(1+12e^{-0.08t})^{-1}$. Hence the rate of change of the population will be $P'(t)=$

$-40(1+12e^{-0.08t})^{-2}(12)(-0.08) = \frac{38.4e^{-0.08t}}{(1+12e^{-0.08t})^2}$ and the rate of change in 1995 is $P''(35) = \frac{38.4e^{-2.8}}{(1+12e^{-2.8})^2} = 0.7805$ billion per year.
b) $P(35) = \frac{40}{1+12e^{-2.8}} = 23.1251$. The percentage rate of change in 1995 is $100[\frac{P'(35)}{P(35)}] = 100(\frac{0.7805}{23.1251}) = 3.375$ % per year.

39. a) The first year sales of the text will be $f(x) = 20 - 15e^{-0.2x}$ thousand copies when x thousand complementary copies are distributed. If the number of complementary copies distributed is increased from 10,000, that is when $x = 10$, by 1,000, that is , $\Delta x = 1$, the approximate change in sales is $\Delta f = f'(10)\Delta x$. Since $f'(x) = 3e^{-0.2x}$ and $\Delta x = 1$, $\Delta f = f'(10) = 3e^{-2} = 0.406$ thousand or 406 copies.
b) The actual change in sales is $\Delta f = f(11) - f(10) = (20 - 15e^{-2.2}) - (20 - 15e^{-2}) = 0.368$ or 368 copies.

41. a) Let $P(x)$ denote the profit. $P(x)$ =(number of radios sold)(profit per radio)= $(1,000e^{-0.1x})(x-5)$. $P'(x) = (1,000e^{-0.1x})(1) + (x-5)(1,000e^{-0.1x})(-0.1) = 1,000e^{-0.1x}[1 - 0.1(x-5)] = 100e^{-0.1x}(15-x) = 0$ when $x = 15$. Since $P'(x) > 0$ (P is increasing) for $x < 15$ and $P'(x) < 0$ (P is decreasing) for $x > 15$, $P(x)$ is the absolute maximum when $x = \$15$.

43. $f(x) = xe^{-x}$, $f'(x) = -xe^{-x} + e^{-x} = e^{-x}(1-x) = 0$ when $x = 1$. $f''(x) = -e^{-x} + (1-x)e^{-x}(-1) = xe^{-x} - 2e^{-x} = e^{-x}(x-2) = 0$ at $x = 2$. $f(1) = e^{-1} = \frac{1}{e}$ and $f(2) = 2e^{-2} = \frac{2}{e^2}$.

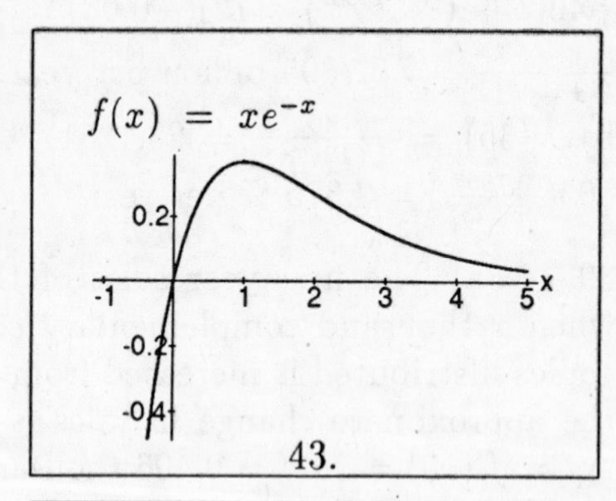

43.

x	$-\infty$		1		2		∞
$f(x)$	$-\infty$		$\frac{1}{e}$		$\frac{2}{e^2}$		0
$f'(x)$	∞	+	0	−	−	−	0
		inc	0	dec	dec	dec	
$f''(x)$	$-\infty$	−	−	−	0	+	0
		down	down	down		up	

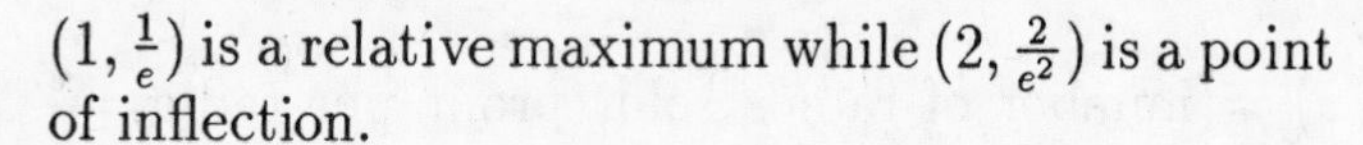
$(1, \frac{1}{e})$ is a relative maximum while $(2, \frac{2}{e^2})$ is a point of inflection.

45. $f(x) = e^{-x^2}$, $f'(x) = -2xe^{-x^2} = 0$ when $x = 0$. $f''(x) = -2xe^{-x^2}(-2x) + e^{-x^2}(-2) = 2e^{-x^2}(2x^2-1) = 0$ at $x = \pm\frac{\sqrt{2}}{2}$. $f(0) = 1$ and $f(\pm\frac{\sqrt{2}}{2}) = e^{-1/2} = \frac{1}{e^{1/2}}$.

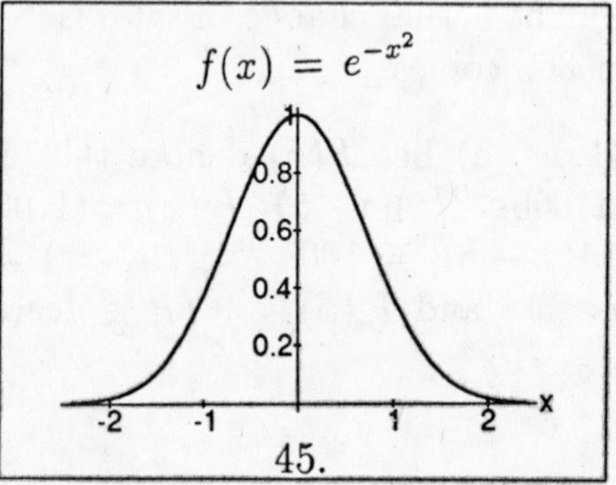

45.

x	$-\infty$		$-\frac{\sqrt{2}}{2}$		0		$\frac{\sqrt{2}}{2}$		∞
$f(x)$	0		$\frac{1}{e^{1/2}}$		1		$\frac{1}{e^{1/2}}$		0
$f'(x)$	∞	+	+	+	0	−	−	−	0
		inc	inc	inc	0	dec	dec	dec	
$f''(x)$		+	0	−	−	−	0	+	
		up		down	down	down		up	

$(0, 1)$ is a relative maximum while $(\frac{\pm\sqrt{2}}{2}, e^{-1/2})$ are points of inflection.

47. $f(x) = e^x + e^{-x}$, $f'(x) = e^x - e^{-x} = 0$ when $x = 0$. $f''(x) = e^x + e^{-x} > 0$.

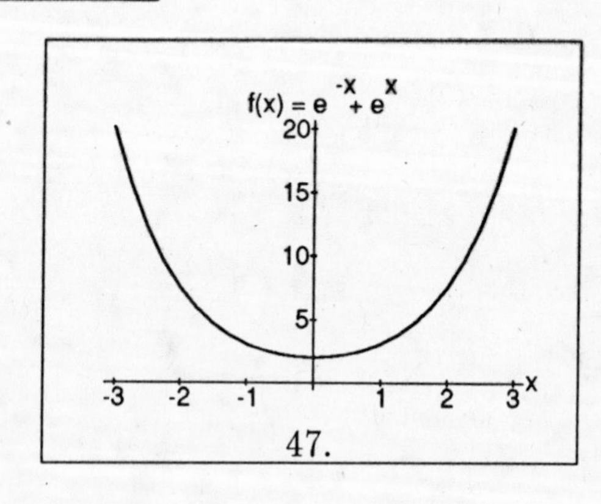

47.

x	$-\infty$		0		∞
$f(x)$	∞		2		∞
$f'(x)$	$-\infty$	−	0	+	∞
		dec		inc	
$f''(x)$		+	+	+	
		up	up	up	

$(0,2)$ is the absolute minimum while the graph is always concave upward.

49. $f(x) = x - \ln x$ with $x > 0$, $f'(x) = 1 - \frac{1}{x} = \frac{x-1}{x} = 0$ when $x = 1$. $f''(x) = \frac{1}{x^2} > 0$.

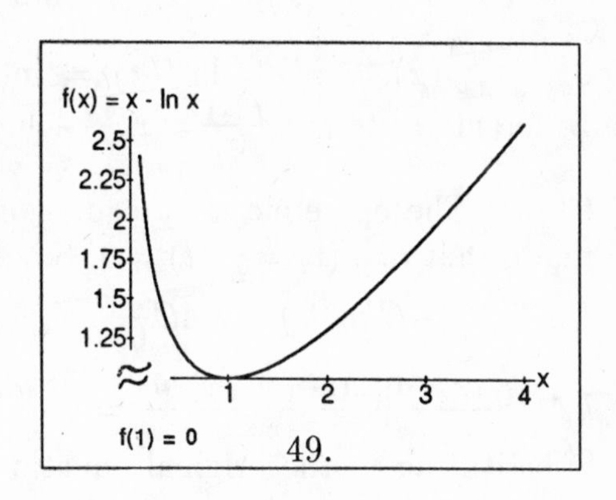

49.

x	0		1		∞
$f(x)$	$-\infty$		1		∞
$f'(x)$	$-\infty$	$-$	0	$+$	1
		dec		inc	
$f''(x)$	0	$+$	$+$	$+$	0
		up	up	up	

$(1,1)$ is the absolute minimum while the graph is always concave upward.

51. $f(x) = \frac{\ln x}{x}$ with $x > 0$, $f'(x) = \frac{x(1/x) - \ln x}{x^2} = \frac{1 - \ln x}{x^2} = 0$ when $\ln x = 1$ or $x = e$. $f''(x) = \frac{x^2(-1/x) - (1 - \ln x)(2x)}{x^4} = \frac{-x - 2x + 2x\ln x}{x^4} = \frac{2\ln x - 3}{x^3} = 0$ when $\ln x = \frac{3}{2}$ or $x = e^{3/2}$.

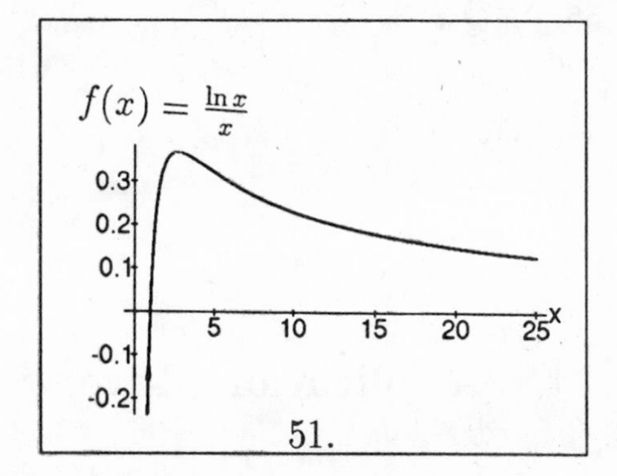

51.

x	0		e		$e^{2/3}$		∞
$f(x)$	∞		$\frac{1}{e}$		$\frac{3}{2e^{3/2}}$		∞
$f'(x)$		$+$	0	$-$	$-$	$-$	
		inc		dec	dec	dec	
$f''(x)$		$-$	$-$	$-$	0	$+$	
		down	down	down		up	

(e, e^{-1}) is the absolute maximum while $(e^{3/2}, \frac{3}{2e^{3/2}})$ is a point of inflection.

53. $f(x) = \sqrt[4]{\frac{2x+1}{1-3x}} = (\frac{2x+1}{1-3x})^{1/4}$. $\ln f(x) = \ln(\frac{2x+1}{1-3x})^{1/4} = (\frac{1}{4})\ln\frac{2x+1}{1-3x} = (\frac{1}{4})[\ln(2x+1) - \ln(1-3x)]$. Thus $\frac{f'(x)}{f(x)} = (\frac{1}{4})[\frac{2}{2x+1} - \frac{-3}{1-3x}]$ and $f'(x) = (\frac{2x+1}{1-3x})^{1/4}(\frac{1}{4})(\frac{2}{2x+1} + \frac{3}{1-3x})$.

55. $f(x) = \frac{e^{-3x}\sqrt{2x-5}}{(6-5x)^4}$. $\ln f(x) = \ln[\frac{e^{-3x}\sqrt{2x-5}}{(6-5x)^4}] = \ln e^{-3x} + \ln(2x-5)^{1/2} - \ln(6-5x)^4 = -3x + (\frac{1}{2})\ln(2x-5) - 4\ln(6-5x)$. $\frac{f'(x)}{f(x)} = -3 + (\frac{1}{2})\frac{2}{2x-5} - \frac{4(-5)}{6-5x}$, $f'(x) = \frac{e^{-3x}\sqrt{2x-5}}{(6-5x)^4}[-3 + \frac{1}{2x-5} + \frac{20}{6-5x}]$.

57 $f(x) = a^x$, $\ln f(x) = \ln a^x = (\ln a)x$. Differentiate to get $\frac{f'(x)}{f(x)} = \ln a$, $f'(x) = f(x)\ln a = a^x \ln a$.

59. $f(x) = \frac{e^{2x}}{x^x}$. $\ln f(x) = \ln \frac{e^{2x}}{x^x} = \ln e^{2x} - \ln x^x = 2x - x\ln x$. $\frac{f'(x)}{f(x)} = 2 - [x(\frac{1}{x}) + (\ln x)(1)] = 1 - \ln x$. Thus $f'(x) = (\frac{e^{2x}}{x^x})(1 - \ln x)$.

61. $f(x) = x^x e^x$, $\ln f(x) = \ln(x^x e^x) = \ln x^x + \ln e^x = x\ln x + x\ln e = x\ln x + x$. Differentiate to get $\frac{f'(x)}{f(x)} = x(\frac{1}{x}) + \ln x + 1 = 2 + \ln x$, $f'(x) = x^x e^x(2 + \ln x)$.

63. The epidemic is spreading most rapidly when the rate of change $R(t)$ is a maximum, that is $R'(t) = f''(t) = 0$. Since $f(t) = A\frac{1}{1+Ce^{-kt}} = A(1+Ce^{-kt})^{-1}$, $f'(t) = A(-1)(1+Ce^{-kt})^{-2}(-Cke^{-kt}) = kAC\frac{e^{-kt}}{(1+Ce^{-kt})^2}$. $f''(t) = kAC\frac{(1+Ce^{-kt})^2(-ke^{-kt})-(e^{-kt})(2)(1+Ce^{-kt})(-Cke^{-kt})}{(1+Ce^{-kt})^4} = kAC\frac{(1+Ce^{-kt})(-ke^{-kt})+2Cke^{-kt}}{(1+Ce^{-kt})^3} = \frac{k^2ACe^{-kt}(Ce^{-kt}-1)}{(1+Ce^{-kt})^3} = 0$ if $Ce^{-kt} = 1$, $e^{-kt} = \frac{1}{C}$, $t = \frac{\ln C}{k}$. Substituting in the original equation leads to $f(\frac{\ln C}{k}) = \frac{A}{1+Ce^{-\ln C}} = \frac{A}{1+Ce^{\ln C^{-1}}} = \frac{A}{2}$ which means half of the total number of susceptible residents. $R'(t) < 0$ when $t > \frac{\ln C}{k}$ (the graph of the curve is decreasing) and $R'(t) > 0$ when $t < \frac{\ln C}{k}$ (the graph of the curve is increasing)

65. $Q(t) = Q_0e^{kt}$. $Q'(t) = kQ_0e^{kt}$. The percentage rate of change is $\frac{100Q'(t)}{Q(t)} = \frac{100kQ_0e^{kt}}{Q_0e^{kt}} = 100k$.

4.5 Applications Involving Compound Interest

1. a) Simple interest: $B(t) = (P)(1+rt)^t$. If $P = 5,000$, $r = 0.05$, and $t = 20$, then $B(20) = 5,000[1+0.05]^{20} = \$13,266.49$.
b) Interest compounded k times per year: $B(t) = (P)(1+\frac{r}{k})^{kt}$. If $P = 5,000$, $r = 0.05$, $t = 20$, and $k = 2$, then $B(20) = 5,000[1+\frac{0.05}{2}]^{40} = \$13,425.32$.
c) Interest compounded continuously: $B(t) = (P)e^{rt}$. If $P = 5,000$ $r = 0.05$, and $t = 20$, then $B(20) = 5,000e^{0.05(20)} = \$13,591.41$.

3. a) If the interest rate is 12 percent and the interest is compounded quarterly, the balance after t years is $B(t) = (P)(1+\frac{0.12}{4})^{4t} = (P)(1.03)^{4t}$. The doubling time is the value of t for which $B(t) = 2P$, that is, $2P = (P)(1.3)^{4t}$, $2 = (1.3)^{4t}$, $\ln 2 = 4t\ln 1.03$, or $t = \frac{\ln 2}{4\ln 1.03} = 5.86$ years.
b) If the interest rate is 12 percent and the interest is compounded continuously, the balance after t years is $B(t) = (P)e^{0.12t}$. The doubling time is the value of t for which $B(t) = 2P$, that is $2P = (P)e^{0.12t}$, $2 = e^{0.12t}$, $\ln 2 = 0.12t$ or $t = \frac{\ln 2}{0.12} = 5.78$ years.

5. The balance after t years is $B(t) = (P)(1+\frac{k}{t})^{kt}$. The doubling time is the value of t for which $B(t) = 2P$, that is, $2P = (P)(1+\frac{r}{k})^{kt}$, $2 = (1+\frac{r}{k})^{kt}$, $\ln 2 = kt\ln(1+\frac{r}{k})$ or $t = \frac{\ln 2}{k\ln(1+r/k)}$ years.

7. If P dollars is invested at an annual rate of 6 percent compounded semiannually, the balance after t years will be $B(t) = (P)(1+\frac{0.06}{2})^{2t} = (P)(1.03)^{2t}$. The tripling time is the value of t for which $B(t) = 3P$, that is, for which $3P = (P)(1.03)^{2t}$, $\ln 3 = 2t\ln 1.03$ or $t = \frac{\ln 3}{2\ln 1.03} = 18.58$ years.

9. If P dollars is invested at an annual rate of r percent compounded k times per year, the balance after t years will be $B(t) = (P)(1+\frac{r}{k})^{kt}$. The tripling time is the value of t for which $B(t) = 3P$, that is, for which $3P = (P)(1+\frac{r}{k})^{kt}$, $3 = (1+\frac{r}{k})^{kt}$, $\ln 3 = kt\ln(1+\frac{r}{k})$ or $t = \frac{\ln 3}{k\ln(1+r/k)}$ years.

11. a) Use $B(t) = (P)(1+\frac{r}{k})^{kt}$ with $P = 1,000$, $B = 2,500$, $r = 0.06$, and $k = 4$ to get $2,500 = 1,000(1+\frac{0.06}{4})$, $(1.015)^{4t} = 2.5$, $4t\ln 1.015 = \ln 2.5$, or $t = \frac{\ln 2.5}{4\ln 1.015} = 15.39$ years.
b) Use $B(t) = Pe^{rt}$ with $P = 1,000$, $B = 2,500$, and $r = 0.06$ to get $2,500 = 1,000e^{0.06t}$, $0.06t = \ln 2.5$, or $t = \frac{\ln 2.5}{0.06} = 15.27$ years.

13. a) With $r = 0.06$ and $k = 4$, the effective rate is $(1+\frac{r}{k})^k - 1 = (1+\frac{0.06}{4})^4 - 1 = 0.0614$ or 6.14 percent.
b) With $r = 0.06$ the effective rate is $e^r - 1 = e^{0.06} - 1 = 0.0618$ or 6.18 percent.

15. Compare effective interest rates. The effective rate corresponds to 10.25 % compounded semiannually is $(1+\frac{r}{k})^k - 1 = (1+\frac{0.1025}{2})^2 - 1 = 0.1051$ or 10.51 %.
The effective rate corresponds to 10.20 % compounded continuously is $e^r - 1 = e^{0.1020} - 1 = 0.1074$ or 10.74 %.
Hence 10.20 compounded continuously is the better investment.

17. Use $B(t) = (P)(1+\frac{r}{k})^{kt}$ with $k = 4$, $P = 1,000$, $t = 8$, and $B(8) = 2,203.76$ to get $2,203.76 = 1,000(1+\frac{r}{4})^{32}$, $(2.20376)^{1/32} = 1+\frac{r}{4}$, $r = 4(2.20376)^{1/32} - 4 = 0.10$ or 10 %.

19. At 6% compounded annually, the effective interest rate is $(1+\frac{r}{k})^k - 1 = (1+\frac{0.06}{1})^1 - 1 = 0.06$. At r % compounded continuously, the effective interest rate is $e^r - 1$. Setting the two effective rates equal to each other yields $e^r - 1 = 0.06$, $e^r = 1.06$, $r = \ln 1.06 = 0.0583$ or 5.83 %.

21. a) Using $P(t) = (B)(1+\frac{r}{k})^{-kt}$ with $B = 20,000$, $r = 0.08$, $t = 15$, and $k = 4$, $P(15) = 20,000(1+\frac{0.08}{4})^{-4(15)} = \$6,095.65$.
b) Using $P(t) = Be^{-rt}$ with $B = 20,000$, $r = 0.08$, and $t = 15$, $P(15) = 20,000e^{-0.08(15)} = \$6,023.88$.

23. a) The three deposits of \$2,000 and the value of each (at 10 % compounded continuously) at the end of the term is shown in the accompanying picture. The amount of the annuity at the end of the term is the corresponding sum $2,000 + 2,000e^{0.1} + 2,000e^{0.2} = \$6,653.15$.

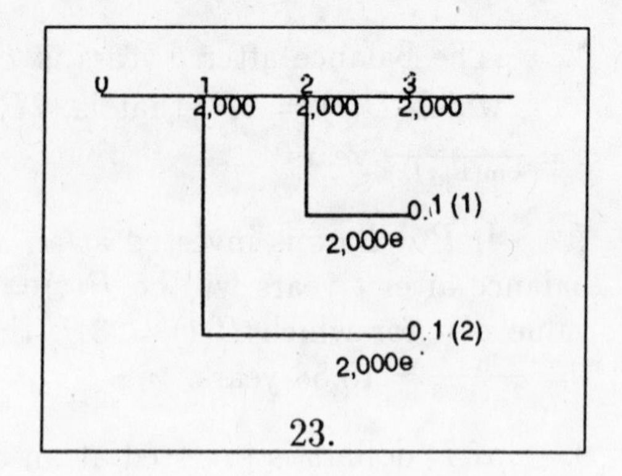

23.

25. The figure shows the sequence of \$50 deposits and the value of that money (at 6% interest compounded semiannually) at the time of the fourth deposit.

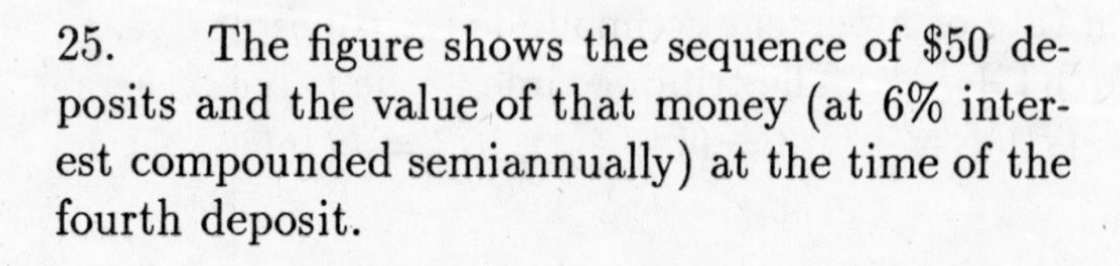

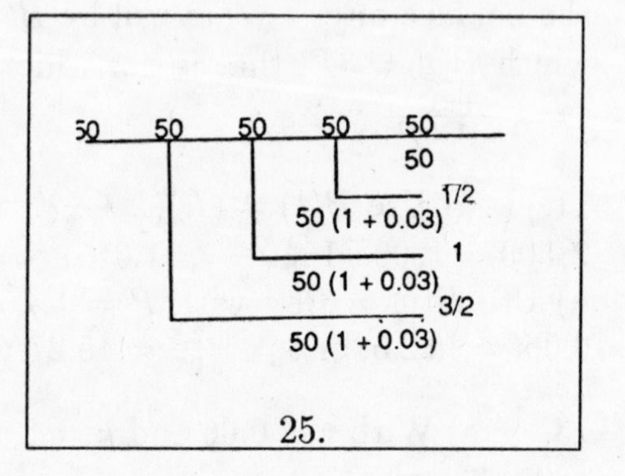

25.

The amount after the fourth deposit is $50(1+\frac{0.06}{2})^3+50(1+\frac{0.06}{2})^2+50(1+\frac{0.06}{2})^1+50(1+\frac{0.06}{2})^0 =$ \$209.18.

27. The figure shows the sequence of \$500 withdrawals and the present value of each withdrawal at 6% interest compounded annually.

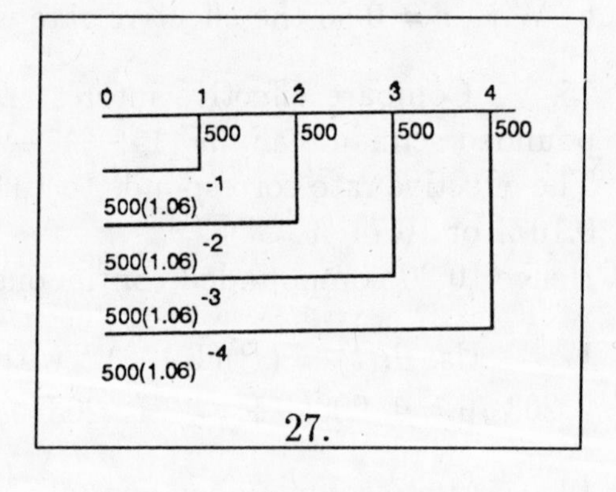

27.

The money to be invested now (that is, the present value of the sequence of payments) is $500(1.06)^{-1} + 500(1.06)^{-2} + 500(1.06)^{-3} + 500(1.06)^{-4} = \$1,732.55$.

29. The percentage rate of change of the market price $V(t) = 8,000e^{\sqrt{t}}$ of the land (expressed in decimal form) is $\frac{V'(t)}{V(t)} = \frac{8,000e^{\sqrt{t}}}{8,000e^{\sqrt{t}}}(\frac{1}{2\sqrt{t}}) = \frac{1}{2\sqrt{t}}$ which will be equal to the prevailing interest rate of 6 % when $\frac{1}{2\sqrt{t}} = 0.06$ or $t = (\frac{1}{0.12})^2 = 69.44$. Moreover, $\frac{1}{2\sqrt{t}} > 0.06$ when $0 < t < 69.44$ and $\frac{1}{2\sqrt{t}} < 0.06$ when $69.44 < t$. Hence the percentage rate of growth of the value of the land is greater than the prevailing interest rate when $0 < t < 69.44$ and less than the prevailing interest rate when $69.44 < t$. Thus the land should be sold in 69.44 years.

31. Since the stamp collection is currently worth \$1,200 and its value is increases linearly at the rate of \$200 per year, its value t years from now is $V(t) = 1,200+200t$. The percentage rate of change of the value (expressed in decimal form) is $\frac{V'(t)}{V(t)} = \frac{200}{1,200+200t} = \frac{200}{200(6+t)} = \frac{1}{6+t}$

which will be equal to the prevailing interest rate of 8% when $\frac{1}{6+t} = 0.08$ or $t = \frac{1}{0.08} - 6 = 6.5$. Moreover, $\frac{1}{6+t} > 0.08$ when $0 < t < 6.5$ and $\frac{1}{6+t} < 0.08$ when $6.5 < t$. Hence the percentage rate of growth of the value of the collection is greater than the prevailing interest rate when $0 < t < 6.5$ and less than the prevailing interest rate when $t > 6.5$. Thus the collection should be sold in 6.5 years.

Review Problems

1. a) If $f(x) = 5e^{-x}$, then $f(x)$ approaches 0 as x increases without bound, and $f(x)$ increases without bound as x decreases without bound. The y–intercept is $f(0) = 5$.

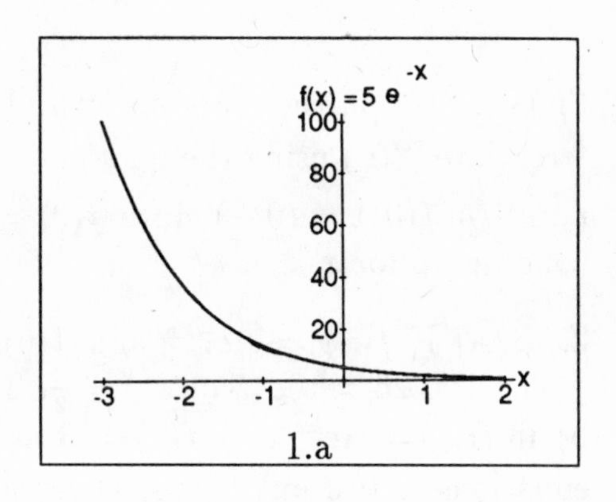

1.a

b) If $f(x) = 5 - e^{-x}$, then $f(x)$ approaches 5 as x increases without bound, and $f(x)$ decreases without bound as x decreases without bound. The y–intercept is $f(0) = 4$.

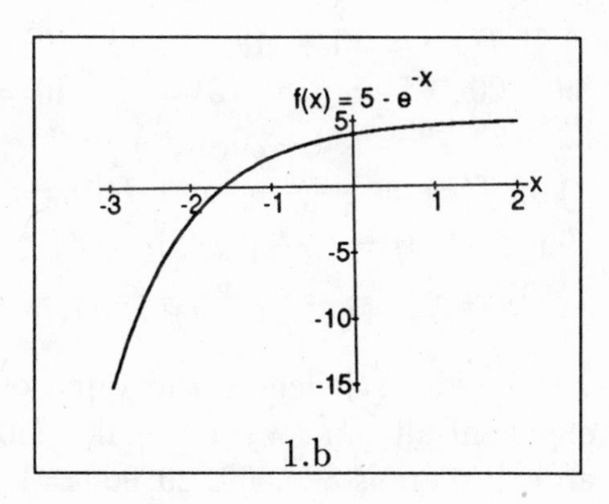

1.b

c) If $f(x) = 1 - \frac{6}{2+e^{-3x}}$, the y–intercept is $f(0) = 1 - \frac{6}{2+e^0} = 1 - \frac{6}{2+1} = -1$. As x increases without bound, e^{-3x} approaches 0, and so $\lim_{x \to \infty} f(x) = 1 - \frac{6}{2+0} = -2$. As x decreases without bound, e^{-3x} increases without bound, and so $\lim_{x \to -\infty} f(x) = 1 - 0 = 1$.

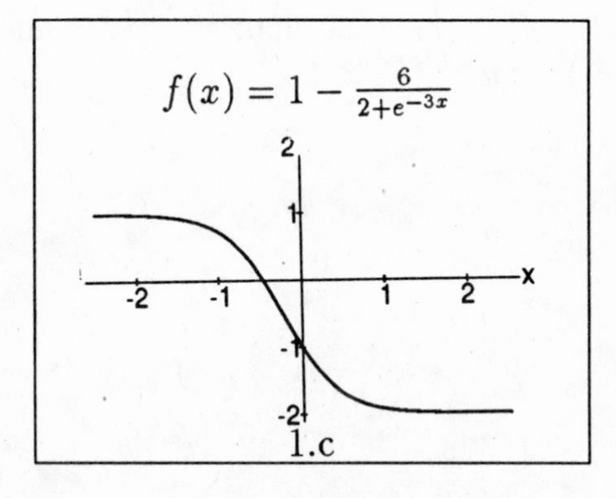

1.c

d) If $f(x) = \frac{3+2e^{-2x}}{1+e^{-2x}}$, then $f(x)$ approaches 3 as x increases without bound, since e^{-2x} approaches 0. Rewriting the function, by multiplying numerator and denominator by e^{2x} yields $f(x) = \frac{3e^{2x}+2}{e^{2x}+1}$. As x decreases without bound, e^{2x} approaches 0 and thus $f(x)$ approaches 2. The y−intercept is $f(0) = \frac{3+2}{1+1} = \frac{5}{2}$. Applying the quotient rule to determine $f'(x)$ yields $f'(x) = \frac{2e^{2x}}{(e^{2x}+1)^2}$.

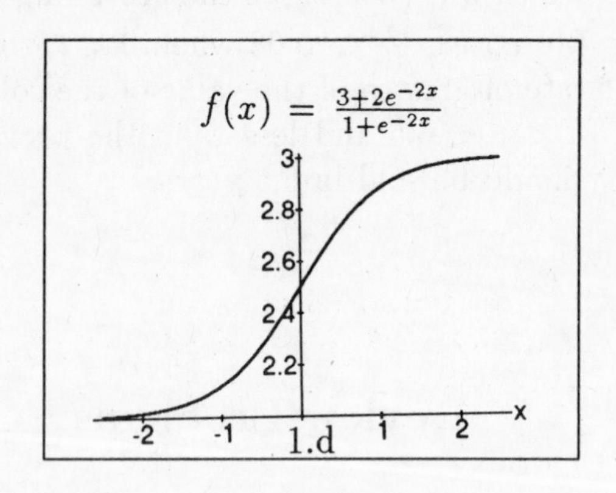

Since $2e^{2x} > 0$, there are no critical points and $f(x)$ increases for all x. To find $f''(x)$, differentiate $f'(x)$ using the quotient rule obtaining $f''(x) = \frac{4e^{2x}(1-e^{2x})}{(e^{e^{2x}}+1)^3}$. There is one inflection point $(0, f(0)) = (0, \frac{5}{2})$ since $f(0) = 0$. The function $f(x)$ is concave down for $x > 0$ and concave up for $x < 0$.

2. a) If $f(x) = Ae^{-kx}$ and $f(0) = 10$, then $10 = Ae^0$. Hence $f(x) = 10e^{-kx}$. Since $f(1) = 25$, $25 = 10e^{-k}$ or $e^{-k} = \frac{5}{2}$. Then $f(4) = 10e^{-4k} = 10(e^{-k})^4 = 10(\frac{5}{2})^4 = 390.625$.
b) If $f(x) = Ae^{kx}$ and $f(1) = 3$ as well as $f(2) = 10$, then $3 = Ae^k$ and $10 = Ae^{2k}$, two equations in two unknowns. Division eliminates A, so $\frac{3}{10} = \frac{Ae^k}{Ae^{2k}}$, $\frac{3}{10} = \frac{e^k}{e^{2k}}$, or $e^k = \frac{10}{3}$. Since $e^k = \frac{10}{3}$, $3 = A(\frac{10}{3})$ and so $A = \frac{9}{10}$. Thus $f(3) = \frac{9}{10}e^{3k} = \frac{9}{10}(e^k)^3 = (\frac{9}{10})(\frac{10}{3})^3 = \frac{100}{3}$.
c) If $f(x) = 30 + Ae^{-kx}$ and $f(0) = 50$, then $50 = 30 + Ae^0$ or $A = 20$. Hence, $f(x) = 30 + 20e^{-kx}$. Since $f(3) = 40$, $40 = 30 + 20e^{-3k}$, $10 = 20e^{-3k}$, or $e^{-3k} = \frac{1}{2}$. Thus $f(9) = 30 + 20e^{-9k} = 30 + 20(e^{-3k})^3 = 30 + 20(\frac{1}{2})^3 = 32.5$.
d) If $f(x) = \frac{6}{1+Ae^{-kx}}$ and $f(0) = 3$ then $3 = \frac{6}{1+Ae^0}$, $3 = \frac{6}{1+A}$, $3 + 3A = 6$, or $A = 1$. Hence, $f(x) = \frac{6}{1+e^{-kx}}$. Since $f(5) = 2$, $2 = \frac{6}{1+e^{-5k}}$, $2 + 2e^{-5k} = 6$, or $e^{-5k} = 2$. Then, $f(10) = \frac{6}{1+e^{-10k}} = \frac{6}{1+(e^{-5k})^2} = \frac{6}{1+(2)^2} = \frac{6}{5}$.

3. Let $V(t)$ denote the value of the machine after t years. Since the value decreases exponentially and was originally \$50,000, it follows that $V(t) = 50,000e^{-kt}$. Since the value after 5 years is \$20,000, $20,000 = V(5) = 50,000e^{-5k}$, $e^{-5k} = \frac{2}{5}$, $-5k = \ln\frac{2}{5}$, or $k = -\frac{1}{5}\ln\frac{2}{5}$. Hence $V(t) = 50,000e^{[1/5\ln(2/5)]t}$ and so $V(10) = 50,000e^{2\ln(2/5)}$, $V(10) = 50,000e^{2\ln(2/5)} = 50,000e^{\ln(4/25)} = 50,000(\frac{4}{25}) = \$8,000$.

4. The sales function is $Q(x) = 50 - 40e^{-0.1x}$ units, where x is the amount (in thousands) spent on advertising.
a) As x increases without bound, $Q(x)$ approaches 50. The vertical axis intercept is $Q(0) = 10$. The graph is like that of a learning curve.
b) If no money is spent on advertising, sales will be $Q(0) = 10$ thousand units.
c) If \$8,000 is spent on advertising, sales will be $Q(8) = 50 - 40e^{-0.8} = 32.027$ thousand or 32,027 units.

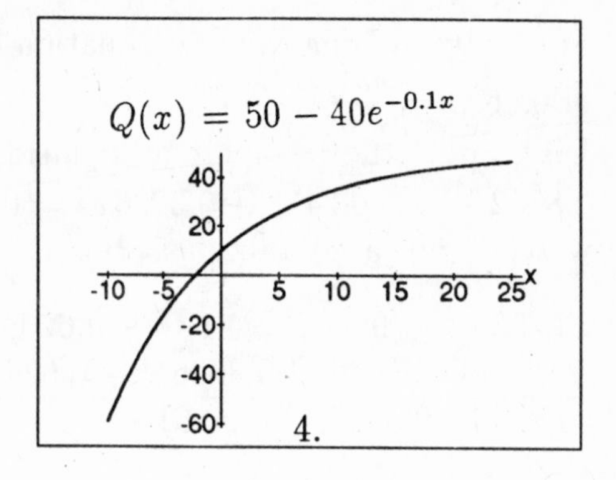

d) Sales will be 35 thousand if $Q(x) = 35$, that is, if $50 - 40e^{-0.1x} = 35$, $-40e^{-0.1x} = -15$, $e^{-0.1x} = \frac{3}{8}$, $-0.1x = \ln\frac{3}{8}$, $x = -\frac{\ln(3/8)}{0.1} = 9.808$ thousand or \$9,808.
e) Since $Q(x)$ approaches 50 as x increases without bound, the most optimistic sales projection is 50,000 units.

5. The output function is $Q(t) = 120 - Ae^{-kt}$. Since $Q(0) = 30$, $30 = 120 - A$ or $A = 90$. Since $Q(8) = 80$, $80 = 120 - 90e^{-8k}$, $-40 = -90e^{-8k}$ or $e^{-8k} = \frac{4}{9}$. Hence, $Q(4) = 120 - 90e^{-4k} = 120 - 90(e^{-8k})^{1/2} = 120 - 90(\frac{4}{9})^{1/2} = 60$ units.

6. The population t years from now will be $P(t) = \frac{30}{1+2e^{-0.05t}}$.
a) The vertical axis intercept is $P(0) = \frac{30}{1+2} = 10$ million. As t increases without bound, $e^{-0.05t}$ approaches 0. Hence, $\lim_{t\to\infty} P(t) = \lim_{t\to\infty} \frac{30}{1+2e^{-0.05t}} = 30$. As t decreases without bound, $e^{-0.05t}$ increases without bound. Hence, the denominator $1 + 2e^{-0.05t}$ increases without bound and $\lim_{t\to-\infty} P(t) = \lim_{t\to-\infty} \frac{30}{1+2e^{-0.05t}} = 0$.

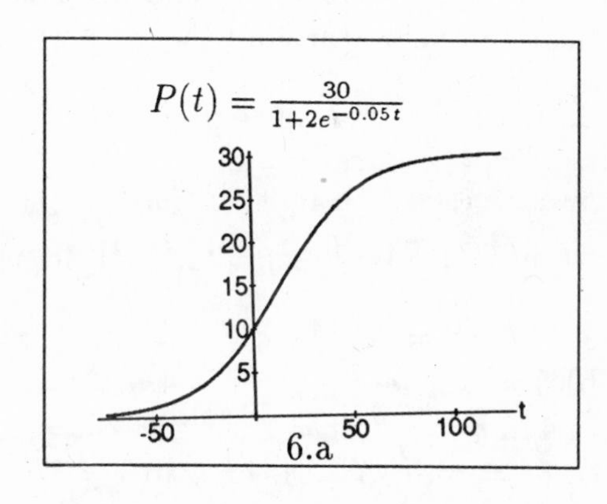

b) The current population is $P(0) = 10$ million.
c) The population in 20 years will be $P(20) = \frac{30}{1+2e^{-0.05(20)}} = \frac{30}{1+2e^{-1}} = 17.28$ million.
d) In the long run (as t increases without bound), $e^{-0.05t}$ approaches 0 and so the population $P(t)$ approaches 30 million.

7. Let $P(t) = \frac{B}{1+Ae^{-Bkt}}$ be the population at time t, where $t = 0$ corresponds to the year 1940. $P(0) = \frac{B}{1+A} = 132$, $P(10) = \frac{B}{1+Ae^{-10Bk}} = 151$, and $P(30) = \frac{B}{1+Ae^{-30Bk}} = 199$. Dividing the first equation by the second and using $x = e^{-10Bk}$ yields $\frac{132}{151} = \frac{1+Ae^{-10Bk}}{1+A}$ or $0.8742 = \frac{1+Ax}{1+A}$. This reduces to $0.8742 + 0.8742A - 1 - Ax = 0$ or $(0.8742 - x)A = 0.1258$.

Similarly dividing the first equation by the third and using $x = e^{-10Bk}$ yields $\frac{132}{199} = \frac{1+Ae^{-30Bk}}{1+A}$ or $0.6633 = \frac{1+Ax^3}{1+A}$. This reduces to $0.6633+0.6633A-1-Ax^3 = 0$ or $(0.6633-x^3)A = 0.3367$. Division of the last pair of equations in x and A eliminates A. Thus $\frac{0.8742-x}{0.6633-x^3} = 0.3736$, $0.8742 - x - 0.2478 + 0.3736x^3 = 0$, or $0.3736x^3 - x + 0.6264 = 0$. $x = 1$ is one solution, which is not acceptable as $kB \neq 0$. By synthetic division,

0.3736	0	−1	0.6264	1
	0.3736	0.3736	−0.6264	
0.3736	0.3736	−0.6264	0	

The resulting quadratic is $0.3736x^2 + 0.3736x - 0.6264 = 0$. From the quadratic formula $x = \frac{-0.3736\pm\sqrt{(0.3736)^2+4(0.3736)(0.6264)}}{2(0.3736)} = \frac{-0.3736\pm1.03715}{2(0.3736)} \to 0.88804$ for $x > 0$.
Substituting in the linear equation in x leads to $(0.8742 - 0.88804)A = 0.1258$ or $A = -9.0885$. Now $P(0) = \frac{B}{1+A} = \frac{B}{1-9.0885} = 132$, $132 + 132(-9.0885) = B$, or $B = -1067.7$.
Since $x = e^{-10Bk} = e^{10677k} = 0.88804$, $10677 = \ln 0.88804 = -0.11874$, or $k = -0.000\,011\,12$.
Thus $kB = 0.011874$ and $P(t) = \frac{-1067.7}{1-9.0885e^{-0.011874t}}$.
This model satisfies $P(0) = 132$, $P(10) = 151$, and $P(30) = 199$
b) In the year 2,000, that is when $t = 60$, the population will be $P(60) = \frac{-1067.7}{1-9.0885e^{-0.71244}} = 308.8 \approx 309$ billion people.
c) This model does not lend itself to the long run predictions, since it is not valid when t is near 186 years (around the year 2126), because $9.0885e^{-(0.011874)(186)} = 1$.

8. a) $\ln e^5 = 5$ since $n = \ln e^n$.
b) $e^{\ln 2} = 2$ since $n = e^{\ln n}$.
c) $e^{3\ln 4 - \ln 2} = e^{\ln 4^3 - \ln 2} = e^{\ln \frac{64}{2}} = e^{\ln 32} = 32$.
d) $\ln(9e^2) + \ln(3e^{-2}) = \ln[(9e^2)(3e^{-2})] = \ln 27 = \ln 3^3 = 3\ln 3$.

9. a) $8 = 2e^{0.04x}$, $e^{0.04x} = 4$, $0.04x = \ln 4$, or $x = \frac{\ln 4}{0.04} = 34.66$.
b) $5 = 1 + 4e^{-6x}$, $4e^{-6x} = 4$, $e^{-6x} = 1$, $-6x = \ln 1 = 0$, or $x = 0$.
c) $4\ln x = 8$, $\ln x = 2$, or $x = e^2 = 7.39$.
d) $5^x = e^3$, $\ln 5^x = \ln e^3$, $x\ln 5 = 3$, or $x = \frac{3}{\ln 5} = 1.86$.

10. Let $Q(t)$ denote the number of bacteria after t minutes. Since $Q(t)$ grows exponentially and 5,000 bacteria were present initially, $Q(t) = 5{,}000e^{kt}$. Since 8,000 bacteria were present after 10 minutes, $8{,}000 = Q(10) = 5{,}000e^{10k}$, $e^{10k} = \frac{8}{5}$, or $k = \frac{1}{10}\ln\frac{8}{5}$. The bacteria will double when $Q(t) = 10{,}000$, that is, when $5{,}000e^{kt} = 10{,}000$, $e^{kt} = 2$, $kt = \ln 2$, or $t = \frac{\ln 2}{k} = \frac{10\ln 2}{\ln(8/5)} = 14.75$ minutes.

11. a) $f(x) = 2e^{3x+5}$, $f'(x) = 2e^{3x+5}\frac{d}{dx}(3x+5) = 6e^{3x+5}$.
b) $f(x) = x^2e^{-x}$, $f'(x) = e^{-x}\frac{d}{dx}(x^2) + x^2\frac{d}{dx}e^{-x} = 2xe^{-x} - x^2e^{-x} = x(2-x)e^{-x}$.
c) $f(x) = \ln\sqrt{x^2+4x+1} = \ln(x^2+4x+1)^{1/2} = (\frac{1}{2})\ln(x^2+4x+1)$, $f'(x) = (\frac{1}{2})(\frac{1}{x^2+4x+1})\frac{d}{dx}(x^2+$

$4x+1) = (\frac{1}{2})(\frac{2x+4}{x^2+4x+1}) = \frac{x+2}{x^2+4x+1}$.
d) $f(x) = x \ln x^2 = 2x \ln x$, $f'(x) = 2(x\frac{d}{dx}\ln x + \ln x \frac{d}{dx}x) = 2(x\frac{1}{x} + \ln x) = 2(1+\ln x)$.
e) $f(x) = \frac{x}{\ln 2x}$, $f'(x) = \frac{(\ln 2x)(1) - x(\frac{1}{2x})(2)}{(\ln 2x)^2} = \frac{\ln 2x - 1}{(\ln 2x)^2}$.

12. a) Use implicit differentiation on $e^{y/x} = 3xy^2$. $e^{y/x}\frac{d}{dx}(\frac{y}{x}) = 6xy\frac{dy}{dx} + 3y^2$, $e^{y/x}\frac{x\frac{dy}{dx} - y}{x^2} = 6xy\frac{dy}{dx} + 3y^2$, $(x\frac{dy}{dx} - y)e^{y/x} = 6x^3y\frac{dy}{dx} + 3x^2y^2$, $(xe^{y/x} - 6x^3y)\frac{dy}{dx} = 3x^2y^2 + ye^{y/x}$, $\frac{dy}{dx} = \frac{3x^2y^2 + ye^{y/x}}{xe^{y/x} - 6x^3y} = \frac{y(3x^2y + e^{y/x})}{x(e^{y/x} - 6x^2y)}$. Simplify further by substituting $e^{y/x} = 3xy^2$ (from the original equation) to get $\frac{dy}{dx} = \frac{y(3x^2y + 3xy^2)}{x(3xy^2 - 6x^2y)} = \frac{3xy^2(x+y)}{3x^2y(y-2x)} = \frac{y(x+y)}{x(y-2x)}$.
b) To simplify the calculation, rewrite the equation $\ln\frac{x}{y} = x^3y^2$ as $\ln x - \ln y = x^3y^2$. Use logarithmic differentiation to get $\frac{1}{x} - \frac{1}{y}\frac{dy}{dx} = 2x^3y\frac{dy}{dx} + 3x^2y^2$, $y - x\frac{dy}{dx} = 2x^4y^2\frac{dy}{dx} + 3x^3y^3$, $(2x^4y^2 + x)\frac{dy}{dx} = y - 3x^3y^3$, $\frac{dy}{dx} = \frac{y - 3x^3y^3}{2x^4y^2 + x} = \frac{y(1-3x^3y^2)}{x(2x^3y^2+1)}$.

13. The average level of carbon monoxide in the air t years from now is $Q(t) = 4e^{0.03t}$ parts per million.
a) The rate of change of the carbon monoxide level t years from now is $Q'(t) = 0.12e^{0.03t}$, and the rate two years from now is $Q'(2) = 0.12e^{0.06} = 0.13$ parts per million per year.
b) The percentage rate of change of the carbon monoxide level t years from now is $100[\frac{Q'(t)}{Q(t)}] = 100(\frac{0.12e^{0.03t}}{4e^{0.03t}}) = 3$ % per year which is a constant, independent of time.

14. Let $F(p)$ denote the profit, where p is the price per camera. Then $F(p)$ =(number of cameras sold)(profit per camera)$= 800e^{-0.01p}(p-40) = 800(p-40)e^{-0.01p}$. $F'(p) = 800[e^{-0.01p}(1) + (p-40)e^{-0.01p}(-0.01)] = 800e^{-0.01p}(1.4 - 0.01p) = 8e^{-0.01p}(140-p) = 0$ when $p = 140$. Since $F'(p) > 0$ (and F is increasing) for $0 < p < 140$, and $F'(p) < 0$ (and F is decreasing) for $p > 140$, it follows that $F(p)$ has its absolute maximum at $p = 140$. Thus the cameras should be sold for \$140 apiece to maximize the profit.

15. a) $f(x) = xe^{-2x}$, $f'(x) = xe^{-2x}(-2) + e^{-2x}(1) = e^{-2x}(1-2x) = 0$ when $x = \frac{1}{2}$. $f(\frac{1}{2}) = \frac{1}{2e}$. $f''(x) = e^{-2x}(-2) + (1-2x)e^{-2x}(-2) = 4e^{-2x}(x-1) = 0$ when $x = 1$. $f(1) = \frac{1}{e^2}$.

x	$-\infty$		$\frac{1}{2}$		1		∞
$f(x)$	$-\infty$		$\frac{1}{2e}$		$\frac{1}{e^2}$		0
$f'(x)$	∞	$+$	0	$-$	$-$	$-$	0
		inc		dec	dec	dec	
$f''(x)$		$-$	$-$	$-$	0	$+$	
		down	down	down		up	

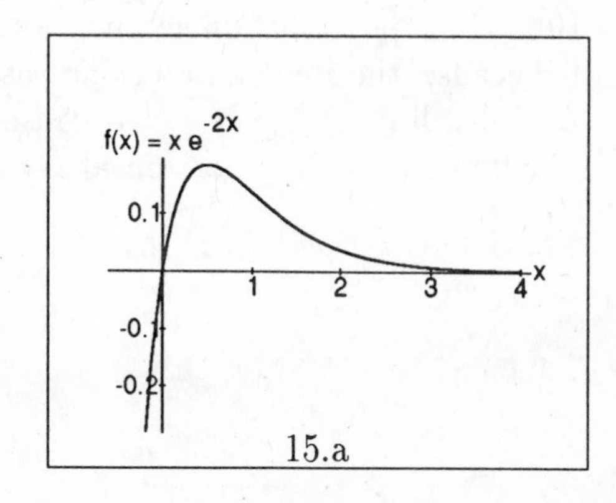

15.a

$(\frac{1}{2}, \frac{1}{2e})$ is the absolute maximum while $(1, \frac{1}{e^2})$ is a point of inflection.

b) $f(x) = e^x - e^{-x}$, $f'(x) = e^x + e^{-x} > 0$, so there are no extrema.

x	$-\infty$		0		∞
$f(x)$	$-\infty$		0		∞
$f'(x)$	∞	$+$	$+$	$+$	
		inc	inc	inc	
$f''(x)$		$-$	0	$+$	
		down		up	

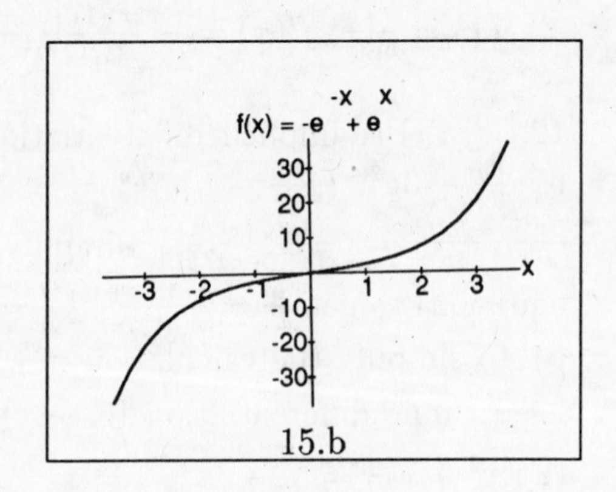

15.b

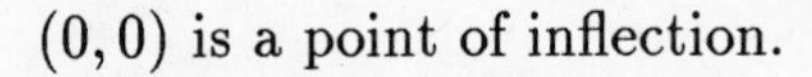

$(0, 0)$ is a point of inflection.

c) $f(x) = \frac{4}{1+e^{-x}} = 4(1+e^{-x})^{-1}$, $f'(x) = 4(-1)(1+e^{-x})^{-2}\frac{d}{dx}(1+e^{-x}) = 4(-1)(1+e^{-x})^{-2}(-e^{-x}) = \frac{4e^{-x}}{(1+e^{-x})^2} > 0$.

$f''(x) = \frac{(1+e^{-x})^2(4e^{-x})(-1) - 4e^{-x}(2)(1+e^{-x})(e^{-x})(-1)}{(1+e^{-x})^4} = \frac{4e^{-x}(1+e^{-x})(-1-e^{-x}+2e^{-x})}{(1+e^{-x})^4} = \frac{4e^{-x}(e^{-x}-1)}{(1+e^{-x})^3} = 0$ when $x = 0$.

x	$-\infty$		0		∞
$f(x)$	0		2		4
$f'(x)$	0	$+$	$+$	$+$	$\frac{1}{2}$
		inc	inc	inc	
$f''(x)$		$+$	0	$-$	
		up		down	

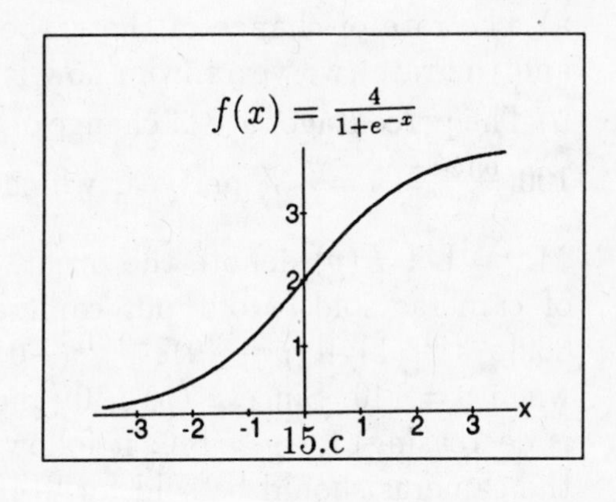

15.c

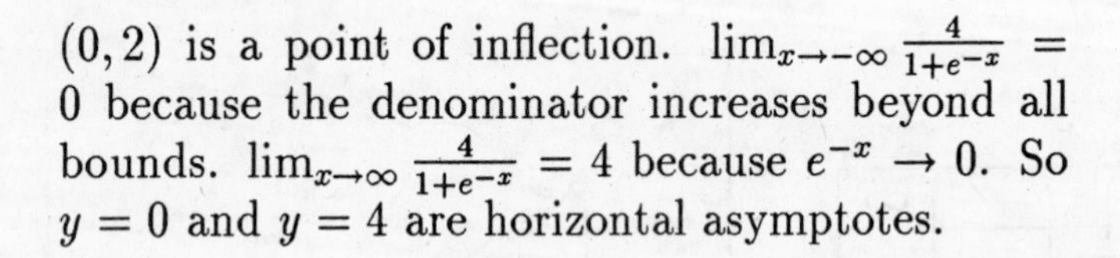

$(0, 2)$ is a point of inflection. $\lim_{x\to-\infty} \frac{4}{1+e^{-x}} = 0$ because the denominator increases beyond all bounds. $\lim_{x\to\infty} \frac{4}{1+e^{-x}} = 4$ because $e^{-x} \to 0$. So $y = 0$ and $y = 4$ are horizontal asymptotes.

d) $f(x) = \ln(x^2+1)$, $f'(x) = \frac{2x}{x^2+1} = 0$ when $x = 0$. $f(0) = 0$. $f''(x) = \frac{(x^2+1)(2)-2x(2x)}{(x^2+1)^2} = \frac{2-2x^2}{(x^2+1)^2} = \frac{2(1-x)(1+x)}{(x^2+1)^2} = 0$ when $x = \pm 1$. $f(\pm 1) = \ln 2$.

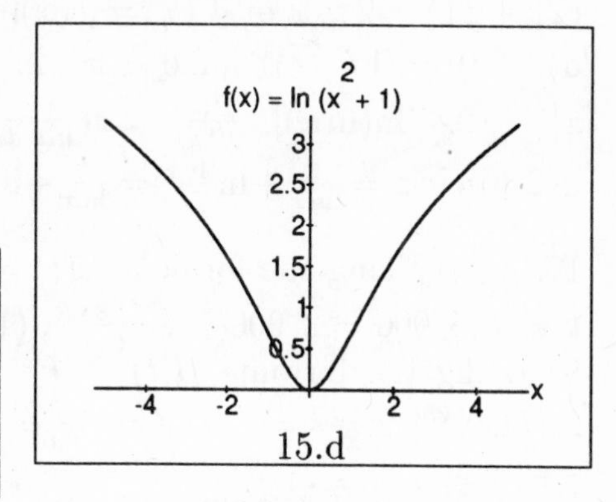

15.d

x	$-\infty$		-1		0		1		∞
$f(x)$	∞		$\ln 2$		0		$\ln 2$		$-\infty$
$f'(x)$	0	$-$	$-$	$-$	0	$+$	$+$	$+$	0
		dec	dec	dec		inc	inc	inc	
$f''(x)$	0	$-$	0	$+$	2	$+$	0	$-$	0
		down		up		up		down	

$(0,0)$ is the absolute minimum while $(\pm 1, \ln 2)$ are points of inflection.

e) $f(x) = \frac{\ln \sqrt[3]{x}}{x} = (\frac{1}{3})(\frac{\ln x}{x})$, $f'(x) = \frac{x(1/x)-\ln x}{x^2} = \frac{1-\ln x}{x^2} = 0$ when $x = e$. $f(e) = \frac{1}{3e}$. $f''(x) = \frac{x^2(-1/x)-(1-\ln x)(2x)}{x^4} = \frac{-1-2+2\ln x}{x^3} = \frac{2\ln x - 3}{x^3} = 0$ when $\ln x = \frac{3}{2}$, or $x = e^{1.5} = 4.4817$. $f(4.4817) = 0.1116$. Note that the domain is $0 < x$.

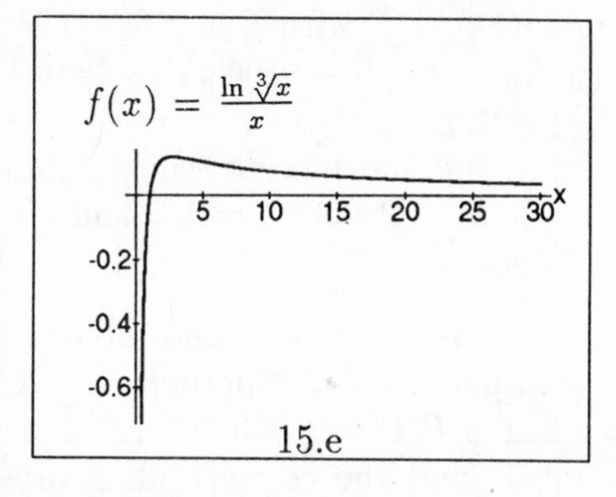

15.e

x	0		e		4.4817		∞
$f(x)$	$-\infty$		0.12		0.11		0
$f'(x)$		$+$	0	$-$	$-$	$-$	
		inc		dec	dec	dec	
$f''(x)$		$-$	$-$	$-$	0	$+$	
		down	down	down		up	

$(e, \frac{1}{3e})$ is the absolute maximum, while $(e^{1.5}, 0.1132)$ is a point of inflection. $\lim_{x\to 0^+} \frac{\ln x}{x} = -\infty$ (you can get a "feeling" for the result from using a calculator or a computer program). Similarly it can be shown that $\lim_{x\to 0^+} \frac{\ln x}{x} = 0$. So $y = 0$ is a vertical asymptote.

16. a) $f(x) = \sqrt{(x^2+1)(x^2+x+2)}$, $\ln f(x) = \ln[(x^2+1)(x^2+x+2)]^{1/2} = (\frac{1}{2})[\ln(x^2+1) + \ln(x^2+x+2)]$. $\frac{f'(x)}{f(x)} = (\frac{1}{2})[\frac{2x}{x^2+1} + \frac{2x+1}{x^2+x+2}]$ or $f'(x) = f(x)[\frac{x}{x^2+1} + \frac{2x+1}{2(x^2+x+2)}]$.

b) $f(x) = \frac{e^{-3x}\sqrt{1-2x}}{(x^2+2x-3)^2}$. $\ln f(x) = \ln[\frac{e^{-3x}\sqrt{1-2x}}{(x^2+2x-3)^2}] = \ln e^{-3x} + \ln(1-2x)^{1/2} - \ln(x^2+2x-3)^2 = -3x + \frac{1}{2}\ln(1-2x) - 2\ln(x^2+2x-3)$.. Thus $\frac{f'(x)}{f(x)} = -3 + \frac{-2}{2(1-2x)} - \frac{2(2x+2)}{x^2+2x-3}$ or $f'(x) = f(x)[-3 - \frac{1}{1-2x} - \frac{4(x+1)}{x^2+2x-3}]$.

c) $f(x) = x^{2x+1}$ for $0 < x$. $\ln f(x) = \ln x^{2x+1} = (2x+1)\ln x$. $\frac{f'(x)}{f(x)} = (2x+1)\frac{d}{dx}\ln x + 2\ln x =$

$(2x+1)\frac{1}{x}+2\ln x$ or $f'(x)=f(x)[\frac{2x+1}{x}+2\ln x]$.
d) $f(x)=(\ln\sqrt{x})^x$ for $0<x$. $\ln f(x)=\ln(\ln x^{1/2})^x=x\ln(\frac{1}{2}\ln x)=x[\ln(\frac{1}{2})+\ln(\ln x)]=x[-\ln 2+\ln(\ln x)]$. $\frac{f'(x)}{f(x)}=x(\frac{1}{\ln x}\frac{d}{dx}\ln x)-\ln 2+\ln\ln x)=x(\frac{1}{x\ln x})-\ln 2+\ln\ln x=\frac{1}{\ln x}-\ln 2+\ln\ln x=\frac{1}{\ln x}+\ln\frac{\ln x}{2}=\frac{1}{\ln x}+\ln\frac{1}{2}\ln x=\frac{1}{\ln x}+\ln\sqrt{\ln x}$. Thus $f'(x)=f(x)[\frac{1}{\ln x}+\ln\sqrt{\ln x}]$

17. a) Using the formula $B(t)=(P)(1+\frac{r}{k})^{kt}$ with $P=2,000$, $B=5,000$, $r=0.08$, and $k=4$, $5,000=2,000(1+\frac{0.08}{4})^{4t}$, $(1.02)^{4t}=\frac{5}{2}$, $4t\ln 1.02=\ln\frac{5}{2}$, or $t=\frac{1}{4}\frac{\ln(5/2)}{\ln 1.02}=11.57$ years.
b) Using the formula $B(t)=Pe^{rt}$ with $P=2,000$, $B=5,000$, and $r=0.08$, $5,000=2,000e^{0.08t}$, $e^{0.08t}=\frac{5}{2}$, or $t=\frac{\ln(5/2)}{0.08}=11.45$.

18. Compare the effective interest rates. The effective interest rate for 8.25 % compounded quarterly is $(1+\frac{r}{k})^k-1=(1+\frac{0.0825}{4})^4-1=0.0851$ or 8.51 %. The effective interset rate for 8.20 compounded continuously is $e^r-1=e^{0.082}-1=0.0855$ or 8.55 %.

19. a) Using the present value formula $P=(B)(1+\frac{r}{k})^{-kt}$ with $B=2,000$, $t=10$, $r=0.0625$, and $k=12$, $P=2,000(1+\frac{0.0625}{12})^{-120}=\frac{2,000}{1.8652182}=\$1,072.26$.
b) Using the present value formula $P=Be^{-rt}$ with $B=2,000$, $t=10$, and $r=0.0625$, $P=2,000e^{-0.0625(10)}=\$1,070.52$.

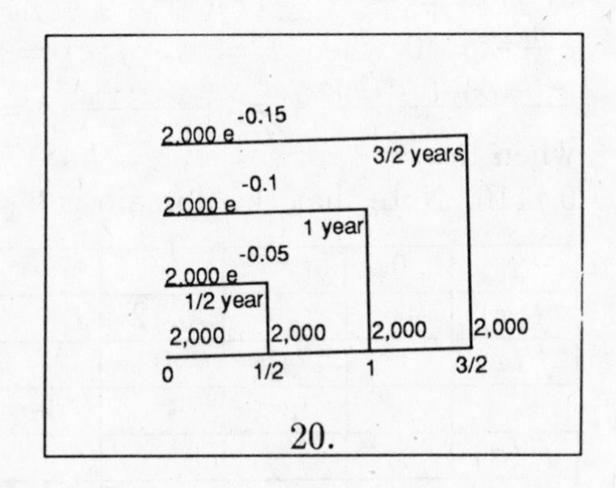

20.

20. Since $r=0.10$ and interest is compounded continuously, the appropriate present value formula is $P(t)=2,000e^{-0.1t}$. The sequence of payments and the corresponding present values are shown in the accompanying figure.

The present value of the annuity is the corresponding sum $2,000+2,000e^{-0.05}+2,000e^{-0.1}+2,000e^{-0.15}=\$7,433.55$.

21. The value of the coin collection in t years is $V(t)=2,000e^{\sqrt{t}}$. Hence, the percentage rate of change of the value of the collection (expressed in decimal form) is $\frac{V'(t)}{V(t)}=\frac{2,000e^{\sqrt{t}}}{2,000e^{\sqrt{t}}}(\frac{1}{2\sqrt{t}})=\frac{1}{2\sqrt{t}}$ which will be equal to the prevailing interest rate of 7 % when $\frac{1}{2\sqrt{t}}=0.07$ or $t=(\frac{1}{0.14})^2=51.02$.
Moreover, $\frac{1}{2\sqrt{t}}>0.07$ when $0<t<51.02$ and $\frac{1}{2\sqrt{t}}<0.07$ when $51.02<t$. Hence the percentage rate of growth of the collection is greater than the prevailing interest rate when $0<t<51.02$ and less than the prevailing interest rate when $51.02<t$. Thus the coin collection should be sold in 51.02 years.

22. Using the present value formula $P = Be^{-rt}$ with $B = (375)(10^9)$ and $t = 1,986 - 1,626 = 360$, $P = (375)(10^9)e^{-360r} = 24$, $e^{-360r} = \frac{24}{(375)(10^9)} = 6.4(10^{-11})$, $-360r = \ln 6.4 - 11 \ln 10 = -23.4721$, $r = \frac{23.4721}{360} = 0.0652$ or 6.52%.

Chapter 5

Antidifferentiation

5.1 Antidifferentiation; the Indefinite Integral

1. $I = \int x^5 dx = \frac{x^6}{6} + C.$

3. $I = \int \frac{1}{x^2} dx = \int x^{-2} dx = -x^{-1} + C = -\frac{1}{x} + C.$

5. $I = \int 5dx = 5x + C.$

7. $I = \int (3x^2 - \sqrt{5x} + 2)dx = 3\int x^2 dx - \sqrt{5}\int x^{1/2} dx + 2\int dx = 3(\frac{x^3}{3}) - \sqrt{5}(\frac{x^{3/2}}{3/2}) + 2x + C = x^3 - \frac{2\sqrt{5}x^{3/2}}{3} + 2x + C.$

9. $I = \int (3\sqrt{x} - \frac{2}{x^3} + \frac{1}{x})dx = 3\int x^{1/2} dx - 2\int x^{-3} dx + \int x^{-1} dx = 3(\frac{x^{3/2}}{3/2}) - 2(\frac{x^{-2}}{-2}) + \ln|x| + C = 2x^{3/2} + x^{-2} + \ln|x| + C = 2x^{3/2} + \frac{1}{x^2} + \ln|x| + C.$

11. $I = \int (\frac{e^x}{2} + x\sqrt{x})dx = \frac{1}{2}\int e^x dx + \int x^{3/2} dx = (\frac{1}{2})e^x + \frac{x^{5/2}}{5/2} + C = \frac{e^x}{2} + \frac{2x^{5/2}}{5} + C.$

13. $I = \int (\frac{1}{3x} - \frac{3}{2x^2} + e^2 + \frac{\sqrt{x}}{2})dx = \frac{1}{3}\int x^{-1} dx - \frac{3}{2}\int x^{-2} dx + e^2 \int dx + \frac{1}{2}\int x^{1/2} dx = \frac{1}{3}\ln|x| - \frac{3}{2}(\frac{x^{-1}}{-1}) + e^2 x + (\frac{1}{2})(\frac{x^{3/2}}{3/2}) + C = \frac{1}{3}\ln|x| + \frac{3}{2x} + e^2 x + \frac{x^{3/2}}{3} + C.$

15. $I = \int \frac{x^2+2x+1}{x^2} dx = \int (1 + \frac{2}{x} + \frac{1}{x^2})dx = x + 2\ln|x| - x^{-1} + C = x + 2\ln|x| - \frac{1}{x} + C.$

17. $I = \int (x^3 - 2x^2)(\frac{1}{x} - 5)dx = \int (x^2 - 2x - 5x^3 + 10x^2)dx = \int (11x^2 - 2x - 5x^3)dx = 11(\frac{x^3}{3}) - x^2 - 5(\frac{x^4}{4}) + C = \frac{11x^3}{3} - x^2 - \frac{5x^4}{4} + C.$

19. $I = \int \sqrt{x}(x^2 - 1)dx = \int x^{5/2} dx - \int x^{1/2} dx = \frac{2x^{7/2}}{7} - \frac{2x^{3/2}}{3} + C.$

21. Let $P(t)$ denote the population of the town t months from now. Since $\frac{dP}{dt} = 4 + 5t^{2/3}$, $P(t)$ is an antiderivative of $4 + 5t^{2/3}$. Thus $P(t) = \int(4 + 5t^{2/3})dt = 4t + 5(\frac{3t^{5/3}}{5}) + C = 4t + 3t^{5/3} + C$. Since the current population is 10,000, it follows that $10,000 = P(0) = 4(0) + 3(0)^{5/3} + C$ or $C = 10,000$. $P(t) = 4t + 3t^{5/3} + 10,000$ and $P(8) = 4(8) + 3(8^{5/3}) + 10,000 = 32 + 3(2^5) + 10,000 = 32 + 96 + 10,000 = 10,128$ people.

23. $\frac{dy}{dt} = 3 + 2t + 6t^2$. The distance is $y(t) = 3t + t^2 + 2t^3 + S_0$. $y(2) = 6 + 4 + 16 + S_0$, $y(1) = 3 + 1 + 2 + S_0$, so the distance traveled during the second minute is $|y(2) - y(1)| = 26 - 6 = 20$ meters.

25. The rate is $r(t)$ people per hour. In a time period dt the number of people is $r(t)dt$. The number of people entering the fair from 11:00 a.m. to 1:00 p.m. is $\int_2^4 r(t)dt$.

27. The rate is $f(x) = 5,000 + 60\sqrt{x}$ bicycles per month and the price $p(x) = 80 + 3\sqrt{x}$ dollars per bicycle. Since revenue per month is $R(x) = f(x)p(x) = (5,000 + 60\sqrt{x})(80 + 3\sqrt{x}) = 400,000 + 19,800\sqrt{x} + 180x$. The total revenue for 16 months is $\int_0^{16}(400,000 + 19,800\sqrt{x} + 180x)dx = (400,000x + \frac{2\times 19,800x^{3/2}}{3} + 90x^2)|_0^{16} = 6,400,000 + 844,800 + 23,040 = \$7,267,840$.

29. The rate at which population changes is $r(t) = 0.6t^2 + 0.2t + 0.5$ thousand people per year and the rate of pollution 5 units per thousand people. In a time period dt the pollution increases by $5(0.6t^2 + 0.2t + 0.5)dt$ and in two years $\int_0^2 5(0.6t^2 + 0.2t + 0.5)dt = (t^3 + \frac{t^2}{2} + 2.5t)|_0^2 = 15$ units.

31. Let $C(q)$ denote the total cost of producing q units. Since marginal cost is $\frac{dC}{dq} = 6q + 1$ dollars per unit, $C(q)$ is an antiderivative of $6q + 1$. $C(q) = \int(6q + 1)dq = 3q^2 + q + C_1$. Since the cost of producing the first unit is \$130, $130 = C(1) = 3 + 1 + C_1$ or $C_1 = 126$. $C(q) = 3q^2 + q + 126$. $C(10) = 3(10)^2 + 10 + 126 = \$436$.

33. Let $P(q)$ denote the profit from the production and sale of q units. Since $\frac{dP}{dq} = 100 - 2q$ dollars per unit, $P(q)$ is an antiderivative of $100 - 2q$. $P(q) = \int(100 - 2q)dq = 100q - q^2 + C$. Since the profit is \$700 when 10 units are produced, $700 = P(10) = 100(10) - (10)^2 + C$ or $C = -200$. $P(q) = 100q - q^2 - 200$ which attains its maximum value when its derivative, $\frac{dP}{dq} = 100 - 2q = 0$ or $q = 50$. Hence the maximum profit is $P(50) = 100(50) - (50)^2 - 200 = 5,000 - 2,500 - 200 = \$2,300$.

35. $\frac{dy}{dx} = 3x^2 + 6x - 2$, $y = x^3 + 3x^2 - 2x + C$. Since $(0,6)$ lies on the graph its coordinates satisfy the equation of the curve, that is $C = 6$. Thus $y = f(x) = x^3 + 3x^2 - 2x + 6$.

37. Let's assume that the function is continuous at $x = 1$ and at $x = 4$. Then $f'(1) = f'(4) = 0$ and $f'(x) = (x - 4)(x - 1) = x^2 - 5x + 4$ will do the trick. Integrating leads to $f(x) = \frac{x^3}{3} - \frac{5x^2}{2} + 4x + C$. Note that $f(1) = \frac{1}{3} - \frac{5}{2} + 4 + C = \frac{11}{6} + C < f(4) = \frac{64}{3} - \frac{16}{2} + 16 + C =$

$\frac{44}{3} + C$ and $x = 4$ corresponds to a maximum while $x = 1$ flags a minimum.

39. $I = \int 3e^{5x}dx = \frac{3}{5}\int e^{5x}d(5x) = \frac{3}{5}e^{5x} + C.$

41. $I = \int \frac{2}{3x+2}dx = \frac{2}{3}\int \frac{d(3x+2)}{3x+2} = \frac{2}{3}\ln|3x+2| + C.$

43. $I = \int(2x+3)^5dx$. One guess is that $I = \frac{(2x+3)^6}{6} + C_1$. If this is correct, its derivative must be $(2x+3)^5$. Now $\frac{d}{dx}[\frac{(2x+3)^6}{6} + C_1] = 2(2x+3)^5$, which is twice the desired result. Thus $I = \frac{(2x+3)^6}{12} + \frac{C_1}{2} = \frac{(2x+3)^6}{12} + C$. Note that C and C_1 are arbitrary constants.

45. The rate of revenue is $R'(x) =$ (demand per month at time x)(price at time x) $=$ $[n(x)][p(x)]$. Hence the total revenue is an antiderivative of $[n(x)][p(x)]$ or $R(x) = \int n(x)p(x)dx$. The total revenue over over 12 months is $R(12)$. This can be written in the form of a definite integral, which will be covered in the next chapter. The form is $R(12) = \int_0^{12} n(x)p(x)dx$.

47. The deceleration is $a(t) = -25\ \frac{\text{ft}}{\text{sec}^2}$. The velocity $v(t)$ is an antiderivative of the deceleration, so $v(t) = -25t + C_1$, where C_1 is the initial velocity (when $t = 0$). To stay with the units of feet and seconds, let's convert $60\ \frac{\text{miles}}{\text{hour}}$ to $\frac{\text{ft}}{\text{sec}}$. Thus $\frac{60\text{ miles}}{\text{hour}} = (\frac{60\text{ miles}}{\text{hour}})(\frac{5{,}280\text{ ft}}{\text{mile}})(\frac{1\text{ hr}}{3{,}600\text{ sec}}) = 88\ \frac{\text{ft}}{\text{sec}}$. Thus $v(t) = -25t + 88$.
The car will stop when the velocity is 0, or $t = \frac{88}{25} = 3.52$ seconds. The distance traveled, $s(t)$, is an antiderivative of the velocity, so $s(t) = -\frac{25t^2}{2} + 88t + C_2$. Let the point at which the brakes are applied correspond to $s(0) = 0$ so that $C_2 = 0$. Then the braking distance is $s(3.52) = -\frac{25(3.52)^2}{2} + 88(3.52) = 154.88$ ft.

49. The distance covered during the reaction time of 0.7 seconds is $\frac{60\text{miles}}{\text{hr}}\frac{5{,}280\text{ft}}{\text{mile}}\frac{\text{hr}}{\text{sec.}} \times$ 0.7sec. = 61.6 ft.
The speed is $\frac{dD}{dt} = -28t + 88 = 0$ ft/sec. when the car comes to rest, $t = \frac{22}{7}$ sec. The distance traveled in this time will be $D(\frac{22}{7}) = -14 \times (\frac{22}{7})^2 + \frac{88\times 22}{7} + 61.6 = 199.89$ ft.
The camel may or may not get hit. If the camel sits on the road in the car's path, it will make contact with the car. If the camel is standing and its legs are in the car's path, the car will hit it. If, on the other hand, the car is positioned between the camel's front and rear legs, and if the hood of the car is more than 0.89 ft. (10.7 inches) in length, the camel will escape undamaged. (The camel's stomach is assumed to above the car's hood. Defining a probability function for the camel getting hit and computing this probability would make a neat additional problem in another chapter/section. Cross referencing could become a little cumbersome.)

5.2 Integration by Substitution

1. Let $u = 2x+6$. Then $du = 2dx$ or $dx = \frac{du}{2}$. Hence, $\int(2x+6)^5dx = \frac{1}{2}\int u^5du = \frac{u^6}{12} + C = \frac{(2x+6)^6}{12} + C =$

3. Let $u = 4x-1$. Then $du = 4dx$ or $dx = \frac{du}{4}$. Hence, $\int\sqrt{4x-1}dx = \frac{1}{4}\int u^{1/2}du = (\frac{1}{4})\frac{2u^{3/2}}{3} + C = \frac{(4x-1)^{3/2}}{6} + C$

5. Let $u = 1-x$. Then $du = -dx$ or $dx = -du$. Hence, $\int e^{1-x}dx = -\int e^u du = -e^u + C = -e^{1-x} + C$.

7. Let $u = x^2$. Then $du = 2xdx$ or $xdx = \frac{1}{2}du$. Hence, $\int xe^{x^2}dx = \frac{1}{2}\int e^u du = \frac{1}{2}e^u + C = \frac{1}{2}e^{x^2} + C$.

9. Let $u = x^2+1$. Then $du = 2xdx$ or $xdx = \frac{1}{2}du$. Hence, $\int x(x^2+1)^5dx = \frac{1}{2}\int u^5du = \frac{u^6}{12} + C = \frac{(x^2+1)^6}{12} + C$.

11. Let $u = x^3+1$. Then $du = 3x^2dx$ or $x^2dx = \frac{1}{3}du$. Hence, $\int x^2(x^3+1)^{3/4}dx = \frac{1}{3}\int u^{3/4}du = (\frac{1}{3})(\frac{4u^{7/4}}{7}) + C = \frac{4(x^3+1)^{7/4}}{21} + C$.

13. Let $u = x^5+1$. Then $du = 5x^4dx$ or $x^4dx = \frac{1}{5}du$. Hence, $\int\frac{2x^4}{x^5+1}dx = \frac{2}{5}\int\frac{du}{u} = \frac{2}{5}\ln|u| + C = \frac{2}{5}\ln|x^5+1| + C$.

15. Let $u = x^2+2x+5$. Then $du = (2x+2)dx$ or $(x+1)dx = \frac{1}{2}du$. Hence, $\int(x+1)(x^2+2x+5)^{12}dx = \frac{1}{2}\int u^{12}du = \frac{u^{13}}{26} + C = \frac{(x^2+2x+5)^{13}}{26} + C$.

17. Let $u = x^5+5x^4+10x+12$. Then $du = (5x^4+20x^3+10)dx$ or $(x^4+4x^3+2)dx = \frac{1}{5}du$. Hence, $\int\frac{3x^4+12x^3+6}{x^5+5x^4+10x+12}dx = 3\int\frac{x^4+4x^3+2}{x^5+5x^4+10x+12}dx = \frac{3}{5}\int\frac{1}{u}du = \frac{3}{5}\ln|u| + C = \frac{3}{5}\ln|x^5+5x^4+10x+12| + C$.

19. Let $u = x^2-2x+6$. Then $du = (2x-2)dx$ or $(3x-3)dx = \frac{3}{2}du$. Hence, $\int\frac{3x-3}{(x^2-2x+6)^2}dx = \frac{3}{2}\int\frac{1}{u^2}du = \frac{3}{2}\int u^{-2}du = (\frac{3}{2})(-\frac{1}{u}) + C = -\frac{3}{2(x^2-2x+6)} + C$.

21. Let $u = \ln 5x$. Then $du = \frac{5}{5x}dx = \frac{1}{x}dx$. Hence, $\int\frac{\ln 5x}{x}dx = \int udu = \frac{u^2}{2} + C = \frac{(\ln 5x)^2}{2} + C$.

23. Let $u = \ln x$. Then $du = \frac{1}{x}dx$. Hence, $\int\frac{1}{x(\ln x)^2}dx = \int\frac{1}{u^2}du = -\frac{1}{u} + C = -\frac{1}{\ln x} + C$.

25. Let $u = \ln(x^2+1)$. Then $du = \frac{2x}{x^2+1}dx$. Hence, $\int\frac{2x\ln(x^2+1)}{x^2+1}dx = \int udu = \frac{u^2}{2} + C =$

$\frac{[\ln(x^2+1)]^2}{2} + C$.

27. Let $u = x-1$. Then $du = dx$ and $x = u+1$. Hence, $\int \frac{x}{x-1}dx = \int \frac{u+1}{u}du = \int(1+\frac{1}{u})du = u + \ln|u| + C_1 = x - 1 + \ln|x-1| + C_1 = x + \ln|x-1| + C$. (Note that $C = C_1 - 1$.)

29. Let $u = x - 5$. Then $du = dx$ and $x = u + 5$. Hence, $\int \frac{x}{(x-5)^6}dx = \int \frac{u+5}{u^6}du = \int(u^{-5} + 5u^{-6})du = -\frac{u^{-4}}{4} - u^{-5} + C = -\frac{1}{4(x-5)^4} - \frac{1}{(x-5)^5} + C$.

31. Let $u = x - 4$. Then $du = dx$ and $x = u + 4$ Hence, $\int \frac{x+3}{(x-4)^2}dx = \int \frac{(u+4)+3}{u^2}du = \int(u^{-1} + 7u^{-2})du = \ln|u| - 7u^{-1} + C = \ln|x-4| - \frac{7}{x-4} + C$.

33. Let $u = 2x - 1$. Then $2x + 3 = u + 4$ and $du = 2dx$ or $dx = \frac{1}{2}du$. Hence, $\int(2x+3)\sqrt{2x-1}dx = \frac{1}{2}\int(u+4)u^{1/2}du = \frac{1}{2}\int(u^{3/2} + 4u^{1/2})du = \frac{1}{2}(\frac{2u^{5/2}}{5} + \frac{8u^{3/2}}{3}) + C = \frac{1}{2}[\frac{2(2x-1)^{5/2}}{5} + \frac{8(2x-1)^{3/2}}{3}] + C$.

35. The slope of the tangent is the derivative of f. Thus, $f'(x) = \frac{2x}{1-3x^2}$ and so f is an antiderivative of $\frac{2x}{1-3x^2}$. That is, $f(x) = \int \frac{2x}{1-3x^2}dx = \frac{-1}{3}\ln|1-3x^2| + C$. Since the graph of f passes through the point (0,5), it follows that $5 = f(0) = -\frac{1}{3}\ln 1 + C$ or $C = 5$. Hence, $f(x) = -\frac{1}{3}\ln|1-3x^2| + 5$.

37. Let $V(t)$ denote the value of the machine after t years. Since $\frac{dV}{dt} = -960e^{-t/5}$ dollars per year, $V(t)$ is an antiderivative of $-960e^{-t/5}$. Thus, $V(t) = \int(-960e^{-t/5})dt = -960(-5)e^{-t/5} + C = 4,800e^{-t/5} + C$. Since the value of the machine was originally \$5,000, it follows that $5,000 = V(0) = 4,800e^0 + C = 4,800 + C$ or $C = 200$. The value of the machine after t years is $V(t) = 4,800e^{-t/5} + 200$ and the value after 10 years is $V(10) = 4,800e^{-2} + 200 = \$849.61$.

39. Let $V(x)$ denote the value of the farm land x years from now. Since $\frac{dV}{dx} = \frac{0.4x^3}{\sqrt{0.2x^4+8,000}}$ dollars per year. $V(x)$ is an antiderivative of this expression. Let $u = 0.2x^4 + 8,000$. Then $du = 0.8x^3dx$ or $0.4x^3dx = \frac{1}{2}du$. Hence, $V(x) = \int \frac{0.4x^3}{\sqrt{0.2x^4+8,000}}dx = \frac{1}{2}\int u^{-1/2}du = u^{1/2} + C = (0.2x^4 + 8,000)^{1/2} + C$. Since the land is currently worth \$500 per acre, it follows that $500 = V(0) = (8,000)^{1/2} + C$ or $C = 500 - (8,000)^{1/2}$. Hence the value of the land 10 years from now will be $V(10) = [0.2(10^4) + 8,000]^{1/2} + 500 - (8,000)^{1/2} = (10,000)^{1/2} + 500 - (8,000)^{1/2} = 600 - (8,000)^{1/2} = \$510.56$ per acre.

5.3 Integration by Parts

1. $I = \int xe^{-x}dx.$

$f(x) = x$	$g(x) = e^{-x}$
$f'(x) = 1$	$G(x) = -e^{-x}$

$I = -xe^{-x} + \int e^{-x}dx = -xe^{-x} - e^{-x} + C = -(x+1)e^{-x} + C.$

3. $I = \int xe^{-x/5}dx.$

$f(x) = x$	$g(x) = e^{-x/5}$
$f'(x) = 1$	$G(x) = -5e^{-x/5}$

$I = -5xe^{-x/5} - \int(-5e^{-x/5})dx = -5xe^{-x/5} + 5\int e^{-x/5}dx = -5xe^{-x/5} - 25e^{-x/5} + C = -5(x+5)e^{-x/5} + C.$

5. $I = \int(1-x)e^x dx.$

$f(x) = 1 - x$	$g(x) = e^x$
$f'(x) = -1$	$G(x) = e^x$

$I = (1-x)e^x - \int(-e^x)dx = (1-x)e^x + \int e^x dx = (1-x)e^x + e^x + C = (2-x)e^x + C.$

7. $I = \int x \ln 2x dx.$

$f(x) = \ln 2x$	$g(x) = x$
$f'(x) = \frac{1}{x}$	$G(x) = \frac{x^2}{2}$

$I = \frac{x^2}{2}\ln 2x - \int(\frac{1}{x})(\frac{x^2}{2})dx = \frac{x^2}{2}\ln 2x - \frac{1}{2}\int x dx = \frac{x^2}{2}\ln 2x - \frac{1}{2}(\frac{x^2}{2}) + C = \frac{x^2}{2}(\ln 2x - \frac{1}{2}) + C.$

9. $I = \int x\sqrt{x-6}dx.$

$f(x) = x$	$g(x) = \sqrt{x-6} = (x-6)^{1/2}$
$f'(x) = 1$	$G(x) = \frac{2(x-6)^{3/2}}{3}$

$I = \frac{2}{3}x(x-6)^{3/2} - \int(\frac{2}{3})(x-6)^{3/2}dx = \frac{2}{3}x(x-6)^{3/2} - \frac{2}{3}(\frac{2}{5})(x-6)^{5/2} + C = \frac{2}{3}x(x-6)^{3/2} - \frac{4(x-6)^{5/2}}{15} + C.$

11. $I = \int x(x+1)^8 dx.$

$f(x) = x$	$g(x) = (x+1)^8$
$f'(x) = 1$	$G(x) = \frac{(x+1)^9}{9}$

$I = \frac{1}{9}x(x+1)^9 - \int(\frac{1}{9})(x+1)^9 dx = \frac{1}{9}x(x+1)^9 - \frac{1}{9}(\frac{1}{10})(x+1)^{10} + C = \frac{1}{9}x(x+1)^9 - \frac{(x+1)^{10}}{90} + C.$

13. $I = \int \frac{x}{\sqrt{x+2}}dx.$

$f(x) = x$	$g(x) = \frac{1}{\sqrt{x+2}} = (x+2)^{-1/2}$
$f'(x) = 1$	$G(x) = 2(x+2)^{1/2}$

$I = 2x(x+2)^{1/2} - \int 2(x+2)^{1/2}dx = 2x(x+2)^{1/2} - 2(\frac{2}{3})(x+2)^{3/2} + C = 2x(x+2)^{1/2} - (\frac{4}{3})(x+2)^{3/2} + C.$

15. $I = \int x^2 e^{-x}dx.$

$f_1(x) = x^2$	$g_1(x) = e^{-x}$
$f_1'(x) = 2x$	$G_1(x) = -e^{-x}$

$I = -x^2e^{-x} - \int 2x(-e^{-x})dx = -x^2e^{-x} + 2\int xe^{-x}dx.$

$f_2(x) = x$	$g_2(x) = e^{-x}$
$f_2'(x) = 1$	$G_2(x) = -e^{-x}$

$I = -x^2e^{-x} + 2[-xe^{-x} - \int(-e^{-x})dx] = -x^2e^{-x} + 2(-xe^{-x} - e^{-x}) + C.$

17. $I = \int x^3 e^x dx.$

$f_1(x) = x^3$	$g_1(x) = e^x$
$f_1'(x) = 3x^2$	$G_1(x) = e^x$

$I = x^3e^x - \int 3x^2e^x dx = x^3e^x - 3\int x^2e^x dx.$

$f_2(x) = x^2$	$g_2(x) = e^x$
$f_2'(x) = 2x$	$G_2(x) = e^x$

$I = x^3e^x - 3[x^2e^x - \int 2xe^x dx] = x^3e^x - 3[x^2e^x - 2\int xe^x dx] = x^3e^x - 3[x^2e^x - 2(xe^x - e^x)] + C = (x^3 - 3x^2 + 6x - 6)e^x + C.$

19. $I = \int x^2 \ln x dx.$

$f_1(x) = \ln x$	$g_1(x) = x^2$
$f_1'(x) = \frac{1}{x}$	$G_1(x) = \frac{x^3}{3}$

$I = \frac{x^3 \ln x}{3} - \int (\frac{1}{x})(\frac{x^3}{3})dx = \frac{x^3 \ln x}{3} - \frac{1}{3}\int x^2 dx = \frac{x^3 \ln x}{3} - \frac{x^3}{9} + C.$

21. $I = \int \frac{\ln x}{x^2} dx = \int (\frac{1}{x^2}) \ln x dx.$

$f(x) = \ln x$	$g(x) = \frac{1}{x^2}$
$f'(x) = \frac{1}{x}$	$G(x) = -\frac{1}{x}$

$I = -\frac{1}{x}\ln x - \int (-\frac{1}{x})(\frac{1}{x})dx = -\frac{1}{x}\ln x + \int \frac{1}{x^2}dx = -\frac{1}{x}\ln x - \frac{1}{x} + C = -\frac{1}{x}(\ln x + 1) + C.$

23. $I = \int x^3 e^{x^2} dx = \int x^2(xe^{x^2})dx.$

$f(x) = x^2$	$g(x) = xe^{x^2}$
$f'(x) = 2x$	$G(x) = \frac{1}{2}e^{x^2}$

$I = \frac{1}{2}x^2e^{x^2} - \int 2x(\frac{1}{2}e^{x^2})dx = \frac{1}{2}x^2e^{x^2} - \int xe^{x^2}dx = \frac{x^2}{2}e^{x^2} - \frac{1}{2}e^{x^2} + C.$

25. $I = \int x^7(x^4 + 5)^8 dx = \int x^4[x^3(x^4 + 5)^8]dx.$

$f(x) = x^4$	$g(x) = x^3(x^4 + 5)^8$
$f'(x) = 4x^3$	$G(x) = \frac{(x^4+5)^9}{36}$

$I = \frac{x^4(x^4+5)^9}{36} - \int 4x^3 \frac{(x^4+5)^9}{36}dx = \frac{x^4(x^4+5)^9}{36} - \frac{4}{36}\int (x^4 + 5)^9 x^3 dx = \frac{x^4(x^4+5)^9}{36} - \frac{(x^4+5)^{10}}{360} + C.$

27. Let $f(x)$ be the function whose tangent has slope $x \ln \sqrt{x}$. Then $f'(x) = x \ln \sqrt{x}$ for $x > 0$ and $f(x) = \int x \ln \sqrt{x} dx = \int x \ln x^{1/2} dx = \int x(\frac{1}{2}) \ln x dx = \frac{1}{2}\int x \ln x dx$. To integrate by parts,

$f(x) = \ln x$	$g(x) = x$
$f'(x) = \frac{1}{x}$	$G(x) = \frac{x^2}{2}$

$f(x) = \int x \ln x dx = \frac{1}{2}[(\frac{x^2}{2}) \ln x - \int (\frac{x^2}{2})(\frac{1}{x})dx] = \frac{1}{2}[\frac{x^2 \ln x}{2} - \frac{1}{2}\int x dx] = \frac{1}{2}[\frac{x^2 \ln x}{2} - \frac{x^2}{4}] + C = \frac{x^2 \ln x}{4} - \frac{x^2}{8} + C$. Since $(2, f(2)) = (2, -3)$, that is, when $x = 2$, $f = -3$, $-3 = \frac{2^2 \ln 2}{4} - \frac{2^2}{8} + C$, $-3 = \ln 2 - \frac{1}{2} + C$, or $C = -\frac{5}{2} - \ln 2$. Thus $f(x) = \frac{x^2 \ln x}{4} - \frac{x^2}{8} - \frac{5}{2} - \ln 2.$

29. Let t denote time and $Q(t)$ the number of units produced. Then, $\frac{dQ}{dt} = 100te^{-0.5t}$ and $Q(t) = 100\int te^{-0.5t}dt.$

$f(t) = t$	$g(t) = e^{-0.5t}$
$f'(t) = 1$	$G(t) = -\frac{e^{-0.5t}}{0.5} = -2e^{-0.5t}$

Thus $Q(t) = 100[-2te^{-0.5t} - \int (-2)e^{-0.5t}dt] = -200te^{-0.5t} - 400e^{-0.5t} + C = -200(t + 2)e^{-0.5t} + C$. Since no units are produced when $t = 0$, $Q(0) = 0 = -200(2) + C$ or $C = 400$. Hence, $Q(t) = -200(t+2)e^{-0.5t} + 400$, and the number of units produced during the first three hours is $Q(3) = 176.87$.

31. Let q denote the number of units produced and $C(t)$ the cost of producing the first q units. Then, $\frac{dC}{dq} = 0.5(q + 1)e^{0.1q}$ and $C(q) = 0.5\int (q + 1)e^{0.1q}dq = \frac{1}{2}\int (q + 1)e^{q/10}dq.$

$f(q) = q + 1$	$g(q) = e^{q/10}$
$f'(q) = 1$	$G(q) = 10e^{q/10}$

Thus $C(q) = \frac{1}{2}[10(q + 1)e^{q/10} - 10\int e^{q/10}dq] = \frac{1}{2}[10(q + 1)e^{q/10} - 100e^{q/10}] + C = 5(q+1)e^{q/10} - 50e^{q/10} + C$. When $q = 10$, $C = 200$ and $200 = 5(10+$

$1)e^{10/10}-50e^{10/10}+C = 5e+C$, or $C = 200-5e$. Thus $C(q) = 5(q+1)e^{q/10}-50e^{q/10}+200-5e$. and $C(20) = 5(20+1)e^{20/10}-50e^{20/10}+200-5e = 105e^2-50e^2+200-5e = 65e^2+200-5e =$ \$666.70 is the total cost of producing the first 20 units.

33. a)

$f(x) = x^n$	$g(x) = e^{ax}$
$f'(x) = nx^{n-1}$	$G(q) = \frac{1}{a}e^{ax}$

$\int x^n e^{ax}dx = \frac{1}{a}x^n e^{ax} - \int (nx^{n-1})(\frac{1}{a}e^{ax})dx = \frac{1}{a}x^n e^{ax} - \frac{n}{a}\int x^{n-1}e^{ax}dx$.

b) Let $n = 3$, then $n = 2$, $n = 1$, and $a = 5$. Then, by by the formula from part a), $\int x^3e^{5x}dx = \frac{1}{5}x^3e^{5x} - \frac{3}{5}\int x^2e^{5x}dx = \frac{1}{5}x^3e^{5x} - \frac{3}{5}[\frac{1}{5}x^2e^{5x} - \frac{2}{5}\int xe^{5x}dx] = \frac{1}{5}x^3e^{5x} - \frac{3}{5}[\frac{1}{5}x^2e^{5x} - \frac{2}{5}(\frac{1}{5}xe^{5x} - \frac{1}{5}\int e^{5x}dx) = \frac{1}{5}x^3e^{5x} - \frac{3}{5}[\frac{1}{5}x^2e^{5x} - \frac{2}{5}(\frac{1}{5}xe^{5x} - \frac{1}{25}e^{5x})] + C = \frac{1}{5}x^3e^{5x} - \frac{3}{25}x^2e^{5x} + \frac{6}{25}(\frac{1}{5}xe^{5x} - \frac{1}{25}e^{5x}) + C = \frac{1}{5}x^3e^{5x} - \frac{3}{25}x^2e^{5x} + \frac{6}{125}xe^{5x} - \frac{6}{1225}e^{5x} + C = (\frac{1}{5}x^3 - \frac{3}{25}x^2 + \frac{6}{125}x - \frac{6}{625})e^{5x} + C$.

5.4 The Use of Integral Tables

The following formulas will be referenced below:
(1) $\int \frac{1}{p^2-x^2}dx = \frac{1}{2p}\ln|\frac{p+x}{p-x}|$
(2) $\int \frac{1}{x(ax+b)}dx = \frac{1}{b}\ln|\frac{x}{ax+b}|$
(3) $\int \frac{1}{\sqrt{x^2\pm p^2}}dx = \ln|x+\sqrt{x^2\pm p^2}|$
(4) $\int x^n e^{ax}dx = \frac{1}{a}x^n e^{ax} - \frac{n}{a}\int x^{n-1}e^{ax}dx$.

1. Use (2) with $a = 2$ and $b = -3$ to get $I = \int \frac{1}{x(2x-3)}dx = -\frac{1}{3}\ln|\frac{x}{2x-3}| + C$.

3. Use (3) with $p = 5$ to get $I = \int \frac{1}{\sqrt{x^2+25}}dx = \ln|x+\sqrt{x^2+25}| + C$.

5. Use (1) with $p = 2$ to get $I = \int \frac{1}{4-x^2}dx = \frac{1}{4}\ln|\frac{2+x}{2-x}| + C$.

7. Use (2) with $a = 3$ and $b = 2$ to get $I = \int \frac{1}{(3x^2+2x)}dx = \int \frac{1}{x(3x+2)}dx = \frac{1}{2}\ln|\frac{x}{3x+2}| + C$.

9. Use (4) with $n = 2$, then $n = 1$, and $a = 3$ to get $I = \int x^2e^{3x}dx = \frac{1}{3}x^2e^{3x} - \frac{2}{3}\int xe^{3x}dx = \frac{1}{3}x^2e^{3x} - \frac{2}{3}(\frac{1}{3}xe^{3x} - \frac{1}{3}\int e^{3x}dx) = \frac{1}{3}x^2e^{3x} - \frac{2}{9}xe^{3x} + \frac{2}{27}e^{3x} + C = (\frac{x^2}{3} - \frac{2x}{9} + \frac{2}{27})e^{3x} + C$.

11. Use $\int \frac{x}{ax^2+c}dx = \frac{1}{2a}\ln|ax^2+c|$ with $a = -1$ and $c = 2$ to get $\int \frac{x}{2-x^2}dx = -\frac{1}{2}\ln|2-x^2| + C$.

13. Use $\int(\ln ax)^2dx = x(\ln ax)^2 - 2x\ln ax + 2x$ with $a = 2$ to get $\int(\ln 2x)^2dx = x(\ln 2x)^2 - 2x\ln 2x + 2x + C$.

15. Use $\int \frac{1}{x\sqrt{ax+b}}dx = \frac{1}{\sqrt{b}} \ln |\frac{\sqrt{ax+b}-\sqrt{b}}{\sqrt{ax+b}+\sqrt{b}}|$ with $a = 2$ and $b = 5$ to get $\int \frac{1}{3x\sqrt{2x+5}}dx = \frac{1}{3} \int \frac{1}{\sqrt{2x+5}}dx = \frac{1}{3\sqrt{5}} \ln |\frac{\sqrt{2x+5}-\sqrt{5}}{\sqrt{2x+5}+\sqrt{5}}| + C$

17. Use $\int \frac{1}{b+ce^{ax}}dx = \frac{1}{ab}[ax - \ln |b + ce^{ax}|]$ with $a = -1$, $b = 2$, and $c = -3$ to get $\int \frac{1}{2-3e^{-x}}dx = -\frac{1}{2}[-x - \ln |2 - 3e^{-x}|] + C = \frac{x}{2} + \frac{1}{2} \ln |2 - 3e^{-x}| + C.$

19. Rewrite the two integrals with the constant of integration:
(1) $\int \frac{1}{\sqrt{x^2 \pm p^2}}dx = \ln |\frac{x+\sqrt{x^2 \pm p^2}}{p}| + C_1$ and
(2) $\int \frac{1}{\sqrt{x^2 \pm p^2}}dx = \ln |x + \sqrt{x^2 \pm p^2}| + C_2$

Now rewrite the right hand side of (1) using properties of logarithms to get $\ln |\frac{x+\sqrt{x^2 \pm p^2}}{p}| + C_1 = \ln |x + \sqrt{x^2 \pm p^2}| - \ln |p| + C_1 = \ln |x + \sqrt{x^2 \pm p^2}| + C_2$ where C_2 replaces the constant $-\ln |p| + C_1$. In other words, the two given antiderivatives of $\frac{1}{\sqrt{x^2 \pm p^2}}$ differ by only a constant.

21. $\int (\ln x)^n dx = x(\ln x)^n - n \int (\ln x)^{n-1} dx$, so $\int (\ln x)^3 dx = x(\ln x)^3 - 3 \int (\ln x)^2 dx = x(\ln x)^3 - 3[x(\ln x)^2 - 2 \int \ln x dx] = x(\ln x)^3 - 3x(\ln x)^2 + 6[x \ln x - \int dx] = x(\ln x)^3 - 3x(\ln x)^2 + 6x \ln x - 6x + C.$

Review Problems

1. $\int (x^5 - 3x^2 + \frac{1}{x^2})dx = \int (x^5 - 3x^2 + x^{-2})dx = \frac{x^6}{6} - x^3 - \frac{1}{x} + C.$

2. $\int (x^{2/3} - \frac{1}{x} + 5 + \sqrt{x})dx = \frac{3x^{5/3}}{5} - \ln |x| + 5x + \frac{2x^{3/2}}{3} + C.$

3. Let $u = 3x + 1$. Then $du = 3dx$ or $dx = \frac{1}{3}du$. $\int \sqrt{3x + 1}dx = \frac{1}{3} \int u^{1/2} du = \frac{1}{3}(\frac{2u^{3/2}}{3}) + C = \frac{2(3x+1)^{3/2}}{9} + C.$

4. Let $u = 3x^2 + 2x + 5$. Then $du = (6x + 2)dx$ or $(3x + 1)dx = \frac{1}{2}du$. Hence $I = \int (3x + 1)\sqrt{3x^2 + 2x + 5}dx = \frac{1}{2} \int u^{1/2} du = \frac{1}{2}(\frac{2u^{3/2}}{3}) + C = \frac{(3x^2+2x+5)^{3/2}}{3} + C.$

5. Let $u = x^2 + 4x + 2$. Then $du = (2x + 4)dx$ or $(x + 2)dx = \frac{1}{2}du$. Hence $I = \int (x + 2)(x^2 + 4x + 2)^5 dx = \frac{1}{2} \int u^5 du = \frac{1}{2}(\frac{u^6}{6}) + C = \frac{(x^2+4x+2)^6}{12} + C.$

6. Let $u = x^2 + 4x + 2$. Then $du = (2x + 4)dx$ or $(x + 2)dx = \frac{1}{2}du$. Hence $I = \int \frac{x+2}{x^2+4x+2}dx =$

$\frac{1}{2}\int \frac{1}{u}du = \frac{1}{2}\ln|u| + C = \frac{1}{2}\ln|x^2+4x+2| + C$.

7. Let $u = 2x^2 + 8x + 3$. Then $du = (4x+8)dx$ or $(3x+6)dx = \frac{3}{4}du$. Hence $I = \int \frac{3x+6}{(2x^2+8x+3)^2}dx = \frac{3}{4}\int \frac{1}{u^2}du = \frac{3}{4}\int u^{-2}du = -\frac{3}{4}u^{-1} + C = -\frac{3}{4(2x^2+8x+3)} + C$.

8. Let $u = x-5$. Then $du = dx$. Hence $I = \int (x-5)^{12}dx = \int u^{12}du = \frac{u^{13}}{13} + C = \frac{(x-5)^{13}}{13} + C$.

9. Method 1:
Let $u = x - 5$. Then $du = dx$ and $x = u + 5$. Hence $I = \int x(x-5)^{12}dx = \int (u+5)u^{12}du = \int (u^{13} + 5u^{12})du = \frac{u^{14}}{14} + 5\frac{u^{13}}{13} + C = \frac{(x-5)^{14}}{14} + \frac{5(x-5)^{13}}{13} + C$.
Method 2:

$f(x) = x$	$g(x) = (x-5)^{12}$
$f'(x) = 1$	$G(x) = \frac{(x-5)^{13}}{13}$

Hence $I = \int x(x-5)^{12}dx = \frac{x(x-5)^{13}}{13} - \int \frac{(x-5)^{13}}{13}dx = \frac{x(x-5)^{13}}{13} - \frac{1}{13}[\frac{(x-5)^{14}}{14}] + C = \frac{x(x-5)^{13}}{13} - \frac{(x-5)^{14}}{182} + C$.

10. Let $u = 3x$. Then $du = 3dx$ or $dx = \frac{1}{3}du$. Hence $I = 5\int e^{3x}dx = \frac{5}{3}\int e^u du = \frac{5e^u}{3} + C = \frac{5e^{3x}}{3} + C$.

11. $I = \int 5xe^{3x}dx$.

$f(x) = x$	$g(x) = e^{3x}$
$f'(x) = 1$	$G(x) = \frac{e^{3x}}{3}$

Hence $I = 5(\frac{xe^{3x}}{3} - \int \frac{e^{3x}}{3}dx) = 5(\frac{xe^{3x}}{3} - \frac{e^{3x}}{9}) + C = (\frac{5}{3}x - \frac{5}{9})e^{3x} + C$.

12. $I = \int xe^{-x/2}dx$.

$f(x) = x$	$g(x) = e^{-x/2}$
$f'(x) = 1$	$G(x) = -2e^{-x/2}$

Hence $I = -2xe^{-x/2} - \int(-2e^{-x/2})dx) = -2xe^{-x/2} + 2\int e^{-x/2}dx = -2xe^{-x/2} - 4e^{-x/2} + C = -2(x+2)e^{-x/2} + C$.

13. Rewrite $I = \int x^5 e^{x^3}dx = \int x^3(x^2 e^{x^3})dx$.

$f(x) = x^3$	$g(x) = x^2 e^{x^3}$
$f'(x) = 3x^2$	$G(x) = \frac{e^{x^3}}{3}$

Hence $I = \frac{x^3 e^{x^3}}{3} - \frac{1}{3}\int 3x^2 e^{x^3}dx = \frac{x^3 e^{x^3}}{3} - \int x^2 e^{x^3}dx = \frac{x^3 e^{x^3}}{3} - (\frac{1}{3})e^{x^3} + C = \frac{1}{3}(x^3 - 1)e^{x^3} + C$.

14. $I = \int (2x+1)e^{0.1x}dx$.

$f(x) = 2x+1$	$g(x) = e^{0.1x}$
$f'(x) = 2$	$G(x) = 10e^{0.1x}$

Hence $I = 10e^{0.1x}(2x+1) - 20\int e^{0.1x}dx = 10(2x+1)e^{0.1x} - 200e^{0.1x} + C = 10(2x-19)e^{0.1x} + C$.

15. $I = \int x\ln 3x dx$.

$f(x) = \ln 3x$	$g(x) = x$
$f'(x) = \frac{1}{x}$	$G(x) = \frac{x^2}{2}$

Hence $I = \frac{x^2\ln 3x}{2} - \int(\frac{1}{x})(\frac{x^2}{2})dx = \frac{x^2\ln 3x}{2} - \frac{1}{2}\int x dx = \frac{x^2\ln 3x}{2} - \frac{x^2}{4} + C$.

16. $I = \int \ln 3x dx$.

$f(x) = \ln 3x$	$g(x) = 1$
$f'(x) = \frac{1}{x}$	$G(x) = x$

Hence $I = x\ln 3x - \int x(\frac{1}{x})dx = x\ln 3x -$

$\int dx = x\ln 3x - x + C$.

17. Let $u = \ln 3x$. Then $du = \frac{1}{x}dx$ and so $I = \int \frac{\ln 3x}{x}dx = \int u\,du = \frac{u^2}{2} + C = \frac{(\ln 3x)^2}{2} + C$.

18. $I = \int \frac{\ln 3x}{x^2}dx$.

$f(x) = \ln 3x$	$g(x) = \frac{1}{x^2}$
$f'(x) = \frac{1}{x}$	$G(x) = -\frac{1}{x}$

Hence $I = -\frac{\ln 3x}{x} - \int(-\frac{1}{x})(\frac{1}{x})dx = -\frac{\ln 3x}{x} + \int \frac{1}{x^2}dx = -\frac{\ln 3x}{x} - \frac{1}{x} + C = -\frac{1}{x}(\ln 3x + 1) + C$.

19. Rewrite $I = \int x^3(x^2+1)^8 dx = \int x^2[x(x^2+1)^8]dx$.

$f(x) = x^2$	$g(x) = x(x^2+1)^8$
$f'(x) = 2x$	$G(x) = \frac{(x^2+1)^9}{18}$

Hence $I = \frac{x^2(x^2+1)^9}{18} - \int 2x[\frac{(x^2+1)^9}{18}]dx = \frac{x^2(x^2+1)^9}{18} - \frac{1}{9}\int x(x^2+1)^9 dx = \frac{x^2(x^2+1)^9}{18} - \frac{(x^2+1)^{10}}{180} + C$.

20. Let $u = x^2+1$. Then $du = 2x\,dx$ and so $I = \int 2x\ln(x^2+1)dx = \int \ln u\,du = \int(1)(\ln u)du$.

$f(u) = \ln u$	$g(u) = 1$
$f'(u) = \frac{1}{u}$	$G(u) = u$

Hence $I = u\ln u - \int u(\frac{1}{u})du = u\ln u - \int du = u\ln u - u + C = u(\ln u - 1) + C = (x^2+1)[\ln(x^2+1) - 1] + C$.

21. The slope of the tangent is the derivative. Hence, $f'(x) = x(x^2+1)^3$ and so f is an antiderivative of $x(x^2+1)^3$. That is, $f(x) = \int x(x^2+1)^3 dx = \frac{(x^2+1)^4}{8} + C$. Since the graph of f passes through the point $(1,5)$, $5 = f(1) = \frac{2^4}{8} + C = 2 + C$ or $C = 3$. Hence, $f(x) = \frac{(x^2+1)^4}{8} + 3$.

22. Let $Q(x)$ denote the number of commuters using the new subway line x weeks from now. It is given that $\frac{dQ}{dx} = 18x^2 + 500$ commuters per week. Hence, $Q(x)$ is an antiderivative of $18x^2 + 500$. That is, $Q(x) = \int(18x^2+500)dx = 6x^3 + 500x + C$. Since 8,000 commuters currently use the subway, $8{,}000 = Q(0) = C$. Hence, $Q(x) = 6x^3 + 500x + 8{,}000$, and the number of commuters who will be using the subway in 5 years is $Q(5) = 11{,}250$.

23. Let $Q(x)$ denote the number of inmates in county prisons x years from now. It is given that $\frac{dQ}{dx} = 280e^{0.2x}$ inmates per year. Hence, $Q(x)$ is an antiderivative of $280e^{0.2x}$. That is, $Q(x) = \int(280e^{0.2x})dx = \frac{280}{0.2}e^{0.2x} + C = 1{,}400e^{0.2x} + C$. Since the prisons currently house 2,000 inmates, $2{,}000 = Q(0) = 1{,}400 + C$ or $C = 600$. Hence, $Q(x) = 1{,}400e^{0.2x} + 600$, and the number of inmates 10 years from now will be $Q(10) = 1{,}400e^2 + 600 = 10{,}945$.

24. Let $P(q)$ denote the profit, $R(q)$ the revenue, and $C(q)$ the cost when the level of production is q units. Since the marginal revenue is $R'(q) = 200q^{-1/2}$, $C'(q) = 0.4q$, and profit is revenue minus cost, $\frac{dP}{dq} = \frac{dR}{dq} - \frac{dC}{dq} = 200q^{-1/2} - 0.4q$ dollars per unit. The profit function $P(q)$ is an antiderivative of the marginal profit. That is, $P(q) = \int(200q^{-1/2} - 0.4q)dq = 400q^{1/2} - 0.2q^2 + C$. Since profit is \$2,000 when the level of production is 25 units, $2{,}000 = P(25) = 400(5) - 0.2(25)^2 + C$ or $C = 125$. Hence, $P(q) = 400q^{1/2} - 0.2q^2 + 125$, and the profit when 36 units are produced is $P(36) = \$2{,}265.80$.

25. The rate of change of price is $P'(x) = 0.2 + 0.003x^2$. $P(x) = 0.2x + 0.001x^2 + C = 0.2x + 0.001x^2 + 160$ where $C = 160$ cents ($1.60) since "now" means $x = 0$. $P(10) = 2 + 0.1 + 160 = 162.1$ or $1.62.

26. Let $V(t)$ denote the value of the machine t years from now. Since $\frac{dV}{dt} = 220(t-10)$ dollars per year, the function $V(t)$ is an antiderivative of $220(t-10)$. Thus, $V(t) = \int 220(t-10)dt = 110t^2 - 2,200t + C$. Since the machine was originally worth $12,000, it follows that $V(0) = 12,000 = C$. Thus, the value of the machine after t years will be $V(t) = 110t^2 - 2,200t + 12,000$ and the value after 10 years will be $V(10) = \$1,000$.

27. $f(x) = 0.5 + \frac{1}{(x+1)^2}$ meters per year. The growth per year is $G(x) = \int f(x)dx = 0.5x - \frac{1}{x+1} + G_0$. During the second year the tree will grow $G(2) - G(1) = 0.5\times 2 - \frac{1}{2+1} - 0.5\times 1 + \frac{1}{2} = \frac{2}{3}$ meter.

28. Let $N(t) =$ denote the number of bushels t days from now. The number of bushels will be an antiderivative of $\frac{dN}{dt} = 0.3t^2 + 0.6t + 1$. So $N(t) = 0.1t^3 + 0.3t^2 + t + C$ and the revenue is $R(t) = 3N(t)$. No revenue is generated initially, so $R(0) = 0 = C$. In 5 days, the revenue will be $N(5) = 3[0.1(5^3) + 0.3(5^2) + 5] = 3(12.5 + 7.5 + 5) = \$75$.

29. Use $\int \frac{1}{p^2 - x^2}dx = \frac{1}{2p}\ln\left|\frac{p+x}{p-x}\right|$ with $p = 2$ to get $I = \int \frac{5}{8-2x^2}dx = \frac{5}{2}\int \frac{1}{4-x^2}dx = \frac{5}{2}\int \frac{1}{2^2-x^2}dx = (\frac{5}{2})(\frac{1}{4})\ln\left|\frac{2+x}{2-x}\right| + C = \frac{5}{8}\ln\left|\frac{2+x}{2-x}\right| + C$.

30. Use $\int \frac{1}{\sqrt{x^2 \pm p^2}}dx = \ln|x + \sqrt{x^2 \pm p^2}|$ with $p = \frac{4}{3}$ to get $\int \frac{2}{\sqrt{9x^2+16}}dx = \frac{2}{3}\int \frac{1}{\sqrt{x^2+(16/9)}}dx = \frac{2}{3}\ln|x + \sqrt{x^2 + (\frac{4}{3})^2}| + C$.

31. Use $\int x^n e^{ax}dx = \frac{1}{a}x^n e^{ax} - \frac{n}{a}\int x^{n-1}e^{ax}dx$ with $n = 2$, then $n = 1$, and $a = -\frac{1}{2}$ to get $\int x^2 e^{-x/2}dx = -2x^2e^{-x/2} + 4\int xe^{-x/2}dx = -2x^2e^{-x/2} + 4(-2xe^{-x/2} + 2\int e^{-x/2}dx) = -2x^2e^{-x/2} + 4(-2xe^{-x/2} - 4e^{-x/2}) + C = -2x^2e^{-x/2} - 8xe^{-x/2} - 16e^{-x/2} + C = (-2x^2 - 8x - 16)e^{-x/2} + C$

32. Use $\int \frac{1}{x(ax+b)}dx = \frac{1}{b}\ln\left|\frac{x}{ax+b}\right|$ with $a = 5$ and $b = 9$ to get $I = \int \frac{4}{x(5x+9)}dx = 4\int \frac{1}{x(5x+9)}dx = \frac{4}{9}\ln\left|\frac{x}{5x+9}\right| + C$.

Chapter 6

Further Topics in Integration

6.1 The Definite Integral

1. $\int_0^1(x^4-3x^3+1)dx = (\frac{x^5}{5}-\frac{3x^4}{4}+x)|_0^1 = (\frac{1}{5}-\frac{3}{4}+1)-0 = \frac{9}{20}$.

3. $\int_2^5(2+2t+3t^2)dt = (2t+t^2+t^3)|_2^5 = (10+25+125)-(4+4+8) = 144$.

5. $\int_1^3(1+\frac{1}{x}+\frac{1}{x^2})dx = (x+\ln|x|-\frac{1}{x})|_1^3 = (3+\ln 3-\frac{1}{3})-(1+\ln 1-1) = \frac{8}{3}+\ln 3$.

7. $\int_{-3}^{-1}\frac{t+1}{t^3}dt = \int_{-3}^{-1}(t^{-2}+t^{-3})dt = (-\frac{1}{t}-\frac{1}{2t^2})|_{-3}^{-1} = (1-\frac{1}{2})-(\frac{1}{3}-\frac{1}{18}) = \frac{2}{9}$.

9. Let $u = 2x-4$. Then $du = 2dx$ or $dx = \frac{1}{2}du$. When $x = 1$, $u = -2$, and when $x = 2$, $u = 0$. Hence, $\int_1^2(2x-4)^5dx = \frac{1}{2}\int_{-2}^0 u^5du = \frac{u^6}{12}|_{-2}^0 = -\frac{64}{12} = -\frac{16}{3}$.

11. Let $u = 6t+1$. Then $du = 6dt$ or $dt = \frac{1}{6}du$. When $t = 0$, $u = 1$, and when $t = 4$, $u = 25$. Hence, $\int_0^4\frac{1}{\sqrt{6t+1}}dt = \frac{1}{6}\int_1^{25}u^{-1/2}du = \frac{u^{1/2}}{3}|_1^{25} = \frac{5}{3}-\frac{1}{3} = \frac{4}{3}$.

13. Let $u = t^4+2t^2+1$. Then $du = (4t^3+4t)dt$ or $(t^3+t)dt = \frac{1}{4}du$. When $t = 0$, $u = 1$, and when $t = 1$, $u = 4$. Hence, $\int_0^3(t^3+t)\sqrt{t^4+2t^2+1}dt = \frac{1}{4}\int_1^4 u^{1/2}du = \frac{u^{3/2}}{6}|_1^4 = \frac{1}{6}(4^{3/2}-1^{3/2}) = \frac{1}{6}(8-1) = \frac{7}{6}$.

15. Let $u = x-1$. Then $du = dx$ and $x = u+1$. When $x = 2$, $u = 1$, and when $x = e+1$, $u = e$. Hence, $\int_2^{e+1}\frac{x}{x-1}dx = \int_1^e\frac{u+1}{u}du = \int_1^e(1+\frac{1}{u})du = (u+\ln|u|)|_1^e = (e+\ln e)-(1+\ln 1) = e$.

17. Let $g(t) = 1$ and $f(t) = \ln t$. Then $G(t) = t$ and $f'(t) = \frac{1}{t}$. So $\int_1^{e^2}\ln t\,dt = (t\ln t)|_1^{e^2} - \int_1^{e^2}(\frac{1}{t})(t)dt = (t\ln t)|_1^{e^2} - \int_1^{e^2}dt = (t\ln t - t)|_1^{e^2} = (e^2\ln e^2 - e^2)-(\ln 1 - 1) = e^2+1$

19. Let $g(x) = e^{-x}$ and $f(x) = x$. Then $G(x) = -e^{-x}$ and $f'(x) = 1$. So $\int_{-2}^{2} xe^{-x}dx = (-xe^{-x})|_{-2}^{2} - \int_{-2}^{2}(-e^{-x})dx = (-xe^{-x} - e^{-x})|_{-2}^{2} = (-2e^{-2} - e^{-2}) - (2e^{2} - e^{2}) = -3e^{-2} - e^{2}$

21. Let $u = \ln x$. Then $du = \frac{1}{x}dx$. When $x = 1$, $u = 0$, and when $x = e^2$, $u = 2$. Hence $\int_1^2 \frac{(\ln x)^2}{x}dx = \int_0^2 u^2 du = \frac{u^3}{3}|_0^2 = \frac{8}{3}$.

23. Let $g(t) = e^{-0.1t}$ and $f(t) = 20 + t$. Then $G(t) = -10e^{-0.1t}$ and $f'(t) = 1$. So $\int_0^{10}(20+t)e^{-0.1t}dt = -10(20+t)e^{-0.1t}|_0^{10} + 10\int_0^{10} e^{-0.1t}dt = [-10(20+t)e^{-0.1t} - 100e^{-0.1t}]|_0^{10} = (-300 - 10t)e^{-0.1t}|_0^{10} = (-300 - 100)e^{-1} + (300 + 0)e^{0} = -400e^{-1} + 300 = 152.85$.

25. Let $P(x)$ denote the population of the town x months from now. Then $\frac{dP}{dx} = 5 + 3x^{2/3}$, and the amount by which the population will increase during the next 8 months is $P(8) - P(0) = \int_0^8(5 + 3x^{2/3})dx = (5x + \frac{9x^{5/3}}{5})|_0^8 = [5(8) + \frac{9(8^{5/3})}{5}] - 0 = 97.6 \approx 98$ people.

27. Let $V(x)$ denote the value of the machine after x years. Then $\frac{dV}{dx} = 220(x - 10)$, and the amount by which the machine depreciates during the 2nd year is $V(2) - V(1) = \int_1^2 220(x - 10)dx = 220(\frac{x^2}{2} - 10x)|_1^2 = 220[(2 - 20) - (\frac{1}{2} - 10)] = -\$1,870$ where the negative sign indicates that the value of the machine has decreased.

29. Let $C(q)$ denote the total cost of producing q units. Then the marginal cost is $\frac{dC}{dq} = 6(q - 5)^2$, and the increase in cost is $C(13) - C(10) = \int_{10}^{13} 6(q - 5)^2 dq = 2(q - 5)^3|_{10}^{13} = 2(8^3 - 5^3) = \$774$.

31. Let $N(t)$ denote the number of bushels that are produced over the next t days. Then $\frac{dN}{dt} = 0.3t^2 + 0.6t + 1$, and the increase in the crop over the next five days is $N(5) - N(0) = \int_0^5(0.3t^2 + 0.6t + 1)dt = (0.1t^3 + 0.3t + t)|_0^5 = [0.1(125) + 0.3(25) + 5] - 0 = 25$ bushels. If the price remains fixed at \$3 per bushel, the corresponding increase in the value of the crop is \$75.

33. Let $D(t)$ denote the demand for the product. Since the current demand is 5,000 and the demand increases exponentionally, $D(t) = 5,000e^{0.02t}$ units per year. Let $R(t)$ denote the total revenue t years from now. Then the rate of change of revenue is $\frac{dR}{dt} = \frac{\text{dollars}}{\text{year}} = \frac{\text{dollars}}{\text{unit}}\frac{\text{units}}{\text{year}} = 400D(t) = 400(5,000e^{0.02t}) = 2,000,000e^{0.02t}$.The increase in revenue over the next 2 years is $R(2) - R(0) = \int_0^2 2,000,000e^{0.02t}dt = 100,000,000e^{0.02t}|_0^2 = 100,000,000(e^{0.04} - 1) = \$4,081,077$.

35. a) If $F(x)$ is an antiderivative of $f(x)$, then $\int_a^b f(x)dx + \int_b^c f(x)dx = [F(b) - F(a)] + [F(c) - F(b)] = F(c) - F(a) = \int_a^c f(x)dx$.
b) Since $|x| = -x$ when $x < 0$ and $|x| = x$ when $0 \leq x$, write the integral as $\int_{-1}^{1} |x|dx = \int_{-1}^{0}(-x)dx + \int_0^1 xdx = (-\frac{x^2}{2})|_{-1}^{0} + (\frac{x^2}{2})|_0^1 = -(0 - \frac{1}{2}) + (\frac{1}{2} - 0) = 1$.
c) Since $|x - 3| = -(x - 3)$ when $x < 3$ and $|x - 3| = x - 3$ when $3 \leq x$, write the integral as $\int_0^4(1 + |x - 3|)^2dx = \int_0^3[1 - (x - 3)]^2dx + \int_3^4[1 + (x - 3)]^2dx = \int_0^3(4 - x)^2dx + \int_3^4(x - 2)^2dx =$

$-\frac{(4-x)^3}{3}|_0^3 + \frac{(x-2)^3}{3}|_3^4 = -(\frac{1}{3})[1^3 - 4^3] + (\frac{1}{3})[2^3 - 1^3] = \frac{70}{3}.$

6.2 Area and Integration

1. The element of area has a height y and a base dx. Thus $dA = ydx = (4-3x)dx$. Summing up these infinitesimal elements leads to the integral $A = \int_0^{4/3}(4-3x)dx = (4x - \frac{3x^2}{2})|_0^{4/3} = \frac{4(4)}{3} - \frac{3(4^2)}{2(3^2)} = \frac{8}{3}$.

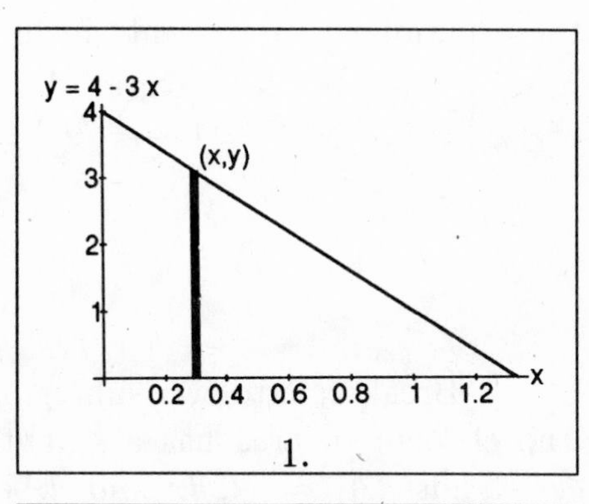

1.

3. The element of area has a height y and a base dx. Thus $dA = ydx = 5dx$. Summing up these infinitesimal elements in $-2 < x < 1$ leads to the integral $A = \int_{-2}^{1}(5)dx = 5x|_{-2}^{1} = 5-(-10) = 15$.

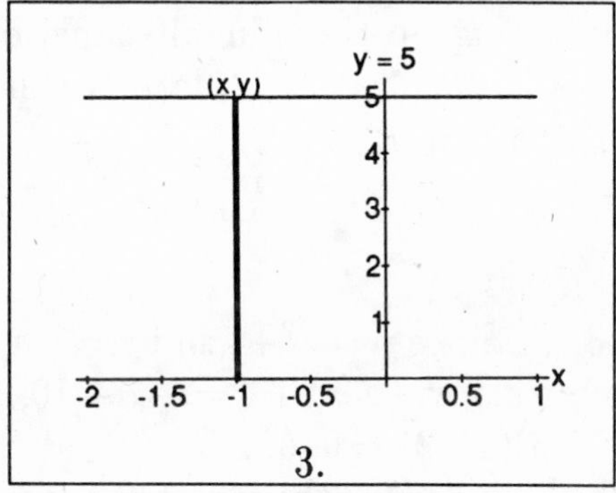

3.

5. The element of area has a height y and a base dx. Thus $dA = ydx = \sqrt{x}dx$. Summing up these infinitesimal elements in $4 < x < 9$ leads to the integral $A = \int_4^9 \sqrt{x}dx = \int_4^9 x^{1/2}dx = \frac{2x^{3/2}}{3}|_4^9 = \frac{2}{3}(27-8) = \frac{38}{3}$.

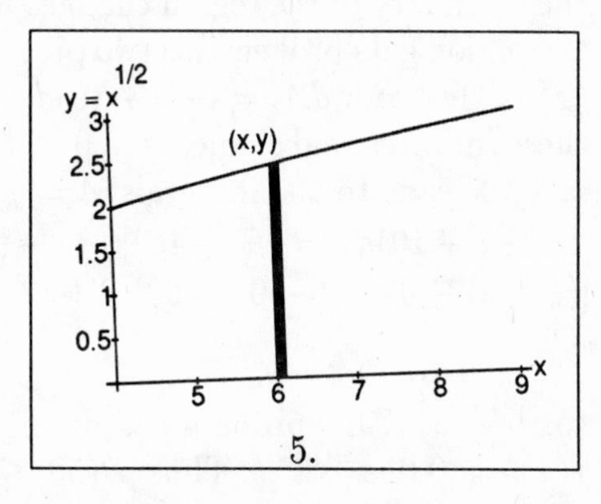

5.

7. The element of area has a height y and a base dx. Thus $dA = ydx = (1 - x^2)dx$. Summing up these infinitesimal elements in $-1 < x < 1$ leads to the integral $A = \int_{-1}^{1}(1 - x^2)dx = (x - \frac{x^3}{3})|_{-1}^{1} = (1 - \frac{1}{3}) - (-1 - \frac{-1}{3}) = \frac{4}{3}$.

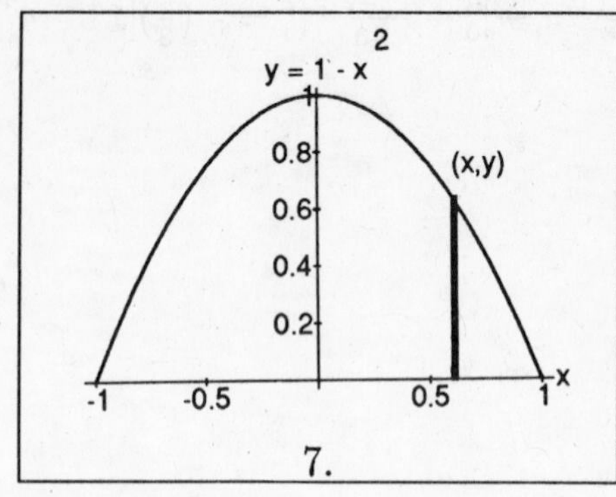

7.

9. The element of area has a height y and a base dx. Thus $dA = ydx = e^x dx$. Summing up these infinitesimal elements in $\ln(\frac{1}{2}) = -\ln 2 < x < 0$ leads to the integral $A = \int_{-\ln 2}^{0} e^x dx = e^x|_{-\ln 2}^{0} = 1 - e^{-\ln 2} = 1 - e^{\ln(1/2)} = 1 - \frac{1}{2} = \frac{1}{2}$.

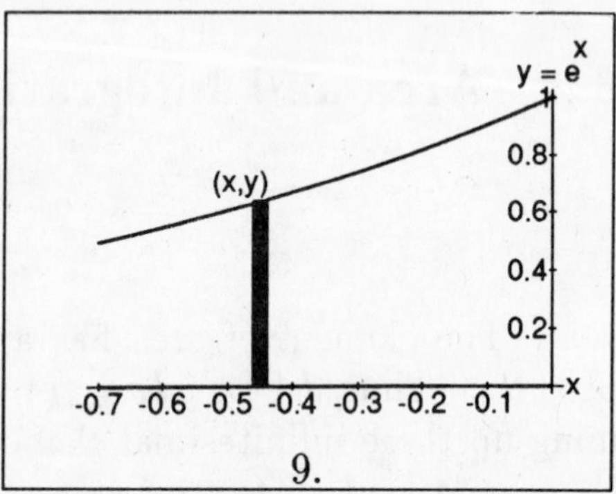

9.

11. Break R into two subregions R_1 and R_2. The element of area has a height y and a base dx. Thus $dA_1 = \sqrt{x}dx$ and $dA_2 = (2 - x)dx$. Summing up these infinitesimal elements in $0 < x_1 < 1$ and $1 < x < 2$ leads to the integrals $A = \int_0^1 x^{1/2}dx + \int_1^2 (2 - x)dx = \frac{2x^{3/2}}{3}|_0^1 + (2x - \frac{x^2}{2})|_1^2 = \frac{2}{3} + [(4 - 2) - (2 - \frac{1}{2})] = \frac{7}{6}$.

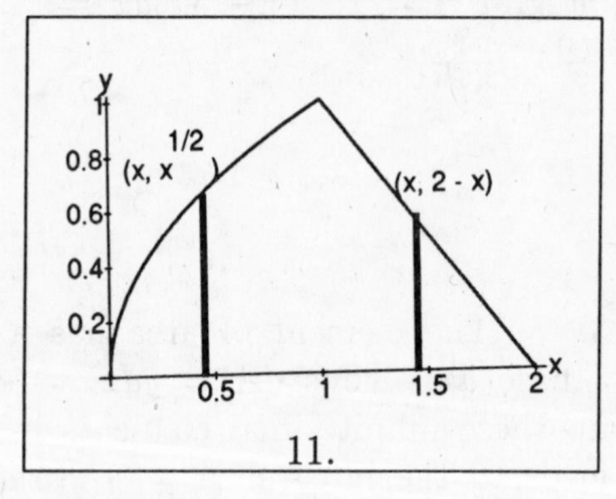

11.

13. Since $y = x^2 + 4$ and $y = -x + 10$, the curves intersect when $x^2 + 4 = -x + 10$, $x^2 + x - 6 = 0$, $(x+3)(x-2) = 0$, or $x = -3$ and $x = 2$. Note that the boundary of the region changes when $x = 2$, so the region gets broken into two pieces to get $dA_1 = (x^2 + 4)dx$ and $dA_2 = (-x + 10)dx$. Summing up these infinitesimal elements in $0 < x_1 < 2$ and $2 < x < 10$ leads to the integrals $A = \int_0^2 (x^2 + 4)dx + \int_2^{10}(-x + 10)dx = (\frac{x^3}{3} + 4x)|_0^2 + (-\frac{x^2}{2} + 10x)|_2^{10} = [(\frac{8}{3} + 8) - 0] + [(-50 + 100) - (-2 + 20)] = \frac{128}{3}$.

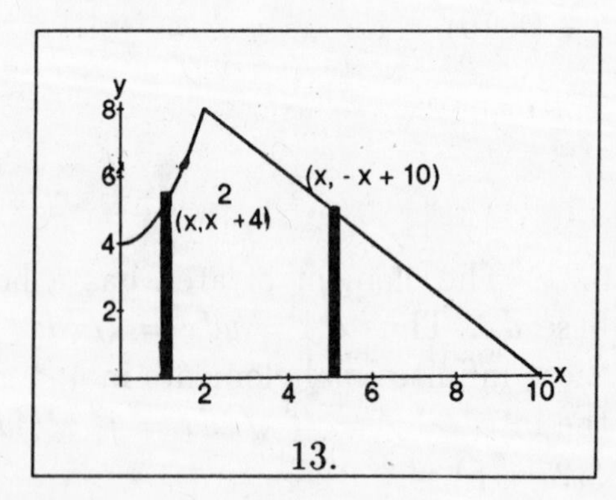

13.

15. a) The probability density function is $f(x) = 0.01e^{-0.01x}$. Thus $P(50 \leq x \leq 60) = \int_{50}^{60} 0.01e^{-0.01x}dx = -e^{-0.01x}|_{50}^{60} = -e^{-0.6} + e^{-0.5} = 0.0577$.

b) $P(0 \le x \le 60) = \int_0^{60} 0.01e^{-0.01x}dx = -e^{-0.01x}|_0^{60} = -e^{-0.6} + 1 = 0.4512.$
c) Using the previous result, $P(60 \le x) = 1 - P(0 \le x \le 60) = 1 - 0.4512 = 0.5488.$

17. The area under the graph of $f(x) = 0.5e^{-0.5x}$ for $0 < x < \infty$ is 1 or 100%.
a) For calls between 2 and 3 minutes, $A = \int_2^3 0.5e^{-0.5x}dx = -e^{-0.5x}|_2^3 = e^{-1} - e^{-1.5} = 0.1447$ or 14.47%.
b) For calls of at most 2 minutes, $A = \int_0^2 0.5e^{-0.5x}dx = -e^{-0.5x}|_0^2 = 1 - e^{-1} = 0.6321$ or 63.21%.

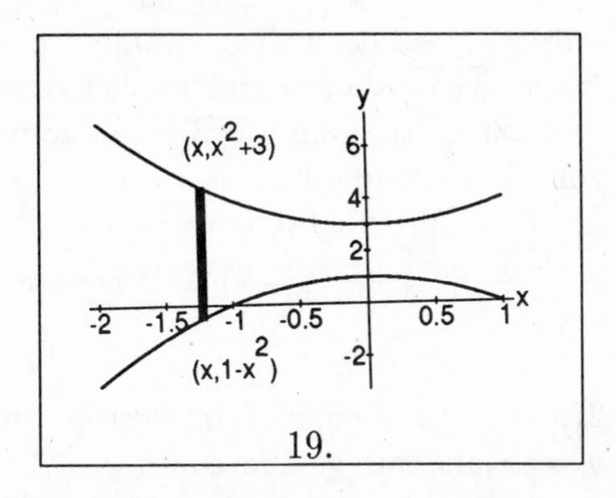

19.

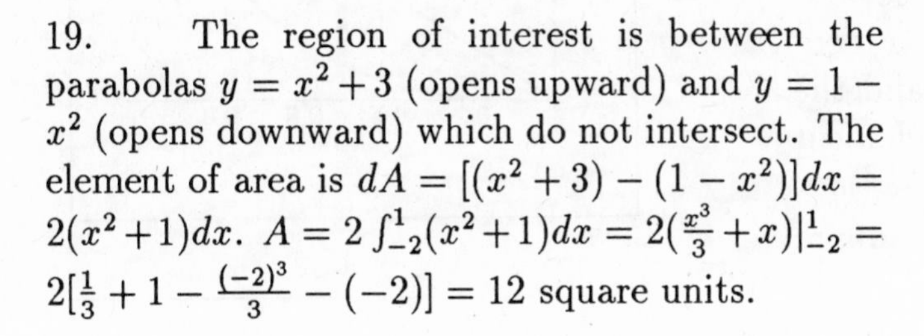
c) For calls of at least 2 minutes, $A = \int_2^\infty 0.5e^{-0.5x}dx = -e^{-0.5x}|_2^\infty = e^{-1} - 0 = 0.3679$ or 36.79%.
Note that 63.21+36.79=1.0, which should not be surprising because the area for $0 \le x < 2$ and that for $2 \le x$ constitutes the entire area under the curve. $f(x) = 0$ when $x < 0$.

19. The region of interest is between the parabolas $y = x^2 + 3$ (opens upward) and $y = 1 - x^2$ (opens downward) which do not intersect. The element of area is $dA = [(x^2+3) - (1-x^2)]dx = 2(x^2+1)dx$. $A = 2\int_{-2}^1 (x^2+1)dx = 2(\frac{x^3}{3} + x)|_{-2}^1 = 2[\frac{1}{3} + 1 - \frac{(-2)^3}{3} - (-2)] = 12$ square units.

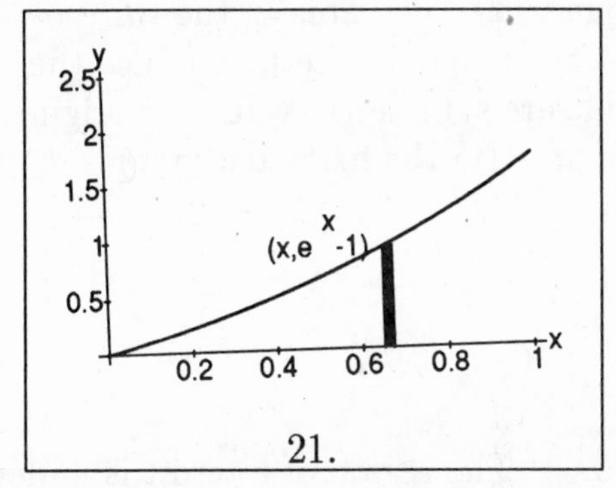

21.

21. The region of interest is under the curve $y = e^x$, above the line $y = 1$ and to the left of the line $x = 1$. The element of area is $dA = (e^x - 1)dx$. $A = \int_0^1 (e^x - 1)dx = (e^x - x)|_0^1 = e - 1 - 1 = e - 2 = 0.72$ square units.

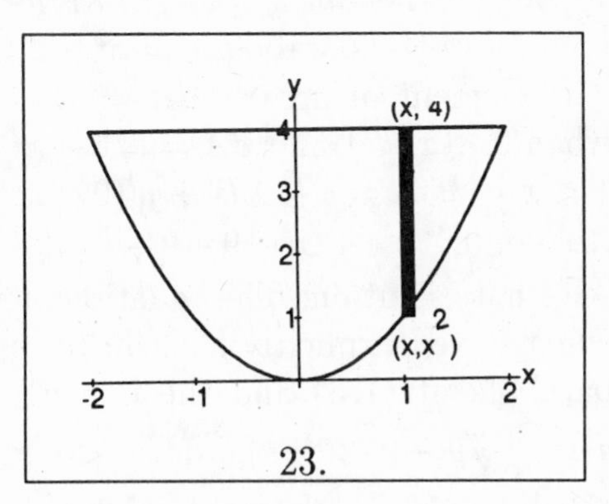

23.

23. The region of interest is under the line $y = 4$ and above the parabola $y = x^2$. The element of area is $dA = (4 - x^2)dx$. $A = 2\int_0^2 (4 - x^2)dx = 2(4x - \frac{x^3}{3})|_0^2 = \frac{32}{3}$ square units. Note that the region is symmetric WRT the y-axis.

25. a) The region of interest is under the parabola $y = 4 - x^2$ and above the line $y = 3$ for $0 < x < 1$. The element of area is $dA = (4 - x^2 - 3)dx = (1 - x^2)dx$. $A = \int_0^1 (1 - x^2)dx = (x - \frac{x^3}{3})|_0^1 = \frac{2}{3}$ square units.

b) The region of interest is under the line $y = 3$ for $0 < x < 1$ and under the parabola $y = 4 - x^2$ for $1 < x < 2$. The element of area is $dA = 3dx$ when $0 < x < 1$ and $dA = (4 - x^2)dx$ when $1 < x < 2$. $A = \int_0^1 3dx + \int_1^2 (4 - x^2)dx = 3x|_0^1 + (4x - \frac{x^3}{3})|_1^2 = 3 + 8 - \frac{8}{3} - 4 + \frac{1}{3} = \frac{14}{3}$ square units.
Alternate solution: The equation of the parabola $x = \sqrt{4 - y}$ when solved explicitly for x in terms of y. Since the line $y = 3$ is horizontal, the (horizontal) element of area is $dA = (\sqrt{4 - y} - 0)dy$ for $0 < y < 3$. Thus $A = \int_0^3 \sqrt{4 - y}dy = -\frac{(4-y)^{3/2}}{3/2}|_0^3 = -\frac{2}{3}(1 - 8) = \frac{14}{3}$ square units.

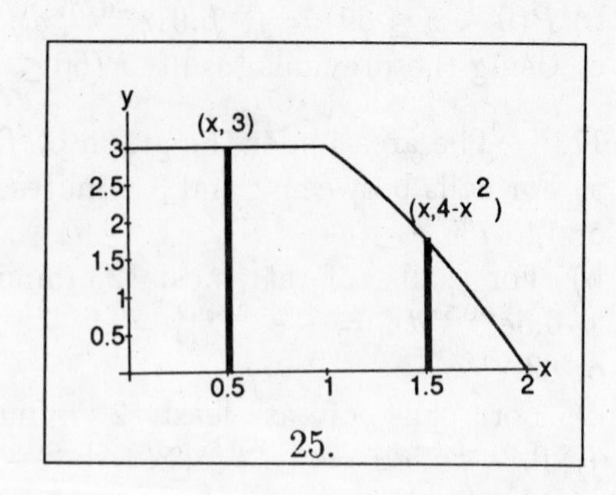

25.

27. The region of interest is under the line $y = x$ and above the curve $y = x^3$ in the first quadrant. The area in the third quadrant is identical to that in the first (since the curve and the line are symmetric WRT the origin.) The element of area (in the first quadrant) is $dA = (x - x^3)dx$. $A = 2\int_0^1 (x - x^3)dx = 2(\frac{x^2}{2} - \frac{x^4}{4})|_0^1 = \frac{1}{2}$ square units.

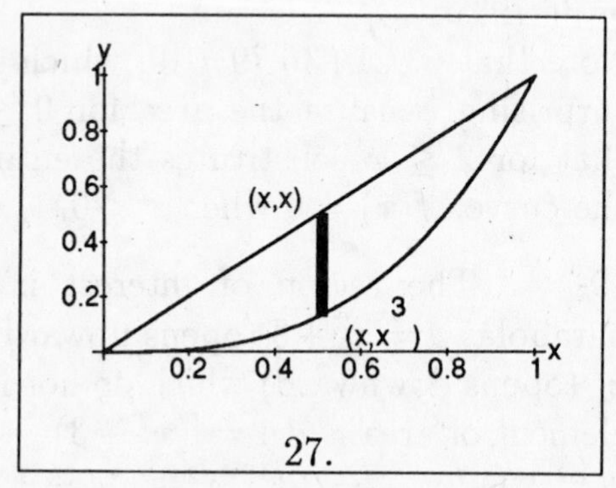

27.

29. The region of interest is under the line $y = 11$, above the line $y = 11 - 8x$ for $0 < x < 1$, and above the parabola $y = x^2 + 2$ for $1 < x < 3$. The element of area is $dA = (11 - 11 + 8x)dx$ when $0 < x < 1$ and $dA = (11 - x^2 - 2)dx$ when $1 < x < 3$. $A = 8\int_0^1 xdx + \int_1^3 (9 - x^2)dx = 4x^2|_0^1 + (9x - \frac{x^3}{3})|_1^3 = 4 + 27 - 9 - 9 + \frac{1}{3} = \frac{40}{3}$ square units.

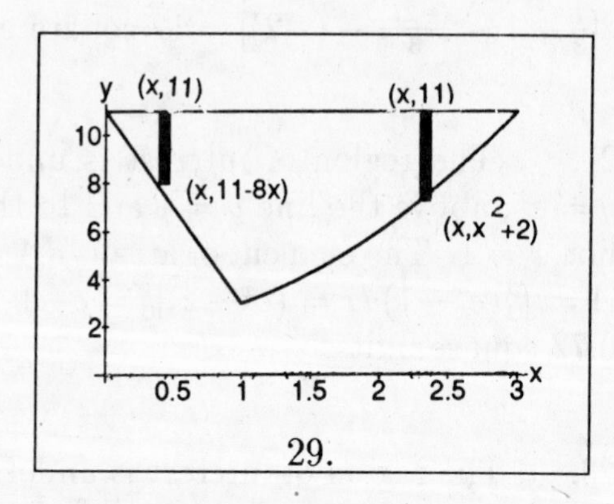

29.

Alternate solution: The equation of the parabola is $x = \sqrt{y - 2}$ and that of the line $x = \frac{11-y}{8}$ when solved explicitly for x in terms of y. Since the oblique line $y = 11 - 8x$ intersects the parabola at $(1, 3)$ and the line $y = 11$ is horizontal, the (horizontal) element of area is $dA = (\sqrt{y - 2} - \frac{11-y}{8})dy$ for $3 < y < 11$. Thus $A = \int_3^{11} (\sqrt{y - 2} - \frac{11-y}{8})dy = [\frac{(y-2)^{3/2}}{3/2} - \frac{11y}{8} + \frac{y^2}{16}]|_3^{11} = 18 - \frac{2}{3} - \frac{121}{8} + \frac{33}{8} + \frac{121}{16} - \frac{9}{16} = \frac{40}{3}$ square units.

31. Since $y = x^2 - 3x + 1$ and $y = -x^2 + 2x + 2$, the curves intersect when $x^2 - 3x + 1 = -x^2 + 2x + 2$, $2x^2 - 5x - 1 = 0$, $x = \frac{5\pm\sqrt{25+8}}{4}$, or $x_1 = 2.6861$, $y_1 = 0.1568$, and $x_2 = -0.1861$, $y_2 = 1.5932$. $dA = [(-x^2+2x+2)-(x^2-3x+1)]dx = (-2x^2+5x+1)dx$. Summing up these infinitesimal elements in $-0.1861 < x < 2.6861$ leads to the integral $A = \int_{-0.1861}^{2.6861}(-2x^2 + 5x + 1)dx = (\frac{-2x^3}{3} + \frac{5x^2}{2} + x)|_{-0.1861}^{2.6861} = \frac{-2(2.6861)^3}{3} + \frac{5(2.6861)^2}{2} + 2.6861 - [(\frac{-2(-0.1861)^3}{3} + \frac{5(-0.1861)^2}{2} - 0.1861)] = -12.9204 + 18.0378 + 2.6861 - 0.000004 - 0.0866 + 0.1861 = 7.99]$.

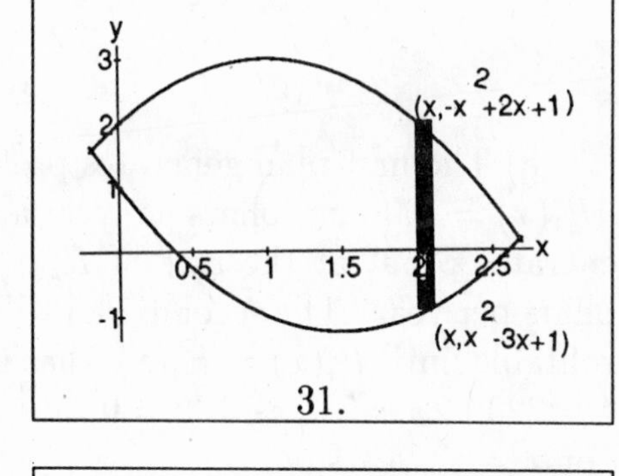

31.

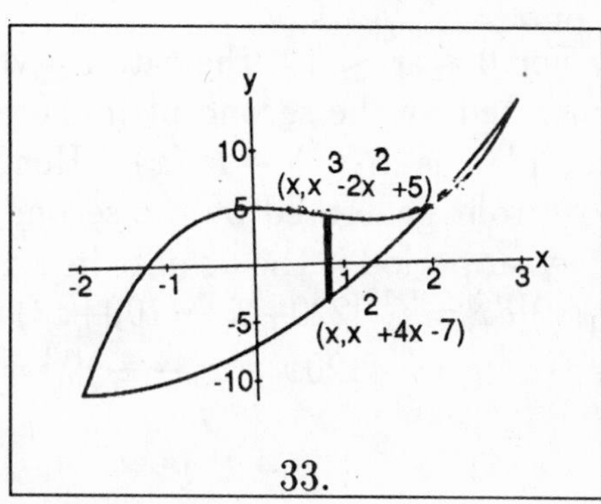

33.

33. Since $y = x^3 - 2x^2 + 5$ and $y = x^2 + 4x - 7$, the curves intersect when $x^3 - 2x^2 + 5 = x^2 + 4x - 7$, $f(x) = x^3 - 3x^2 - 4x + 12 = 0$. Candidates for rational roots are:
$x = \pm1, \pm2, \pm3, \pm4, \pm6, \pm12$. Since $f(2) = 0$, $x = 2$ is a root. Division of $x^3 - 3x^2 - 4x + 12$ by $x - 2$ yields $x^2 - x - 6$, which has roots $x = -2$ and $x = 3$.
Note that the boundary of the region changes when $x = 2$, so the region gets broken into two pieces to get $dA_1 = [(x^3 - 2x^2 + 5) - (x^2 + 4x - 7)]dx = (x^3 - 3x^2 - 4x + 12)dx$ and $dA_2 = (-x^3 + 3x^2 + 4x - 12)dx$.
Summing up these infinitesimal elements in $-2 < x_2 < 2$ and $2 < x < 3$ leads to the integrals $A = \int_{-2}^{2}(x^3 - 3x^2 - 4x + 12)dx + \int_{2}^{3}(-x^3 + 3x^2 + 4x - 12)dx = (\frac{x^4}{4} - x^3 - 2x^2 + 12x)|_{-2}^{2} + (-\frac{x^4}{4} + x^3 + 2x^2 - 12x)|_{2}^{3} = \frac{2^4}{4} - 2^3 - 2(2)^2 + 12(2) - [\frac{(-2)^4}{4} - (-2)^3 - 2(-2)^2 + 12(-2)] + [-\frac{3^4}{4} + 3^3 + 2(3)^2 - 12(3)] - [(-\frac{2^4}{4} + 2^3 + 2(2)^2 - 12(2))] = 53 - \frac{81}{4} = 53 - 20 - \frac{1}{4} = \frac{131}{4}$.

6.3 Applications to Business and Economics

1. a) The first plan generates profit at the rate of $R_1(x) = 100 + x^2$ dollars per year and the second generates profit at the rate of $R_2(x) = 220 + 2x$ dollars per year. The second plan will be the more profitable until $R_1(x) = R_2(x)$, that is, until $100 + x^2 = 220 + 2x$, $x^2 - 2x - 120 = 0$, $(x-12)(x+10) = 0$, or $x = 12$ years.
b) For $0 \le x \le 12$, the rate at which the profit generated by the second plan exceeds that of the first plan is $R_2(x) - R_1(x)$. Hence the net excess profit generated by the second plan over the 12-year period is the definite integral $\int_0^{12}[R_2(x) - R_1(x)]dx = \int_0^{12}[220 + 2x - (100 + x^2)dx = \int_0^{12}(120 + 2x - x^2)dx = (120x + x^2 - \frac{x^3}{3})|_0^{12} = \$1,008.00$.

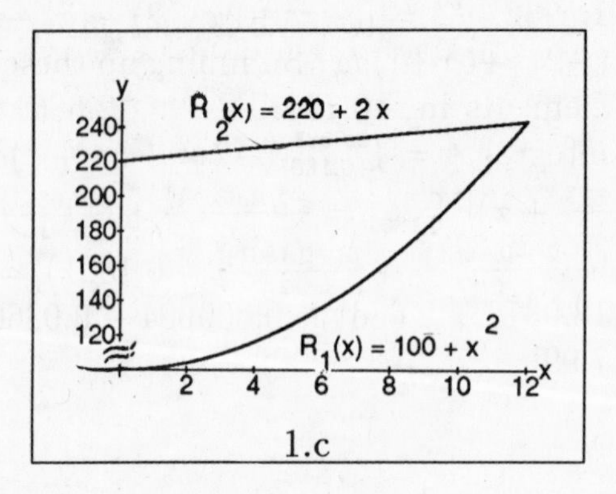

1.c

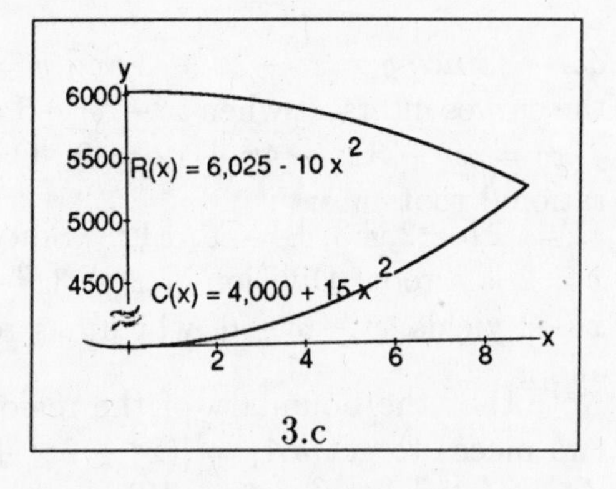

3.c

c) In geometric terms, the net excess profit generated by the second plan is the area of the region between the curves $y = R_2(x)$ and $y = R_1(x)$ from $x = 0$ to $x = 12$.

3. a) The machine generates revenue at the rate of $R(x) = 6,025 - 10x^2$ dollars per year and results in costs that accumulate at the rate of $C(x) = 4,000 + 15x^2$ dollars per year. The use of the machine will be profitable as long as the rate at which revenue is generated is greater than the rate at which costs accumulate, that is, until $R(x) = C(x)$, $6,025 - 10x^2 = 4,000 + 15x^2$, $25x^2 = 2,025$, $x^2 = 81$, or $x = 9$ years.
b) The difference $R(x) - C(x)$ represents the rate of change of the net earnings generated by the machine. Hence, the net earnings over the next 9 years is the definite integral $\int_0^9[R(x) - C(x)]dx = \int_0^9[(6,025 - 10x^2) - (4,000 + 15x^2)]dx = \int_0^9(2,025 - 25x^2)dx = (2,025x - \frac{25x^3}{3})|_0^9 = \$12,150$.
c) In geometric terms, the net earnings in part b) is the area of the region between the curves $y = R(x)$ and $y = C(x)$ from $x = 0$ to $x = 9$.

5. a) $Q_1(t) = 60 - 2(t-1)^2$, $Q_2(t) = 50 - 5t$. a) The excess production is $I = \int_0^4[60 - 2(t-1)^2 - 50 + 5t]dt = [10t - \frac{2(t-1)^3}{3} + \frac{5t^2}{2}]|_0^4 = 40 - 18 + 40 - 0 - \frac{2}{3} = \frac{184}{3}$.
b) The excess production is the difference between the areas under the production curves.

7. a) The contributions are $R(t) = 6,537e^{-0.3t}$ dollars per week t weeks from now. The expenses accumulate at $E(t) = 593$ dollars per week. The drive is profitable as long as rate of revenue exceeds weekly expenses. Thus $e^{-0.3t} = \frac{593}{6,537} = 0.090714$, $-0.3t = \ln 0.090714$, $t = 8$ weeks.
b) The net earnings during the first 8 weeks are $N = \int_0^8 (6,537e^{-0.3t} - 593)dt = (-\frac{6,537}{0.3}e^{-0.3t} - 593t)|_0^8 = \frac{6,537}{0.3}(1 - 0.09072) - (593)(8) = 19,813.26 - 4,744 = 15,069.26$.

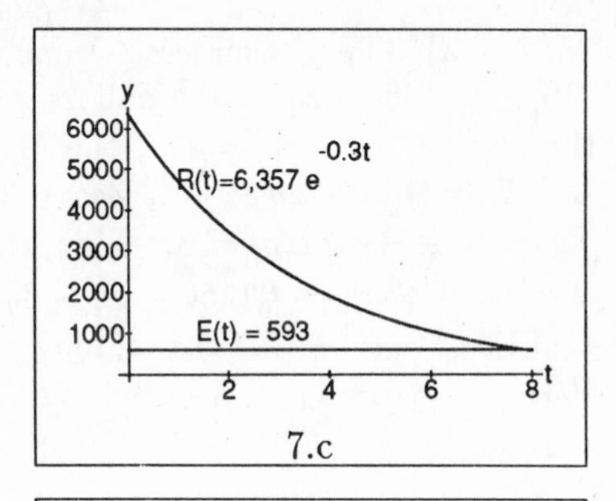

7.c

c)In geometric terms, the net earnings in part b) is the area of the region between the curves $y = R(t)$ and $y = E(t)$ from $t = 0$ to $t = 8$.

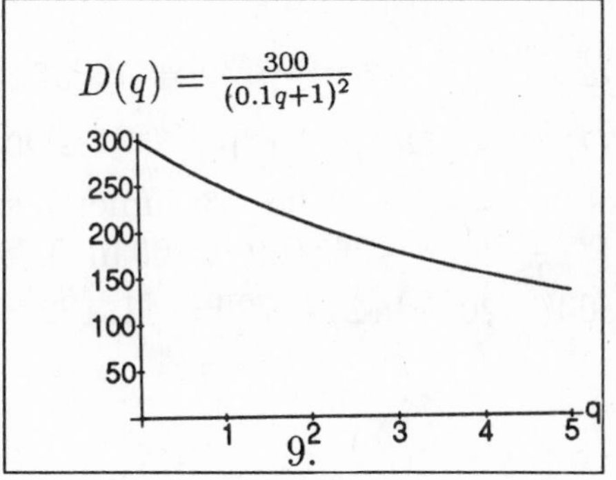

9.

9. a) If the consumers' demand function is $D(q) = \frac{300}{(0.1q+1)^2}$ dollars per unit, the total amount that consumers are willing to spend to get 5 units is the definite integral $\int_0^5 D(q)dq = 300\int_0^5 (0.1q + 1)^{-2}dq = -3,000(0.1q + 1)^{-1}|_0^5 = -3,000(\frac{1}{1.5} - 1) = \$1,000$.

b) The total willingness to spend in part a) is the area of the region under the demand curve from $q = 0$ to $q = 5$.

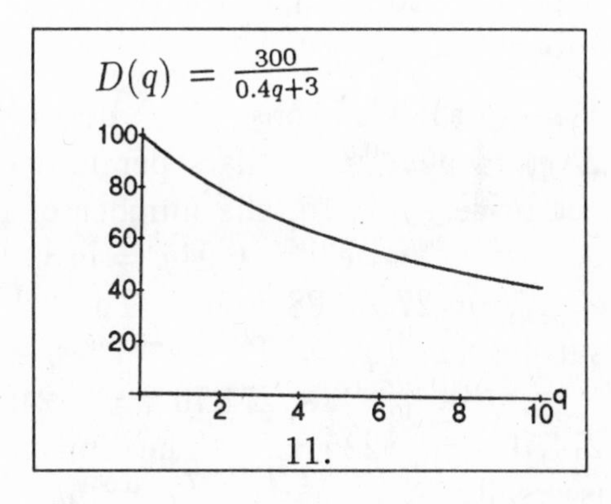

11.

11. a) If the consumers' demand function is $D(q) = \frac{300}{0.4q+3}$ dollars per unit, the total amount that consumers are willing to spend to get 10 units is the definite integral $\int_0^{10} D(q)dq = 300\int_0^{10}(0.4q+3)^{-1}dq = \frac{300}{0.4}\ln|0.4q + 3||_0^{10} = 750(1.9459 - 1.0986) = 635.47$.

b) The total willingness to spend in part a) is the area of the region under the demand curve from $q = 0$ to $q = 10$.

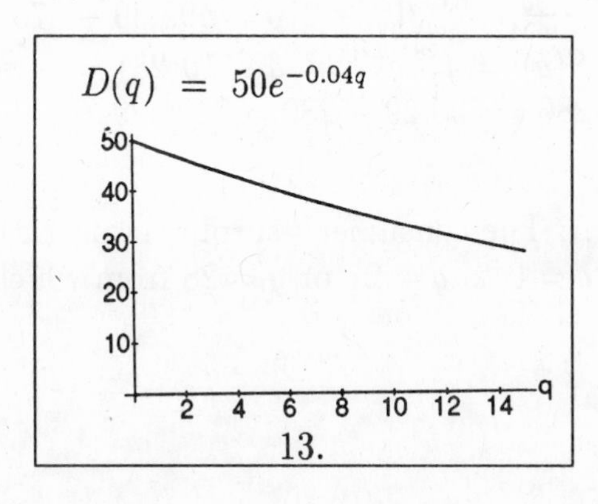

13.

13. a) If the consumers' demand function is $D(q) = 50e^{-0.04q}$ dollars per unit, the total amount that consumers are willing to spend to get 15 units is the definite integral $\int_0^{15} D(q)dq = 50\int_0^{15} e^{-0.04q}dq = -\frac{50}{0.04}e^{-0.04q}|_0^{15} = 1,250(1 - 0.5488) = \$563.99$.
b) The total willingness to spend in part a) is the area of the region under the demand curve from $q = 0$ to $q = 15$.

15. a) The consumers' demand function is $D(q) = 150 - 2q - 3q^2$ dollars per unit. For the market price $p_0 = 117$, the number of units is $117 = 150 - 2q - 3q^2$, $3q^2 + 2q - 33 = 0$, $q = \frac{-1 \pm \sqrt{1+99}}{3}$, or $q = 3$. Thus the consumer's surplus is $S(q) = \int_0^3 (150 - 2q - 3q^2 - 117)dq = 150(3) - 3^2 - 3^3 - 117(3) = 3(150 - 3 - 9 - 117) =$ \$63.

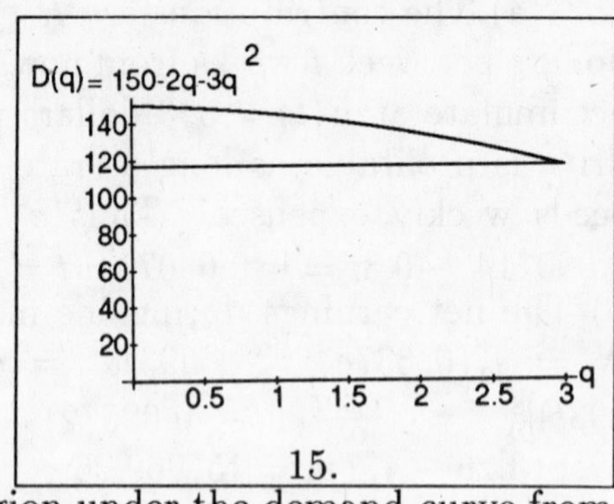

15.

b) The consumer's surplus in part a) is the area of the region under the demand curve from $q = 0$ to $q = 3$ from which the actual spending is subtracted.

17. a) $D(q) = p_0$ if $\frac{400}{0.5q+2} = 20$, $20 = 0.5q + 2$ or $q = 36$. The consumer's surplus is $I = \int_0^{36} \frac{400}{0.5q+2} dq - 36 \times 20 = 800 \ln |0.5q + 2| |_0^{36} - 720 = 800(\ln 20 - \ln 2) - 720 = \$1,122.07$

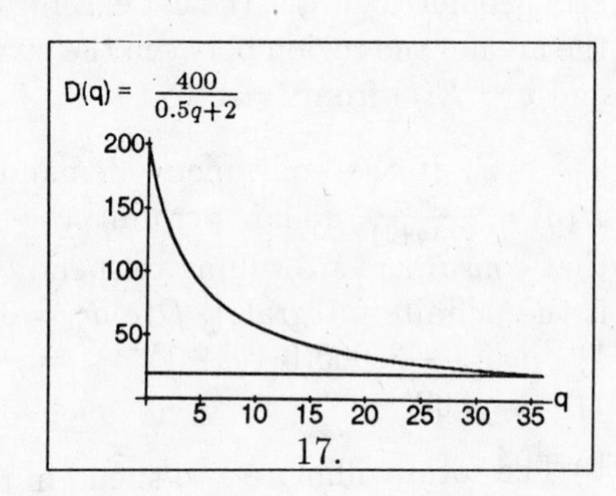

17.

b) The consumer's surplus in part a) is the area of the region under the demand curve from $q = 0$ to $q = 36$ from which the actual spending is subtracted.

19. a) The consumers' demand function is $D(q) = 30e^{-0.04q}$ dollars per unit. For the market price $p_0 = 10$, the number of units is $10 = 30e^{-0.04q}$, $3 = e^{0.04q}$, $0.04q = \ln 3$, or $q = \frac{\ln 3}{0.04} = 27.47$, or 27 or 28 units. Thus the consumer's surplus is $S(q) = \int_0^{27.47} e^{-0.04q} dq - 10(27.47) = -\frac{30}{0.04} e^{-0.04q} |_0^{27.47} - 274.70 = -249.95 + 750 - 274.70 = \$225.35$. The consumer's surplus is really $S(q) = \int_0^{27} e^{-0.04q} dq - 10(27) = -\frac{30}{0.04} e^{-0.04q} |_0^{27} - 270 = 495.30 - 270 = \$225.30$ or $S(q) = \int_0^{28} e^{-0.04q} dq - 10(28) = -\frac{30}{0.04} e^{-0.04q} |_0^{28} - 280 = 505.29 - 280 = \$225.29$

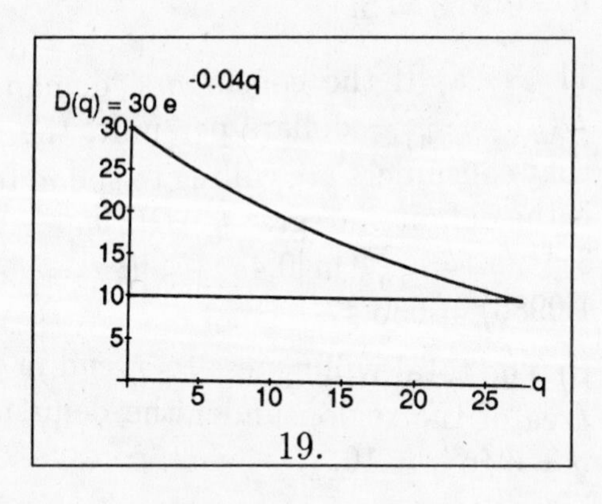

19.

b) The consumer's surplus in part a) is the area of the region under the demand curve from $q = 0$ to $q = 27$ or $q = 28$ from which the actual spending is subtracted.

21. The dollar price per unit is $p = 124-2q$ and the cost function is $C(q) = 2q^3-59q^2+4q+7,600$. The revenue function is $R(q) = (124-2q)q-(2q^3-59q^2+4q+7,600) = 124q-2q^2-2q^3+59q^2-4q-7,600 = -2q^3+57q^2+120q-7,600$. $R'(q) = -6q^2+114q+120 = -(6q^2-114q-120) = 0$ if $q = \frac{57\pm\sqrt{57^2+6(120)}}{6} = 20$ (since $q > 0$). $R''(q) = -12q+114$ and $R(20) < 0$, so $q = 20$ is a maximum.

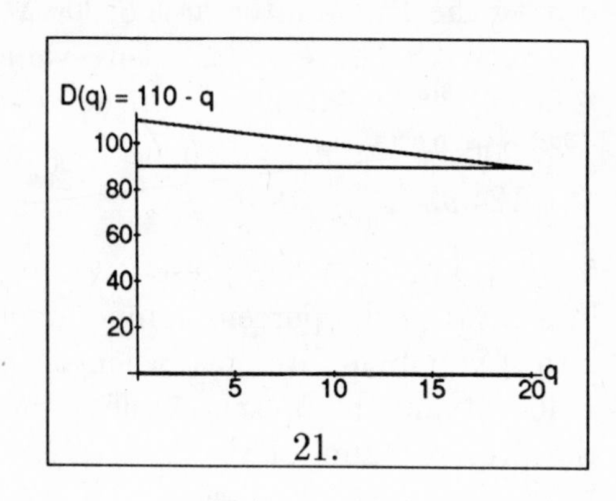

21.

b) The corresponding consumer's surplus is $S(q) = \int_0^{20}(110-q)dq - 20(110-20) = (110q - \frac{q^2}{2})|_0^{20} - (20)(90) = 110(20) - \frac{20^2}{2} - 1,800 = 2,200-200-1,800 = \$200$.

23. The supply function for a certain commodity is $S(p) = \frac{q+1}{3}$ and the demand function is $D(q) = \frac{16}{q+2} - 3$. The supply equals the demand if $\frac{16}{q+2} - 3 = \frac{q+1}{3}$, $16(3) - 3(3)(q+2) = (q+1)(q+2)$, $48-9q-18 = q^2+3q+2$, $q^2+12q-28 = 0$, or $q = -6 \pm \sqrt{36+28} = 2$ (since $q > 0$.)

b) The corresponding consumer's surplus is $S(q) = \int_0^2(\frac{16}{q+2} - 3)dq - 2(1) = (16\ln|q+2| - 3q)|_0^2 - 2 = 16\ln 4 - 6 - 16\ln 2 - 2 = \$3.09$.

6.4 The Definite Integral as the Limit of a Sum

1. Recall that P dollars invested at an annual interest rate of 6 % compounded continuously will be worth $Pe^{0.06t}$ dollars t years later. To approximate the future value of the income stream, divide the 5-year time interval $0 \le t \le 5$ into n equal subintervals of length Δt years and let t_j denote the beginning of the jth subinterval. Then, the money deposited during the jth subinterval is $2,400\Delta t$. This money will remain in the account approximately $5-t_j$ years hence. The future value of the money deposited during the jth subinterval is $2,400e^{0.06(5-t_j)}\Delta t$. The future value of the income stream is $\lim_{n\to\infty}\sum_{j=1}^n 2,400e^{0.06(5-t_j)}\Delta t = \int_0^5 2,400e^{0.06(5-t)}dt = 2,400e^{0.3}\int_0^5 e^{-0.06t}dt = \frac{2,400}{-0.06}e^{0.3}e^{-0.06t}|_0^5 = \frac{2,400}{0.06}(e^{0.3}-1) = \$13,994.35$.

3. Recall that P dollars invested at an annual interest rate of 8 % compounded continuously will be worth $Pe^{0.08t}$ dollars t years later. To approximate the future value of the income stream, divide the 3-year time interval $0 \le t \le 3$ into n equal subintervals of length Δt years and let t_j denote the beginning of the jth subinterval. Then, the money deposited

during the jth subinterval is $8,000\Delta t$. This money will remain in the account approximately $3-t_j$ years hence. The future value of the money deposited during the jth subinterval is $8,000e^{0.08(3-t_j)}\Delta t$. The future value of the income stream is $\lim_{n\to\infty}\sum_{j=1}^{n} 8,000e^{0.08(3-t_j)}\Delta t = \int_0^3 8,000e^{0.08(3-t)}dt = 8,000e^{0.24}\int_0^3 e^{-0.08t}dt = \frac{8,000}{-0.08}e^{0.24}e^{-0.08t}|_0^3 = \frac{8,000}{0.08}e^{0.24}(-e^{-0.24}+1) =$ $27,124.92 .

5. Recall that the present value of B dollars payable t years from now with an annual interest rate of 6 percent compounded continuously is $Be^{-0.06t}$. Divide the interval $0\le t\le 5$ into n equal subintervals of length Δt years. Then, the income from the j_{th} subinterval is $2,400\Delta t$ and the present value of income from the j_{th} subinterval is $2,400e^{-0.06t_j}\Delta t$. Hence, the present value of the investment is $\lim_{n\to\infty}\sum_{j=1}^{n} 2,400e^{-0.06t_j}\Delta t = 2,400\int_0^5 e^{-0.06t}dt = \frac{2,400e^{-0.06t}}{-0.06}|_0^5 = -\frac{2,400}{0.06}(e^{-0.3}-1) = \$10,367.27$.

7. Recall that the present value of B dollars payable t years from now with an annual interest rate of 12 percent compounded continuously is $Be^{-0.12t}$. Divide the interval $0\le t\le 5$ into n equal subintervals of length Δt years. Then, the income from the j_{th} subinterval is $1,200\Delta t$ and the present value of income from the j_{th} subinterval is $1,200e^{-0.12t_j}\Delta t$. Hence, the present value of the investment is $\lim_{n\to\infty}\sum_{j=1}^{n} 1,200e^{-0.12t_j}\Delta t = 1,200\int_0^5 e^{-0.12t}dt = \frac{1,200e^{-0.12t}}{-0.12}|_0^5 = -10,000(e^{-0.6}-1) = \$4,511.88$.

9. $P(t) = \int_0^{10}(10,000+500t)e^{-0.1t}dt.$

$f(t) = 10,000+500t$	$g(t) = e^{-0.1t}$
$f'(t) = 500$	$G(t) = -10e^{-0.1t}$

$P(t) = -10e^{-0.1t}(10,000+500t)|_0^{10} + 5,000\int_0^{10} e^{-0.1t}dt =$
$-10[15,000e^{-1} - 10,000] - 50,000e^{-0.1t}|_0^{10} = 44,818.08 - 50,000(e^{-1}-1) = \$76,424.11.$

11. $f(t) = e^{-0.2t}$. Of the 200 present members, $200f(8)$ will still be in the club in 8 months. Of the 10 new members picked up t months from now, $10f(8-t)$ will still be members in 8 months. Thus $P(t) = 200e^{-1.6} + \int_0^8 10e^{-0.2(8-t)}dt = 200e^{-1.6} + 10e^{-1.6}\int_0^8 e^{0.2t}dt = 40.379 + 50e^{-1.6}e^{0.2t}|_0^8 = 80.28$. The club will consist of approximately 80 members in 8 months.

13. Since $f(t) = e^{-t/10}$ is the fraction of members active after t months, and since there were 8,000 charter members, the number of charter members still active at the end of 10 months is $8,000f(10) = 8,000e^{-1}$. Now, divide the interval $0\le t\le 10$ into n equal subintervals of length Δt months and let t_j denote the beginning of the jth subinterval. During this j^{th} subinterval, $200\Delta t$ new members join, and at the end of the 10 months ($10-t_j$ months later), the number of these retaining membership is $200f(10-t_j) = 200e^{-(10-t_j)/10}\Delta t$. Hence the number of new members still active 10 months from now is approximately $\lim_{n\to\infty}\sum_{j=1}^{n} 200e^{-(10-t_j)/10}\Delta t = \int_0^{10} 200e^{-(10-t)/10}dt$. Hence, the total number N of active members 10 months from now is $N = 8,000e^{-1} + \int_0^{10} 200e^{-(10-t)/10}dt =$

$8,000e^{-1}+200e^{-1}\int_0^{10}e^{t/10}dt = 8,000e^{-1}+200e^{-1}(10e^{t/10})|_0^{10} = 8,000e^{-1}+2,000e^{-1}(e-1) = 8,000e^{-1}+2,000-2,000e^{-1} = 4,207$.

15. Divide the interval $0 \le r \le 3$ into n equal subintervals of length Δr, and let r_j denote the beginning of the jth subinterval. This divides the circular disc of radius 3 into n concentric circles, as shown in the figure. If Δr is small, the area of the jth ring is $2\pi r_j\Delta r$ where $2\pi r_j$ is the circumference of the circle of radius r_j that forms the inner boundary of the ring and Δr is the width of the ring. Then, since $D(r) = 5,000e^{-0.1r}$ is the population density (people per square mile) r miles from the center, it follows that the number of people in the jth ring is $D(r_j)$(area of the jth ring)$= 5,000e^{-0.1r_j}(2\pi r_j\Delta r)$.

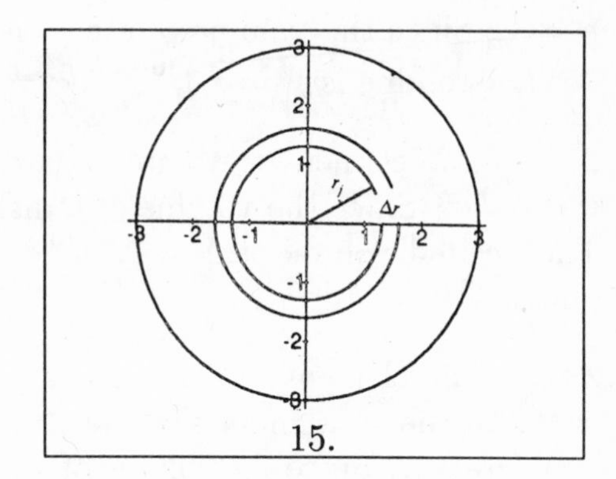

15.

Hence, if N is the total number of people within 3 miles of the center of the city, $N = \lim_{n\to\infty}\sum_{j=1}^{n}5,000e^{-0.1r_j}(2\pi r_j\Delta r) = \int_0^3 5,000(2\pi)re^{-0.1r}dr = 10,000\pi\int_0^3 re^{-0.1r}dr$. Applying integration by parts, $\int_0^3 re^{-0.1r}dr = -10re^{-0.1r}|_0^3 - \int_0^3(-10)e^{-0.1r}dr = (-10re^{-0.1r} - 100e^{-0.1r})|_0^3 = (-30e^{-0.3}-100e^{-0.3})-(-100) = -130e^{-0.3}+100 = 3.6936$. Hence, the total number of people within 10 miles of the center of the city is $N = 10,000\pi\int_0^3 re^{-0.1r}dr = 10,000\pi(3.6936) = 116,039$. (Your answer may differ slightly due to round-off errors.)

17. Let $S(r) = k(R^2-r^2)$ denote the speed of the blood in centimeters per second at a distance r from the central axis of the artery of (fixed) radius R. The area of a small circular ring at a distance r_j is (approximately) $2\pi r_j\Delta r$ square centimeters, so the amount of blood passing through the ring is $V(r) = 2\pi r_j\Delta r[k(R^2-r_j^2)] = 2\pi k(R^2r_j - r_j^3)\Delta r$ cubic centimeters per second. Hence, the total quantity of blood flowing through the artery per second is $\lim_{n\to\infty}\sum_{j=1}^{n}2\pi k(R^2r_j-r_j^3)\Delta r) = 2\pi k\int_0^R(rR^2-r^3)dr = 2\pi k(\frac{R^2r^2}{2}-\frac{r^4}{4})|_0^R = \frac{\pi kR^4}{2}$. The area of the artery is πR^2 and the average velocity of the blood through the artery is $V_{ave} = \frac{\pi kR^4/2}{\pi R^2} = \frac{kR^2}{2}$. The maximum speed for the blood occurs at $r=0$, so $S(0) = kR^2$. Thus $V_{ave} = \frac{1}{2}S(0)$.

19. The element of volume is a disk (cylinder) of height (thickness) $h = dx$ and radius $r = \sqrt{x}$. Since the volume of a disk is $V_d = \pi r^2 h$, the element of volume $dV = \pi(\sqrt{x})^2dx$. The desired volume is $V = \pi\int_1^4 x\,dx = \frac{\pi x^2}{2}|_1^4 = \frac{15\pi}{2}$ cubic units.

21. The element of volume is a disk (cylinder) of height (thickness) $h = dx$ and radius $r = 4-x^2$. Since the volume of a disk is $V_d = \pi r^2 h$, the element of volume $dV = \pi(4-x^2)^2dx$. The desired volume is $V = 2\pi\int_0^2(16-8x^2+x^4)dx = 2\pi(16x-\frac{8x^3}{3}+\frac{x^5}{5})|_0^2 = \frac{512\pi}{15}$ cubic units.

Note symmetry WRT the y-axis.

23. The element of volume is a disk (cylinder) of height (thickness) $h = dx$ and radius $r = \frac{1}{x}$. Since the volume of a disk is $V_d = \pi r^2 h$, the element of volume $dV = \pi \frac{1}{x^2} dx$. The desired volume is $V = \pi \int_1^{10} x^{-2} dx = -\frac{\pi}{x}\Big|_1^{10} = 0.9\pi$ cubic units.

25. The element of volume is a disk (cylinder) of height (thickness) $h = dx$ and radius $r = e^{-0.1x}$. Since the volume of a disk is $V_d = \pi r^2 h$, the element of volume $dV = \pi e^{-0.2x} dx$. The desired volume is $V = \pi \int_0^{10} e^{-0.2x} dx = -\frac{\pi}{0.2} e^{-0.2x}\Big|_0^{10} = -\frac{\pi}{0.2}(e^{-2} - 1) = 13.58$ cubic units.

27. The element of volume is a disk (cylinder) of height (thickness) $h = dx$ and radius $= \frac{rx}{h}$. (If the area under the line $y = \frac{rx}{h}$ is rotated about the x-axis with $0 < x < h$, the cone is generated.) Since the volume of a disk is $V_d = \pi r^2 h$, the element of volume $dV = \pi (\frac{rx}{h})^2 dx$. The desired volume is $V = \frac{\pi r^2}{h^2} \int_0^h x^2 dx = \frac{\pi r^2}{h^2} \frac{x^3}{3}\Big|_0^h = \frac{\pi r^2 h}{3}$ cubic units.

29. $f(x) = 2x - x^2$. $f_{av} = \frac{1}{2-0} \int_0^2 (2x - x^2) dx = \frac{1}{2}(x^2 - \frac{x^3}{3})\Big|_0^2 = \frac{2}{3}$.

31. $f(x) = \frac{1}{x}$. $f_{av} = \frac{1}{2-1} \int_1^2 x^{-1} dx = \ln|x|\Big|_1^2 = \ln 2$.

33. $P(t) = 0.09t^2 - 0.2t + 1.6$. $P_{av} = \frac{1}{3-0} \int_0^3 (0.09t^2 - 0.2t + 1.6) dt = \frac{1}{3}(0.03t^3 - 0.1t^2 + 1.6t)\Big|_0^3 = \frac{1}{3}(0.81 - 0.9 + 4.8) = 1.57$ dollars per pound.

35. $Q(t) = 2{,}000e^{0.05t}$. $Q_{av} = \frac{1}{5} \int_0^5 2{,}000 e^{0.05t} dt = \frac{400}{0.05} e^{0.05t}\Big|_0^5 = 8{,}000(e^{0.25} - 1) = 2{,}272$ bacteria.

37. The equation of the line indicating the number of kilograms left at time t is $y = 60{,}000(-t + 1)$. $y_{av} = 60{,}000 \int_0^1 (-t + 1) dt = 60{,}000(-\frac{t^2}{2} + t)\Big|_0^1 = 30{,}000$ kilograms.

39. The area of a ring of thickness dr is $2\pi r dr$. The area of a circle of radius R is $A = 2\pi \int_0^R r dr = \pi R^2$ square units.

41. Let $f(t)$ denote the fraction of the membership of the group that will remain active for at least t years, P_0 the initial membership, and $r(t)$ the rate per year at which additional members are added to the group. Then, the size of the group N years from now is the number of initial members still active plus the number of new members still active. Of the P_0 initial members, $f(N)$ is the fraction remaining active for N years. Hence, the number of initial members still active after N years is $P_0 f(N)$. To find the number of new members still active after N years, divide the interval $0 \le t \le N$ into n equal subintervals of length Δt years and let t_j denote the beginning of the jth subinterval. During the jth subinterval, approximately $r(t_j)\Delta t$ new members joined the group. Of these, the fraction still active $t = N$ (that is $N - t_j$ years later) is $f(N - t_j)$, and so the number of these still active after N years is $\lim_{n\to\infty} \sum_{j=1}^n r(t_j) f(N - t_j)\Delta t = \int_0^N r(t) f(N - t) dt$. Putting it all together, the

total number of active members N years from now is $P_0 f(N) + \int_0^N r(t)f(N-t)dt$.

43. a) Let t denote the time from now in months and $R(t)$ the total revenue generated. The price is $P(t) = 18 + 0.3\sqrt{t}$ dollars per barrel and the revenue is generated at a rate of $\frac{dR}{dt} = 300(18 + 0.3\sqrt{t})$. Then $R(t) = \int_0^{36} 300(18 + 0.3\sqrt{t})dt = 300(18t + \frac{0.6t^{3/2}}{3}))|_0^{36} = 300[18(36) + 0.2(6^3)] = 300(648 + 43.2) = \$207,360$.
b) Divide the interval $0 \le t \le 36$ into n equal subintervals of length Δt years. Then, the quantity of oil during the jth subinterval is $300\Delta t$ and the revenue generated by that much oil is $(18 + 0.3\sqrt{t_j})300\Delta t$. Hence, the total revenue is $\lim_{n\to\infty} \sum_{j=1}^{n}(18 + 0.3\sqrt{t_j})300\Delta t = \int_0^{36}(18 + 0.3\sqrt{t})300dt = \$207,360$ (computed in part a) above.).

6.5 Numerical Integration

1. For $\int_1^2 x^2dx$ with $n = 4$, $\Delta x = \frac{2-1}{4} = 0.25$, and $x_1 = 1$, $x_2 = 1.25$, $x_3 = 1.50$, $x_4 = 1.75$, $x_5 = 2$.
By the trapezoidal rule, $\int_1^2 x^2dx = \frac{\Delta x}{2}[f(x_1) + 2f(x_2) + 2f(x_3) + 2f(x_4) + f(x_5)] = \frac{0.25}{2}[1^2 + 2(1.25)^2 + 2(1.5)^2 + 2(1.75)^2 + 2^2] = 2.3438$.
b) By Simpson's rule, $\int_1^2 x^2dx = \frac{\Delta x}{3}[f(x_1) + 4f(x_2) + 2f(x_3) + 4f(x_4) + f(x_5)] = \frac{0.25}{3}[1^2 + 4(1.25)^2 + 2(1.5)^2 + 4(1.75)^2 + 2^2] = 2.3333$.

3. For $\int_0^1 \frac{1}{1+x^2}dx$ with $n = 4$, $\Delta x = \frac{1-0}{4} = 0.25$, and $x_1 = 0$, $x_2 = 0.25$, $x_3 = 0.50$, $x_4 = 0.75$, $x_5 = 1$.
By the trapezoidal rule, $\int_0^1 \frac{1}{1+x^2}dx = \frac{\Delta x}{2}[f(x_1) + 2f(x_2) + 2f(x_3) + 2f(x_4) + f(x_5)] = \frac{0.25}{2}[1 + \frac{2}{1+(0.25)^2} + \frac{2}{1+(0.5)^2} + \frac{2}{1+(0.75)^2} + \frac{1}{2}] = 0.7828$.
b) By Simpson's rule, $\int_0^1 \frac{1}{1+x^2}dx = \frac{\Delta x}{3}[f(x_1) + 4f(x_2) + 2f(x_3) + 4f(x_4) + f(x_5)] = \frac{0.25}{3}[1 + \frac{4}{1+(0.25)^2} + \frac{2}{1+(0.5)^2} + \frac{4}{1+(0.75)^2} + \frac{1}{2}] = 0.7854$.

5. For $\int_{-1}^0 \sqrt{1+x^2}dx$ with $n = 4$, $\Delta x = \frac{0-(-1)}{4} = 0.25$, and $x_1 = -1$, $x_2 = -0.75$, $x_3 = -0.5$, $x_4 = -0.25$, $x_5 = 0$.
By the trapezoidal rule, $\int_1^2 \sqrt{1+x^2}dx = \frac{\Delta x}{2}[f(x_1) + 2f(x_2) + 2f(x_3) + 2f(x_4) + f(x_5)] = \frac{0.25}{2}[\sqrt{1+(-1)^2} + 2\sqrt{1+(-0.75)^2} + 2\sqrt{1+(-0.5)^2} + 2\sqrt{1+(-0.25)^2} + \sqrt{1+(0)^2}] = 1.1515$.
b) By Simpson's rule, $\int_1^2 \sqrt{1+x^2}dx = \frac{\Delta x}{3}[f(x_1) + 4f(x_2) + 2f(x_3) + 4f(x_4) + f(x_5)] = \frac{0.25}{3}[\sqrt{1+(-1)^2} + 4\sqrt{1+(-0.75)^2} + 2\sqrt{1+(-0.5)^2} + 4\sqrt{1+(-0.25)^2} + \sqrt{1+(0)^2}] = 1.1478$.

7. For $\int_0^1 e^{-x^2}dx$ with $n = 4$, $\Delta x = \frac{1-0}{4} = 0.25$, and $x_1 = 0$, $x_2 = 0.25$, $x_3 = 0.50$,

$x_4 = 0.75$, $x_5 = 1$.
By the trapezoidal rule, $\int_1^2 e^{-x^2}dx = \frac{\Delta x}{2}[f(x_1)+2f(x_2)+2f(x_3)+2f(x_4)+f(x_5)] = \frac{0.25}{2}[1+2e^{-(0.25)^2}+2e^{-(0.5)^2}+2e^{-(0.75)^2}+e^{-1}] = 0.7430$.
b) By Simpson's rule, $\int_1^2 e^{-x^2}dx = \frac{\Delta x}{3}[f(x_1)+4f(x_2)+2f(x_3)+4f(x_4)+f(x_5)] = \frac{0.25}{3}[1+4e^{-(0.25)^2}+2e^{-(0.5)^2}+4e^{-(0.75)^2}+e^{-1}] = 0.7469$.

9. For $\int_1^2 \frac{1}{x^2}dx$ with $n=4$, $\Delta x = \frac{2-1}{4} = 0.25$, and $x_1 = 1$, $x_2 = 1.25$, $x_3 = 1.50$, $x_4 = 1.75$, $x_5 = 2$.
By the trapezoidal rule, $\int_1^2 \frac{1}{x^2}dx = \frac{\Delta x}{2}[f(x_1)+2f(x_2)+2f(x_3)+2f(x_4)+f(x_5)] = \frac{0.25}{2}[1+\frac{2}{(1.25)^2}+\frac{2}{(1.5)^2}+\frac{2}{(1.75)^2}+\frac{1}{2^2}] = 0.5090$.
The error estimate is $|E_n| \le \frac{M(b-a)^3}{12n^2}$. For $n=4$, $a=1$ and $b=2$, $|E_4| \le \frac{M(2-1)^3}{12(4^2)} = \frac{M}{192}$ where M is the maximum value of $|f''(x)|$ on $1 \le x \le 2$. Now $f(x) = x^{-2}$, $f'(x) = -2x^{-3}$, and $f''(x) = 6x^{-4}$. For $1 \le x \le 2$, $|f''(x)| = \frac{6}{x^4} \le \frac{6}{1^4} = 6$. Hence, $|E_4| = \frac{6}{192} = 0.0313$.
b) By Simpson's rule, $\int_1^2 \frac{1}{x^2}dx = \frac{\Delta x}{3}[f(x_1)+4f(x_2)+2f(x_3)+4f(x_4)+f(x_5)] = \frac{0.25}{3}[1+\frac{4}{(1.25)^2}+\frac{2}{(1.5)^2}+\frac{4}{(1.75)^2}+\frac{1}{2^2}] = 0.5004$.
The error estimate is $|E_n| \le \frac{M(b-a)^5}{180n^4}$. For $n=4$, $a=1$ and $b=2$, $|E_4| \le \frac{M(2-1)^5}{180(4^4)} = \frac{M}{46{,}080}$ where M is the maximum value of $|f^{(4)}(x)|$ on $1 \le x \le 2$. Now $f''(x) = 6x^{-4}$, $f^{(3)}(x) = -24x^{-5}$, and $f^{(4)}(x) = 120x^{-6}$. For $1 \le x \le 2$, $|f^{(4)}(x)| = \frac{120}{x^6} \le \frac{120}{1^6} = 120$. Hence, $|E_4| \le \frac{120}{46{,}080} = 0.0026$.

11. For $\int_1^3 \sqrt{x}dx$ with $n=10$, $\Delta x = \frac{3-1}{10} = 0.2$, and $x_1 = 1$, $x_2 = 1.2$, $x_3 = 1.4$, ..., $x_{10} = 2.8$, $x_{11} = 3$.
By the trapezoidal rule, $\int_1^3 \sqrt{x}dx = \frac{\Delta x}{2}[f(x_1)+2f(x_2)+2f(x_3)+\cdots+2f(x_{10})+f(x_{11})] = \frac{0.2}{2}[1+2\sqrt{1.2}+2\sqrt{1.4}+2\sqrt{1.6}+2\sqrt{1.8}+2\sqrt{2}+2\sqrt{2.2}+2\sqrt{2.4}+2\sqrt{2.6}+2\sqrt{2.8}+\sqrt{3}] = 2.7967$.
The error estimate is $|E_n| \le \frac{M(b-a)^3}{12n^2}$. For $n=10$, $a=1$ and $b=3$, $|E_{10}| \le \frac{M(3-1)^3}{12(10^2)} = \frac{8M}{1200} = \frac{M}{150}$ where M is the maximum value of $|f''(x)|$ on $1 \le x \le 3$. Now $f(x) = x^{1/2}$, $f'(x) = \frac{1}{2}x^{-1/2}$, and $f''(x) = -\frac{1}{4}x^{-3/2}$. For $1 \le x \le 3$, $|f''(x)| = |-\frac{1}{4}x^{-3/2}| \le \frac{1}{4}(1^{-3/2}) = \frac{1}{4}$. Hence, $|E_{10}| \le \frac{1}{150}(\frac{1}{4}) = 0.0017$.
b) By Simpson's rule, $\int_1^3 \sqrt{x}dx = \frac{\Delta x}{3}[f(x_1)+4f(x_2)+2f(x_3)+\cdots+4f(x_{10})+f(x_{11})] = \frac{0.2}{3}[1+4\sqrt{1.2}+2\sqrt{1.4}+4\sqrt{1.6}+2\sqrt{1.8}+4\sqrt{2}+2\sqrt{2.2}+4\sqrt{2.4}+2\sqrt{2.6}+4\sqrt{2.8}+\sqrt{3}] = 2.7974$.
The error estimate is $|E_n| \le \frac{M(b-a)^5}{180n^4}$. For $n=10$, $a=1$ and $b=3$, $|E_{10}| \le \frac{M(3-1)^5}{180(10^4)} = \frac{32M}{180(10)^4}$ where M is the maximum value of $|f^{(4)}(x)|$ on $1 \le x \le 3$. Now $f''(x) = -\frac{1}{4}x^{-3/2}$, $f^{(3)}(x) = \frac{3}{8}x^{-5/2}$, and $f^{(4)}(x) = -\frac{15}{16}x^{-7/2}$. For $1 \le x \le 3$, $|f^{(4)}(x)| = |-\frac{15}{16}x^{-7/2}| \le \frac{15}{16}(1^{-7/2}) = \frac{15}{16}$. Hence, $|E_{10}| \le \frac{32}{180(10{,}000)}(\frac{15}{16}) = 0.000\,02$.

13. For $\int_0^1 e^{x^2}dx$ with $n=4$, $\Delta x = \frac{1-0}{4} = 0.25$, and $x_1 = 0$, $x_2 = 0.25$, $x_3 = 0.50$, $x_4 = 0.75$, $x_5 = 1$.
By the trapezoidal rule, $\int_1^2 e^{x^2}dx = \frac{\Delta x}{2}[f(x_1)+2f(x_2)+2f(x_3)+2f(x_4)+f(x_5)] = \frac{0.25}{2}[1+$

$2e^{(0.25)^2} + 2e^{(0.5)^2} + 2e^{(0.75)^2} + e^1] = 1.4907$.
The error estimate is $|E_n| \leq \frac{M(b-a)^3}{12n^2}$. For $n = 4$, $a = 0$ and $b = 1$, $|E_4| \leq \frac{M(1-0)^3}{12(4^2)} = \frac{M}{192}$ where M is the maximum value of $|f''(x)|$ on $0 \leq x \leq 1$. Now $f(x) = e^{x^2}$, $f'(x) = -2xe^{-x^2}$, and $f''(x) = (4x^2+2)e^{x^2}$. For $0 \leq x \leq 1$, $|f''(x)| = [4(1^2)+2]e^{1^2} = 6e$. Hence, $|E_4| = \frac{6e}{192} = 0.0849$.
b) By Simpson's rule, $\int_1^2 e^{x^2}\,dx = \frac{\Delta x}{3}[f(x_1) + 4f(x_2) + 2f(x_3) + 4f(x_4) + f(x_5)] = \frac{0.25}{3}[1 + 4e^{(0.25)^2} + 2e^{(0.5)^2} + 4e^{(0.75)^2} + e^1] = 1.4637$.
The error estimate is $|E_n| \leq \frac{M(b-a)^5}{180n^4}$. For $n = 4$, $a = 0$ and $b = 1$, $|E_4| \leq \frac{M(2-1)^5}{180(4^4)} = \frac{M}{46{,}080}$ where M is the maximum value of $|f^{(4)}(x)|$ on $0 \leq x \leq 1$. Now $f''(x) = (4x^2+2)e^{x^2}$, $f^{(3)}(x) = (8x^3+12x)e^{x^2}$, and $f^{(4)}(x) = (16x^4+48x^2+12)e^{x^2}$. For $0 \leq x \leq 1$, $|f^{(4)}(x)| = [16(1^4)+48(1^2)+12]e^{1^2} = 76e$. Hence, $|E_4| \leq \frac{76e}{46{,}080} = 0.0045$.

15. The integral to be approximated is $\int_1^3 \frac{1}{x}dx$. The derivatives of $f(x) = \frac{1}{x} = x^{-1}$ are $f'(x) = -x^{-2}$, $f''(x) = 2x^{-3}$, $f^{(3)}(x) = -6x^{-4}$, and $f^{(4)}(x) = 24x^{-5}$.
a) For the trapezoidal rule, $|E_n| \leq \frac{M(b-a)^3}{12n^2}$, where M is the maximum value of $|f''(x)|$ on $1 \leq x \leq 3$. Now $|f''(x)| = \frac{2}{x^3} \leq \frac{2}{1^3} = 2$ on $1 \leq x \leq 3$. $|E_n| \leq \frac{2(3-1)^3}{12n^2} = \frac{4}{3n^2}$ which is less than 0.000 05 if $4 < 3(0.000\ 05)n^2$ or $n > \sqrt{\frac{4}{3(0.000\ 05)}} \approx 163.3$. Hence, 164 intervals should be used.
b) For Simpson's rule, $|E_n| \leq \frac{M(b-a)^5}{180n^4}$, where M is the maximum value of $|f^{(4)}(x)|$ on $1 \leq x \leq 3$. Now $|f^{(4)}(x)| = |\frac{24}{x^5}| \leq \frac{24}{1^5} = 24$ on $1 \leq x \leq 3$. $|E_n| \leq \frac{24(3-1)^5}{180n^4} = \frac{768}{180n^4}$ which is less than 0.000 05 if $768 < 180(0.000\ 05)n^4$ or $n > \sqrt[4]{\frac{768}{180(0.000\ 05)}} \approx 17.1$. Hence, 18 subintervals should be used.

17. The integral to be approximated is $\int_1^2 \frac{1}{\sqrt{x}}dx$. The derivatives of $f(x) = \frac{1}{\sqrt{x}} = x^{-1/2}$ are $f'(x) = -\frac{1}{2}x^{-3/2}$, $f''(x) = \frac{3}{4}x^{-5/2}$, $f^{(3)}(x) = -\frac{15}{8}x^{-7/2}$, and $f^{(4)}(x) = \frac{105}{16}x^{-9/2}$.
a) For the trapezoidal rule, $|E_n| \leq \frac{M(b-a)^3}{12n^2}$, where M is the maximum value of $|f''(x)|$ on $1 \leq x \leq 2$. Now $|f''(x)| = \frac{3}{4}x^{-5/2} \leq \frac{3}{4}$ on $1 \leq x \leq 2$. $|E_n| \leq \frac{3}{4}\frac{(2-1)^3}{12n^2} = \frac{1}{16n^2}$ which is less than 0.000 05 if $1 < 16(0.000\ 05)n^2$ or $n > \sqrt{\frac{1}{16(0.000\ 05)}} \approx 35.4$. Hence, 36 intervals should be used.
b) For Simpson's rule, $|E_n| \leq \frac{M(b-a)^5}{180n^4}$, where M is the maximum value of $|f^{(4)}(x)|$ on $1 \leq x \leq 2$. Now $|f^{(4)}(x)| = |\frac{105}{16}x^{-9/2}| \leq \frac{105}{16}$ on $1 \leq x \leq 2$. $|E_n| \leq \frac{105(2-1)^5}{16(180)n^4} = \frac{7}{192n^4}$ which is less than 0.000 05 if $7 < 192(0.000\ 05)n^4$ or $n > \sqrt[4]{\frac{7}{192(0.000\ 05)}} \approx 5.2$. Hence, 6 subintervals should be used.

19. The integral to be approximated is $\int_{1.2}^{2.4} e^x dx$.
a) For the trapezoidal rule, $|E_n| \leq \frac{M(b-a)^3}{12n^2}$, where M is the maximum value of $|f''(x)|$ on $1.2 \leq x \leq 2.4$. Now $|f''(x)| = |e^x| \leq e^{2.4}$ on $1.2 \leq x \leq 2.4$. $|E_n| \leq \frac{e^{2.4}(2.4-1.2)^3}{12n^2} = \frac{1.728e^{2.4}}{12n^2}$

which is less than 0.000 05 if $1.728e^{2.4} < 12(0.000\ 05)n^2$ or $n > \sqrt{\frac{1.728e^{2.4}}{12(0.000\ 05)}} \approx 178.2$. Hence, 179 intervals should be used.
b) For Simpson's rule, $|E_n| \leq \frac{M(b-a)^5}{180n^4}$, where M is the maximum value of $|f^{(4)}(x)|$ on $1.2 \leq x \leq 2.4$. Now $|f^{(4)}(x)| = |e^x| \leq e^{2.4}$ on $1.2 \leq x \leq 2.4$. $|E_n| \leq \frac{e^{2.4}(2.4-1.2)^5}{180n^4}$ which is less than 0.000 05 if $e^{2.4}(1.2)^5 < 180(0.000\ 05)n^4$ or $n > \sqrt[4]{\frac{e^{2.4}(1.2)^5}{180(0.000\ 05)}} \approx 7.4$. Hence, 8 subintervals should be used.

21. The integral to be approximated is $\int_0^1 \sqrt{1-x^2}dx$.
a) For the trapezoidal rule, $|E_n| \leq \frac{M(b-a)^3}{12n^2}$, where M is the maximum value of $|f''(x)|$ on $0 \leq x \leq 1$. Now $f(x) = (1-x^2)^{1/2}$, $f'(x) = -x(1-x^2)^{-1/2}$, $f''(x) = -(1-x^2)^{-3/2}$, $f^{(3)}(x) = -3(1-x^2)^{-5/2}$, and $f^{(4)}(x) = -\frac{3(4x^2-5)}{(1-x^2)^{7/2}}$. Unfortunately there is no maximum value for $f''(x)$ and $f^{(4)}(x)$ because they grow without bound on the interval $0 \leq x \leq 1$. A BASIC computer program was wriiten (using Microsoft Quickbasic) to approximate the answer. Both the basic program and the output (with reduced precision) are shown below:

```
     'program trapez.bas
     input "nstart ";nstart%
     kill"trapez21.dat" :  blank$ = " " '(20 blanks)
     open "trapez21.dat" as # 1 len = 80
     field #1, 80 as dat$ :  n% = nstart%
beg:
     lset dat$ = chr$(13)+chr$(10) :  put #1, 1
     rec% = 2 :  prepi# = 4*delx#*sum#/2
     xa#=0 :  xb#=1 :  print " n = ",n% :  sum#=0
     delx#=(xb#-xa#)/n% :  x#=xa#-delx#
     if n% < 1 then end
     print " y=sqr(1-x^2), 0<x<1, n = ";n%
     for i=1 to n%+1
         x#=x#+delx# :  if 1-x#^2 > 0 then y#=sqr(1-x#^2)
         if 1<i and i<n%+1 then
            sum# = sum# + 2 * y#
            lset dat$ = "x="+left$(str$(x#)+blank$,18)+"2y="+left$(str$(2*y#)+blank$,18)_
                        +"s="+left$(str$(sum#)+blank$,18)+chr$(13)+chr$(10)
            put #1, rec% :  rec% = rec% + 1
         end if
         if i=1 or i=n%+1 then
            sum# = sum# + y#
            lset dat$ = "x="+left$(str$(x#)+blank$,18)+"2y="+left$(str$(2*y#)+blank$,18)_
                        +"s="+left$(str$(sum#)+blank$,18)+chr$(13)+chr$(10)
            put #1, rec% :  rec% = rec% + 1
         end if
     next i
```

```
print "n=";n%;"pi = 4*delx*sum/2 = ";4*delx#*sum#/2 :  print
lset dat$ = "pi="+left$(str$(2*delx#*sum#)+blank$,17)+ "n="+str$(n%)+chr$(13)+chr$(10)
put #1, rec% :  rec% = rec% + 1
if (2*delx#*sum# < 3.1415 or 2*delx#*sum# > 3.1425) then
    n% = n% + 20
    go to beg
end if
end
```

```
x= 0                    y= 1                s = 1
x= .00625               2 y= 1.9999609      s = 2.9999609
x= .0125                2 y= 1.9998437      s = 4.9998046
x= .01875               2 y= 1.9996484      s = 6.9994530
x= .025                 2 y= 1.9993749      s = 8.9988279
x= .03125               2 y= 1.9990231      s = 10.997851
x= .0375                2 y= 1.9985932      s = 12.996444
x= .04375               2 y= 1.9980850      s = 14.994529
x= .05                  2 y= 1.9974984      s = 16.992027
x= .05625               2 y= 1.9968334      s = 18.988861
x= .0625                2 y= 1.9960899      s = 20.984951
x= .06875               2 y= 1.9952678      s = 22.980219
x= .075                 2 y= 1.9943670      s = 24.974586
x= .08125               2 y= 1.9933875      s = 26.967973
x= .0875                2 y= 1.9923290      s = 28.960302
x= .09375               2 y= 1.9911915      s = 30.951494
x= .1                   2 y= 1.9899748      s = 32.941469
x= .10625               2 y= 1.9886788      s = 34.930148
x= .1125                2 y= 1.9873034      s = 36.917451
x= .11875               2 y= 1.9858483      s = 38.903299
x= .125                 2 y= 1.9843134      s = 40.887613
x= .13125               2 y= 1.9826986      s = 42.870311
x= .1375                2 y= 1.9810035      s = 44.851315
x= .14375               2 y= 1.9792280      s = 46.830543
x= .15                  2 y= 1.9773719      s = 48.807915
x= .15625               2 y= 1.9754350      s = 50.783350
x= .1625                2 y= 1.9734170      s = 52.756767
x= .16875               2 y= 1.9713177      s = 54.728085
x= .175                 2 y= 1.9691368      s = 56.697222
x= .18125               2 y= 1.9668741      s = 58.664096
x= .1875                2 y= 1.9645292      s = 60.628625
x= .19375               2 y= 1.9621018      s = 62.590727
x= .2                   2 y= 1.9595917      s = 64.550319
x= .20625               2 y= 1.9569986      s = 66.507317
x= .2125                2 y= 1.9543221      s = 68.461640
x= .21875               2 y= 1.9515618      s = 70.413201
x= .225                 2 y= 1.9487175      s = 72.361919
```

```
x= .23125          2 y= 1.9457887        s = 74.307708
x= .2375           2 y= 1.9427750        s = 76.250483
x= .24375          2 y= 1.9396761        s = 78.190159
x= .25             2 y= 1.9364916        s = 80.126651
x= .25625          2 y= 1.9332210        s = 82.059872
x= .2625           2 y= 1.9298639        s = 83.989736
x= .26875          2 y= 1.9264199        s = 85.916156
x= .275            2 y= 1.9228884        s = 87.839044
x= .28125          2 y= 1.9192690        s = 89.758313
x= .2875           2 y= 1.9155612        s = 91.673874
x= .29375          2 y= 1.9117645        s = 93.585639
x= .3              2 y= 1.9078784        s = 95.493517
x= .30625          2 y= 1.9039022        s = 97.397420
x= .3125           2 y= 1.8998355        s = 99.297255
x= .31875          2 y= 1.8956776        s = 101.19293
x= .325            2 y= 1.8914280        s = 103.08436
x= .33125          2 y= 1.8870860        s = 104.97144
x= .3375           2 y= 1.8826510        s = 106.85409
x= .34375          2 y= 1.8781224        s = 108.73222
x= .35             2 y= 1.8734993        s = 110.60572
x= .35625          2 y= 1.8687813        s = 112.47450
x= .3625           2 y= 1.8639675        s = 114.33846
x= .36875          2 y= 1.8590572        s = 116.19752
x= .375            2 y= 1.8540496        s = 118.05157
x= .38125          2 y= 1.8489439        s = 119.90051
x= .3874999        2 y= 1.8437394        s = 121.74425
x= .3937499        2 y= 1.8384351        s = 123.58269
x= .39999999       2 y= 1.8330302        s = 125.41572
x= .40624999       2 y= 1.8275239        s = 127.24324
x= .41249999       2 y= 1.8219152        s = 129.06516
x= .41874999       2 y= 1.8162031        s = 130.88136
x= .42499999       2 y= 1.8103866        s = 132.69175
x= .43124999       2 y= 1.8044649        s = 134.49621
x= .43749999       2 y= 1.7984368        s = 136.29465
x= .44374999       2 y= 1.7923012        s = 138.08695
x= .44999999       2 y= 1.7860571        s = 139.87301
x= .45624999       2 y= 1.7797032        s = 141.65271
x= .46249999       2 y= 1.7732385        s = 143.42595
x= .46874999       2 y= 1.7666617        s = 145.19261
x= .47499999       2 y= 1.7599715        s = 146.95258
x= .48124999       2 y= 1.7531667        s = 148.70575
x= .48749999       2 y= 1.7462459        s = 150.45200
x= .49374999       2 y= 1.7392077        s = 152.19120
x= .49999999       2 y= 1.7320508        s = 153.92326
x= .50624999       2 y= 1.7247735        s = 155.64803
x= .51249999       2 y= 1.7173744        s = 157.36540
x= .51874999       2 y= 1.7098519        s = 159.07526
x= .52499999       2 y= 1.7022044        s = 160.77746
```

```
x= .53124999        2 y= 1.6944302      s = 162.47189
x= .5375            2 y= 1.6865274      s = 164.15842
x= .54375           2 y= 1.6784944      s = 165.83691
x= .55              2 y= 1.6703293      s = 167.50724
x= .55625           2 y= 1.6620300      s = 169.16927
x= .5625            2 y= 1.6535945      s = 170.82287
x= .56875           2 y= 1.6450208      s = 172.46789
x= .575             2 y= 1.6363068      s = 174.10419
x= .58125           2 y= 1.6274500      s = 175.73164
x= .5875            2 y= 1.6184483      s = 177.35009
x= .59375           2 y= 1.6092991      s = 178.95939
x= .6               2 y= 1.6            s = 180.55939
x= .60625           2 y= 1.5905482      s = 182.14994
x= .6125            2 y= 1.5809411      s = 183.73088
x= .61875           2 y= 1.5711759      s = 185.30206
x= .625             2 y= 1.5612494      s = 186.86331
x= .63125           2 y= 1.5511588      s = 188.41446
x= .6375            2 y= 1.5409007      s = 189.95537
x= .64375           2 y= 1.5304717      s = 191.48584
x= .65000000        2 y= 1.5198684      s = 193.00571
x= .65625000        2 y= 1.5090870      s = 194.51479
x= .66250000        2 y= 1.4981238      s = 196.01292
x= .66875000        2 y= 1.4869746      s = 197.49989
x= .67500000        2 y= 1.4756354      s = 198.97553
x= .68125000        2 y= 1.4641016      s = 200.43963
x= .68750000        2 y= 1.4523687      s = 201.89200
x= .69375000        2 y= 1.4404317      s = 203.33243
x= .70000000        2 y= 1.4282856      s = 204.76071
x= .70625000        2 y= 1.4159250      s = 206.17664
x= .71250000        2 y= 1.4033442      s = 207.57998
x= .71875000        2 y= 1.3905372      s = 208.97052
x= .72500000        2 y= 1.3774977      s = 210.34802
x= .73125000        2 y= 1.3642190      s = 211.71224
x= .73750000        2 y= 1.3506942      s = 213.06293
x= .74375000        2 y= 1.3369157      s = 214.39985
x= .75000000        2 y= 1.3228756      s = 215.72272
x= .75625000        2 y= 1.3085655      s = 217.03129
x= .76250000        2 y= 1.2939764      s = 218.32527
x= .76875000        2 y= 1.2790988      s = 219.60436
x= .77500000        2 y= 1.2639224      s = 220.86829
x= .78125000        2 y= 1.2484365      s = 222.11672
x= .78750000        2 y= 1.2326293      s = 223.34935
x= .79375000        2 y= 1.2164882      s = 224.56584
x= .80000000        2 y= 1.2            s = 225.76584
x= .80625000        2 y= 1.1831499      s = 226.94899
x= .81250000        2 y= 1.1659223      s = 228.11491
x= .81875000        2 y= 1.1483003      s = 229.26321
x= .82500000        2 y= 1.1302654      s = 230.39348
```

```
x= .83125000          2 y= 1.1117975       s = 231.50528
x= .83750000          2 y= 1.0928746       s = 232.59815
x= .84375000          2 y= 1.0734727       s = 233.67162
x= .85000000          2 y= 1.0535653       s = 234.72519
x= .85625000          2 y= 1.0331232       s = 235.75831
x= .86250000          2 y= 1.0121141       s = 236.77043
x= .86875000          2 y= .99050176       s = 237.76093
x= .87500000          2 y= .96824583       s = 238.72917
x= .88125000          2 y= .94530087       s = 239.67448
x= .88750000          2 y= .92161542       s = 240.59609
x= .89375000          2 y= .89713084       s = 241.49322
x= .90000000          2 y= .87177978       s = 242.36500
x= .90625000          2 y= .84548432       s = 243.21049
x= .91250000          2 y= .81815340       s = 244.02864
x= .91875000          2 y= .78967952       s = 244.81832
x= .92500000          2 y= .75993420       s = 245.57825
x= .93125000          2 y= .72876179       s = 246.30701
x= .93750000          2 y= .69597054       s = 247.00299
x= .94375000          2 y= .66131970       s = 247.66430
x= .95000000          2 y= .62449979       s = 248.28880
x= .95625000          2 y= .58510148       s = 248.87391
x= .96250000          2 y= .54256336       s = 249.41647
x= .96875000          2 y= .49607837       s = 249.91255
x= .97500000          2 y= .44440972       s = 250.35696
x= .98125000          2 y= .38547859       s = 250.74244
x= .98750000          2 y= .31523800       s = 251.05767
x= .99375000          2 y= .22325713       s = 251.28093
x= 1                  2 y= .22325713       s = 251.39256
pi= 3.14240706231660n= 160
```

b) For Simpson's rule, the following program was used:

```
     'program simpson.bas
     open "sim.dat" as # 1 len = 80
     field #1,80 as dat$ :  rec% = 1
     input " start count ";nstart% :   n = nstart%
     b$ = " " '(20)
beg:
     xa=0 :  xb=1 :  sum=0 :  rec% = 1
     delx=(xb-xa)/n :  x=xa-delx
     for i=1 to n+1
        x=x+delx :  if 1-x^2 > 0 then y=sqr(1-x^2)
        if 1<i and i<n+1 then
           if i mod 2 = 1 then
              sum = sum + 2 * y
```

```
       lset dat$ = "x="+left$(str$(x)+b$,18)+_
                   "y="+left$(str$(y)+b$,18)+_
                   "s="+left$(str$(sum)+b$,18)+chr$(13)+chr$(10)
       put #1,rec% :  rec% = rec% + 1
     end if
     if i mod 2 = 0 then
       sum = sum + 4 * y
       lset dat$ = "x="+left$(str$(x)+b$,18)+_
                   "y="+left$(str$(y)+b$,18)+_
                   "s="+left$(str$(sum)+b$,18)+chr$(13)+chr$(10)
       put #1,rec% :  rec% = rec% + 1
     end if
   end if
   if i=1 or i=n+1 then
     sum = sum + y
     lset dat$ = str$(x)+","+str$(y)+","+str$(sum)+chr$(13)+chr$(10)
     put #1, rec% :  rec% = rec% + 1
   end if
next i
lset dat$ = "n = "+str$(n)+str$(delx*sum/3)+" , pi = "+str$(4*delx*sum/3)
put #1, rec% :  rec% = rec% + 1
print "pi = ";4*delx*sum/3,"n = ";n
if (4*delx*sum/3 < 3.1415 or 4*delx*sum/3 > 3.1425) then
   n = n + 20
   go to beg
end if
close 1
end
```

```
x= .00625             y= .9999805          s= 4.999922
x= .0125              y= .9999219          s= 6.999765
x= .01875             y= .9998242          s= 10.99906
x= .025               y= .9996874          s= 12.99844
x= .03125             y= .9995116          s= 16.99648
x= .0375              y= .9992966          s= 18.99508
x= .04375             y= .9990425          s= 22.99125
x= .05                y= .9987492          s= 24.98874
x= 5.625001E-02       y= .9984167          s= 28.98241
x= 6.250001E-02       y= .998045           s= 30.9785
x= 6.875001E-02       y= .9976339          s= 34.96904
x= 7.500001E-02       y= .9971835          s= 36.96341
x= 8.125001E-02       y= .9966937          s= 40.95018
x= 8.750001E-02       y= .9961645          s= 42.94251
x= 9.375002E-02       y= .9955958          s= 46.92489
x= .1                 y= .9949874          s= 48.91487
x= .10625             y= .9943395          s= 52.89223
```

```
x= .1125         y= .9936517      s= 54.87953
x= .11875        y= .9929242      s= 58.85123
x= .125          y= .9921567      s= 60.83554
x= .13125        y= .9913493      s= 64.80094
x= .1375         y= .9905018      s= 66.78194
x= .14375        y= .9896141      s= 70.7404
x= .15           y= .988686       s= 72.71777
x= .15625        y= .9877176      s= 76.66864
x= .1625         y= .9867086      s= 78.64206
x= .16875        y= .9856589      s= 82.58469
x= .175          y= .9845684      s= 84.55383
x= .18125        y= .9834371      s= 88.48758
x= .1875         y= .9822646      s= 90.45211
x= .1937499      y= .9810509      s= 94.37631
x= .1999999      y= .9797959      s= 96.33591
x= .2062499      y= .9784994      s= 100.2499
x= .2124999      y= .9771611      s= 102.2042
x= .2187499      y= .975781       s= 106.1073
x= .2249999      y= .9743588      s= 108.0561
x= .2312499      y= .9728944      s= 111.9476
x= .2374999      y= .9713876      s= 113.8904
x= .2437499      y= .9698381      s= 117.7698
x= .2499999      y= .9682459      s= 119.7063
x= .2562499      y= .9666106      s= 123.5727
x= .2624999      y= .964932       s= 125.5026
x= .2687499      y= .96321        s= 129.3554
x= .2749999      y= .9614443      s= 131.2783
x= .2812499      y= .9596345      s= 135.1168
x= .2874999      y= .9577807      s= 137.0324
x= .2937499      y= .9558823      s= 140.8559
x= .2999999      y= .9539393      s= 142.7638
x= .3062499      y= .9519512      s= 146.5716
x= .3124999      y= .9499178      s= 148.4714
x= .3187499      y= .9478388      s= 152.2628
x= .3249998      y= .9457141      s= 154.1542
x= .3312498      y= .9435431      s= 157.9284
x= .3374998      y= .9413256      s= 159.811
x= .3437498      y= .9390613      s= 163.5673
x= .3499998      y= .9367498      s= 165.4408
x= .3562498      y= .9343908      s= 169.1783
x= .3624998      y= .9319838      s= 171.0423
x= .3687498      y= .9295287      s= 174.7604
x= .3749998      y= .9270249      s= 176.6145
x= .3812498      y= .924472       s= 180.3123
x= .3874998      y= .9218698      s= 182.1561
x= .3937498      y= .9192177      s= 185.8329
x= .3999998      y= .9165152      s= 187.666
x= .4062498      y= .9137621      s= 191.321
```

```
x= .4124998          y= .9109577          s= 193.1429
x= .4187498          y= .9081017          s= 196.7753
x= .4249997          y= .9051935          s= 198.5857
x= .4312497          y= .9022326          s= 202.1947
x= .4374997          y= .8992186          s= 203.9931
x= .4437497          y= .8961508          s= 207.5777
x= .4499997          y= .8930287          s= 209.3638
x= .4562497          y= .8898518          s= 212.9232
x= .4624997          y= .8866195          s= 214.6964
x= .4687497          y= .8833311          s= 218.2297
x= .4749997          y= .879986           s= 219.9897
x= .4812497          y= .8765836          s= 223.496
x= .4874997          y= .8731232          s= 225.2423
x= .4937497          y= .8696041          s= 228.7207
x= .4999997          y= .8660256          s= 230.4528
x= .5062497          y= .8623869          s= 233.9023
x= .5124997          y= .8586874          s= 235.6197
x= .5187497          y= .8549262          s= 239.0394
x= .5249997          y= .8511024          s= 240.7416
x= .5312498          y= .8472152          s= 244.1305
x= .5374998          y= .8432639          s= 245.817
x= .5437498          y= .8392474          s= 249.174
x= .5499998          y= .8351648          s= 250.8443
x= .5562499          y= .8310151          s= 254.1684
x= .5624999          y= .8267974          s= 255.822
x= .5687499          y= .8225105          s= 259.112
x= .5749999          y= .8181534          s= 260.7484
x= .58125            y= .8137251          s= 264.0033
x= .5875             y= .8092242          s= 265.6217
x= .59375            y= .8046496          s= 268.8403
x= .6                y= .8                s= 270.4403
x= .6062501          y= .7952741          s= 273.6214
x= .6125001          y= .7904705          s= 275.2023
x= .6187501          y= .7855879          s= 278.3447
x= .6250001          y= .7806247          s= 279.9059
x= .6312501          y= .7755793          s= 283.0083
x= .6375002          y= .7704502          s= 284.5492
x= ,6437502          y= .7652357          s= 287.6101
x= .6500002          y= .7599341          s= 289.13
x= .6562502          y= .7545433          s= 292.1481
x= .6625003          y= .7490616          s= 293.6463
x= .6687503          y= .7434871          s= 296.6202
x= .6750003          y= .7378175          s= 298.0959
x= .6812503          y= .7320505          s= 301.0241
x= .6875004          y= .7261841          s= 302.4764
x= .6937504          y= .7202155          s= 305.3573
x= .7000004          y= .7141424          s= 306.7856
x= .7062504          y= .7079621          s= 309.6174
```

```
x= .7125005         y= .7016717          s= 311.0208
x= .7187505         y= .6952681          s= 313.8018
x= .7250005         y= .6887484          s= 315.1793
x= .7312505         y= .682109           s= 317.9078
x= .7375006         y= .6753466          s= 319.2585
x= .7437506         y= .6684573          s= 321.9323
x= .7500006         y= .6614372          s= 323.2552
x= .7562506         y= .6542821          s= 325.8723
x= .7625006         y= .6469874          s= 327.1663
x= .7687507         y= .6395486          s= 329.7245
x= .7750007         y= .6319604          s= 330.9884
x= .7812507         y= .6242173          s= 333.4852
x= .7875007         y= .6163137          s= 334.7179
x= .7937508         y= .6082431          s= 337.1509
x= .8000008         y= .599999           s= 338.3509
x= .8062508         y= .5915738          s= 340.7172
x= .8125008         y= .58296            s= 341.8831
x= .8187509         y= .574149           s= 344.1797
x= .8250009         y= .5651314          s= 345.3099
x= .8312509         y= .5558974          s= 347.5335
x= .8375009         y= .5464359          s= 348.6264
x= .843751          y= .5367349          s= 350.7734
x= .850001          y= .5267811          s= 351.8269
x= .856251          y= .51656            s= 353.8932
x= .862501          y= .5060553          s= 354.9053
x= .8687511         y= .4952491          s= 356.8863
x= .8750011         y= .4841209          s= 357.8545
x= .8812511         y= .4726484          s= 359.7451
x= .8875011         y= .4608056          s= 360.6667
x= .8937511         y= .4485631          s= 362.4609
x= .9000012         y= .4358875          s= 363.3327
x= .9062512         y= .4227396          s= 365.0237
x= .9125012         y= .409074           s= 365.8418
x= .9187512         y= .3948369          s= 367.4211
x= .9250013         y= .379964           s= 368.1811
x= .9312513         y= .3643776          s= 369.6386
x= .9375013         y= .3479818          s= 370.3345
x= .9437513         y= .330656           s= 371.6572
x= .9500014         y= .3122458          s= 372.2817
x= .9562514         y= .2925462          s= 373.4518
x= .9625014         y= .2712767          s= 373.9944
x= .9687514         y= .2480335          s= 374.9865
x= .9750015         y= .2221984          s= 375.4309
x= .9812515         y= .1927318          s= 376.2018
x= .9875015         y= .1576096          s= 376.5171
x= .9937515         y= .111615           s= 376.9635
1.000002, .111615, 377.0751
n = 160 .7855732 , pi = 3.142293
```

23. The integral to be approximated is $\int_0^1 \sqrt{1+x^3}dx$.
For the trapezoidal rule, $|E_n| \le \frac{M(b-a)^3}{12n^2}$, where M is the maximum value of $|f''(x)|$ on $0 \le x \le 1$. Now $f(x) = (1+x^3)^{1/2}$, $f'(x) = \frac{1}{2}(1+x^3)^{-1/2}(3x^2) = \frac{3x^2}{2}(1+x^3)^{-1/2}$, $f''(x) = \frac{3x^2}{2}(-\frac{1}{2})(1+x^3)^{-3/2}(3x^2) + 3x(1+x^3)^{-1/2} = (1+x^3)^{-3/2}[(-\frac{9}{4})x^4 + 3x + 3x^4] = (1+x^3)^{-3/2}(3x)(1+\frac{x^3}{4})$. $|E_n| \le \frac{M(b-a)^3}{12n^2}$, where M is the maximum value of $|f''(x)|$ on $0 \le x \le 1$. Now $|f''(x)| = (1+x^3)^{-3/2}(3x)(1+\frac{x^3}{4}) \to (1+0^3)^{-3/2}(\frac{15}{4}) < 4$ as a bound on $0 \le x \le 1$. (Note the inconsistency of $x = 0$ in the denominator and $x = 1$ in the numerator for a worst case condition.) Thus $|E_n| \le \frac{4(1-0)^3}{12n^2} = \frac{1}{3n^2}$ or $n = \sqrt{\frac{1}{3E_n}}$. With $n = 18$, $\int_0^1 \sqrt{1+x^3}dx = \frac{\Delta x}{2}[f(x_1)+2f(x_2)+2f(x_3)+\cdots+2f(x_{17})+f(x_{18})] = [\frac{1-0}{2(18)}][f(0.055)+2f(0.111)+2f(0.166)+\cdots+2f(x0.944)+f(1.0)] = (0.0277)[1+2.000+2.001+2.004+2.010+2.021+2.036+2.057+2.085+2.121+2.164+2.216+2.277+2.346+2.425+2.512+2.609+2.714+1.414] = 1.112$. This is a rough approximation. With $n = 18$, the number extended to more decimal places is 1.1117208. $n = 50$ produces 1114833, $n = 80$ generates 1.1114618, and $n = 130$ yields 1.1114532. Needless to say a computer mechanization was used to obtain these results.

Review Problems

1. $\int_0^1 (5x^4 - 8x^3 + 1)dx = (x^5 - 2x^4 + x)|_0^1 = (1-2+1) - 0 = 0$

2. $\int_1^4 (\sqrt{x} + x^{-3/2})dx = (\frac{2x^{3/2}}{3} - 2x^{-1/2}|_1^4 = [\frac{2(8)}{3} - \frac{2}{2}] - [\frac{2(1)}{3} - \frac{2}{1}] = \frac{17}{3}$

3. Let $u = 5x-2$. Then $du = 5dx$ or $dx = \frac{1}{5}du$. When $x = -1$, $u = -7$, and when $x = 2$, $u = 8$. Hence, $\int_{-1}^2 30(5x-2)^2 dx = 6\int_{-7}^8 u^2 du = 2u^3|_{-7}^8 = 2[(8)^3 - (-7)^3] = 1,710$.

4. Let $u = x^2 - 1$. Then $du = 2xdx$. When $x = 0$, $u = -1$, and when $x = 1$, $u = 0$. Hence, $\int_0^1 2xe^{x^2-1}dx = \int_{-1}^0 e^u du = e^u|_{-1}^0 = e^0 - e^{-1} = 0.3679$.

5. Let $u = x^2 - 6x + 2$. Then $du = (2x-6)dx$ or $(x-3)dx = \frac{1}{2}du$. When $x = 0$, $u = 2$, and when $x = 1$, $u = -3$. Hence, $\int_0^1 (x-3)(x^2-6x+2)^3 dx = \frac{1}{2}\int_2^{-3} u^3 du = \frac{u^4}{8}|_2^{-3} = \frac{1}{8}[(-3)^4 - 2^4] = \frac{65}{8}$.

6. Let $u = x^2+4x+5$. Then $du = (2x+4)dx$ or $(3x+6)dx = \frac{3}{2}du$. When $x = -1$, $u = 2$, and when $x = 1$, $u = 10$. Hence, $\int_{-1}^1 \frac{3x+6}{(x^2+4x+5)^2}dx = \frac{3}{2}\int_2^{10} u^{-2}du = -\frac{3}{2u}|_2^{10} = -\frac{3}{2}(\frac{1}{10} - \frac{1}{2}) = \frac{3}{5}$.

7. Let $g(x) = e^x$ and $f(x) = x$. Then, $G(x) = e^x$ and $f'(x) = 1$. Thus $\int_{-1}^{1} xe^x dx = xe^x|_{-1}^{1} - \int_{-1}^{1} e^x dx = (xe^x - e^x)|_{-1}^{1} = (e - e) - (-e^{-1} - e^{-1}) = 2e^{-1} = 0.7356$

8. Let $u = \ln x$. Then $du = \frac{1}{x}dx$. When $x = e$, $u = 1$, and when $x = e^2$, $u = 2$. Hence, $\int_e^{e^2} \frac{1}{x(\ln x)^2}dx = \int_1^2 u^{-2}du = -\frac{1}{u}|_1^2 = -\frac{1}{2} - (-1) = \frac{1}{2}$.

9. Let $g(x) = x^2$ and $f(x) = \ln x$. Then, $G(x) = \frac{x^3}{3}$ and $f'(x) = \frac{1}{x}$. Thus $\int_1^e x^2 \ln x dx = \frac{x^3}{3} \ln x|_1^e - \frac{1}{3}\int_1^e x^3(\frac{1}{x})dx = \frac{x^3}{3} \ln x|_1^e - (\frac{1}{3})\int_1^e x^2 dx = (\frac{x^3}{3} \ln x - \frac{x^3}{9})|_1^e = \frac{x^3}{3}(\ln x - \frac{1}{3})|_1^e = [\frac{e^3}{3}(1 - \frac{1}{3})] - [\frac{1}{3}(0 - \frac{1}{3})] = \frac{2e^3}{9} + \frac{1}{9} = \frac{2e^3+1}{9}$.

10. Let $g(x) = e^{0.2x}$ and $f(x) = 2x + 1$. Then, $G(x) = 5e^{0.2x}$ and $f'(x) = 2$. Thus $\int_0^{10}(2x+1)e^{0.2x}dx = 5(2x+1)e^{0.2x}|_0^{10} - 10\int_0^{10} e^{0.2x}dx = [5(2x+1)e^{0.2x} - 50e^{0.2x}]|_0^{10} = 5(2x - 9)e^{0.2x}|_0^{10} = [5(11)e^2] - [5(-9)] = 55e^2 + 45$.

11. Let $P(x)$ denote the population x months from now. Then $\frac{dP}{dx} = 10 + 2\sqrt{x}$, and the amount by which the population will increase during the next 9 months is $P(9) - P(0) = \int_0^9 (10 + 2x^{1/2})dx = (10x + \frac{4x^{3/2}}{3})|_0^9 = 90 + \frac{4(27)}{3} = 126$ people.

12. Let $N(t)$ denote the size of the crop (in bushels) t days from now. Then $\frac{dN}{dt} = 0.3t^2 + 0.6t + 1$ bushels per day, and the increase in size of the crop over the next 6 days is $N(6) - N(0) = \int_0^6 (0.3t^2 + 0.6t + 1)dt = (0.1t^3 + 0.3t^2 + t)|_0^6 = 0.1(6^3) + 0.3(6^2) + 6 = 38.4$ bushels. Hence, at \$2 per bushel, the value of the crop will increase by $2(38.4) = \$76.80$.

13. Let $Q(t)$ denote the total consumption (in billion-barrel units) of oil over the next t years. Then the demand (billion barrels per year) is the rate of change $\frac{dQ}{dt}$ of total consumption with respect to time. The fact that this demand is increasing exponentially at the rate of 10 percent per year and is currently equal to 40 (billion barrels per year) implies that the demand is $\frac{dQ}{dt} = 40e^{0.1t}$ billion barrels per year. Hence, the total consumption during the next 5 years is the definite integral $Q(5) - Q(0) = \int_0^5 40e^{0.1t}dt = 400e^{0.1t}|_0^5 = 400(e^{0.5} - 1) = 259.49$ billion barrels per year.

14. $A = \int_{-1}^{3}(3x^2 + 2)dx = (x^3 + 2x)|_{-1}^{3} = (27 + 6) - (-1 - 2) = 36$

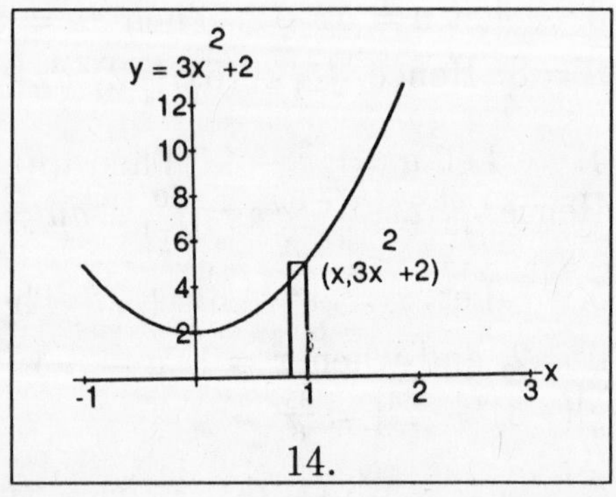

14.

15. $A = \int_1^4 \frac{1}{x^2} dx = -\frac{1}{x}\Big|_1^4 = -\frac{1}{4} - (-1) = \frac{3}{4}$

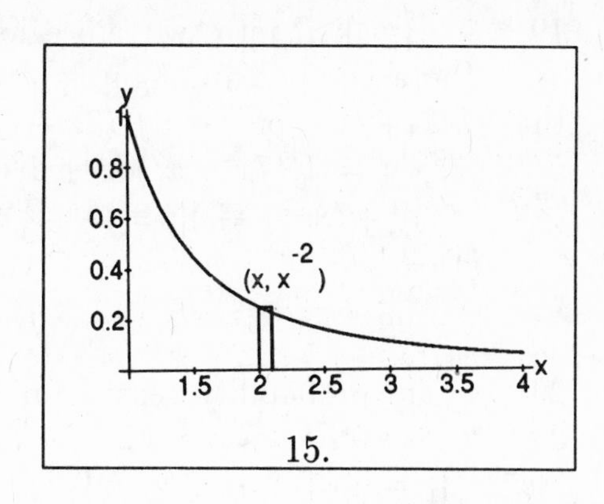

15.

16. $A = \int_{-1}^{2} (2 + x - x^2) dx = (2x + \frac{x^2}{2} - \frac{x^3}{3})\Big|_{-1}^{2} = (4 + 2 - \frac{8}{3}) - (-2 + \frac{1}{2} + \frac{1}{3}) = \frac{9}{2}$

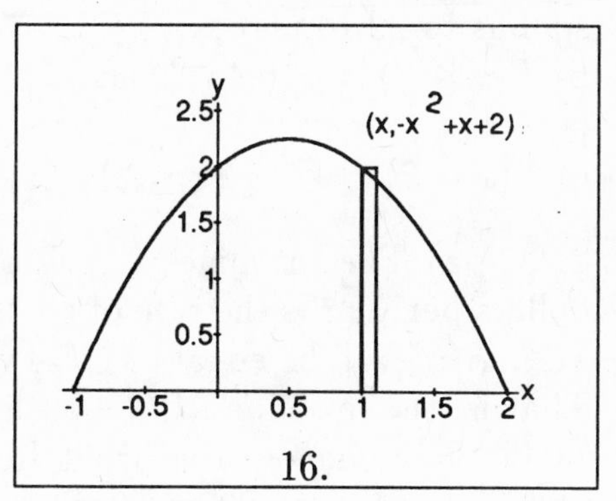

16.

17. Break R into two subregions R_1 and R_2 as in the accompanying figure. The area of $R =$ that of $R_1 +$ that of $R_2 = \int_0^4 \sqrt{x} dx + \int_4^8 \frac{8}{x} dx = \frac{2}{3} x^{3/2}\Big|_0^4 + 8 \ln |x|\Big|_4^8 = \frac{16}{3} + 8 \ln 8 - 8 \ln 4 = \frac{16}{3} + 8 \ln 2.$

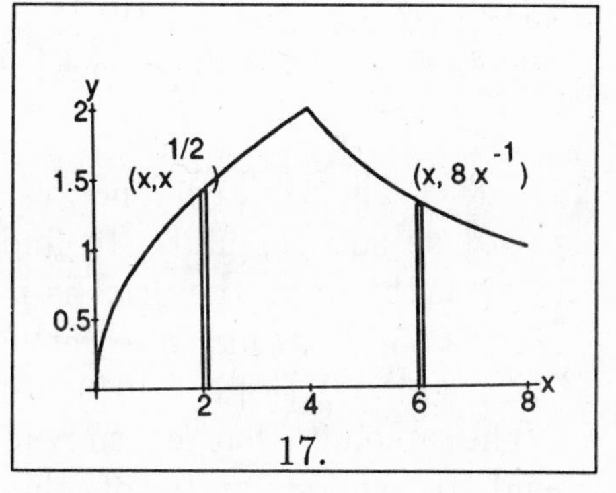

17.

18. $\int_0^1 (x - x^4) dx = (\frac{x^2}{2} - \frac{x^5}{5})\Big|_0^1 = \frac{1}{2} - \frac{1}{5} = \frac{3}{10}.$

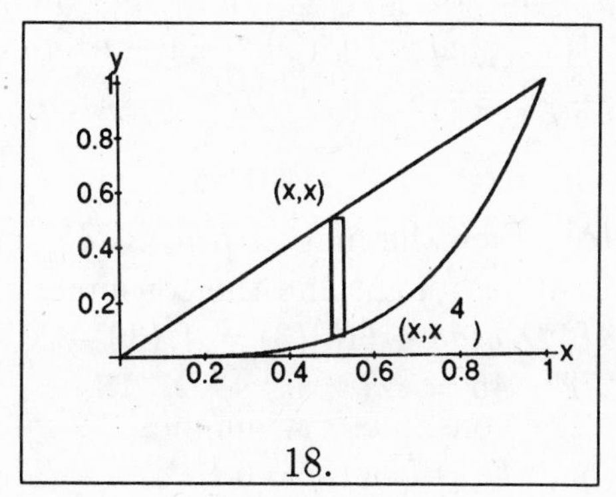

18.

19. Break R into two subregions R_1 and R_2 as in the accompanying figure. The area of $R =$ that of R_1+ that of $R_2 = \int_0^1(7x - x^2)dx + \int_1^2[(8 - x^2) - x^2]dx = \int_0^1(7x - x^2)dx + \int_1^2(8 - 2x^2)dx = (\frac{7x^2}{2} - \frac{x^3}{3})|_0^1 + (8x - \frac{2x^3}{3})|_1^2 = (\frac{7}{2} - \frac{1}{3}) + [(16 - \frac{16}{3}) - (8 - \frac{2}{3})] = \frac{13}{2}$.

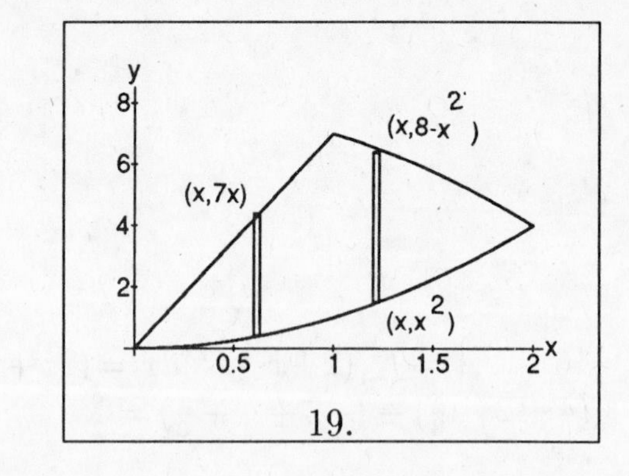

19.

20. The probability density function is $f(x) = 0.4e^{-0.4x}$. a) $P(1 \le x \le 2) = \int_1^2 0.4e^{-0.4x}dx = -e^{-0.4x}|_1^2 = -e^{-0.8} + e^{-0.4} = 0.2210$ which corresponds to 22.10 percent.
b) $P(x \le 2) = \int_0^2 0.4e^{-0.4x}dx = -e^{-0.4x}|_0^2 = -e^{-0.8} + 1 = 0.5507$ which corresponds to 55.07 percent.
c) $P(x \ge 2) = 1 - P(x \le 2) = 1 - 0.5507 = 0.4493$ which corresponds to 44.93 percent.

21. a) The machine is profitable as long as $R(x) \ge C(x)$, where $R(x) = 4,575 - 5x^2$ dollars per year is the rate of revenue and $C(x) = 1,200 + 10x^2$ dollars per year is the rate of cost. $R(x)$ will be equal to $C(x)$ when $4,575 - 5x^2 = 1,200 + 10x^2$, $15x^2 = 3,375$ or $x = 15$. That is, the machine will be profitable for 15 years.
b) Let $P(x)$ denote the profit from the use of the machine. Then, $\frac{dP}{dx} = R(x) - C(x) = (4,575 - 5x^2) - (1,200 + 10x^2) = 3,375 - 15x^2$. The total net profit generated over the next 5 years is $P(15) - P(0) = \int_0^{15}(3,375 - 15x^2)dx = (3,375x - 5x^3)|_0^{15} = 3,375(15) - 5(15)^3 =$ \$33,750.

22. a) The demand function is $D(q) = 30 - 2q - q^2$ dollars per unit. To find the number of units bought when the price is $p = 15$, solve the equation $15 = D(q)$ for q to get $15 = 30 - 2q - q^2$, $q^2 + 2q - 15 = 0$, $(q+5)(q-3) = 0$, or $q = 3$ units.
b) The amount that consumers are willing to spend to get 3 units of the commodity is $\int_0^3 D(q)dq = \int_0^3(30 - 2q - q^2)dq = (30q - q^2 - \frac{q^3}{3})|_0^3 =$ \$72.

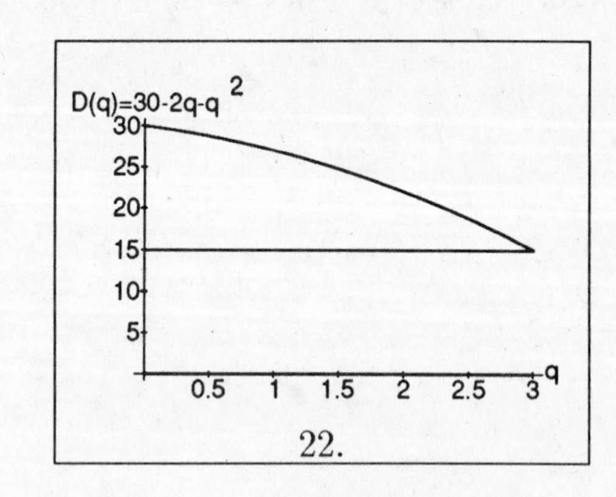

22.

c) When the market price is \$15 per unit, 3 units will be bought and the consumer's surplus will be $\int_0^3 D(q)dq - (15)(3) = \int_0^3(30 - 2q - q^2)dq - 45 = 72 - 45 =$ \$27.
The consumer's willingness to spend in part b) is equal to the area under the demand curve $p = D(q)$ from $q = 0$ to $q = 3$. The consumers' surplus in part c) is equal to the area of the region between the demand curve and the horizontal line $p = 15$.

line $p = 15$.

23. Recall that P dollars invested at an annual interest rate of 8 percent compounded continuously will be worth $Pe^{0.08t}$ dollars after t years. Let t_j denote the time (in years) of the jth deposit of \$1,200. This deposit will remain in the account for $5 - t_j$ years and hence will grow to $1,200e^{0.08(5-t_j)}$ dollars. At the end of 5 years the amount in the account is $\sum_{j=1}^{n} 1,200e^{0.08(5-t_j)}$. Rewrite the sum as $\sum_{j=1}^{n} 1,200e^{0.08(5-t_j)}\Delta t$. This sum can be approximated by the definite integral $\int_0^5 1,200e^{0.08(5-t)}dt = 1,200e^{0.4}\int_0^5 e^{-0.08t}dt = \frac{1,200e^{0.4}}{-0.08}(e^{-0.08t})|_0^5 = -\frac{1,200e^{0.4}}{0.08}(e^{-0.4} - 1) = -\frac{1,200}{0.08}(1 - e^{0.4}) = \$7,377.37$.

24. Recall that the present value of B dollars payable t years from now with an annual interest rate of 7 percent compounded continuously is $Be^{-0.07t}$. Divide the interval $0 \leq t \leq 10$ into n equal subintervals of length Δt years, and let t_j denote the beginning of the jth subinterval. During the jth subinterval, the income will equal (dollars per year)(number of years)$=$ $1,000\Delta t$ and the present value of the income will $=$ $1,000(\Delta t)e^{-0.07t_j} = 1,000e^{-0.07t_j}\Delta t$. Hence, over the entire 10 years, the present value of the investment $= \sum_{j=1}^{n} 1,000e^{-0.07t_j}\Delta t$. Now as n increases without bound, this approximation improves and the sum approaches the corresponding integral. Hence, tye present value is $\int_0^{10} 1,000e^{-0.07t}dt = \frac{1,000}{-0.07}e^{-0.07t}|_0^{10} = -\frac{1,000}{0.07}(e^{-0.7} - 1) = \$7,191.64$.

25. Let $f(t) = e^{-0.2t}$ denote the fraction of the homes that will remain unsold for t weeks. Of the 200 homes currently on the market, the number that will still be on the market 10 weeks from now is $200f(10) = 200e^{-2}$. To find the number of additional homes on the market 10 weeks from now, divide the interval $0 \leq t \leq 10$ into n equal subintervals of length Δt weeks and let t_j denote the beginning of the jth subinterval. During the jth subinterval, $8\Delta t$ additional homes are placed on the market, and 10 weeks from now (that is $10 - t_j$ weeks later) the number of these still on the market will be $8(\Delta t)f(10-t_j)$. Hence, the total number N of homes on the market 10 weeks from now will be approximately $N = 200f(10) + \sum_{j=1}^{n} 8f(10 - t_j)\Delta t$. Now, as n increases without bound, this approximation improves while the sum approaches the corresponding integral. Hence, $N = 200f(10) + \int_0^{10} 8f(10 - t)dt = 200e^{-2} + 8\int_0^{10} e^{-0.2(10-t)}dt = 200e^{-2} + 8e^{-2}\int_0^{10} e^{0.2t}dt = 200e^{-2} + \frac{8e^{-2}}{0.2}e^{0.2t}|_0^{10} = 200e^{-2} + 40e^{-2}(e^2 - 1) = 200e^{-2} + 40(1 - e^{-2}) = 160e^{-2} + 40 = 61.65$ or about 62 homes.

26. Since $y = x^2 + 1$ is non-negative over the interval $x = -1$ to $x = 2$, then the volume of revolution is given by $V = \pi\int_a^b [f(x)]^2dx = \pi\int_{-1}^{2}(x^2 + 1)^2dx = \pi\int_{-1}^{2}(x^4 + 2x^2 + 1)dx = \pi(\frac{x^5}{5} + \frac{2x^3}{3} + x)|_{-1}^{2} = \pi[(\frac{32}{5} + \frac{16}{3} + 2) - (-\frac{1}{5} - \frac{2}{3} - 1)] = \frac{234}{15}\pi = 49.01$

27. Since $y = e^{-x/20}$ is non-negative over the interval $x = 0$ to $x = 10$, then the volume of revolution is given by $V = \pi\int_a^b [f(x)]^2dx = \pi\int_0^{10}(e^{-x/20})^2dx = \pi\int_0^{10} e^{-x/10}dx = \pi(-10)e^{-x/10}|_0^{10} = -10\pi(e^{-1} - e^0) = 10\pi(1 - \frac{1}{e}) = 19.86$.

28. Since the price of chicken t months after the beginning of the year is $P(t) = 0.06t^2 -$

$0.2t + 12$ dollars per pound, the average price during the first six months is $\frac{1}{6-0}\int_0^6 (0.06t^2 - 0.2t + 1.2)dx = (\frac{1}{6})(0.02t^3 - 0.1t^2 + 1.2t)|_0^6 = (\frac{1}{6})[0.02(6)^3 - 0.1(6)^2 + 1.2(6)] = \$1.32$ per pound.

29. For $\int_1^3 \frac{1}{x}dx$ with $n = 10$, $\Delta x = \frac{3-1}{10} = 0.2$, and $x_1 = 1$, $x_2 = 1.2$, $x_3 = 1.4, \cdots$, $x_{10} = 2.8$, $x_{11} = 3$.
a) By the trapezoidal rule, $\int_1^3 \frac{1}{x}dx = \frac{\Delta x}{2}[f(x_1) + 2f(x_2) + 2f(x_3) + \cdots + 2f(x_{10}) + f(x_{11})] = \frac{0.2}{2}[1 + \frac{2}{1.2} + \frac{2}{1.4} + \frac{2}{1.6} + \frac{2}{1.8} + \frac{2}{2.0} + \frac{2}{2.2} + \frac{2}{2.4} + \frac{2}{2.6} + \frac{2}{2.8} + \frac{1}{3}] = 1.1016$.
The error estimate is $|E_n| \le \frac{M(b-a)^3}{12n^2}$. For $n = 10$, $a = 1$ and $b = 3$, $|E_{10}| \le \frac{M(3-1)^3}{12(10^2)} = \frac{8M}{1{,}200} = \frac{M}{150}$ where M is the maximum value of $|f''(x)|$ on $1 \le x \le 3$. Now $f(x) = x^{-1}$, $f'(x) = -\frac{1}{x^2}$, and $f''(x) = \frac{2}{x^3}$. For $1 \le x \le 3$, $|f''(x)| \le 2(1^3) = 2$. Hence, $|E_{10}| \le \frac{2}{150} = 0.0133$.
b) By Simpson's rule, $\int_1^3 \frac{1}{x}dx = \frac{\Delta x}{3}[f(x_1) + 4f(x_2) + 2f(x_3) + \cdots + 4f(x_{10}) + f(x_{11})] = \frac{0.2}{3}[1 + \frac{4}{1.2} + \frac{2}{1.4} + \frac{4}{1.6} + \frac{2}{1.8} + \frac{4}{2} + \frac{2}{2.2} + \frac{4}{2.4} + \frac{2}{2.6} + \frac{4}{2.8} + \frac{1}{3}] = 1.0987$.
The error estimate is $|E_n| \le \frac{M(b-a)^5}{180n^4}$. For $n = 10$, $a = 1$ and $b = 3$, $|E_{10}| \le \frac{M(3-1)^5}{180(10^4)} = \frac{32M}{180(10)^4}$ where M is the maximum value of $|f^{(4)}(x)|$ on $1 \le x \le 3$. Now $f''(x) = \frac{2}{x^3}$, $f^{(3)}(x) = -\frac{6}{x^4}(x)$, and $f^{(4)}(x) = \frac{24}{x^5}$. For $1 \le x \le 3$, $|f^{(4)}(x)| = |\frac{24}{1^5}| \le 24$. Hence, $|E_{10}| \le \frac{(32)(24)}{180(10{,}000)} = 0.0004$.

30. For $\int_0^2 e^{x^2}dx$ with $n = 8$, $\Delta x = \frac{2-0}{8} = 0.25$, and $x_1 = 0$, $x_2 = 0.25$, $x_3 = 0.50, \cdots$, $x_8 = 1.75$, $x_9 = 2$.
a) By the trapezoidal rule, $\int_0^2 e^{x^2}dx = \frac{\Delta x}{2}[f(x_1) + 2f(x_2) + 2f(x_3) + \cdots + 2f(x_8) + f(x_9)] = \frac{0.25}{2}[1 + 2e^{(0.25)^2} + 2e^{(0.5)^2} + 2e^{(0.75)^2} + e^1 + 2e^{(1.25)^2} + 2e^{(1.5)^2} + 2e^{(1.75)^2} + e^2] = 17.5651$. The error estimate is $|E_n| \le \frac{M(b-a)^3}{12n^2}$. For $n = 8$, $a = 0$ and $b = 2$, $|E_8| \le \frac{M(2-0)^3}{12(8^2)} = \frac{M}{96}$ where M is the maximum value of $|f''(x)|$ on $0 \le x \le 2$. Now $f(x) = e^{x^2}$, $f'(x) = 2xe^{x^2}$, and $f''(x) = (2 + 4x^2)e^{x^2}$. For $0 \le x \le 2$, $|f''(x)| \le [2 + 4(2)^2]e^{2^2} = 18e^4$. Hence, $|E_8| \le \frac{18e^4}{96} = 10.2372$.
b) By Simpson's rule, $\int_0^2 e^{x^2}dx = \frac{\Delta x}{3}[f(x_1) + 4f(x_2) + 2f(x_3) + 4f(x_4) + \cdots + f(x_9)] = \frac{0.25}{3}[1 + 4e^{(0.25)^2} + 2e^{(0.5)^2} + 4e^{(0.75)^2} + 2e^1 + 4e^{(1.25)^2} + 2e^{(1.5)^2} + 4e^{(1.75)^2} + 2e^2] = 16.5386$. The error estimate is $|E_n| \le \frac{M(b-a)^5}{180n^4}$. For $n = 8$, $a = 0$ and $b = 2$, $|E_8| \le \frac{M(2-0)^5}{180(8^4)} = \frac{M}{23{,}040}$ where M is the maximum value of $|f^{(4)}(x)|$ on $0 \le x \le 2$. Now $f^{(3)}(x) = (8x^3 + 12x)e^{x^2}$, and $f^{(4)}(x) = (16x^4 + 48x^2 + 12)e^{x^2}$. For $0 \le x \le 2$, $|f^{(4)}(x)| = [16(2)^4 + 48(2)^2 + 12]e^{2^2} \le 460e^4$. Hence, $|E_8| \le \frac{460e^4}{23{,}040} = 1.0901$.

31. The integral to be approximated is $\int_1^3 \sqrt{x}dx$. The derivatives of $f(x) = \sqrt{x} = x^{1/2}$ are $f'(x) = \frac{1}{2}x^{-1/2}$, $f''(x) = -\frac{1}{4}x^{-3/2}$, $f^{(3)}(x) = \frac{3}{8}x^{-5/2}$, and $f^{(4)}(x) = -\frac{15}{16}x^{-7/2}$.
a) For the trapezoidal rule, $|E_n| \le \frac{M(b-a)^3}{12n^2}$, where M is the maximum value of $|f''(x)|$ on $1 \le x \le 3$. Now $|f''(x)| = \frac{1}{4}x^{-3/2} \le \frac{1}{4}$ on $1 \le x \le 3$. $|E_n| \le \frac{1}{4}\frac{(3-1)^3}{12n^2} = \frac{1}{6n^2}$ which is less than $0.000\,05$ if $1 < 6(0.000\,05)n^2$ or $n > \sqrt{\frac{1}{6(0.000\,05)}} \approx 57.7$. Hence, 58 intervals should be used.

b) For Simpson's rule, $|E_n| \le \frac{M(b-a)^5}{180n^4}$, where M is the maximum value of $|f^{(4)}(x)|$ on $1 \le x \le 3$. Now $|f^{(4)}(x)| = |-\frac{15}{16}x^{-7/2}| \le \frac{15}{16}$ on $1 \le x \le 3$. $|E_n| \le \frac{15(3-1)^5}{16(180)n^4} = \frac{1}{6n^4}$ which is less than 0.000 05 if $1 < 6(0.000\ 05)n^4$ or $n > \sqrt[4]{\frac{1}{6(0.000\ 05)}} \approx 7.6$. Hence, 8 subintervals should be used.

32. The integral to be approximated is $\int_{0.5}^{1} e^{x^2}\,dx$. The derivatives of $f(x) = e^{x^2}$ are $f'(x) = 2xe^{x^2}$, $f''(x) = (4x^2+2)e^{x^2}$, $f^{(3)}(x) = (8x^3+12x)e^{x^2}$, and $f^{(4)}(x) = (16x^4+48x^2+12)e^{x^2}$.
a) For the trapezoidal rule, $|E_n| \le \frac{M(b-a)^3}{12n^2}$, where M is the maximum value of $|f''(x)|$ on $0.5 \le x \le 1$. Now $|f''(x)| = |4(1^2)+2|e^1 = 6e$ on $0.5 \le x \le 1$. $|E_n| \le \frac{6e(1-0.5)^3}{12n^2} = \frac{e}{16n^2}$ which is less than 0.000 05 if $e < 16(0.000\ 05)n^2$ or $n > \sqrt{\frac{e}{16(0.000\ 05)}} \approx 58.3$. Hence, 59 intervals should be used.
b) For Simpson's rule, $|E_n| \le \frac{M(b-a)^5}{180n^4}$, where M is the maximum value of $|f^{(4)}(x)|$ on $0.5 \le x \le 1$. Now $|f^{(4)}(x)| = |16(1)^4 + 48(1)^2 + 12|e^1 \le 76e$ on $0.5 \le x \le 1$. $|E_n| \le \frac{76e(1-0.5)^5}{180n^4} = \frac{19e}{1{,}440n^4}$ which is less than 0.000 05 if $19e < 1{,}440(0.000\ 05)n^4$ or $n > \sqrt[4]{\frac{19e}{1{,}440(0.000\ 05)}} \approx 5.2$. Hence, 6 subintervals should be used.

Chapter 7

Differential Equations

7.1 Introduction: Separable Differential Equations

1. Let Q denote the number of bacteria. Then, $\frac{dQ}{dt}$ is the rate of change of Q, and since this rate of change is proportional to Q, it follows that $\frac{dQ}{dt} = kQ$, where k is a positive constant of proportionality.

3. Let Q denote the investment. Then $\frac{dQ}{dt}$ is the rate of change of Q, and since this rate of change is equal to 7 % of the size of Q, it follows that $\frac{dQ}{dt} = 0.07Q$.

5. Let P denote the population. Then $\frac{dP}{dt}$ is the rate of change of P, and since this rate of change is the constant 500, it follows that $\frac{dP}{dt} = 500$.

7. Let Q denote the temperature of the object and M the temperature of the surrounding medium. Then $\frac{dQ}{dt}$ is the rate of change of Q, and since this rate of change is proportional to the difference between the temperature of the object and that of the surrounding medium, it follows that $\frac{dQ}{dt} = k(M - Q)$, where k is a positive constant of proportionality. (Notice that if $M > Q$, $\frac{dQ}{dt}$ is positive and the temperature Q is increasing.)

9. Let Q denote the number of facts recalled and N the total number of relevant facts in the person's memory. Then $\frac{dQ}{dt}$ is the rate of change of Q, and $N - Q$ is the number of relevant facts not recalled. Since the rate of change is proportional to $N - Q$, it follows that $\frac{dQ}{dt} = k(N - Q)$, where k is a positive constant of proportionality.

11. Let Q denote the number of people implicated and N the total number of people not to be implicated. Then $\frac{dQ}{dt}$ is the rate of change of Q, and $N - Q$ is the number involved but not implicated. Since the rate of change is jointly proportional to Q and $N - Q$, it follows

that $\frac{dQ}{dt} = kQ(N-Q)$, where k is a positive constant of proportionality.

13. If $y = Ce^{kx}$, then $\frac{dy}{dx} = kCe^{kx} = ky$.

15. If $y = C_1e^x + C_2xe^x$, then $\frac{dy}{dx} = C_1e^x + C_2(xe^x + e^x) = (C_1+C_2)e^x + C_2xe^x$ and $\frac{d^2y}{dx^2} = (C_1+C_2)e^x + C_2(xe^x+e^x) = (C_1+2C_2)e^x + C_2xe^x$. Thus $\frac{d^2y}{dx^2} - 2\frac{dy}{dx} + y = (C_1+2C_2)e^x + C_2xe^x - 2[(C_1+C_2)e^x + C_2xe^x] + C_1e^x + C_2xe^x = (C_1+2C_2-2C_1-2C_2+C_1)e^x + (C_2-2C_2+C_2)xe^x = 0$

17. If $\frac{dy}{dx} = 3x^2 + 5x - 6$, then $y = \int(3x^2+5x-6)dx = x^3 + \frac{5x^2}{2} - 6x + C$.

19. If $\frac{dV}{dx} = \frac{2}{x+1}$, then $V = \int \frac{2}{x+1}dx = 2\ln|x+1| + C = \ln(x+1)^2 + C$.

21. If $\frac{d^2P}{dt^2} = 50$, then $\frac{dP}{dt} = \int 50dt = 50t + C_1$ and $P = \int(50t + C_1)dt = 25t^2 + C_1t + C_2$.

23. Separate the variables of $\frac{dy}{dx} = 3y$ and integrate to get $\int \frac{1}{y}dy = \int 3dx$, $\ln|y| = 3x + C_1$, $|y| = e^{3x+C_1} = e^{C_1}e^{3x}$ or $y = \pm e^{C_1}e^{3x} = Ce^{3x}$ where C is the constant $\pm e^{C_1}$.

25. Separate the variables of $\frac{dy}{dx} = e^y$ and integrate to get $\int e^{-y}dy = \int dx$, $-e^{-y} = x + C_1$ or $e^{-y} = C - x$, where C is the constant $-C_1$. Hence, $\ln e^{-y} = \ln(C-x)$, $-y = \ln(C-x)$, or $y = -\ln(C-x)$.

27. Separate the variables of $\frac{dy}{dx} = \frac{x}{y}$ and integrate to get $\int ydy = \int xdx$, $\frac{y^2}{2} = \frac{x^2}{2} + C_1$ or $y^2 = x^2 + C$, $y = \pm\sqrt{x^2+C}$, where C is the constant $2C_1$.

29. Separate the variables of $\frac{dy}{dx} = y + 10$ and integrate to get $\int \frac{1}{y+10}dy = \int dx$, $\ln|y+10| = x + C_1$ or $|y+10| = e^{C_1+x}$, $y + 10 = \pm e^{C_1+x} = \pm e^{C_1}e^x$, $y = \pm e^{C_1}e^x - 10 = Ce^x - 10$, where C is the constant $\pm e^{C_1}$.

31. Separate the variables of $\frac{dy}{dx} = y(1-2y)$ and integrate to get $\int \frac{1}{y(1-2y)}dy = \int dx$, $\int \frac{1}{y}dy + 2\int \frac{1}{1-2y}dy = \int dx$, $\ln|y| - \ln|1-2y| = x + C$, $\ln|\frac{y}{1-2y}| = x + C$, $|\frac{y}{1-2y}| = e^{x+C} = e^xe^C$, or $\frac{y}{1-2y} = Ae^x$, where $A = \pm e^C$, $y = Ae^x - 2yAe^x$, $y = \frac{Ae^x}{1+2Ae^x} = \frac{1}{(1/A)e^{-x}+2} = \frac{1}{2+Be^{-x}}$, where $B = \frac{1}{A}$.

33. $\frac{dy}{dx} = 5x^4 - 3x^2 - 2$, $y = x^5 - x^3 - 2x + C$. Since $y = 4$ when $x = 1$, $C = 6$ and $y = x^5 - x^3 - 2x + 6$.

35. If $\frac{d^2A}{dt^2} = e^{-t/2}$, then $\frac{dA}{dt} = \int e^{-t/2}dt = -2e^{-t/2} + C_1$. Since $\frac{dA}{dt} = 1$ when $t = 0$, $1 = -2e^0 + C_1$ or $C_1 = 3$. Hence $\frac{dA}{dt} = -2e^{-t/2} + 3$ and $A = \int(-2e^{-t/2}+3)dt = 4e^{-t/2} + 3t + C_2$. Since $A = 2$ when $t = 0$, $2 = 4e^0 + 3(0) + C_2$ or $C_2 = -2$. Hence, $A = 4e^{-t/2} + 3t - 2$.

37. Separate the variables of $\frac{dy}{dx} = 4x^3y^2$ and integrate to get $\int \frac{1}{y^2}dy = \int 4x^3dx$, $-\frac{1}{y} = x^4 + C$, or $y = -\frac{1}{x^4+C}$. Since $y = 2$ when $x = 1$, $2 = -\frac{1}{1+C}$, $2 + 2C = -1$, or $C = -\frac{3}{2}$.

Hence, $y = -\frac{1}{x^4-3/2} = \frac{2}{3-2x^4}$.

39. $\frac{dy}{dx} = y(1-y)$, $\frac{dy}{y(1-y)} = dx$.
Before proceeding let's break up the fraction (by the method of partial fractions). $\frac{1}{y(1-y)} = \frac{A}{y} + \frac{B}{1-y}$. Now multiply by the least common denominator. $1 = A(1-y) + By$ which is rewritten $1 = (B-A)y + A$. The constants must agree in both members of the equation, so $A = 1$. Since there is no variable in the left member, $B = 1$ to make $(B-A)y = 0$.
$\int \frac{dy}{y(1-y)} = \int(\frac{1}{y} + \frac{1}{1-y})dy = \int dx$. $\ln|y| - \ln|1-y| = x + C$, $\ln|\frac{y}{1-y}| = x + C$. Since $y = \frac{1}{3}$ when $x = 0$, $C = \ln|\frac{1}{2}| = -\ln 2$. Thus $\ln|\frac{y}{1-y}| = x - \ln 2$, $\ln|\frac{y}{1-y}| + \ln 2 = \ln|\frac{2y}{1-y}| = x$, $\frac{2y}{1-y} = e^x$, $2y - e^x + e^x y = 0$, $y(2+e^x) = e^x$, $y = \frac{e^x}{2+e^x} = \frac{1}{1+2e^{-x}}$.

41. a) $\frac{dC}{dq} = 3(q-4)^2$. $C(q) = (q-4)^3 + C_0$.
b) The cost of producing 0 units is \$436, so $436 = (-4)^3 + C_0$. Thus $C_0 = 436 + 64 = 500$ and $C(14) = (14-4)^3 + 500 = 1,000 + 500 = \$1,500$.

43. a) Let $Q(t)$ denote the ozone level t hours after 7:00 a.m. Since $\frac{dQ}{dt} = \frac{0.24-0.03t}{\sqrt{36+16t-t^2}} = (0.24-0.03t)(36+16t-t^2)^{-1/2}$ parts per million per hour, then by substituting $u = 36 + 16t - t^2$, $Q(t) = \int(0.24-0.03t)(36+16t-t^2)^{-1/2}dt = 0.03\int(8-t)(36+16t-t^2)^{-1/2}dt = 0.03(36+16t-t^2)^{1/2} + C$. Since the ozone level was 0.25 at 7:00 a.m., $0.25 = Q(0) = 0.03\sqrt{36} + C = 0.18 + C$ or $C = 0.07$. Hence, $Q(t) = 0.03(36+16t-t^2)^{1/2} + 0.07$.
b) The peak ozone level occurs when $\frac{dQ}{dt} = 0$, that is, when $0.24 - 0.03t = 0$ or $t = 8$. Note that $Q(t)$ has its absolute maximum at this critical point since $\frac{dQ}{dt} > 0$ (Q is increasing) for $0 < t < 8$, and $\frac{dQ}{dt} < 0$ (Q is decreasing) for $8 < t$. Thus the peak ozone level is $Q(8) = 0.03(36 + 16(8) - 8^2)^{1/2} + 0.07 = 0.03(10) + 0.07 = 0.37$ parts per million, which occurs at 3:00 p.m. (8 hours after 7:00 a.m.).

45. $\frac{dP}{dt} = 5(0.6t^2 + 0.2t + 0.5)$ units of pollution per year, $P_0 = 60$, so $P(t) = 5(0.2t^3 + 0.1t^2 + 0.5t) + 60$ and $P(2) = 5(1.6 + 0.4 + 1.0) + 60 = 75$ units.

47. Let $C(t)$ be the concentration of the drug. Then $\frac{dC}{dt} = -kC$, $\int C^{-1}dC = \int k dt$, $\ln|C| = -kt + K_1$, $|C| = e^{-kt+K_1} = e^{K_1}e^{-kt} = Ke^{-kt}$, where $K = e^{K_1} > 0$. Thus $C(t) = Ke^{-kt}$.

49. Let y be the differentiable function. Then $\frac{dy}{dx} = y$, $\int y^{-1}dy = \int dx$, $\ln|y| = x + C_1$, $|y| = e^{x+C_1} = e^{C_1}e^x = Ce^x$, where $C = \pm e^{C_1}$. Thus $y = Ce^x$.

51. Let Q_0 be the original amount of sugar and $s(t)$ the amount thereof that has dissolved. Then $\frac{ds}{dt} = k(Q_0 - s)$, $-\int \frac{d(Q_0-s)}{Q_0-s} = -\int k\,dt$, $\ln|Q_0 - s| = -kt + C_1$, $|Q_0 - s| = e^{C_1 - kt}$. With $C = \pm e^{C_1}$, $Q_0 - s = Ce^{-kt}$, $s = Q_0 - Ce^{-kt}$. At $t = 0$ $s(0) = Q_0$, thus $s(t) = Q_0 - Q_0e^{-kt} = Q_0(1 - e^{-kt})$.

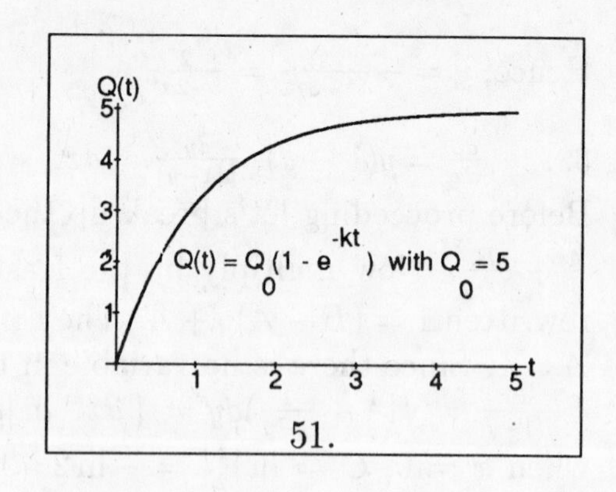

51.

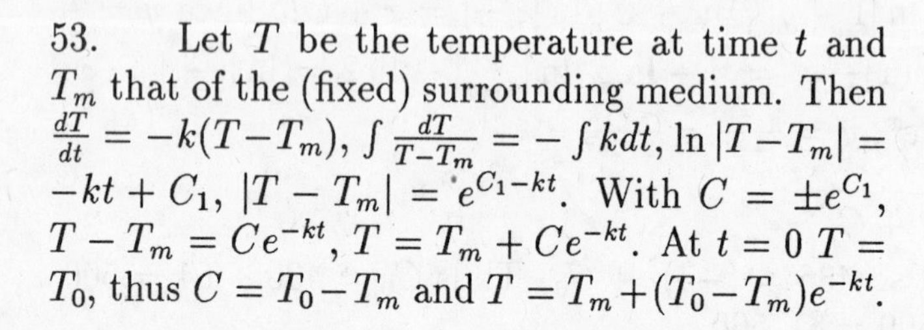
53. Let T be the temperature at time t and T_m that of the (fixed) surrounding medium. Then $\frac{dT}{dt} = -k(T - T_m)$, $\int \frac{dT}{T - T_m} = -\int k\,dt$, $\ln|T - T_m| = -kt + C_1$, $|T - T_m| = e^{C_1 - kt}$. With $C = \pm e^{C_1}$, $T - T_m = Ce^{-kt}$, $T = T_m + Ce^{-kt}$. At $t = 0$ $T = T_0$, thus $C = T_0 - T_m$ and $T = T_m + (T_0 - T_m)e^{-kt}$.

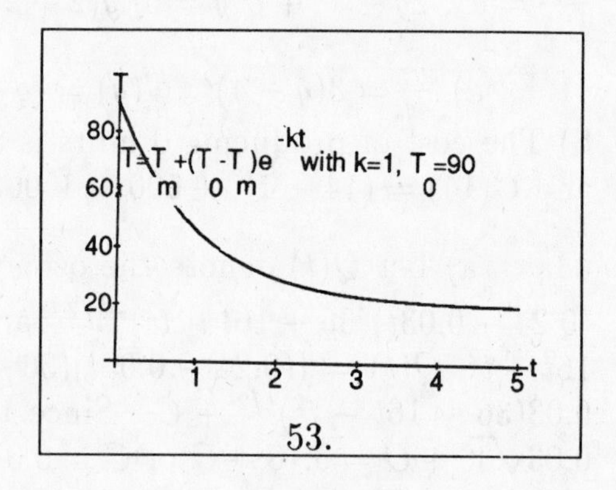

53.

55. Let $Q(t)$ denote the amount of salt in the tank and t the number of minutes that have elapsed. Then, $\frac{dQ}{dt}$ =rate of change of salt with respect to time (pounds per minute)=(rate at which salt enters the tank)−(rate at which salt leaves the tank)= $\frac{\text{pounds entering}}{\text{gallon}}\frac{\text{gallons entering}}{\text{minute}} - \frac{\text{pounds leaving}}{\text{gallon}}\frac{\text{gallons leaving}}{\text{minute}}$.

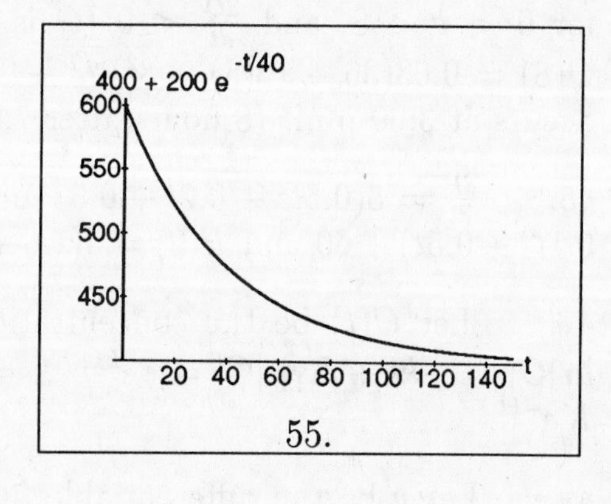

55.

Since $\frac{\text{pounds leaving}}{\text{gallon}} = \frac{\text{pounds of salt in the tank}}{\text{gallon of mixture in the tank}} = \frac{Q}{200}$ it follows that $\frac{dQ}{dt} = 2(5) - (\frac{Q}{200})(5) = 10 - \frac{Q}{40} = \frac{400-Q}{40}$. Separate the variables and integrate to get $\int \frac{1}{400-Q}dQ = \int \frac{1}{40}dx$, $-\ln|400 - Q| = \frac{t}{40} + C_1$, $\ln|400 - Q| = -\frac{t}{40} - C_1$, $400 - Q = \pm e^{-C_1}e^{-t/40}$, or $Q(t) = 400 - Ce^{-t/40}$, where C is the constant $\pm e^{-C_1}$. The absolute value can be dropped since $400 - Q > 0$.
The tank initially held 200 gallons containing 3 pounds of salt per gallon for a total of 600 pounds of salt, $600 = Q(0) = 400 - C$ or $C = -200$. Hence, $Q(t) = 400 - (-200e^{-t/40}) = 400 + 200e^{-t/40}$. As t increases without bound, $e^{-t/40}$ approaches 0, so $Q(0) = 400$.

57. Let t denote time, Q the number of residents who have been infected, and B the

total number of susceptible residents. The differential equation describing the spread of the epidemic is $\frac{dQ}{dt} = kQ(B-Q)$, where k is a positive constant of proportionality. Notice that this is the same differential equation as that in example 1.10 of the text. The solution was found to be $Q(t) = \frac{B}{1+Ae^{-Bkt}}$. Since there are 2,000 susceptible residents, $B = 2,000$. Since 500 residents had the disease initially, $Q(0) = 500$ and so $500 = \frac{2,000}{1+Ae^0}$, $500 + 500A = 2,000$, or $A = 3$. Hence, $Q(t) = \frac{2,000}{1+3e^{-2,000kt}}$. To find k, use the fact that $Q(1) = 855$ to get $855 = \frac{2,000}{1+3e^{-2,000k}}$, $855 + 2,565e^{-2,000k} = 2,000$ or $2,565e^{-2,000k} = 1,145$. Now $-2,000k = \ln\frac{1,145}{2,565} = -0.8066$. Hence the function giving the number of residents who have been infected after t weeks is (approximately) $Q(t) = \frac{2,000}{1+3e^{-0.8066t}}$.

7.1.59 Let P be the number of people involved and Q the number of people implicated. $\frac{dQ}{dt} = kQ(P-Q)$, $\frac{dQ}{Q(P-Q)} = kdt$.
Before proceeding let's break up the fraction (by the method of partial fractions). $\frac{1}{Q(P-Q)} = \frac{A}{Q} + \frac{B}{P-Q}$. Now multiply by the least common denominator. $1 = A(P-Q) + BQ$ which must be an identity (that is true **for all values**), so when $Q = 0$ $A = \frac{1}{P}$, and when $Q = P$ $B = \frac{1}{P}$.
$\int\frac{dQ}{Q(P-Q)} = \frac{1}{P}\int(\frac{dQ}{Q} + \frac{dQ}{P-Q}) = k\int dt$. $\frac{1}{P}[\ln|Q| - \ln|P-Q|] = kt + C_1$, $\ln|\frac{Q}{P-Q}| = kPt + C_1$, $\frac{Q}{P-Q} = Ce^{kPt}$. Since $Q = 7$ when $t = 0$, $C = \frac{7}{P-7}$. When $t = 3$ $Q = 16$, $\frac{16}{P-16} = \frac{7}{P-7}e^{3Pk}$. When $t = 6$ $Q = 28$ and $\frac{28}{P-28} = \frac{7}{P-7}e^{6Pk} = \frac{7}{P-7}[\frac{16(P-7)}{7(P-16)}]^2$ after substituting for $e^{3kP} = \frac{16(P-7)}{7(P-16)}$. $(28)(7)(P^2 - 32P + 256) = 256(P-28)(P-7) = 256(P^2 - 35P + 196)$, $196P^2 - 6,272P + 50,176 = 256P^2 - 8,960P + 50,176$, $(256-196)P^2 + (6,272 - 8,960)P = 0$, and $P = 0$ (to be rejected in the context of this problem) or $P = \frac{2,688}{60} = \frac{672}{15} = 44.8 \approx 45$ people.

7.1.61 Let p be the price of the commodity. Its rate of change is proportional to the surplus, so $\frac{dp}{dt} = k[7 - p - (1+p)] = k(6-2p)$, where k is a positive constant of proportionality. Separation of variables and integration leads to $\int\frac{1}{3-p}dp = \int 2kdt$, $-\ln|3-p| = 2kt + C_1$, $\ln|p-3| = -2kt - C_1$, $p - 3 = e^{C_1}e^{-2kt} = Ce^{-2kt}$. The absolute sign was dropped since C can conveniently be positive or negative as the need prescribes. Since $p = 6$ when $t = 0$, $3 = C$ and $p(t) = 3 + 3e^{-2kt}$. Similarly $p(4) = 4$ so $4 - 3 = 1 = 3e^{-8k}$. $\frac{1}{3} = e^{-8k}$, $-8k = \ln\frac{1}{3} = -\ln 3$, or $k = \frac{\ln 3}{8}$. Thus $p(t) = 3 + 3e^{-2t(\ln 3)/8} = 3 + 3e^{-(\ln 3)t/4}$.
b) $\lim_{t\to\infty} p(t) = 3$ since the exponential shrinks to 0 when t increases beyond all bounds. Also $D = S$ when $7 - p = 1 + p$, $2p = 6$ or $p = 3$.

63. Let S be the concentration of the solute inside the cell, S_0 that of the solute outside the cell, and A the area of the cell wall. The rate of change of the inside solute is jointly proportional to the area of the cell surface and the difference between the solute inside and outside the wall, so $\frac{dS}{dt} = kA(S_0 - S)$, where k is a positive constant of proportionality, S_0 is constant, and so is A. Separation of variables and integration leads to $\int\frac{1}{S_0 - S}dS = \int kAdt$,

$-\ln|S_0 - S| = kAt + C_1$, $\ln|S - S_0| = -Akt - C_1$, $S - S_0 = \pm e^{-C_1}e^{-kAt} = Ce^{-kAt}$. The absolute sign was dropped since C can conveniently be positive or negative as the need prescribes. Thus $S = S_0 + Ce^{-kAt}$.

7.2 First Order Linear Differential Equations

1. $\frac{dy}{dx} + (\frac{3}{x})y = x$. The integrating factor is $I(x) = e^{\int (3/x)dx} = e^{3\ln x} = e^{\ln x^3} = x^3$. The constant of integration is arbitrary in an integrating factor, so we make it easy on ourselves by letting it be 0. Multiplying the original equation by x^3 gives $x^3\frac{dy}{dx} + x^3(\frac{3}{x})y = x(x^3)$ or $x^3\frac{dy}{dx} + 3x^2y = x^4$. The left-hand member can be written as the derivative of a product, so $\frac{d}{dx}(x^3y) = x^4$, $d(x^3y) = x^4dx$, and $\int d(x^3y) = \int x^4dx$. Thus $x^3y = \frac{x^5}{5} + C$ (remember the constant of integration **immediately** after the last integration) or $y = \frac{x^2}{5} + \frac{C}{x^3}$.

3. $\frac{dy}{dx} + (\frac{1}{2x})y = \sqrt{x}e^x$. The integrating factor is $I(x) = e^{\int 1/(2x)dx} = e^{(1/2)\ln x} = e^{\ln x^{1/2}} = x^{1/2} = \sqrt{x}$. The constant of integration is arbitrary in an integrating factor, so we make it easy on ourselves by letting it be 0. Multiplying the original equation by $\sqrt{x}$ gives $\sqrt{x}\frac{dy}{dx} + \sqrt{x}(\frac{1}{2x})y = (\sqrt{x})^2e^x$, $\sqrt{x}\frac{dy}{dx} + (\frac{1}{2\sqrt{x}})y = xe^x$. The left-hand member can be written as the derivative of a product, so $\frac{d}{dx}(\sqrt{x}y) = xe^x$, $d(\sqrt{x}y) = xe^xdx$. Thus $\int d(\sqrt{x}y) = \int xe^xdx$, $\sqrt{x}y = xe^x - e^x + C$. This last result is due to an integration by parts.

$f(x) = x$	$g(x) = e^x$
$f'(x) = 1$	$G(x) = e^x$

$I = xe^x - \int e^xdx = xe^x - e^x + C.$

Thus $y = \sqrt{x}e^x - \frac{e^x}{\sqrt{x}} + \frac{C}{\sqrt{x}}$

5. $x^2\frac{dy}{dx} + xy = 2$ is rewritten so that the coefficient of $\frac{dy}{dx}$ is 1. Thus $\frac{dy}{dx} + (\frac{1}{x})y = \frac{2}{x^2}$. The integrating factor is $I(x) = e^{\int 1/x\,dx} = e^{\ln x} = x$. Multiplying the original equation by x gives $x\frac{dy}{dx} + y = \frac{2}{x}$ The left-hand member can be written as the derivative of a product, so $\frac{d}{dx}(xy) = \frac{2}{x}$, $d(xy) = \frac{2}{x}dx$. Thus $\int d(xy) = \int \frac{2}{x}dx$, $xy = 2\int \frac{1}{x}dx = 2\ln|x| + C$. Thus $y = \frac{2}{x}\ln|x| + \frac{C}{x}$

7. $\frac{dy}{dx} + (\frac{2x+1}{x})y = e^{-2x}$. The integrating factor is $I(x) = e^{\int (2x+1)/x\,dx} = e^{\int (2+1/x)\,dx} = e^{2x+\ln x} = e^{2x}e^{\ln x} = xe^{2x}$. Multiplying the original equation by xe^{2x} gives $xe^{2x}\frac{dy}{dx} + xe^{2x}(\frac{2x+1}{x})y = xe^{2x}e^{-2x}$ or $xe^{2x}\frac{dy}{dx} + (2x+1)e^{2x}y = x$. The left-hand member can be written as the derivative of a product, so $\frac{d}{dx}(xye^{2x}) = x$, $d(xye^{2x}) = xdx$, and $\int d(xye^{2x}) = \int xdx$. Thus $xye^{2x} = \frac{x^2}{2} + C$ or $y = \frac{xe^{-2x}}{2} + \frac{Ce^{-2x}}{x}$

9. $x\frac{dy}{dx} - 2y = 2x^3$ is rewritten so that the coefficient of $\frac{dy}{dx}$ is 1. Thus $\frac{dy}{dx} - \frac{2}{x}y = 2x^2$. The integrating factor is $I(x) = e^{\int(-2/x)dx} = e^{-2\ln x} = e^{\ln x^{-2}} = x^{-2}$. Multiplying the original equation by x^2 gives $x^{-2}\frac{dy}{dx} - x^{-2}\frac{2}{x}y = 2$ or $x^{-2}\frac{dy}{dx} - \frac{2}{x^3}y = 2$ The left-hand member can be written as the derivative of a product, so $\frac{d}{dx}(\frac{y}{x^2}) = 2$, $d(\frac{y}{x^2}) = 2dx$. Thus $\int d(\frac{y}{x^2}) = \int 2dx$, $\frac{y}{x^2} = 2x + C$, or $y = 2x^3 + Cx^2$. Since $y = 2$ when $x = 1$, $2 = 2 + C$, $C = 0$ and thus $y = 2x^3$.

11. $\frac{dy}{dx} + y = x$. The integrating factor is $I(x) = e^{\int dx} = e^x$. Multiplying the original equation by e^x gives $e^x\frac{dy}{dx} + e^x y = xe^x$. The left-hand member can be written as the derivative of a product, so $\frac{d}{dx}(ye^x) = xe^x$, $d(ye^x) = xe^x dx$, and $\int d(ye^x) = \int xe^x dx$. Thus $ye^x = xe^x - e^x + C$ or $y = x - 1 + Ce^{-x}$. Since $y = 4$ when $x = 0$, $4 = 0 - 1 + Ce^0$, $C = 5$ and thus $y = x - 1 + 5e^{-x}$.

13. Method 1, separation of variables:
$\frac{dy}{dx} + 3y = 5$, $\frac{dy}{dx} = 5 - 3y$, $\frac{dy}{5-3y} = dx$, $\int \frac{dy}{5-3y} = \int dx$, $-\frac{1}{3}\ln|5 - 3y| = x + C_1$, $\ln|5 - 3y| = -3x - 3C_1$, $|5 - 3y| = e^{-3x-3C_1} = e^{-3x}e^{-3C_1} = C_2e^{-3x}$, $3y = 5 + C_2e^{-3x}$. Thus $y = \frac{5}{3} + \frac{C_2}{3}e^{-3x} = \frac{5}{3} + C_3e^{-3x}$. ($C_3$ is really $-C_3$ if $5 - 3y < 0$, which is still an arbitrary constant.)
Method 2, first order linear differential equation:
$\frac{dy}{dx} + 3y = 5$. The integrating factor is $I(x) = e^{\int 3dx} = e^{3x}$. Multiplying the original equation by e^{3x} gives $e^{3x}\frac{dy}{dx} + 3e^{3x}y = 5e^{3x}$. The left-hand member can be written as the derivative of a product, so $\frac{d}{dx}(ye^{3x}) = 5e^{3x}$, $d(ye^{3x}) = 5e^{3x}dx$, and $\int d(ye^{3x}) = \int 5e^{3x}dx$. Thus $ye^{3x} = \frac{5}{3}e^{3x} + C_4$ or $y = \frac{5}{3} + C_4e^{-3x}$.
Note that $C_3 = C_4$ in the final answers for the two methods above.

15. The slope m is also the first derivative of the function y. Thus $\frac{dy}{dx} = x + y$ or $\frac{dy}{dx} - y = x$. The integrating factor is $I(x) = e^{\int(-1)dx} = e^{-x}$. Multiplying the original equation by e^{-x} gives $e^{-x}\frac{dy}{dx} - e^{-x}y = xe^{-x}$. The left-hand member can be written as the derivative of a product, so $\frac{d}{dx}(ye^{-x}) = xe^{-x}$, $d(ye^{-x}) = xe^{-x}dx$, and $\int d(ye^{-x}) = \int xe^{-x}dx$. Thus (using integration by parts) $ye^{-x} = -xe^{-x} - e^{-x} + C$ or $y = -x - 1 + Ce^x$. Since the graph passes through the point $(-1, 2)$, $2 = 1 - 1 + Ce^{-1}$ or $2e = C$. Thus $y = -x - 1 + 2ee^x$, $y = -x - 1 + 2e^{x+1}$.

17. Let $P(t)$ be the price of the house t months from now. The rate of change of the price is the first derivative. Thus $\frac{dP}{dt} = 0.01P + 1{,}000t$ or $\frac{dP}{dt} - 0.01P = 1{,}000t$. The integrating factor is $I(x) = e^{\int(-0.01)dt} = e^{-0.01t}$. Multiplying the original equation by $e^{-0.01t}$ gives $e^{-0.01t}\frac{dP}{dt} - 0.01e^{-0.01t}P = 1{,}000te^{-0.01t}$. The left-hand member can be written as the derivative of a product, so $\frac{d}{dx}(Pe^{-0.01t}) = 1{,}000te^{-0.01t}$, $d(Pe^{-0.01t}) = 1{,}000te^{-0.01t}dt$, and $\int d(Pe^{-0.01t}) = \int 1{,}000te^{-0.01t}dt$. Thus (using integration by parts) $Pe^{-0.01t} = 1{,}000[-\frac{t}{0.01}e^{-0.01t} + \frac{1}{0.01}\int e^{-0.01t}dt] = 10^7[-\frac{t}{100}e^{-0.01t} - e^{-0.01t}] + C$ or $P(t) = 10^7[-\frac{t}{100} - 1] + Ce^{0.01t}$. Since the original value of the house is \$200,000 it follows that $200{,}000 = -10{,}000{,}000 + C$ or

$C = 10,200,000$. Thus $P(t) = 10^7[-\frac{t}{100} - 1] + 10,200,000e^{0.01t}$. In 9 months, when $t = 9$, $P(9) = 10^7[-\frac{9}{100} - 1] + 10,200,000e^{0.09} = 10^7[-\frac{9}{100} - 1] + 10,200,000e^{0.09} = -10,900,000 + 11,160,578 = \$260,578$.

19. a) Let $P(t)$ be the amount of salary at time t, $0 \leq t \leq N$ and $S(t)$ the savings accrued due to $P(t)$ from time t to N. $P(t) = 27,000e^{0.09t}$ is the salary at time t (compounded continuously at 0.09%). Five percent of this salary goes into savings, or $0.05P(t) = 1,350e^{0.09t}$. The money is in the savings account for $N-t$ months, so it grows to $S(t) = 1,350e^{0.09t}e^{0.08(N-t)} = 1,350e^{0.08N}e^{0.01t}$ when compounded continuously at 8%. This is the savings per period of time Δt.
b) In 20 years, $N = 20$. The total value accumulated over 20 years is the integral $S(20) = \int_0^{20} 1,350e^{0.08N}e^{0.01t}dt = \frac{1,350}{0.01}e^{0.08(20)}[e^{0.01(20)} - 1] = 135,000e^{1.6}(e^{0.2} - 1) = \$148,043$.

21. Let $Q(t)$ denote the amount of salt in the tank and t the number of minutes that have elapsed. Then, $\frac{dQ}{dt}$ =rate of change of salt with respect to time (pounds per minute)=(rate at which salt enters the tank)−(rate at which salt leaves the tank)= $\frac{\text{pounds entering}}{\text{gallon}}\frac{\text{gallons entering}}{\text{minute}} - \frac{\text{pounds leaving}}{\text{gallon}}\frac{\text{gallons leaving}}{\text{minute}}$. 1 more gallon of brine flows into the tank than flows out every minue. Thus the quantity of solution in the tank after t minutes is $30+t$. Since $\frac{\text{pounds leaving}}{\text{gallon}} = \frac{\text{pounds of salt in the tank}}{\text{gallon of mixture in the tank}} = \frac{Q}{30+t}$ it follows that $\frac{dQ}{dt} = 2 - (\frac{Q}{30+t})(1)$ or $\frac{dQ}{dt} + (\frac{Q}{30+t}) = 2$. The integrating factor is $I(x) = e^{\int 1/(30+t)dt} = e^{\ln(30+t)} = 30 + t > 0$. Multiplying the original equation by $30 + t$ gives $(30 + t)\frac{dQ}{dt} + Q = 2(30 + t)$. The left-hand member can be written as the derivative of a product, so $\frac{d}{dt}(30 + t)Q = 60 + 2t$, $d[(30 + t)Q] = (60 + 2t)dt$, and $\int d[(30 + t)Q] = \int(60 + 2t)dt$. Thus $(30 + t)Q = 60t + t^2 + C$ or $Q(t) = \frac{t(60+t)}{30+t} + \frac{C}{30+t}$. Since the original amount of salt is 10 pounds $10 = \frac{C}{30}$ or $C = 300$. Thus $Q(t) = \frac{t(60+t)}{30+t} + \frac{300}{30+t}$.
b) It will take 20 minutes to fill the tank. Thus $Q(20) = \frac{20(60+20)}{30+20} + \frac{300}{30+20} = \frac{2(80)}{5} + \frac{30}{5} = 32 + 6 = 38$ pounds of salt.

23. The original amount of carbon dioxide in the room is $0.0025(2,500) = 6.25$ cubic feet. The amount of carbon dioxide entering the room every minute is $0.0003(1,000) = 0.3$ cubic feet per minute. Let $y(t)$ be the amount of carbon dioxide present in the room at any time t. The quantity of carbon dioxide leaving the room every minute is $\frac{1,000y(t)}{2,500} = \frac{y(t)}{2.5}$ cubic feet. The rate of change of carbon dioxide is thus $\frac{dy}{dt} = 0.3 - \frac{y(t)}{2.5}$ or $\frac{dy}{dt} + \frac{1}{2.5}y(t) = 0.3$. The integrating factor is $I(x) = e^{\int 1/2.5dt} = e^{t/2.5}$. Multiplying the original equation by $e^{t/2.5}$ gives $e^{t/2.5}\frac{dy}{dt} + \frac{e^{t/2.5}}{2.5}y(t) = 0.3e^{t/2.5}$. The left-hand member can be written as the derivative of a product, so $\frac{d}{dt}[e^{t/2.5}y(t)] = 0.3e^{t/2.5}$, $d(e^{t/2.5}y(t)) = 0.3e^{t/2.5}dt$, and $\int d[e^{t/2.5}y(t)] = \int 0.3e^{t/2.5}dt$. Thus $e^{t/2.5}y(t) = 0.3(2.5)e^{t/2.5} + C = 0.75e^{t/2.5} + C$ or $y(t) = 0.75 + Ce^{-t/2.5}$. Since the original amount of carbon dioxide is 6.25 cubic feet, $6.25 = 0.75 + C$ or $C = 5.5$. Thus $y(t) = 0.75 + 5.5e^{-t/2.5}$. The percentage carbon dioxide in the room at time t is

$\frac{100}{2{,}500} = \frac{1}{25}$ of the amount or $p(t) = \frac{0.75}{25} + \frac{5.5}{25}e^{-t/2.5} = 0.03 + 0.22e^{-t/2.5}$.
b) $p(t) = 0.08 = 0.03{+}0.22e^{-t/2.5}$, $0.22e^{-t/2.5} = 0.05$, $e^{-t/2.5} = 0.22727$, $-t/2.5 = \ln 0.22727 = -1.4816$, or $t = 3.7$ minutes.
c) In the long run, $\lim_{t\to\infty} p(t) = 0.03$ % since the exponential vanishes as t increases beyond all bounds.

25. Let $x(t)$ denote the concentration of the drug after t units of time (let's say hours). $\frac{dx}{dt} = r - sx$, where r and $s > 0$ are assumed constants. Then $\frac{dx}{dt} + sx = r$. The integrating factor is $I(x) = e^{\int s\,dt} = e^{st}$. Multiplying the original equation by e^{st} gives $e^{st}\frac{dx}{dt} + se^{st}x = re^{st}$. The left-hand member can be written as the derivative of a product, so $\frac{d}{dt}[e^{st}x(t)] = re^{st}$, $d(xe^{st}x(t)) = re^{st}dt$, and $\int d[e^{st}x(t)] = \int re^{st}dt$. Thus $e^{st}x(t) = \frac{r}{s}e^{st} + C$ or $x(t) = \frac{r}{s} + Ce^{-st}$. Since the original amount of the drug is 0, $0 = \frac{r}{s} + Ce^0$ or $C = -\frac{r}{s}$. Thus $x(t) = \frac{r}{s} - \frac{r}{s}e^{-st} = \frac{r}{s}(1 - e^{-st})$.
b) In the long run, $\lim_{t\to\infty} x(t) = \frac{r}{s}$ since the exponential vanishes as t increases beyond all bounds (that's why we assumed $s > 0$).

27. Let $P(t)$ denote the population at time t (assumed to be in years). The birth and death rates are functions of the population, but the rate of immigration is not. The rate of change of the population is $\left(\frac{\text{number of infants born}}{\text{person}}\right)$(number of people)$-\left(\frac{\text{number of persons dying}}{\text{person}}\right)$(number of people)+(number of people immigrating), all per unit of time.
b) $\frac{dP}{dt} = (r - s)P(t) + m$ and $\frac{dP}{dt} - (r - s)P = m$. The integrating factor is $I(x) = e^{\int -(r-s)dt} = e^{-(r-s)t}$. Multiplying the original equation by $e^{-(r-s)t}$ gives $e^{-(r-s)t}\frac{dP}{dt} - (r - s)Pe^{-(r-s)t} = me^{-(r-s)t}$. The left-hand member can be written as the derivative of a product, so $\frac{d}{dt}[e^{-(r-s)t}P] = me^{-(r-s)t}$, $d[Pe^{-(r-s)t}] = me^{-(r-s)t}dt$, and $\int d[Pe^{-(r-s)t}] = \int me^{-(r-s)t}dt$. Thus $e^{-(r-s)t}P = -\frac{m}{r-s}e^{-(r-s)t} + C$ or $P(t) = \frac{m}{s-r} + Ce^{(r-s)t}$. Since the original population is P_0 million people, $P_0 = \frac{m}{s-r} + Ce^0$ or $C = P_0 - \frac{m}{s-r}$. Thus $P(t) = \frac{m}{s-r} + (P_0 + \frac{m}{r-s})e^{(r-s)t}$.
c) With $r - s = 0.02$, $P_0 = 100$, $m = 0.3$, $P(10) = \frac{0.3}{-0.02} + (100 + \frac{0.3}{0.02})e^{(0.02)10} = -15 + 115e^{0.2} = 125.4613$ million people.

Review Problems

1. If $\frac{dy}{dx} = x^3 - 3x^2 + 5$, $y = \int(x^3 - 3x^2 + 5)dx = \frac{x^4}{4} - x^3 + 5x + C$.

2. Separate the variables of $\frac{dy}{dx} = 0.02y$ and integrate to get $\int \frac{1}{y}dy = 0.02dx$, $\ln|y| = 0.02x + C_1$, $|y| = e^{0.02x+C_1} = e^{C_1}e^{0.02x}$, or $y = Ce^{0.02x}$ where $C = \pm e^{C_1}$.

3. Separate the variables of $\frac{dy}{dx} = k(80 - y)$ and integrate to get $\int \frac{1}{80-y}dy = \int k dx$, $-\ln|80-y| = kx + C_1$, $\ln|80-y| = -kx - C_1$, $|80-y| = e^{-kx-C_1} = e^{-C_1}e^{-kx}$, $y-80 = Ce^{-kx}$, or $y = 80 + Ce^{-x}$ where $C = \pm e^{-C_1}$.

4. Separate the variables of $\frac{dy}{dx} = y(1-y)$ and integrate to get $\int \frac{1}{y(1-y)}dy = \int dx$. Notice that $\frac{1}{y(1-y)} = \frac{1}{y} + \frac{1}{1-y}$. Hence $\int \frac{1}{y(1-y)}dy = \int \frac{1}{y}dy + \int \frac{1}{1-y}dy = \ln|y| - \ln|1-y| + C_0$ and so $\ln|y| - \ln|1-y| = x + C_1$, $\ln|\frac{y}{1-y}| = x + C_1$, $|\frac{y}{1-y}| = e^{x+C_1} = e^{C_1}e^x$, $\frac{y}{1-y} = C_2e^x$, (where $C_2 = \pm e^{C_1}$), $y = C_2e^x - C_2ye^x$, $(1 + C_2e^x)y = C_2e^x$, $y = \frac{C_2e^x}{1+C_2e^x} = \frac{1}{(1/C_2)e^{-x}+1} = \frac{1}{1+Ce^{-x}}$, where $C = \frac{1}{C_2}$.

5. Separate the variables of $\frac{dy}{dx} = e^{x+y} = (e^x)(e^y)$ and integrate to get $\int e^{-y}dy = \int e^x dx$. $-e^{-y} = e^x + C_1$ or $e^x + e^{-y} + C_1 = 0$. If $C_1 = -C$ the solution can be rewrtitten as $e^x + e^{-y} = C$. It is better to leave the solution in the present form rather than solving explicitly for y.

6. Separate the variables of $\frac{dy}{dx} = ye^{-2x}$ and integrate to get $\int \frac{1}{y}dy = \int e^{-2x}dx$. $\ln|y| = -\frac{1}{2}e^{-2x} + C_1$, $|y| = e^{-\frac{1}{2}e^{-2x}+C_1}$, $|y| = e^{C_1}e^{-\frac{1}{2}e^{-2x}}$, or $y = Ce^{-\frac{1}{2}e^{-2x}}$ where $C = \pm e^{C_1}$.

7. $\frac{dy}{dx} + (\frac{4}{x})y = e^{-x}$. The integrating factor is $I(x) = e^{\int 4/x\, dx} = e^{4\ln x} = e^{\ln x^4} = x^4$. Multiplying the original equation by x^4 gives $x^4\frac{dy}{dx} + x^4(\frac{4}{x})y = x^4e^{-x}$, $x^4\frac{dy}{dx} + 4x^3y = x^4e^{-x}$. The left-hand member can be written as the derivative of a product, so $\frac{d}{dx}(x^4y) = x^4e^{-x}$, $d(x^4y) = x^4e^{-x}dx$. Thus $\int d(x^4y) = \int x^4e^{-x}dx$, $x^4y = (-x^4 - 4x^3 - 12x^2 - 24x - 24)e^{-x} + C$. Thus $y = \frac{-x^4-4x^3-12x^2-24x-24}{x^4}e^{-x} + \frac{C}{x^4}$.

8. $\frac{dy}{dx} + (\frac{1}{x})y = \frac{2}{x+1}$. The integrating factor is $I(x) = e^{\int 1/x\, dx} = e^{\ln x} = x$. Multiplying the original equation by x gives $x\frac{dy}{dx} + x(\frac{1}{x})y = x\frac{2}{x+1}$, $x\frac{dy}{dx} + y = \frac{2x}{x+1}$. The left-hand member can be written as the derivative of a product, so $\frac{d}{dx}(xy) = 2(1 - \frac{1}{x+1})$, $d(xy) = 2(1 - \frac{1}{x+1})dx$. Thus $\int d(xy) = \int 2(1 - \frac{1}{x+1})dx$, $xy = 2(x - \ln|x+1|) + C$ and $y = 2(1 - \frac{\ln|x+1|}{x}) + \frac{C}{x}$

9. $\frac{dy}{dx} + (\frac{2}{x+1})y = \frac{4}{x+2}$. The integrating factor is $I(x) = e^{\int 2/(x+1)dx} = e^{2\ln(x+1)} = e^{\ln(x+1)^2} = (x+1)^2$. Multiplying the original equation by $(x+1)^2$ gives $(x+1)^2\frac{dy}{dx} + (x+1)^2(\frac{2}{x+1})y = (x+1)^2\frac{4}{x+2}$, $(x+1)^2\frac{dy}{dx} + 2(x+1)y = \frac{4(x+1)^2}{x+2}$. The left-hand member can be written as the derivative of a product, so $\frac{d}{dx}[(x+1)^2y] = 4(\frac{x^2+2x+1}{x+2}) = 4(x + \frac{1}{x+2})$, $d[(x+1)^2y] = 4(x + \frac{1}{x+2})dx$. Thus $\int d[(x+1)^2y] = \int 4(x + \frac{1}{x+2})dx$, $(x+1)^2y = 4(\frac{x^2}{2} + \ln|x+2|) + C$, and $y = 4(\frac{2x^2}{(x+1)^2} + \frac{4\ln|x+2|}{(x+1)^2}) + \frac{C}{(x+1)^2}$,

10. $x^3\frac{dy}{dx} + xy = 5$, $\frac{dy}{dx} + \frac{1}{x^2}y = \frac{5}{x^3}$. The integrating factor is $I(x) = e^{\int 1/x^2 dx} = e^{-1/x}$. Multiplying the last equation by $e^{-1/x}$ gives $e^{-1/x}\frac{dy}{dx} + \frac{e^{-1/x}}{x^2}y = e^{-1/x}\frac{5}{x^3}$. The left-hand

member can be written as the derivative of a product, so $\frac{d}{dx}(ye^{-1/x}) = (\frac{5}{x^3})e^{-1/x}$, $d(ye^{-1/x}) = (\frac{5}{x^3})e^{-1/x}dx$. Thus $\int d(ye^{-1/x}) = \int(\frac{5}{x^3})e^{-1/x}dx$, $ye^{-1/x} = 5\int(\frac{1}{x})(\frac{e^{-1/x}}{x^2}))dx$.

$f(x) = \frac{1}{x}$	$g(x) = \frac{1}{x^2}e^{-1/x}$
$f'(x) = -\frac{1}{x^2}$	$G(x) = e^{-1/x}$

$e^{-1/x}y = 5(\frac{1}{x}e^{-1/x} + \int \frac{1}{x^2}e^{-1/x}dx) = 5(\frac{1}{x}e^{-1/x} + e^{-1/x}) + C$, thus $y = 5(\frac{1}{x} + 1) + Ce^{1/x}$.

11. If $\frac{dy}{dx} = 5x^4 - 3x^2 - 2$, $y = \int(5x^4 - 3x^2 - 2)dx = x^5 - x^3 - 2x + C$. Since $y = 4$ when $x = 1$, $4 = 1 - 1 - 2 + C$ or $C = 6$. Hence $y = x^5 - x^3 - 2x + 6$

12. Separate the variables of $\frac{dy}{dx} = 0.06y$ and integrate to get $\int \frac{1}{y}dy = \int 0.06dx$. $\ln|y| = 0.06x + C_1$, $|y| = e^{0.06x+C_1} = e^{C_1}e^{0.06x}$, or $y = Ce^{0.06x}$ where $C = \pm e^{C_1}$. Since $y = 100$ when $x = 0$, $100 = Ce^0$ or $C = 100$. Hence $y = 100e^{0.06x}$

13. Separate the variables of $\frac{dy}{dx} = 3 - y$ and integrate to get $\int \frac{1}{3-y}dy = \int dx$. $-\ln|3-y| = x + C_1$, $\ln|3-y| = -x - C_1$, $|3-y| = e^{-x-C_1} = e^{-C_1}e^{-x}$, $3 - y = Ce^{-x}$, or $y = 3 - Ce^{-x}$ where $C = \pm e^{-C_1}$. Since $y = 2$ when $x = 0$, $2 = 3 - C$ or $C = 1$. Hence $y = 3 - e^{-x}$.

14. If $\frac{d^2y}{dx^2} = 2$, then $\frac{dy}{dx} = \int \frac{d^2y}{dx^2}dx = \int 2dx = 2x + C_1$. Since $\frac{dy}{dx} = 3$ when $x = 0$, $3 = 2(0) + C_1$ or $C_1 = 3$. Hence $\frac{dy}{dx} = 2x + 3$ and $y = \int(2x+3)dx = x^2 + 3x + C$. Since $y = 5$ when $x = 0$, $5 = 0^2 + 3(0) + C$ or $C = 5$. Hence $y = x^2 + 3x + 5$.

15. $\frac{dy}{dx} - (\frac{5}{x})y = x^2$. The integrating factor is $I(x) = e^{\int(-5/x)dx} = e^{-5\ln x} = e^{\ln x^{-5}} = x^{-5}$. Multiplying the original equation by x^{-5} gives $x^{-5}\frac{dy}{dx} - x^{-5}(\frac{5}{x})y = x^{-5}x^2$, $x^{-5}\frac{dy}{dx} - \frac{5}{x^6}y = x^{-3}$. The left-hand member can be written as the derivative of a product, so $\frac{d}{dx}(x^{-5}y) = x^{-3}$, $d(x^{-5}y) = x^{-3}dx$. Thus $\int d(x^{-5}y) = \int x^{-3}dx$, $x^{-5}y = \frac{x^{-2}}{-2} + C$. Thus $y = -\frac{x^3}{2} + Cx^5$. Since $y = 4$ when $x = -1$, $4 = -\frac{(-1)^3}{2} + C(-1)^5$, or $C = -\frac{7}{2}$. Hence $y = -\frac{x^3}{2} - \frac{7}{2}x^5$.

16. $\frac{dy}{dx} + (\frac{1}{x+1})y = x$. The integrating factor is $I(x) = e^{\int 1/(x+1)dx} = e^{\ln(x+1)} = (x+1)$. Multiplying the original equation by $(x+1)$ gives $(x+1)\frac{dy}{dx} + (x+1)(\frac{1}{x+1})y = x(x+1)$, $(x+1)\frac{dy}{dx} + y = x(x+1)$. The left-hand member can be written as the derivative of a product, so $\frac{d}{dx}[(x+1)y] = x^2 + x$, $d[(x+1)y] = (x^2+x)dx$. Thus $\int d[(x+1)y] = \int(x^2+x)dx$, $(x+1)y = \frac{x^3}{3} + \frac{x^2}{2} + C$, and $y = \frac{x^3}{3(x+1)} + \frac{x^2}{2(x+1)} + \frac{C}{x+1}$. Since $y = 0$ when $x = 3$, $0 = \frac{3^3}{3(3+1)} + \frac{3^2}{2(3+1)} + \frac{C}{3+1}$, $0 = \frac{3^2}{4} + \frac{9}{8} + \frac{C}{4}$, $C = -4(\frac{3^2}{4} + \frac{9}{8})$, $C = -(\frac{18}{2} + \frac{9}{2}) = -\frac{27}{2}$ Hence $y = \frac{x^3}{3(x+1)} + \frac{x^2}{2(x+1)} - \frac{27}{2(x+1)}$.

17. $\frac{dy}{dx} - xy = e^{x^2/2}$. The integrating factor is $I(x) = e^{\int(-x)dx} = e^{-x^2/2}$. Multiplying the original equation by $e^{-x^2/2}$ gives $e^{-x^2/2}\frac{dy}{dx} - xe^{-x^2/2}y = e^{-x^2/2}e^{x^2/2} = 1$. The left-hand member can be written as the derivative of a product, so $\frac{d}{dx}(e^{-x^2/2}y) = 1$, $d(e^{-x^2/2}y) = dx$. Thus $\int d(e^{-x^2/2}y) = \int dx$, $e^{-x^2/2}y = x + C$, and $y = xe^{x^2/2} + Ce^{x^2/2}$. Since $y = 4$ when $x = 0$, $4 = 0e^0 + Ce^0$, or $C = 4$. Thus $y = xe^{x^2/2} + 4e^{x^2/2}$.

18. $\frac{dy}{dx} + 3x^2y = x^2$. The integrating factor is $I(x) = e^{\int(3x^2)dx} = e^{x^3}$. Multiplying the original equation by e^{x^3} gives $e^{x^3}\frac{dy}{dx} + 3x^2e^{x^3}y = x^2e^{x^3}$. The left-hand member can be written as the derivative of a product, so $\frac{d}{dx}(e^{x^3}y) = x^2e^{x^3}$, or $d(e^{x^3}y) = x^2e^{x^3}dx$. Thus $\int d(e^{x^3}y) = \int x^2e^{x^3}dx$, $e^{x^3}y = \frac{1}{3}\int 3x^2e^{x^3}dx$, $e^{x^3}y = \frac{1}{3}e^{x^3} + C$ and $y = \frac{1}{3} + Ce^{-x^3}$. Since $y = -2$ when $x = 0$, $-2 = \frac{1}{3} + Ce^0$ or $C = -\frac{7}{3}$. Thus $y = \frac{1}{3} - \frac{7}{3}e^{-x^3} = \frac{1}{3}(1 - 7e^{-x^3})$.

19. Let $V(t)$ denote the value of the machine after t years. The rate of change of V is $\frac{dV}{dt} = k(V - 5,000)$, where k is a positive constant of proportionality. Separate the variables and integrate to get $\int \frac{1}{V-5,000}dV = \int kdt$, $\ln(V - 5,000) = kt + C_1$, $V - 5,000 = e^{kt+C_1} = e^{C_1}e^{kt}$, or $V(t) = 5,000 + Ce^{kt}$ where $C = e^{C_1}$ and the absolute values can be dropped since $V - 5,000 > 0$. Since the machine was originally worth \$40,000, $40,000 = V(0) = 5,000 + C$ or $C = 35,000$. Hence, $V(t) = 5,000 + 35,000e^{kt}$. Since the machine was worth \$30,000 after 4 years, $30,000 = V(4) = 5,000 + 35,000e^{4k}$, $35,000e^{4k} = 25,000$ or $e^{4k} = \frac{25,000}{35,000} = \frac{5}{7}$. The value of the machine after 8 years is $V(8) = 5,000 + 35,000e^{8k} = 5,000 + 35,000(e^k)^2 = 5,000 + 35,000(\frac{5}{7})^2 = \$22,857$.

20. Let $R(t)$ denote the total revenue generated during the next t months and $P(t)$ the price of oil t months from now. Then, $P(t) = 24 + 0.08t$ and $\frac{dR}{dt} = \frac{\text{dollars}}{\text{month}} = \frac{\text{dollars}}{\text{barrel}}\frac{\text{barrels}}{\text{month}} = P(t)(600) = 600(24 + 0.08t)$. $R(t) = 600\int(24 + 0.08t)dt = 600(24t + 0.04t^2) + C$. Since $R(0) = 0$ it follows that $C = 0$ and the appropriate particular solution is $R(t) = 600(24t + 0.04t^2)$. Since the well will run dry in 36 months, the total future revenue will be $R(36) = 600[24(36) + 0.04(36)^2] = \$549,504$.

21. Let $Q(t)$ denote the number of pounds of salt in the tank after t minutes. Then $\frac{dQ}{dt}$ is the rate of change of salt with respect to time (measured in pounds per minute). Thus, $\frac{dQ}{dt}$ =(rate at which salt enters)−(rate at which salt leaves)= $\frac{\text{pounds entering}}{\text{gallon}}\frac{\text{gallons entering}}{\text{minute}} - \frac{\text{pounds leaving}}{\text{gallon}}\frac{\text{gallons leaving}}{\text{minute}}$. Now $\frac{\text{pounds leaving}}{\text{gallon}} = \frac{\text{pounds of salt in the tank}}{\text{gallons of salt in the tank}} = \frac{Q}{200-t}$ since the tank loses 1 gallon of brine per minute. Hence $\frac{dQ}{dt} = 0(4) - \frac{Q}{200-t}(5) = \frac{5Q}{t-200}$. Separate the variables and integrate to get $\int \frac{1}{Q}dQ = \int \frac{5}{t-200}dt$, $\ln|Q| = 5\ln|t - 200| + C_1$, $\ln|Q| = \ln|t - 200|^5 + C_1$, $Q = e^{C_1}e^{\ln(t-200)^5}$, or $Q(t) = C(t - 200)^5$, where $C = e^{C_1}$, and the absolute values are not needed since Q and $t - 200$ are both positive in the context of this problem. Since there are initially 600 pounds of salt in the tank (3 pounds of salt per gallon times 200 gallons), $600 = Q(0) = (-200)^5C$ or $C = \frac{600}{-(200)^5}$. Hence, $Q(t) = \frac{600}{-(200)^5}(t - 200)^5 = 600(\frac{t-200}{-200})^5 = 600(1 - \frac{t}{200})^5$. The amount of salt in the tank after 100 minutes is $Q(100) = 600(1 - \frac{100}{200})^5 = 18.75$ pounds.

22. Let $Q(t)$ denote the population in millions t years after 1980. The differential equation describing the population growth is $\frac{dQ}{dt} = kQ(10 - Q)$, where k is a positive constant of proportionality. This differential equation (with $10 = B$) was solved in example 1.10. Its solution is $Q(t) = \frac{10}{1+Ae^{-10kt}}$. Since the population was 4 million in 1980, $Q(0) = 4$ and so

$4 = \frac{10}{1+A}$, $4+4A = 10$, or $A = 1.5$. Hence, $Q(t) = \frac{10}{1+1.5e^{-10kt}}$. Since the population was 4.74 million in 1985, $Q(5) = 4.74$ and $4.74 = \frac{10}{1+1.5e^{-50k}}$, $4.74 + 7.11e^{-50k} = 10$, $7.11e^{-50k} = 5.26$, $-50k = \ln\frac{5.26}{7.11}$, or $-10k = \frac{1}{5}\ln\frac{5.26}{7.11} = -0.06$. Hence, the population function is $Q(t) = \frac{10}{1+1.5e^{-0.06t}}$.

23. Let $Q(t)$ denote the amount of new currency in circulation at time t. Then $\frac{dQ}{dt}$ is the rate of change of the new currency with respect to time (measured in dollars per day). Thus $\frac{dQ}{dt}$ =(rate at which new currency enters)−(rate at which new currency leaves). Now, the rate at which new currency enters is 18 million per day. The rate at which new currency leaves is $(\frac{\text{new currency at timet}}{\text{total currency}})$(rate at which new currency enters)= $\frac{Q(t)}{5{,}000}(18)$ million per day. Putting it all together, $\frac{dQ}{dt} = 18 - \frac{18Q}{5{,}000} = 18(1 - \frac{Q}{5{,}000})$. Separate variables to obtain $\frac{dQ}{1-Q/5{,}000} = 18dt$, $\int\frac{dQ}{1-Q/5{,}000} = 18\int dt$, $-5{,}000\ln|1 - \frac{Q}{5{,}000}| = 18t + C$. When $t = 0$, $Q(0) = 0$ which yields $-5{,}000\ln|1 - \frac{0}{5{,}000}| = 18(0) + C$ or $C = 0$. Therefore, the solution becomes $-5{,}000\ln|1 - \frac{Q}{5{,}000}| = 18t$, $\ln|1 - \frac{Q}{5{,}000}| = -\frac{18t}{5{,}000}$. Since Q is a part of 5,000, $1 - \frac{Q}{5{,}000} > 0$ and so $\ln(1 - \frac{Q}{5{,}000}) = -\frac{18t}{5{,}000}$, $1 - \frac{Q}{5{,}000} = e^{-18t/5{,}000}$. Now to find t so that $Q(t) = 0.9(5{,}000)$ substitute into the last solution $1 - \frac{4{,}500}{5{,}000} = e^{-18t/5{,}000}$, $\frac{1}{10} = e^{-18t/5{,}000}$, $\ln\frac{1}{10} = -\frac{18t}{5{,}000}$, $\ln 1 - \ln 10 = -\ln 10 = -\frac{18t}{5{,}000}$, or $t = \frac{5{,}000}{18}\ln 10 = 640$ days $= 1.75$ years.

24. a) Let $Q(t)$ denote the number of pounds of salt in the tank after t minutes. Then $\frac{dQ}{dt}$ is the rate of change of salt with respect to time (measured in pounds per minute). Thus, $\frac{dQ}{dt}$ =(rate at which salt enters)−(rate at which salt leaves)= $\frac{\text{pounds entering}}{\text{gallon}}\frac{\text{gallons entering}}{\text{minute}} - \frac{\text{pounds leaving}}{\text{gallon}}\frac{\text{gallons leaving}}{\text{minute}}$. Now $\frac{\text{pounds leaving}}{\text{gallon}} = \frac{\text{pounds of salt in the tank}}{\text{gallons of salt in the tank}} = \frac{Q}{50+t}$ since the tank gains 1 gallon of brine per minute. Hence $\frac{dQ}{dt} = 2(2) - \frac{Q}{50+t}$. $\frac{dQ}{dt} + \frac{Q}{50+t} = 4$. The integrating factor is $I(x) = e^{\int 1/(50+t)dt} = e^{\ln(50+t)} = 50 + t$. Multiplying the original equation by $50+t$ gives $(50+t)\frac{dQ}{dt} + (50+t)\frac{Q}{50+t} = 4(50+t)$, $(50+t)\frac{dQ}{dt} + Q = 4(50+t)$. The left-hand member can be written as the derivative of a product, so $\frac{d}{dt}[(50+t)Q] = 4(50+t)$, or $d[(50+t)Q] = 4(50+t)dt$. Thus $\int d[(50+t)Q] = \int 4(50+t)dt$, $(50+t)Q = 2(50+t)^2 + C$, and $Q = 2(50+t) + \frac{C}{50+t}$. Since $Q = 30$ when $t = 0$, $30 = 2(50) + \frac{C}{50}$, $-70 = \frac{C}{50}$, $C = -3{,}500$. Hence, $Q(t) = 2(50+t) - \frac{3{,}500}{50+t}$.

b) When $Q(t) = 40$, $40 = 100 + 2t - \frac{3{,}500}{50+t}$, $-60(50+t) = 2t(50+t) - 3{,}500$, $-3000 - 60t = 100t + 2t^2 - 3{,}500$, $500 = 160t + 2t^2$, $t^2 + 80t - 250 = 0$, $t = \frac{-80\pm\sqrt{80^2+4(250)}}{2} = \frac{-80\pm\sqrt{7{,}400}}{10} = \frac{-80+86.02}{2} = 3.01$. Since 1 gallon is added to the solution every minute, there will be 50+3 = 53 gallons in the tank.

25. Let P denote the number of people, x the income. The rate of change of the number of people $\frac{dP}{dt} = -kP\frac{1}{x}$ where k is a positive constant of proportionality. Separation of variables leads to $\frac{dP}{P} = \frac{-k}{x}dt$ and integrating yields $\ln P = \frac{-k}{x}t + C_1$, $P = e^{-kt/x+C_1} = e^{-kt/x}e^{C_1} =$

$Ce^{-kt/x}$, where $C = e^{C_1}$. Note That $P > 0$ and $x > 0$ in the context of this problem.

26. $\frac{dP}{dt} = P(\ln P_0)(\ln \beta)\beta^t$, $\frac{dP}{P} = (\ln P_0)(\ln \beta)\beta^t dt$, integrating leads to $\ln P = (\ln P_0)\beta^t + C_1$, $P = e^{(\ln P_0)\beta^t + C_1} = e^{C_1} e^{(\ln P_0)\beta^t} = Ce^{\ln(P_0)^{\beta^t}} = C(P_0)^{\beta^t}$, where $C = e^{C_1}$. Absolute values were dispensed with because by context $P > 0$.

Chapter 8

Limits of Infinity and Improper Integrals

8.1 Limits at Infinity and l'Hôpital's Rule

1. $\lim_{x\to\infty}(x^3 - 4x^2 - 4) = \infty$ because x^3 dominates.

3. $\lim_{x\to\infty}(1-2x)(x+5) = \lim_{x\to\infty}(5 - 9x - 2x^2) = -\infty$ because x^2 dominates and is subtracted.

5. Divide numerator and denominator by x^2 to get $\lim_{x\to\infty}\frac{x^2-2x+3}{2x^2+5x+1} = \lim_{x\to\infty}\frac{1-2/x+3/x^2}{2+5/x+1/x^2} = \frac{1}{2}$.

7. Divide numerator and denominator by x to get $\lim_{x\to\infty}\frac{2x+1}{3x^2+2x-7} = \lim_{x\to\infty}\frac{2+1/x}{3x+2-7/x} = \frac{2}{\infty} = 0$.

9. Divide numerator and denominator by x to get $\lim_{x\to\infty}\frac{3x^2-6x+2}{2x-9} = \lim_{x\to\infty}\frac{3x-6+2/x}{2-9/x} = \frac{\infty}{2} = \infty$.

11. $\lim_{x\to\infty}(2-3e^x) = 2 - 3\lim_{x\to\infty} e^x = 2 - \infty = -\infty$.

13. Since $\lim_{x\to\infty} e^{-8x} = 0$, it follows that $\lim_{x\to\infty}\frac{3}{2+5e^{-8x}} = \frac{3}{2+0} = \frac{3}{2}$.

15. Since the limit is of the form $\frac{\infty}{\infty}$, l'Hôpital's rule gives $\lim_{x\to\infty}\frac{e^{2x}}{3x} = \lim_{x\to\infty}\frac{2e^{2x}}{3} = \frac{\infty}{3} = \infty$.

17. $\lim_{x\to\infty}\frac{2x}{e^{-3x}} = \lim_{x\to\infty}\frac{2}{-3e^{-3x}} = \lim_{x\to\infty} 2e^{3x} = \infty$.

19. Since the limit is of the form $\frac{\infty}{\infty}$, l'Hôpital's rule (applied twice) gives $\lim_{x\to\infty}\frac{x^2}{e^x} = \lim_{x\to\infty}\frac{2x}{e^x} = \lim_{x\to\infty}\frac{2}{e^x} = 0$.

21. Since the limit is of the form $\frac{\infty}{\infty}$, l'Hôpital's rule gives $\lim_{x\to\infty}\frac{\sqrt{x}}{e^x} = \lim_{x\to\infty}\frac{1/(2\sqrt{x})}{e^x} = \lim_{x\to\infty}\frac{1}{2\sqrt{x}e^x} = \frac{1}{\infty} = 0$.

23. Since the limit is of the form $\frac{\infty}{\infty}$, l'Hôpital's rule gives $\lim_{x\to\infty}\frac{\ln x}{x} = \lim_{x\to\infty}\frac{1/x}{1} = \lim_{x\to\infty}\frac{1}{x} = \frac{1}{\infty} = 0$.

25. Since the limit is of the form $\frac{\infty}{\infty}$, l'Hôpital's rule (applied twice) gives $\lim_{x\to\infty}\frac{\ln(x^2+1)}{x} = \lim_{x\to\infty}\frac{2x/(x^2+1)}{1} = \lim_{x\to\infty}\frac{2x}{x^2+1} = \lim_{x\to\infty}\frac{2}{2x} = 0$.

27. Since the limit is of the form $\frac{\infty}{\infty}$, l'Hôpital's rule (applied twice) gives $\lim_{x\to\infty}\frac{\ln(2x+1)}{\ln(3x-1)} = \lim_{x\to\infty}\frac{2/(2x+1)}{3/(3x-1)} = \lim_{x\to\infty}\frac{6x-2}{6x+3} = \lim_{x\to\infty}\frac{6}{6} = 1$.

29. Since the limit is of the form $\frac{\infty}{\infty}$, l'Hôpital's rule (applied twice) gives $\lim_{x\to\infty}\frac{e^x}{x\ln x} = \lim_{x\to\infty}\frac{e^x}{1+\ln x} = \lim_{x\to\infty}\frac{e^x}{1/x} = \lim_{x\to\infty} xe^x = \infty$.

31. Since the limit is of the form $\frac{\infty}{\infty}$, l'Hôpital's rule (applied twice) gives $\lim_{x\to\infty}\frac{x}{e^{\sqrt{x}}} = \lim_{x\to\infty}\frac{1}{e^{\sqrt{x}}/(2\sqrt{x})} = \lim_{x\to\infty}\frac{2\sqrt{x}}{e^{\sqrt{x}}} = \lim_{x\to\infty}\frac{2/(2\sqrt{x})}{[1/(2\sqrt{x})]e^{\sqrt{x}}} = \lim_{x\to\infty}\frac{2}{e^{\sqrt{x}}} = 0$.

33. Since the limit is of the form 0∞, rewrite the product as a quotient and apply l'Hôpital's rule to get $\lim_{x\to\infty} e^{-2x}\ln x = \lim_{x\to\infty}\frac{\ln x}{e^{2x}} = \lim_{x\to\infty}\frac{1/x}{2e^{2x}} = \lim_{x\to\infty}\frac{1}{2xe^{2x}} = 0$.

35. Since the limit is of the form ∞^0, let $y = x^{2/x}$. Take logarithms to get $\ln y = \ln x^{2/x} = \frac{2\ln x}{x}$. Since the limit of $\ln y$ is of the form $\frac{\infty}{\infty}$, apply l'Hôpital's rule to get $\lim_{x\to\infty}\ln y = \lim_{x\to\infty}\frac{2\ln x}{x} = \lim_{x\to\infty}\frac{2/x}{1} = 0$. Since $\ln y = 0$, it follows that $y \to e^0 = 1$.

37. Since the limit is of the form 1^∞, let $y = (1+\frac{2}{x})^x$. Take logarithms to get $\ln y = \ln(1+\frac{2}{x})^x = x\ln(1+\frac{2}{x}) = \frac{\ln(1+2/x)}{1/x}$. Since the limit of $\ln y$ is of the form $\frac{0}{0}$, apply l'Hôpital's rule to get $\lim_{x\to\infty}\ln y = \lim_{x\to\infty}\frac{\ln(1+2/x)}{1/x} = \lim_{x\to\infty}\frac{(-2/x^2)/(1+2/x)}{-1/x^2} = \lim_{x\to\infty}\frac{2}{1+2/x} = \frac{2}{1+0} = 2$. Since $\ln y \to 2$, it follows that $y \to e^2$.

39. If $p < 0$, then $p = -n$, where n is some positive number, and so $\lim_{x\to\infty} x^p e^{-kx} = \lim_{x\to\infty}\frac{1}{x^n e^{kx}} = \frac{1}{\infty} = 0$.
If $p = 0$, $\lim_{x\to\infty} x^p e^{-kx} = \lim_{x\to\infty}\frac{1}{e^{kx}} = \frac{1}{\infty} = 0$.
If $p > 0$, then $\lim_{x\to\infty} x^p e^{-kx} = \lim_{x\to\infty}\frac{x^p}{e^{kx}} = \frac{\infty}{\infty}$. Each time l'Hôpital's rule is applied, the power of x in the numerator decreases by 1 until eventually it becomes 0 or negative. When that occurs, the limit is like the one of the two cases already considered and hence is equal to 0.

8.2 Improper Integrals

1. $\int_1^\infty \frac{1}{x^3}dx = \lim_{N\to\infty} \int_1^N \frac{1}{x^3}dx = \lim_{N\to\infty} \frac{-1}{2x^2}|_1^N = \lim_{N\to\infty}(\frac{-1}{2N^2} + \frac{1}{2}) = \frac{1}{2}$.

3. $\int_1^\infty \frac{1}{\sqrt{x}}dx = \lim_{N\to\infty} \int_1^N x^{-1/2}dx = \lim_{N\to\infty} 2x^{1/2}|_1^N = \lim_{N\to\infty}(2N^{1/2} - 2) = \infty$.

5. $\int_3^\infty \frac{1}{2x-1}dx = \lim_{N\to\infty} \int_3^N \frac{1}{2x-1}dx = \frac{1}{2}\lim_{N\to\infty} \ln|2x-1||_3^N = \frac{1}{2}\lim_{N\to\infty}[\ln(2N-1) - \ln 5] = \infty$.

7. $\int_3^\infty \frac{1}{(2x-1)^2}dx = \lim_{N\to\infty} \int_3^N (2x-1)^{-2}dx = \frac{1}{2}\lim_{N\to\infty} \frac{-1}{2x-1}|_3^N = -\frac{1}{2}\lim_{N\to\infty}(\frac{1}{2N-1} - \frac{1}{5}) = \frac{1}{10}$.

9. $\int_0^\infty 5e^{-2x}dx = 5\lim_{N\to\infty} \int_0^N e^{-2x}dx = -\frac{5}{2}\lim_{N\to\infty} e^{-2x}|_0^N = -\frac{5}{2}\lim_{N\to\infty}(e^{-2N} - 1) = \frac{5}{2}$.

11. $\int_1^\infty \frac{x^2}{(x^3+2)^2}dx = \lim_{N\to\infty} \frac{1}{3}\int_1^N 3x^2(x^3+2)^{-2}dx = \frac{1}{3}\lim_{N\to\infty} \frac{-1}{x^3+2}|_1^N = \frac{1}{3}\lim_{N\to\infty}[\frac{-1}{N^3+1} + \frac{1}{3}] = \frac{1}{9}$.

13. $\int_1^\infty \frac{x^2}{\sqrt{x^3+2}}dx = \lim_{N\to\infty} \frac{1}{3}\int_1^N 3x^2(x^3+2)^{-1/2}dx = \frac{1}{3}\lim_{N\to\infty} 2(x^3+2)^{1/2}|_1^N$
$= \frac{1}{3}\lim_{N\to\infty}[2(N^3+2)^{1/2} - 2(3)^{1/2}] = \infty$.

15. $\int_1^\infty \frac{e^{-\sqrt{x}}}{\sqrt{x}}dx = \lim_{N\to\infty} 2\int_1^N e^{-\sqrt{x}}(\frac{1}{2\sqrt{x}})dx = -2\lim_{N\to\infty} e^{-\sqrt{x}}|_1^N = -2\lim_{N\to\infty}[e^{-\sqrt{N}} - e^{-1}] = \frac{2}{e}$.

17. $\int_0^\infty 2xe^{-3x}dx = \lim_{N\to\infty} \int_0^N 2xe^{-3x}dx = \lim_{N\to\infty}(-\frac{2}{3}xe^{-3x}|_0^N + \frac{2}{3}\int_0^N e^{-3x}dx)$
$= \lim_{N\to\infty}(-\frac{2}{3}xe^{-3x} - \frac{2}{9}e^{-3x})|_0^N = \lim_{N\to\infty}[-\frac{2}{3}e^{-3x}(x+\frac{1}{3})]|_0^N = \lim_{N\to\infty}[-\frac{2}{3}e^{-3N}(N+\frac{1}{3}) + \frac{2}{3}(\frac{1}{3})] = \frac{2}{9}$.

19. $\int_0^\infty 5xe^{10-x}dx = 5e^{10}\lim_{N\to\infty} \int_0^N xe^{-x}dx = 5e^{10}\lim_{N\to\infty}(-xe^{-x})|_0^N + \int_0^N e^{-x}dx) = 5e^{10}\lim_{N\to\infty}(-xe^{-x} - e^{-x})|_0^N = 5e^{10}\lim_{N\to\infty}[(-Ne^{-N} - e^{-N}) - (0-1)] = 5e^{10}$.

21. $\int_2^\infty \frac{1}{x\ln x}dx = \lim_{N\to\infty} \int_2^N \frac{1}{\ln x}(\frac{1}{x})dx = \lim_{N\to\infty} \ln|\ln x||_2^N = \lim_{N\to\infty}(\ln|\ln N| - \ln|\ln 2|) = \infty$.

23. $\int_0^\infty x^2e^{-x}dx = \lim_{N\to\infty} \int_0^N x^2e^{-x}dx = \lim_{N\to\infty}(-x^2e^{-x}|_0^N + 2\int_0^N xe^{-x}dx) = \lim_{N\to\infty}[(-x^2e^{-x} - 2xe^{-x})|_0^N + \int_0^N 2e^{-x}dx)] = \lim_{N\to\infty}[(-x^2-2x-2)e^{-x}|_0^N = \lim_{N\to\infty}[(-N^2 - 2N - 2)e^{-N} - (-2)] = 2$.

25. To find the present value of the investment of \$2,400 per year for N years, divide the N-year interval $0 \le t \le N$ into n equal subintervals of length Δt years, and let t_j denote the beginning of the jth subinterval. Then, during the jth subinterval, the amount generated

is approximately $2,400\Delta t$ and the present value is $2,400e^{-0.12t_j\Delta t}$. Hence, the present value of an $N-$year investment is $\lim_{n\to\infty}\sum_{j=1}^{n} 2,400e^{-0.12t_j}\Delta t = \int_0^N 2,400e^{-0.12t}dt$. To find the present value P of the total investment, let $N\to\infty$ to get $P = \lim_{N\to\infty}\int_0^N 2,400e^{-0.12t}dt = -\frac{2,400}{0.12}\lim_{N\to\infty} e^{-0.12t}|_0^N = -\frac{2,400}{0.12}\lim_{N\to\infty}(e^{-0.12N}-1) = -\frac{2,400}{0.12} = \$20,000$.

27. To find the present value of an apartment complex generating $f(t) = 10,000 + 500t$ dollars per year for N years, divide the $N-$year interval $0 \le t \le N$ into n equal subintervals of length Δt years, and let t_j denote the beginning of the jth subinterval. Then, during the jth subinterval, the amount generated is approximately $f(t_j)\Delta t$ and at the interest rate of 10 %, the present value is $f(t_j)e^{-0.1t_j}\Delta t$. Hence, the present value of the apartment complex over an $N-$year period is $\lim_{n\to\infty}\sum_{j=1}^{n} f(t_j)e^{-0.1t_j}\Delta t = \int_0^N f(t)e^{-0.1t}dt$. To find the present value P of the total income, let $N\to\infty$ to get $P = \lim_{N\to\infty}\int_0^N(10,000+500t)e^{-0.1t}dt = \lim_{N\to\infty}[-10(10,000+500t)e^{-0.1t}|_0^N - \int_0^N(-5,000)e^{-0.1t}dt] = \lim_{N\to\infty}(-100,000-5,000t-50,000)e^{-0.1t}|_0^N = \lim_{N\to\infty}(-150,000-5,000t)e^{-0.1t}|_0^N = -5,000\lim_{N\to\infty}[(30+N)e^{-0.1N} - 30] = -5,000(-30) = \$150,000$.

29. To find the present value of an investment generating $f(t) = A+Bt$ dollars per year for N years, divide the $N-$year time interval $0 \le t \le N$ into n equal subintervals of length Δt years, and let t_j denote the beginning of the jth subinterval. Then, during the jth subinterval, the amount generated is approximately $f(t_j)\Delta t$. Hence the present value of an $N-$year investment is $\lim_{n\to\infty}\sum_{j=1}^{n} f(t_j)e^{rt_j}\Delta t = \int_0^N (A+Bt)e^{-rt}dt$. To find the present value P of the total income, let $N\to\infty$ to get $P = \lim_{N\to\infty}\int_0^N(A+Bt)e^{-rt}dt = \lim_{N\to\infty}[-\frac{1}{r}(A+Bt)e^{-rt}|_0^N - \frac{B}{r}\int_0^N e^{-rt}dt] = \lim_{N\to\infty}[-\frac{1}{r}(A+Bt)e^{-rt} + \frac{B}{r^2}e^{-rt}]|_0^N = -\frac{1}{r}\lim_{N\to\infty}[(A+Bt)+\frac{B}{r}]e^{-rt}|_0^N = -\frac{1}{r}\lim_{N\to\infty}[(A+BN)+\frac{B}{r}]e^{-rN} - (A+\frac{B}{r}) = \frac{1}{r}(A+\frac{B}{r}) = \frac{A}{r}+\frac{B}{r^2}$.

31. To find the number of patients after N months, divide the $N-$month time interval $0 \le t \le N$ into n equal subintervals of length Δt months, and let t_j denote the beginning of the jth subinterval. Then, the number of people starting treatment during the jth subinterval is approximately $10\Delta t$. Of these, the number still receiving treatment at time $t = N$ (that is, $N - t_j$ months later) is approximately $10f(N-t_j)\Delta t$. Hence, the number of patients receiving treatment at time $t = N$ is $\lim_{n\to\infty}\sum_{j=1}^{n} 10f(N-t_j)\Delta t = \int_0^N 10f(N-t)dt$ and the number of patients receiving treatment in the long run is $P = \lim_{N\to\infty} 10\int_0^N e^{-(N-t)/20}dt = \lim_{N\to\infty} 10e^{-N/20}\int_0^N e^{t/20}dt = \lim_{N\to\infty} 200e^{-N/20}e^{t/20}|_0^N = \lim_{N\to\infty} 200e^{-N/20}(e^{N/20}-1) = \lim_{N\to\infty} 200(1-e^{-N/20}) = 200$ patients.

33. To find the number of units of the drug in the patient's body after N hours, divide the $N-$hour time interval $0 \le t \le N$ into n equal subintervals of length Δt hours, and let t_j denote the beginning of the jth subinterval. Then, during the jth subinterval, approximately $5\Delta t$ units of the drug are received. Of these, the number remaining at time $t = N$ (that is, $N - t_j$ hours later) is approximately $5f(N-t_j)\Delta t$. Hence, the number of units of the drug in the patient's body at time $t = N$ is $\lim_{n\to\infty}\sum_{j=1}^{n} 5f(N-t_j)\Delta t = \int_0^N 5f(N-t)dt$

and the number Q of units in the patient's body in the long run is $Q = \lim_{N\to\infty} 5\int_0^N f(N-t)dt = \lim_{N\to\infty} 5\int_0^N e^{-(N-t)/10}dt = \lim_{N\to\infty} 5e^{-N/10}\int_0^N e^{t/10}dt = \lim_{N\to\infty} 50e^{-N/10}e^{t/10}|_0^N = \lim_{N\to\infty} 50e^{-N/10}(e^{N/10}-1) = \lim_{N\to\infty} 50(1-e^{-N/10}) = 50$ units.

8.3 Probability Density Functions

8.3.1 a) $P(2 \le x \le 5) = \int_2^5 f(x)dx = \int_2^5 \frac{1}{3}dx = \frac{x}{3}|_2^5 = 1$.
b) $P(3 \le x \le 4) = \int_3^4 f(x)dx = \int_3^4 \frac{1}{3}dx = \frac{x}{3}|_3^4 = \frac{1}{3}$.
c) $P(4 \le x) = \int_4^\infty f(x)dx = \int_4^5 \frac{1}{3}dx = \frac{x}{3}|_4^5 = \frac{1}{3}$.

8.3.3 a) $P(0 \le x \le 4) = \int_0^4 f(x)dx = \int_0^4 \frac{4-x}{8}dx = \frac{1}{8}(4x - \frac{x^2}{2})|_0^4 = \frac{1}{8}(16 - \frac{16}{2}) = 1$.
b) $P(2 \le x \le 3) = \int_2^3 f(x)dx = \int_2^3 \frac{4-x}{8}dx = \frac{1}{8}(4x - \frac{x^2}{2})|_2^3 = \frac{1}{8}[(12 - \frac{9}{2}) - (8 - \frac{4}{2})] = \frac{3}{16}$.
c) $P(1 \le x) = \int_1^\infty f(x)dx = \int_1^4 \frac{4-x}{8}dx = \frac{1}{8}(4x - \frac{x^2}{2})|_1^4 = \frac{1}{8}[(16 - \frac{16}{2}) - (4 - \frac{1}{2})] = \frac{9}{16}$.

5. a) $P(1 \le x < \infty) = \int_1^\infty f(x)dx = \int_1^\infty \frac{3}{x^4}dx = \lim_{N\to\infty}\int_1^\infty \frac{3}{x^4}dx = \lim_{N\to\infty} \frac{-1}{x^3}|_1^N = \lim_{N\to\infty}(\frac{-1}{N^3} + 1) = 1$.
b) $P(1 \le x \le 2) = \int_1^2 f(x)dx = \int_1^2 \frac{3}{x^4}dx = -\frac{1}{x^3}|_1^2 = \frac{7}{8}$.
c) $P(2 \le x < \infty) = \int_2^\infty f(x)dx = \int_2^\infty \frac{3}{x^4}dx = \lim_{N\to\infty}\int_2^N \frac{3}{x^4}dx = \lim_{N\to\infty} \frac{-1}{x^3}|_2^N = \lim_{N\to\infty}(\frac{-1}{N^3} + \frac{1}{8}) = \frac{1}{8}$.

7. a) $P(0 \le x) = \int_0^\infty f(x)dx =$
$\lim_{N\to\infty}\int_0^\infty 2xe^{-x^2}dx = \lim_{N\to\infty}(-e^{-x^2})|_0^N = \lim_{N\to\infty}(-e^{-N^2} + 1) = 1$.
b) $P(1 \le x \le 2) = \int_1^2 f(x)dx = \int_1^2 2xe^{-x^2}dx = -e^{-x^2}|_1^2 = -e^{-4} + e^{-1} = 0.3496$.
c) $P(x \le 2) = \int_{-\infty}^2 f(x)dx = \int_0^2 2xe^{-x^2}dx = -e^{-x^2}|_0^2 = -e^{-4} + 1 = 0.9817$.

9. Let x denote the time (in seconds) you must wait. The uniform density function for x is

$$f(x) = \begin{cases} \frac{1}{45} & \text{if} \quad 0 \le x \le 45 \\ 0 & \text{otherwise} \end{cases}$$

Hence, the probability that the light turns green within 15 seconds is $P(0 \le x \le 15) = \int_0^{15} \frac{1}{45}dx = \frac{x}{45}|_0^{15} = \frac{1}{3}$.

11. Let x denote the number of minutes since the start of the movie at the time of your arrival. The uniform density function for x is

$f(x) = \begin{cases} \frac{1}{120} & \text{if} \quad 0 \le x \le 120 \\ 0 & \text{otherwise} \end{cases}$

Hence, the probability that you arrive within 10 minutes(before or after) of the start of a movie is $P(0 \le x \le 10) + P(110 \le x \le 120) = 2P(0 \le x \le 10) = 2\int_0^{10} \frac{1}{120}dx = \frac{x}{60}\Big|_0^{10} = \frac{1}{6}$.

13. a) $P(8 \le x) = \int_8^\infty f(x)dx = \lim_{N\to\infty} \int_8^N \frac{1}{4}e^{-x/4}dx = \lim_{N\to\infty}(-e^{-x/4})\Big|_8^N =$ $\lim_{N\to\infty}(-e^{-N/4} + e^{-2}) = e^{-2} = 0.1353$.

15. For the probability density $f(x) = \begin{cases} \frac{1}{3} & \text{if} \quad 2 \le x \le 5 \\ 0 & \text{otherwise} \end{cases}$

$E(x) = \int_{-\infty}^{\infty} xf(x)dx = \int_2^5 \frac{x}{3}dx = \frac{x^2}{6}\Big|_2^5 = \frac{1}{6}(25-4) = \frac{7}{2}$.
$Var(x) = \int_{-\infty}^{\infty} x^2f(x)dx - [E(x)]^2 = \int_2^5 \frac{x^2}{3}dx - (\frac{7}{2})^2 = \frac{x^3}{9}\Big|_2^5 - \frac{49}{4} = \frac{1}{9}(125-8) - \frac{49}{4} = \frac{3}{4}$.

17. For the probability density $f(x) = \begin{cases} \frac{1}{8}(4-x) & \text{if} \quad 0 \le x \le 4 \\ 0 & \text{otherwise} \end{cases}$

$E(x) = \int_{-\infty}^{\infty} xf(x)dx = \int_0^4 \frac{x}{8}(4-x)dx = \frac{1}{8}\int_0^4(4x-x^2)dx = \frac{1}{8}(2x^2 - \frac{x^3}{3})\Big|_0^4 = \frac{1}{8}(32 - \frac{64}{3}) = \frac{4}{3}$.
$Var(x) = \int_{-\infty}^{\infty} x^2f(x)dx - [E(x)]^2 = \int_0^4 \frac{1}{8}x^2(4-x)dx - (\frac{4}{3})^2 = \frac{1}{8}\int_0^4(4x^2 - x^3)dx - \frac{16}{9} =$ $\frac{1}{8}(\frac{4x^3}{3} - \frac{x^4}{4})\Big|_0^4 - \frac{16}{9} = \frac{1}{8}(\frac{256}{3} - \frac{256}{4}) - \frac{16}{9} = \frac{8}{9}$.

19. For the probability density $f(x) = \begin{cases} \frac{1}{10}e^{-x/10} & \text{if} \quad 0 \le x \\ 0 & \text{otherwise} \end{cases}$

$E(x) = \int_{-\infty}^{\infty} xf(x)dx = \lim_{N\to\infty} \int_0^N \frac{x}{10}e^{-x/10}dx = \frac{1}{10}\lim_{N\to\infty}(-10xe^{-x/10} - 100e^{-x/10})\Big|_0^N =$ $\frac{1}{10}\lim_{N\to\infty}(-10Ne^{-N/10} - 100e^{-N/10} - (-100)] = 10$.
$Var(x) = \int_{-\infty}^{\infty} x^2f(x)dx - [E(x)]^2 = \lim_{N\to\infty} \int_0^N \frac{1}{10}x^2e^{-x/10}dx - 100 = \lim_{N\to\infty} \frac{1}{10}(-10x^2 -$ $200x - 2,000)e^{-x/10}\Big|_0^N - 100 = \lim_{N\to\infty}[(-10N^2 - 200N - 2,000)e^{-N/10} - (-2,000)] - 100 =$ $1,900$.

21. Let x denote your waiting time (in minutes). The probability density function for x is

$f(x) = \begin{cases} \frac{1}{20} & \text{if} \quad 0 \le x \le 20 \\ 0 & \text{otherwise} \end{cases}$

Hence, the average waiting time is
$E(x) = \int_{-\infty}^{\infty} xf(x)dx = \int_0^{20} \frac{x}{20}dx = \frac{1}{40}x^2\Big|_0^{20} = 10$ minutes.

23. From problem 13, the probability density function is

$f(x) = \begin{cases} \frac{1}{4}e^{-x/4} & \text{if} \quad 0 \le x \\ 0 & \text{otherwise} \end{cases}$

where x is the waiting time (in minutes). Hence, the average waiting time is
$E(x) = \int_{-\infty}^{\infty} xf(x)dx = \lim_{N\to\infty} \int_0^N \frac{1}{4}xe^{-x/4}dx = \frac{1}{4}\lim_{N\to\infty}(-4xe^{-x/4} - 16e^{-x/4})\Big|_0^N$
$= \frac{1}{4}\lim_{N\to\infty}[-4Ne^{-N/4} - 16e^{-N/4} - (-16)] = 4$ minutes.

25. For the probability density function $f(x) = \begin{cases} \frac{1}{B-A} & \text{if} \quad A \le x \le B \\ 0 & \text{otherwise} \end{cases}$

It can be shown (see problem 24) that $E(x) = \frac{A+B}{2}$. Hence, $Var(x) = \int_{-\infty}^{\infty} x^2 f(x)dx - [E(x)]^2 = \int_A^B x^2(\frac{1}{B-A})dx - (\frac{A+B}{2})^2 = \frac{1}{B-A}\int_A^B x^2 dx - (\frac{A+B}{2})^2 = \frac{1}{B-A}\frac{x^3}{3}\Big|_A^B - (\frac{A+B}{2})^2 = \frac{B^3-A^3}{3(B-A)} - (\frac{A+B}{2})^2 = \frac{(B-A)(B^2+AB+A^2)}{3(B-A)} - \frac{(A+B)^2}{4} = \frac{B^2+AB+A^2}{3} - \frac{A^2+2AB+B^2}{4} = \frac{B^2-2AB+A^2}{12} = \frac{(B-A)^2}{12}$.

27. For the exponential probability density function $f(x) = \begin{cases} ke^{-kx} & \text{if} \quad 0 \le x \\ 0 & \text{otherwise} \end{cases}$

It can be shown (see problem 26) that $E(x) = \frac{1}{k}$. Hence, $Var(x) = \int_{-\infty}^{\infty} x^2 f(x)dx - [E(x)]^2 = \lim_{N\to\infty} \int_0^N kx^2 e^{-kx}dx - (\frac{1}{k})^2 = k\lim_{N\to\infty}(-\frac{x^2}{k} - \frac{2x}{k^2} - \frac{2}{k^3})e^{-kx}\Big|_0^N - (\frac{1}{k})^2$
$= k\lim_{N\to\infty}[(-\frac{N^2}{k} - \frac{2N}{k^2} - \frac{2}{k^3})e^{-kN} - (-\frac{2}{k^3})] - \frac{1}{k^2} = k\frac{2}{k^3} - \frac{1}{k^2} = \frac{1}{k^2}$.

Review Problems

1. $\lim_{x\to\infty}(x^2+1)(2-x^4) = \lim_{x\to\infty}(-x^6+\cdots) = -\infty$.

2. Divide numerator and denominator by x^3 to get $\lim_{x\to\infty} \frac{1+x^2-3x^3}{2x^3+5} = \lim_{x\to\infty} \frac{1/x^3+1/x-3}{2+5/x^3} = -\frac{3}{2}$.

3. Divide numerator and denominator by x to get $\lim_{x\to\infty} \frac{(x+1)(2-x)}{x+3} = \lim_{x\to\infty} \frac{-x^2+x+2}{x+3} = \lim_{x\to\infty} \frac{2/x+1-x}{1+3/x} = -\infty$.

4. $\lim_{x\to\infty} \frac{2}{3+4e^{-5x}} = \frac{2}{3+0} = \frac{2}{3}$.

5. Since the limit is of the form $\frac{\infty}{\infty}$, l'Hôpital's rule (applied twice) gives $\lim_{x\to\infty} \frac{x^2+1}{e^{2x}} = \lim_{x\to\infty} \frac{2x}{2e^{2x}} = \frac{1}{2e^{2x}} = 0$.

6. There is no use for l'Hôpital's rule here. $\lim_{x\to\infty} \frac{x^2+1}{e^{-2x}} = \lim_{x\to\infty}(x^2+1)e^{2x} = (\infty)(\infty) = \infty$.

7. Multiply numerator and denominator by e^x to get $\lim_{x\to\infty} \frac{1+e^{-x}}{e^{-x}} = \lim_{x\to\infty}(e^x+1) = \infty$.

8. Since the limit is of the form $\frac{\infty}{\infty}$, l'Hôpital's rule gives $\lim_{x\to\infty} \frac{\ln(3x+1)}{\ln(x^2-4)} = \lim_{x\to\infty} \frac{3/(3x+1)}{2x/(x^2-4)} = \lim_{x\to\infty} \frac{3x^2-12}{6x^2+2x} = \frac{3}{6} = \frac{1}{2}$.

9. Since the limit is of the form $\frac{\infty}{\infty}$, l'Hôpital's rule (applied twice) gives $\lim_{x\to\infty} \frac{e^{\sqrt{x}}}{2x} = \lim_{x\to\infty} \frac{e^{\sqrt{x}}/(2\sqrt{x})}{2} = \lim_{x\to\infty} \frac{e^{\sqrt{x}}}{4\sqrt{x}} = \lim_{x\to\infty} \frac{e^{\sqrt{x}}/(2\sqrt{x})}{4/(2\sqrt{x})} = \lim_{x\to\infty} \frac{e^{\sqrt{x}}}{4} = \infty$.

10. Since the limit is of the form $(\infty)(0)$ rewrite the product as a quotient in the form $\frac{0}{0}$ and apply l'Hôpital's rule to get $\lim_{x\to\infty} \frac{e^{1/x}-1}{1/x^2} = \lim_{x\to\infty} \frac{e^{1/x}(-1/x^2)}{-2/x^3} = \lim_{x\to\infty} \frac{xe^{1/x}}{2} = \infty$.

11. Since the limit is of the form ∞^0, let $y = x^{1/x^2}$, take logarithms to get $\ln y = \ln x^{1/x^2} = \frac{\ln x}{x^2}$. Since the limit of $\ln y$ is of the form $\frac{\infty}{\infty}$, apply l'Hôpital's rule to get $\lim_{x\to\infty} \ln y = \lim_{x\to\infty} \frac{\ln x}{x^2} = \lim_{x\to\infty} \frac{1/x}{2x} = \lim_{x\to\infty} \frac{1}{2x^2} = 0$. Since $\ln y \to 0$, it follows that $y \to e^0 = 1$.

12. Since the limit is of the form 1^∞, let $y = (1+\frac{2}{x})^{3x}$. Take logarithms to get $\ln y = \ln(1+\frac{2}{x})^{3x} = 3x\ln(1+\frac{2}{x}) = \frac{\ln(1+2/x)}{1/(3x)}$. Since the limit of $\ln y$ is of the form $\frac{0}{0}$, apply l'Hôpital's rule to get $\lim_{x\to\infty} \ln y = \lim_{x\to\infty} \frac{\ln(1+2/x)}{1/(3x)} = \lim_{x\to\infty} \frac{(-2/x^2)/(1+2/x)}{-1/(3x^2)} = \lim_{x\to\infty} \frac{6}{1+2/x} = \frac{6}{1+0} = 6$. Since $\ln y \to 6$, it follows that $y \to e^6$.

13. $\int_0^\infty \frac{1}{\sqrt[3]{1+2x}}dx = \lim_{N\to\infty} \int_0^N (1+2x)^{-1/3}dx = \lim_{N\to\infty} \frac{3}{4}(1+2x)^{2/3}|_0^N = \frac{3}{4}\lim_{N\to\infty}[(1+2N)^{2/3} - 1] = \infty$.

14. $\int_0^\infty (1+2x)^{-3/2}dx = \lim_{N\to\infty} \int_0^N (1+2x)^{-3/2}dx = -\lim_{N\to\infty}(1+2x)^{-1/2}|_0^N$ $= -\lim_{N\to\infty}[(1+2N)^{-1/2} + 1] = 1$.

15. $\int_0^\infty \frac{3x}{x^2+1}dx = 3\lim_{N\to\infty} \int_0^N x^{-1}(x^2+1)^{-1}dx = 3\lim_{N\to\infty} \frac{1}{2}\ln(x^2+1)|_0^N$ $= \frac{3}{2}\lim_{N\to\infty}[\ln(N^2+1) - \ln 1] = \infty$.

16. $\int_0^\infty 3e^{-5x}dx = 3\lim_{N\to\infty} \int_0^N e^{-5x}dx = -\frac{3}{5}\lim_{N\to\infty} e^{-5x}|_0^N = -\frac{3}{5}\lim_{N\to\infty}(e^{-5N} - 1) = \frac{3}{5}$.

17. $\int_0^\infty xe^{-2x}dx = \lim_{N\to\infty} \int_0^N xe^{-2x}dx = \lim_{N\to\infty}(-\frac{1}{2}xe^{-2x})|_0^N + \frac{1}{2}\int_0^N e^{-2x}dx$ $= \lim_{N\to\infty}(-\frac{1}{2}xe^{-2x} - \frac{1}{4}e^{-2x})|_0^N = \lim_{N\to\infty}[-\frac{1}{2}e^{-2N}(N+\frac{1}{2}) + \frac{1}{4}] = \frac{1}{4}$.

18. $\int_0^\infty 2x^2e^{-x^3}dx = 2\lim_{N\to\infty} \int_0^N x^2e^{-x^3}dx = 2\lim_{N\to\infty}(-\frac{1}{3})e^{-x^3}|_0^N$ $= \frac{-2}{3}\lim_{N\to\infty}(e^{-N^3} - 1) = \frac{2}{3}$.

19. $\int_0^\infty x^2e^{-2x}dx = \lim_{N\to\infty} \int_0^N x^2e^{-2x}dx = \lim_{N\to\infty}(-\frac{1}{2}x^2e^{-2x}|_0^N + \int_0^N xe^{-2x}dx)$ $= \lim_{N\to\infty}[(-\frac{1}{2}x^2e^{-2x} - \frac{1}{2}xe^{-2x} - \frac{1}{4}e^{-2x})|_0^N = \lim_{N\to\infty}[(-\frac{1}{2}N^2 - \frac{1}{2}N - \frac{1}{4})e^{-2N} + \frac{1}{4}] = \frac{1}{4}$.

20. $\int_2^\infty \frac{1}{x(\ln x)^2}dx = \lim_{N\to\infty} \int_2^N \frac{1}{(\ln x)^2}(\frac{1}{x})dx = \lim_{N\to\infty}(-\frac{1}{\ln x})|_2^N = \lim_{N\to\infty}(-\frac{1}{\ln N} + \frac{1}{\ln 2}) = \frac{1}{\ln 2}$.

21. $\int_0^\infty \frac{x-1}{x+2}dx = \lim_{N\to\infty} \int_0^N (1 - \frac{3}{x+2})dx = \lim_{N\to\infty}(x - 3\ln|x+2|)|_0^N = \infty$. (Note: x

grows much more quickly than $\ln x$.)

22. $\int_0^\infty x^5 e^{-x^3}dx = \lim_{N\to\infty}\int_0^N x^3(x^2e^{-x^3})dx = \lim_{N\to\infty}(-\frac{1}{3}x^3e^{-x^3})|_0^N + \int_0^N x^2e^{-x^3}dx = \lim_{N\to\infty}(-\frac{1}{3}x^3e^{-x^3} - e^{-x^3} - \frac{1}{3}e^{-x^3})|_0^N = \lim_{N\to\infty}[(-\frac{1}{3}N^3e^{-N^3} - e^{-N^3} - \frac{1}{3}e^{-N^3}) + \frac{1}{3}] = \frac{1}{3}$.

23. It was determined in exercise 41 of section 6.4 that the number of subscribers in N years will be $P_0f(N) + \int_0^N r(t)f(N-t)dt$ where $P_0 = 20,000$ is the current number of subscribers, $f(t) = e^{-t/10}$ is the fraction of subscribers remaining at least t years, and $r(t) = 1,000$ is the rate at which new subscriptions are sold. Hence, the number of subscribers in the long run is $\lim_{N\to\infty}[(20,000e^{-N/10}) + \int_0^N 1,000e^{-(N-t)/10}dt] = \lim_{N\to\infty}[(20,000e^{-N/10}) + 1,000e^{-N/10}\int_0^N e^{t/10}dt] = 0 + 10,000\lim_{N\to\infty}e^{-N/10}e^{t/10}|_0^N = 10,000\lim_{N\to\infty}e^{-N/10}(e^{N/10} - 1) = 10,000\lim_{N\to\infty}(1-e^{-N/10}) = 10,000$.

24. To find the present value of the investment in N years, divide the N-year interval $0 \le t \le N$ into n equal subintervals of length Δt years, and let t_j denote the beginning of the jth subinterval. Then, during the jth subinterval, the amount generated is approximately $f(t_j)\Delta t$ and the present value is $f(t_j)e^{-0.1t_j}\Delta$. Hence, the present value of an N-year investment is $\lim_{n\to\infty}\sum_{j=1}^n f(t_j)e^{-0.1t_j}\Delta t = \int_0^N f(t)e^{-0.1t}dt$. To find the present value P of the total investment, let $P = \lim_{N\to\infty}\int_0^N f(t)e^{-0.1t}dt = \lim_{N\to\infty}\int_0^N(8,000 + 400t)e^{-0.1t}dt = \lim_{N\to\infty}[-10(8,000+400t)e^{-0.1t}|_0^N + 4,000\int_0^N e^{-0.1t}dt = \lim_{N\to\infty}[-10(8,000+400t)e^{-0.1t} - 40,000e^{-0.1t}]_0^N = \lim_{N\to\infty}(-120,000 - 4,000t)e^{-0.1t}|_0^N = \lim_{N\to\infty}[(-120,000 - 4,000N)e^{-0.1N} + 120,000] = \$120,000$.

25. As in problem 23, in N years the population of the city will be $P_0f(N) + \int_0^N r(t)f(N-t)dt$ where $P_0 = 100,000$ is the current population, $f(t) = e^{-t/20}$ is the fraction of the residents remaining for at least t years, and $r(t) = 100t$ is the rate of new arrivals. Hence, in the long run, the number of residents will be $\lim_{N\to\infty}[100,000e^{-N/20} + \int_0^N 100te^{-(N-t)/20}dt] = 0 + \lim_{N\to\infty}100e^{-N/20}\int_0^N te^{t/20}dt] =$
$\lim_{N\to\infty}100e^{-N/20}[20te^{t/20} - 400e^{t/20}]|_0^N = \lim_{N\to\infty}100e^{-N/20}[(20Ne^{N/20} - 400e^{N/20}) + 400]_0^N] = \lim_{N\to\infty}100(20N - 400 + 400e^{-N/20}) = \infty$. Thus the population will increase without bound.

26. a) $P(1 \le x \le 4) = \int_1^4 f(x)dx = \int_1^4 \frac{1}{3}dx = \frac{x}{3}|_1^4 = 1$.
b) $P(2 \le x \le 3) = \int_2^3 f(x)dx = \int_2^3 \frac{1}{3}dx = \frac{x}{3}|_2^3 = \frac{1}{3}$.
c) $P(x \le 2) = \int_{-\infty}^2 f(x)dx = \int_1^2 \frac{1}{3}dx = \frac{x}{3}|_1^2 = \frac{1}{3}$.

27. a) $P(0 \le x \le 3) = \int_0^3 f(x)dx = \int_0^3 \frac{2(3-x)}{9}dx = \frac{2}{9}(3x - \frac{x^2}{2})|_0^3 = \frac{2}{9}(9 - \frac{9}{2} - 0) = 1$.
b) $P(1 \le x \le 2) = \int_1^2 f(x)dx = \int_1^2 \frac{2(3-x)}{9}dx = \frac{2}{9}(3x - \frac{x^2}{2})|_1^2 = \frac{2}{9}[(6 - \frac{4}{2}) - (3 - \frac{1}{2})] = \frac{1}{3}$.

28. a) $P(0 \le x) = \int_0^\infty f(x)dx = \lim_{N\to\infty}\int_0^\infty 0.2e^{-0.2x}dx = \lim_{N\to\infty}(-e^{-0.2x})|_0^N = \lim_{N\to\infty}(-e^{-0.2N} + 1) = 1$.
b) $P(1 \le x \le 4) = \int_1^4 f(x)dx = \int_1^4 0.2e^{-0.2x}dx = -e^{-0.2x}|_1^4 = -e^{-0.4} + e^{-0.2} = 0.1484$.
c) $P(5 \le x) = \int_5^\infty f(x)dx = \lim_{N\to\infty}\int_5^N 0.2e^{-0.2x}dx = -\lim_{N\to\infty}e^{-0.2x}|_5^N$

$= \lim_{N\to\infty}[-e^{-0.2N} + e^{-1}] = 0.3679.$

29. For the probability density function $f(x) = \begin{cases} \frac{1}{3} & \text{if} \quad 1 \le x \le 4 \\ 0 & \text{otherwise} \end{cases}$

$E(x) = \int_{-\infty}^{\infty} xf(x)dx = \int_1^4 \frac{x}{3}dx = \frac{x^2}{6}|_1^4 = \frac{1}{6}(16-1) = \frac{5}{2}.$
$Var(x) = \int_{-\infty}^{\infty} x^2 f(x)dx - [E(x)]^2 = \int_1^4 \frac{x^2}{3}dx - (\frac{5}{2})^2 = \frac{x^3}{9}|_1^4 - \frac{25}{4} = \frac{1}{9}(64-1) - \frac{25}{4} = \frac{3}{4}.$

30. For the probability density $f(x) = \begin{cases} \frac{2}{9}(3-x) & \text{if} \quad 0 \le x \le 3 \\ 0 & \text{otherwise} \end{cases}$

$E(x) = \int_{-\infty}^{\infty} xf(x)dx = \int_0^3 \frac{2x}{9}(3-x)dx = \frac{2}{9}\int_0^3 (3x - x^2)dx = \frac{2}{9}(\frac{3}{2}x^2 - \frac{x^3}{3})|_0^3 = \frac{2}{9}(\frac{27}{2} - \frac{27}{3}) = 1.$
$Var(x) = \int_{-\infty}^{\infty} x^2 f(x)dx - [E(x)]^2 = \int_0^3 \frac{2}{9}x^2(3-x)dx - 1 = \frac{2}{9}\int_0^3 (3x^2 - x^3)dx - 1 = \frac{2}{9}(x^3 - \frac{x^4}{4})|_0^3 - 1 = \frac{2}{9}(27 - \frac{81}{4}) - 1 = \frac{1}{2}.$

31. For the probability density $f(x) = \begin{cases} 0.2e^{-0.2x} & \text{if} \quad 0 \le x \\ 0 & \text{otherwise} \end{cases}$

$E(x) = \int_{-\infty}^{\infty} xf(x)dx = \lim_{N\to\infty} \int_0^N 0.2e^{-0.2x}dx = \lim_{N\to\infty}(-xe^{-0.2x}|_0^N + \int_0^N e^{-0.2x})$
$= \lim_{N\to\infty}[-(x+5)e^{-0.2x}|_0^N] = \lim_{N\to\infty}[-(N+5)e^{-0.2N} + 5] = 5.$
$Var(x) = \int_{-\infty}^{\infty} x^2 f(x)dx - [E(x)]^2 = \lim_{N\to\infty} \int_0^N 0.2x^2 e^{-0.2x}dx - 25 = \lim_{N\to\infty}(-x^2 e^{-0.2x}|_0^N + \int_0^N 2xe^{-0.2x}dx) - 25 = \lim_{N\to\infty}[(-x^2 e^{-0.2x} - 10xe^{-0.2x})|_0^N + \int_0^N 10e^{-0.2x}dx] - 25 = \lim_{N\to\infty}[-(x^2 + 10x + 50)e^{-0.2x}]|_0^N - 25 = \lim_{N\to\infty}[-(N^2 + 10N + 50)e^{-0.2N} + 50] - 25 = 25.$

32. Let x denote the time (in minutes) between your arrival and the next batch of cookies. Then x is uniformly distributed with probability density function
$f(x) = \begin{cases} \frac{1}{45} & \text{if} \quad 0 \le x \le 45 \\ 0 & \text{otherwise} \end{cases}$
Hence, the probability that you arrive within 5 minutes (before or after) the cookies were baked is $P(0 \le x \le 5) + P(40 \le x \le 45) = 2P(0 \le x \le 5) = 2\int_0^5 \frac{1}{45}dx = \frac{2x}{45}|_0^5 = \frac{2}{9}.$

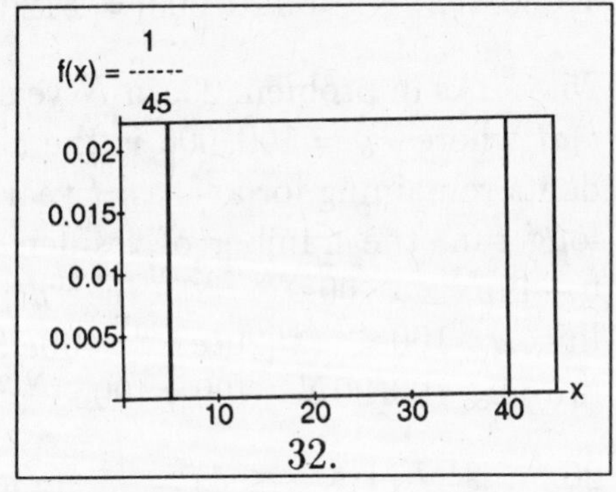

32.

33. Let x denote the time (in minutes) between the arrivals of successive cars. Then the probability density function is
$f(x) = \begin{cases} 0.5e^{-0.5x} & \text{if} \quad 0 \le x \\ 0 & \text{otherwise} \end{cases}$
a) The probability that two cars will arrive at least 6 minutes apart is
$P(6 \le x < \infty) = \int_6^{\infty} f(x)dx = \lim_{N\to\infty} \int_6^N 0.5e^{-0.5x}dx = \lim_{N\to\infty}(-e^{-0.5x})|_6^N$
$= \lim_{N\to\infty}(-e^{-0.5N} + e^{-3}) = 0.0498.$
b)The average time interval between successive arrivals is

$E(x) = \int_{-\infty}^{\infty} xf(x)dx = \lim_{N\to\infty} \int_0^N 0.5xe^{-0.5x}dx = \lim_{N\to\infty}(-xe^{-0.5x}|_0^N + \int_0^N e^{-0.5x}dx) = \lim_{N\to\infty}(-x-2)e^{-0.5x}|_0^N = \lim_{N\to\infty}(-N-2)e^{-0.5N} + 2 = 2$ minutes.

Chapter 9

Functions of Two Variables

9.1 Functions of Two Variables: Surfaces and Level Curves

1. $f(x,y) = (x-1)^2 + 2xy^3$. The domain consists of all ordered pairs (x,y) of real numbers. Moreover $f(2,-1) = (2-1)^2 + 2(2)(-1)^3 = -3$ and $f(1,2) = (1-1)^2 + 2(1)(2)^3 = 16$.

3. $f(x,y) = \sqrt{y^2 - x^2}$. The domain consists of all ordered pairs (x,y) of real numbers for which $y^2 - x^2 \geq 0$, or equivalently, for which $|y| \geq |x|$. Moreover $f(4,5) = \sqrt{5^2 - 4^2} = \sqrt{9} = 3$ and $f(-1,2) = \sqrt{2^2 - (-1)^2} = \sqrt{3} = 1.732$.

5. $f(x,y) = \frac{x}{\ln y}$. The domain consists of all ordered pairs (x,y) of real numbers for which $y > 0$ (since $\ln y$ is defined only for positive values of y) and $y \neq 1$ (since $\ln 1 = 0$). Moreover $f(-1,e^3) = \frac{-1}{\ln e^3} = -\frac{1}{3}$ and $f(\ln 9, e^2) = \frac{\ln 9}{\ln e^2} = \frac{\ln 9}{2} = \ln 9^{1/2} = \ln 3 = 1.099$.

7. $f(x,y) = x + 2y$. With $C = 1$, $C = 2$, and $C = 3$, the three sketched level curves have equations $x + 2y = 1$, $x + 2y = 2$, and $x + 2y = 3$.

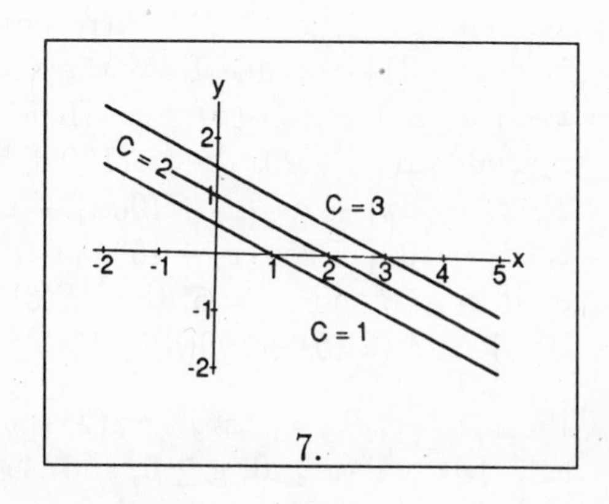

7.

9. $f(x,y) = x^2 - 4x - y$. With $C = -4$ and $C = 5$, the two sketched level curves have equations $x^2 - 4x - y = -4$ and $x^2 - 4x - y = 5$.

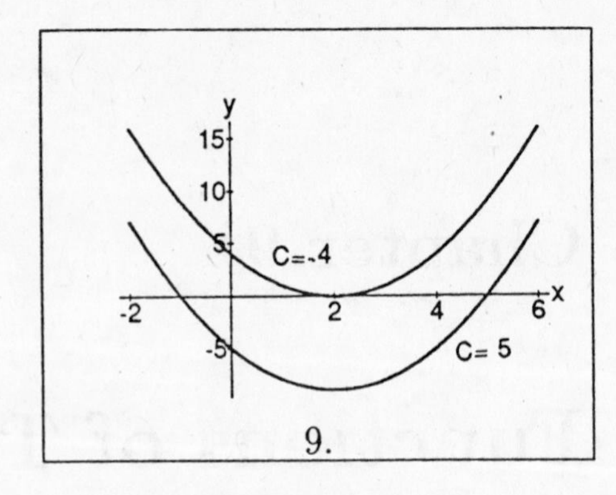

9.

11. $f(x,y) = xy$. With $C = 1, C = -1, C = 2$, and $C = -2$, the four sketched level curves have equations $xy = 1$, $xy = -1$, $xy = 2$, and $xy = -2$.

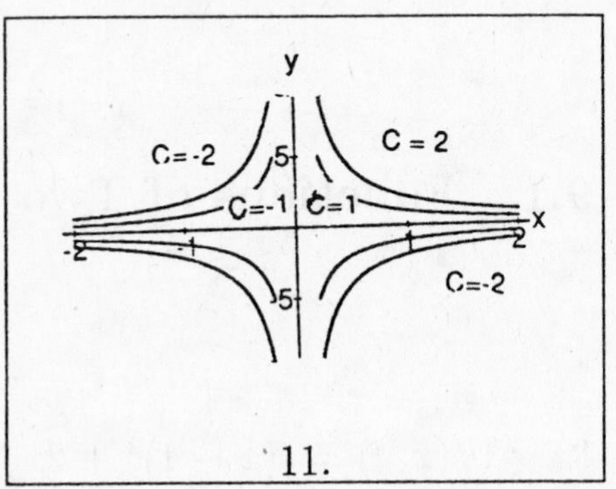

11.

13. $f(x,y) = xe^y$. With $C = 1$ and $C = e$, the two sketched level curves have equations $xe^y = 1$ and $xe^y = e$.

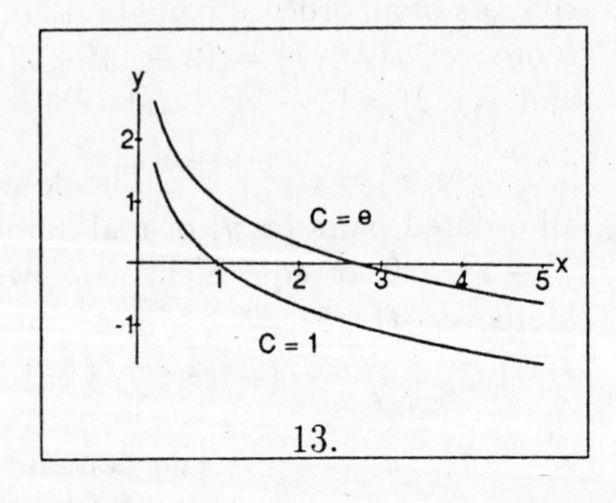

13.

15. a) $Q(x,y) = 10x^2y$ and $x = 20$, $y = 40$. $Q(20,40) = 10(20)^2(40) = 160{,}000$ units.
b) With one more skilled worker, $x = 21$ and the additional output is $Q(21,40) - Q(20,40) = 10(21)^2(40) - 10(20)^2(40) = 16{,}400$ units.
c) With one more unskilled worker, $y = 41$ and the additional output is $Q(20,41) - Q(20,40) = 10(20)^2(41) - 10(20)^2(40) = 4{,}000$ units.
d) With one more skilled worker and one more unskilled worker, $x = 21$ and $y = 41$, so the additional output is $Q(21,41) - Q(20,40) = 10(21)^2(41) - 10(20)^2(40) = 20{,}810$ units.

17. a) Let R denote the total monthly revenue. Then, R =(revenue from the first brand)+(revenue from the second brand)= $x_1D_1(x_1,x_2) + x_2D_2(x_1,x_2)$. Hence, $R(x_1,x_2) = x_1(200 - 10x_1 + 20x_2) + x_2(100 + 5x_1 - 10x_2) = 200x_1 - 10x_1^2 + 20x_1x_2 + 100x_2 + 5x_1x_2 - 10x_2^2 = 200x_1 - 10x_1^2 + 25x_1x_2 + 100x_2 - 10x_2^2$.
b) If $x_1 = 6$ and $x_2 = 5$, then $R(6,5) = 200(6) - 10(6)^2 + 25(6)(5) + 100(5) - 10(5)^2 = \$1{,}840$.

19. $f(x,y) = Ax^ay^b$. $f(2x,2y) = A(2x)^a(2y)^b = A(2)^ax^a(2)^by^b = (2)^a(2)^bAx^ay^b = (2^{a+b})Ax^ay^b$. ($x \geq 0$, $y \geq 0$, and $A > 0$.)
a) If $a + b > 1$, $2^{a+b} > 2$ and f more than doubles.
b) If $a + b < 1$, $2^{a+b} < 2$ and f increases but does not double.
c) If $a + b = 1$, $2^{a+b} = 2$ and f doubles (exactly).

21. Let y denote the number of machines sold in foreign markets and x the number of machines sold domestically. The sales price domestically is $60 - \frac{x}{5} + \frac{y}{20}$ and $50 - \frac{y}{10} + \frac{x}{20}$ in foreign markets. The revenue is $R(x,y) = x(60 - \frac{x}{5} + \frac{y}{20}) + y(50 - \frac{y}{10} + \frac{x}{20}) = 60x - \frac{x^2}{5} + \frac{xy}{20} + 50y - \frac{y^2}{10} + \frac{xy}{20} = 60x + 50y - \frac{x^2}{5} - \frac{y^2}{10} + \frac{xy}{10}$. If the cost of manufacturing, storing, and selling a machine is a constant a for foreign markets and b at home, the profit is $P(x,y) = 60x + 50y - \frac{x^2}{5} - \frac{y^2}{10} + \frac{xy}{10} - bx - ay$.

23. If $Q(K,L) = AK^\alpha L^{1-\alpha}$, then $Q(mK,mL) = A(mK)^\alpha(mL)^{1-\alpha} = Am^\alpha K^\alpha m^{1-\alpha}L^{1-\alpha} = Am^{1+\alpha-\alpha}K^\alpha L^{1-\alpha} = mAK^\alpha L^{1-\alpha} = mQ(K,L)$.

25. If $U(x,y) = 2x^3y^2$, then $U(5,4) = 2(5)^3(4)^2 = 4,000$. The graph of the level curve $U = 4,000$ is shown in the figure.

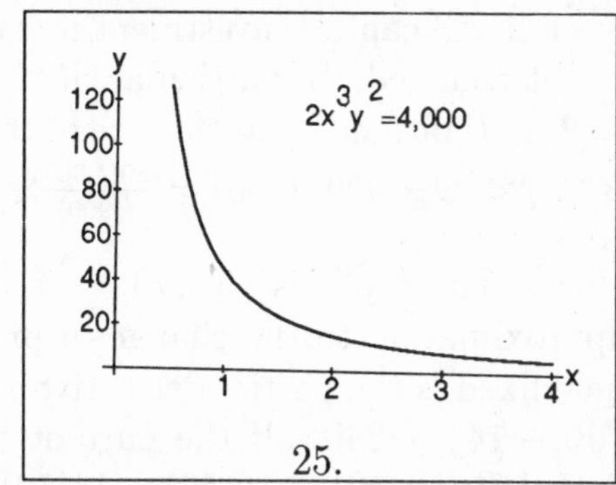

25.

9.2 Partial Derivatives

1. If $f(x,y) = 2xy^5 + 3x^2y + x^2$, then $f_x(x,y) = 2y^5 + 3(2x)y + 2x = 2y^5 + 6xy + 2x$ and $f_y(x,y) = 2x(5y^4) + 3x^2 = 10xy^4 + 3x^2$

3. If $z = (3x+2y)^5$, then $\frac{\partial z}{\partial x} = 5(3x+2y)^4\frac{\partial}{\partial x}(3x+2y) = 15(3x+2y)^4$ and $\frac{\partial z}{\partial y} = 5(3x+2y)^4\frac{\partial}{\partial y}(3x+2y) = 10(3x+2y)^4$

5. If $f(x,y) = \frac{3y}{2x} = \frac{3}{2}x^{-1}y$, then $f_x(x,y) = \frac{3}{2}(-1)x^{-2}y = -\frac{3y}{2x^2}$ and $f_y(x,y) = \frac{3}{2}x^{-1} = \frac{3}{2x}$.

7. If $z = xe^{xy}$, then $\frac{\partial z}{\partial x} = x(ye^{xy}) + e^{xy}(1) = (xy+1)e^{xy}$ and $\frac{\partial z}{\partial y} = x(e^{xy})(x) = x^2e^{xy}$.

9. If $f(x,y) = \frac{e^{2-x}}{y^2} = e^{2-x}y^{-2}$, then $f_x(x,y) = -e^{2-x}y^{-2} = -\frac{e^{2-x}}{y^2}$ and $f_y(x,y) = e^{2-x}(-2y^{-3}) = -\frac{2e^{2-x}}{y^3}$.

11. If $f(x,y) = \frac{2x+3y}{y-x}$, then $f_x(x,y) = \frac{(y-x)(2)-(2x+3y)(-1)}{(y-x)^2} = \frac{5y}{(y-x)^2}$ and $f_y(x,y) = \frac{(y-x)(3)-(2x+3y)(1)}{(y-x)^2} = \frac{-5x}{(y-x)^2}$.

13. If $z = x\ln y$, then $\frac{\partial z}{\partial x} = (1)\ln y$ and $\frac{\partial z}{\partial y} = x(\frac{1}{y}) = \frac{x}{y}$.

15. If $f(x,y) = \frac{\ln(x+2y)}{y^2}$, then $f_x(x,y) = \frac{(y^2)[1/(x+2y)]-\ln(x+2y)(0)}{y^4} = \frac{1}{(x+2y)y^2}$ and $f_y(x,y) = \frac{(y^2)[2/(x+2y)]-\ln(x+2y)(2y)}{y^4} = \frac{(y)(2)-(x+2y)\ln(x+2y)(2)}{(x+2y)y^3} = \frac{2[y-(x+2y)\ln(x+2y)]}{(x+2y)y^3}$.

17. Since $Q = 60K^{1/2}L^{1/3}$, the partial derivative $Q_K = \frac{\partial Q}{\partial K} = 30K^{-1/2}L^{1/3} = \frac{30L^{1/3}}{K^{1/2}}$ is the rate of change of the output with respect to the capital investment. For any values of K and L, this is an approximation to the additional number of units that will be produced each week if the capital investment is increased from K to $K+1$ while the size of the labor force is not changed. In particular, if the capital investment K is increased from 900 (thousand) to 901 (thousand) and the size of the labor force is $L = 1,000$, the resulting change in output is $\Delta Q = Q_K(900, 1000) = \frac{30(1,000)^{1/3}}{(900)^{1/2}} = \frac{30(10)}{30} = 10$ units.

19. The profit is $P(x,y) = (x-30)(70-5x+4y)+(y-40)(80+6x-7y)$ cents. An approximation to the change in profit that will result if y is increased by 1 cent while x is held fixed is the partial derivative $P_y(x,y) = (x-30)(4)+(y-40)(-7)+(1)(80+6x-7y) = 10x - 14y + 240$. If the current prices are $x = 50$ and $y = 52$, the change in profit is $\Delta P = P_y(50, 52) = 10(50) - 14(52) + 240 = 12$ cents.

21. a) Sugar and artificial sweetener are substitute commodities since an increase in the price of one would cause an increase in the demand for the other. Similarly, butter and margerine are substitute commodities.
b) If two commodities are substituite commodities and Q_1 is the demand for the first commodity and Q_2 is the the demand for the second commodity, then $\frac{\partial Q_1}{\partial p_2} \geq 0$ and $\frac{\partial Q_2}{\partial p_1} \geq 0$ since the demand for each is an increasing function of the price of the other.
c) If $Q_1 = 3,000 + \frac{400}{p_1+3} + 50p_2$, then $\frac{\partial Q_1}{\partial p_2} = 50 \geq 0$, and if $Q_2 = 2,000 - 100p_1 + \frac{500}{p_2+4}$, then $\frac{\partial Q_2}{\partial p_1} = -100 < 0$. Hence the commodities are not substitutes.

23. If $f(x,y) = 5x^4y^3 + 2xy$, then $f_x = 5(4x^3)y^3 + 2y = 20x^3y^3 + 2y$ and $f_y = 5x^4(3y^2) + 2x = 15x^4y^2 + 2x$. Hence, $f_{xx} = \frac{\partial}{\partial x}(f_x) = 20(3x^2)y^3 + 0 = 60x^2y^3$, $f_{yy} = \frac{\partial}{\partial y}(f_y) = 15x^4(2y) + 0 = 30x^4y$, $f_{xy} = \frac{\partial}{\partial y}(f_x) = 20x^3(3y^2) + 2(1) = 60x^3y^2 + 2$, $f_{yx} = \frac{\partial}{\partial x}(f_y) = 15(4x^3)y^2 + 2(1) = 60x^3y^2 + 2$.

25. If $f(x,y) = e^{x^2y}$, then $f_x = 2xye^{x^2y}$ and $f_y = x^2e^{x^2y}$. Hence, $f_{xx} = \frac{\partial}{\partial x}(f_x) = 2xy(e^{x^2y})(2xy) + e^{x^2y}(2y) = 2y(2x^2y+1)e^{x^2y}$, $f_{yy} = \frac{\partial}{\partial y}(f_y) = x^2(e^{x^2y})(x^2) = x^4e^{x^2y}$, $f_{xy} = \frac{\partial}{\partial y}(f_x) = 2xy(e^{x^2y})(x^2) + e^{x^2y}(2x) = 2x(x^2y+1)e^{x^2y}$, $f_{yx} = \frac{\partial}{\partial x}(f_y) = x^2(e^{x^2y})(2xy) + e^{x^2y}(2x) = 2x(x^2y+1)e^{x^2y}$.

27. If $f(x,y) = \sqrt{x^2+y^2}$, then $f_x = \frac{1}{2}(x^2+y^2)^{-1/2}(2x) = x(x^2+y^2)^{-1/2}$. Interchanging x and y in the work done from the beginning of the exercise to this point leads to $f_y = y(x^2+y^2)^{-1/2}$. Hence, $f_{xx} = \frac{\partial}{\partial x}(f_x) = x(-\frac{1}{2})(x^2+y^2)^{-3/2}(2x) + (x^2+y^2)^{-1/2} = -\frac{x^2}{(x^2+y^2)^{3/2}} +$

$\frac{x^2+y^2}{(x^2+y^2)^{3/2}} = \frac{y^2}{(x^2+y^2)^{3/2}}$, $f_{yy} = \frac{x^2}{(x^2+y^2)^{3/2}}$ (just interchange x and y), $f_{xy} = \frac{\partial}{\partial y}(f_x) = x(-\frac{1}{2})(x^2 + y^2)^{-3/2}(2y) = -\frac{xy}{(x^2+y^2)^{3/2}} = f_{yx}$.

29. If $\frac{\partial^2 Q}{\partial K^2} < 0$, the marginal product of capital decreases as K increases. This implies that for a fixed level of labor, the effect on output of the addition of \$1,000 of capital is greater when capital investment is small than when capital investment is large. If $\frac{\partial^2 Q}{\partial K^2} > 0$, it follows that for a fixed level of labor, the effect on output of the addition of \$1,000 of capital is greater when capital investment is large than when it is small.

31. a) According to the law of diminishing returns, $\frac{\partial Q}{\partial L}$ is increasing if $L < L_0$ and $\frac{\partial Q}{\partial L}$ is decreasing if $L > L_0$. Rephrased in terms of the derivative of $\frac{\partial Q}{\partial L}$, the law states that $\frac{\partial^2 Q}{\partial L^2} > 0$ if $L < L_0$ and $\frac{\partial^2 Q}{\partial L^2} < 0$ if $L > L_0$.

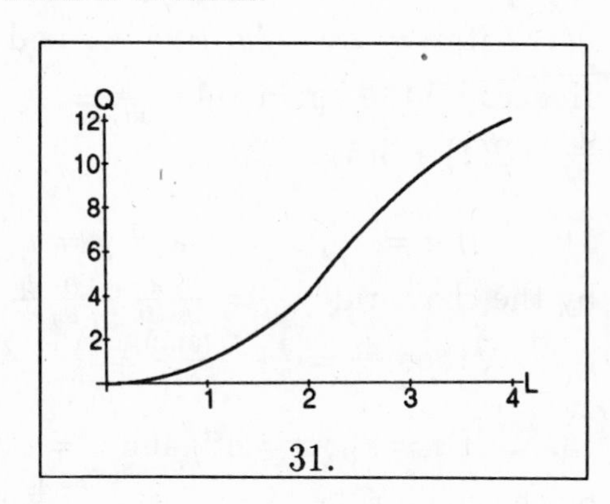

31.

b) A typical graph of Q as a function of L that reflects this situation is shown in the picture. When the size of the labor force is very small, (inadequate for the utilization of existing equipment, for example), the addition of one worker will have little impact on output. When the size of the labor force is very large, (close to the saturation level), the addition of one worker will again have little impact on output.

9.3 The Chain Rule; Approximation by the Total Differential

1. If $z = x + 2y$, $x = 3t$, and $y = 2t + 1$, then $\frac{\partial z}{\partial x} = 1$, $\frac{\partial z}{\partial y} = 2$, $\frac{dx}{dt} = 3$, and $\frac{dy}{dt} = 2$. Hence, by the chain rule, $\frac{dz}{dt} = \frac{\partial z}{\partial x}\frac{dx}{dt} + \frac{\partial z}{\partial y}\frac{dy}{dt} = 1(3) + 2(2) = 7$.

3. If $z = \frac{x}{y}$, $x = t^2$, and $y = 3t$, then $\frac{\partial z}{\partial x} = \frac{1}{y}$, $\frac{\partial z}{\partial y} = -\frac{x}{y^2}$, $\frac{dx}{dt} = 2t$, and $\frac{dy}{dt} = 3$. Hence, by the chain rule, $\frac{dz}{dt} = \frac{\partial z}{\partial x}\frac{dx}{dt} + \frac{\partial z}{\partial y}\frac{dy}{dt} = (\frac{1}{y})(2t) - \frac{x}{y^2}(3) = (\frac{1}{3t})(2t) - \frac{t^2}{(3t)^2}(3) = \frac{2}{3} - \frac{3}{9} = \frac{1}{3}$.

5. If $z = \frac{x+y}{x-y}$, $x = t^3 + 1$, and $y = 1 - t^3$, then $\frac{\partial z}{\partial x} = \frac{(x-y)(1)-(x+y)(1)}{(x-y)^2} = \frac{-2y}{(x-y)^2}$, $\frac{\partial z}{\partial y} =$

$\frac{(x-y)(1)-(x+y)(-1)}{(x-y)^2} = \frac{2x}{(x-y)^2}$, $\frac{dx}{dt} = 3t^2$, and $\frac{dy}{dt} = -3t^2$. Hence, by the chain rule, $\frac{dz}{dt} = \frac{\partial z}{\partial x}\frac{dx}{dt} + \frac{\partial z}{\partial y}\frac{dy}{dt} = [\frac{-2y}{(x-y)^2}](3t^2) + [\frac{2x}{(x-y)^2}](-3t^2) = \frac{-2(1-t^3)(3t^2)+2(t^3+1)(-3t^2)}{[t^3+1-(1-t^3)]^2} = \frac{-6t^2+6t^5-6t^5-6t^2}{(2t^3)^2} = -\frac{12t^2}{4t^6} = -\frac{3}{t^4}$.

7. If $z = (x-y^2)^3$, $x = t^2$, and $y = 2t$, then $\frac{\partial z}{\partial x} = 3(x-y^2)^2$, $\frac{\partial z}{\partial y} = -6y(x-y^2)^2$, $\frac{dx}{dt} = 2t$, and $\frac{dy}{dt} = 2$. Hence, by the chain rule, $\frac{dz}{dt} = \frac{\partial z}{\partial x}\frac{dx}{dt} + \frac{\partial z}{\partial y}\frac{dy}{dt} = 3(x-y^2)^2(2t) - 6y(x-y^2)^2(2) = (6t-12y)(x-y^2)^2 = (6t-24t)(t^2-4t^2)^2 = -18t(9t^4) = -162t^5$.

9. If $z = 2x+3y$, $x = t^2$, and $y = 5t$, then $\frac{\partial z}{\partial x} = 2$, $\frac{\partial z}{\partial y} = 3$, $\frac{dx}{dt} = 2t$, and $\frac{dy}{dt} = 5$. Hence, by the chain rule, $\frac{dz}{dt} = \frac{\partial z}{\partial x}\frac{dx}{dt} + \frac{\partial z}{\partial y}\frac{dy}{dt} = 2(2t) + 3(5) = 4t + 15$, and when $t = 2$, $\frac{dz}{dt} = 2(4) + 3(5) = 23$.

11. If $z = \frac{3x}{y}$, $x = t$, and $y = t^2$, then $\frac{\partial z}{\partial x} = \frac{3}{y}$, $\frac{\partial z}{\partial y} = -\frac{3x}{y^2}$, $\frac{dx}{dt} = 1$, and $\frac{dy}{dt} = 2t$. Hence, by the chain rule, $\frac{dz}{dt} = \frac{\partial z}{\partial x}\frac{dx}{dt} + \frac{\partial z}{\partial y}\frac{dy}{dt} = (\frac{3}{y})(1) - \frac{3x}{y^2}(2t) = (\frac{3}{y}) - \frac{6xt}{y^2}$. When $t = 3$, $x = 3$, and $y = 3^2$, then $\frac{dz}{dt} = \frac{3}{9} - \frac{6(3)(3)}{81} = \frac{1}{3} - \frac{2}{3} = -\frac{1}{3}$

13. If $z = xy$, $x = e^{2t}$, and $y = e^{3t}$, then $\frac{\partial z}{\partial x} = y$, $\frac{\partial z}{\partial y} = x$, $\frac{dx}{dt} = 2e^{2t}$, and $\frac{dy}{dt} = 3e^{3t}$. Hence, by the chain rule, $\frac{dz}{dt} = \frac{\partial z}{\partial x}\frac{dx}{dt} + \frac{\partial z}{\partial y}\frac{dy}{dt} = y(2e^{2t}) + x(3e^{3t})$. When $t = 0$, $x = 1$, and $y = 1$, then $\frac{dz}{dt} = 1(2) + 1(3) = 5$.

15. If $f(x,y) = \frac{x}{y}$, then $f_x = \frac{1}{y}$, $f_y = -\frac{x}{y^2}$, and $\frac{dy}{dx} = -\frac{f_x}{f_y} = -\frac{1/y}{-x/y^2} = \frac{y}{x}$.
If $\frac{x}{y} = C$, then $Cy - x = 0$ and by implicit differentiation with y regarded as a function of x gives $C\frac{dy}{dx} - 1 = 0$ or $\frac{dy}{dx} = \frac{1}{C} = \frac{1}{x/y} = \frac{y}{x}$ (since $C = \frac{x}{y}$).

17. If $f(x,y) = x^2y + 2y^3 - 3x - 2e^{-x}$, then $f_x = 2xy - 3 + 2e^{-x}$, $f_y = x^2 + 6y^2$, and $\frac{dy}{dx} = -\frac{f_x}{f_y} = -\frac{2xy-3+2e^{-x}}{x^2+6y^2} = \frac{3-2xy-2e^{-x}}{x^2+6y^2}$.
If $x^2y + 2y^3 - 3x - 2e^{-x} = C$, then and by implicit differentiation with y regarded as a function of x gives $x^2\frac{dy}{dx} + y(2x) + 2(3y^2\frac{dy}{dx}) - 3 + 2e^{-x} = 0$, $[x^2 + 2(3y^2)]\frac{dy}{dx} = 3 - 2e^{-x} - y(2x)$, and $\frac{dy}{dx} = \frac{3-2e^{-x}-2xy}{x^2+6y^2}$

19. If $f(x,y) = x\ln y$, then $f_x = \ln y$, $f_y = \frac{x}{y}$, and $\frac{dy}{dx} = -\frac{f_x}{f_y} = -\frac{\ln y}{x/y} = -\frac{y\ln y}{x}$.
If $x\ln y = C$, then and by implicit differentiation with y regarded as a function of x gives $x(\frac{1}{y}\frac{dy}{dx}) + (\ln y)(1) = 0$, $\frac{dy}{dx} = -\frac{y\ln y}{x}$

21. If $f(x,y) = x^2 + xy + y^3$, then $f_x = 2x + y$, $f_y = x + 3y^2$, and $\frac{dy}{dx} = -\frac{f_x}{f_y} = -\frac{2x+y}{x+3y^2}$.
When $x = 0$ and $f = 8$, it follows that $8 = y^3$ or $y = 2$. The slope of the tangent at the point $(0,2)$ is the corresponding value of $\frac{dy}{dx}$, that is, $m = \frac{dy}{dx} = -\frac{2(0)+2}{0+3(4)} = -\frac{1}{6}$.

23. If $f(x,y) = (x^2+y)^3$, then $f_x = 6x(x^2+y)^2$, $f_y = 3(x^2+y)^2$, and $\frac{dy}{dx} = -\frac{f_x}{f_y} = -\frac{6x(x^2+y)^2}{3(x^2+y)^2} = -2x$. When $x = -1$ and $f = 8$, it follows that $8 = (1+y)^3$ or $y = 1$. The slope of the tangent at the point $(-1,1)$ is the corresponding value of $\frac{dy}{dx}$, that is, $m = \frac{dy}{dx} = -2(-1) = 2$.

25. If $f(x,y) = e^{2x}\ln\frac{y}{x} = e^{2x}(\ln y - \ln x)$, then $f_x = e^{2x}[-\frac{1}{x} + 2(\ln y - \ln x)] = e^{2x}(-\frac{1}{x} + 2\ln\frac{y}{x})$, $f_y = e^{2x}(\frac{1}{y})$, and $\frac{dy}{dx} = -\frac{f_x}{f_y} = -\frac{e^{2x}(-\frac{1}{x}+2\ln\frac{y}{x})}{e^{2x}(\frac{1}{y})} = -\frac{-y+2xy\ln\frac{y}{x}}{x}$. When $x = 1$ and $f = e^2$, it follows that $e^2 = e^2\ln y$, $\ln y = 1$, or $y = e$. The slope of the tangent at the point $(1,e)$ is the corresponding value of $\frac{dy}{dx}$, that is, $m = \frac{dy}{dx} = e - 2e\ln e = e(1-2) = -e$.

27 $10xy^{1/2} = C$. $\frac{dx}{dy} = -\frac{(10x)(\frac{1}{2}y^{-1/2})}{10y^{1/2}} = -\frac{x}{2y}$. With $x = 30$ and $y = 36$, $\frac{dx}{dy} = -\frac{5}{12} = -0.42$ hours of skilled labor.

29. $2x^3 + 3x^2y + y^3 = 0$. $\frac{dy}{dx} = -\frac{6x^2+6xy}{3x^2+3y^2} = -\frac{2x(x+y)}{x^2+y^2}$. With $x = 20$ and $y = 10$, $dy = 0.5$, $\frac{dy}{dx} = -\frac{2(20)(20+10)}{20^2+10^2} = -\frac{12}{5}$, $\frac{dx}{dy} = -\frac{5}{12} = -0.42$ and $dx = -0.42 \times 0.5 = -0.21$ unit.

31. $U = (x+1)(y+2)$. $\frac{dy}{dx} = -\frac{y+2}{x+1}$. With $x = 25$ and $y = 8$, $dy = -1$, $\frac{dy}{dx} = -\frac{5}{13}$ and $dx = (\frac{dx}{dy})dy = (-\frac{13}{5})(-1) = 2.6$ units.

33. The demand for bicycles is $z = f(x,y) = 200 - 24\sqrt{x} + 4(0.1y+5)^{3/2}$, $x = 129 + 5t$ is the price of bicycles and $y = 80 + 10\sqrt{3t}$ is the price of gasoline. Then $\frac{\partial z}{\partial x} = -\frac{12}{x^{1/2}}$, $\frac{\partial z}{\partial y} = 0.6(0.1y+5)^{1/2}$, $\frac{dx}{dt} = 5$, and $\frac{dy}{dt} = \frac{15}{(3t)^{1/2}}$. Hence, by the chain rule, $\frac{dz}{dt} = \frac{\partial z}{\partial x}\frac{dx}{dt} + \frac{\partial z}{\partial y}\frac{dy}{dt} = (-\frac{12}{x^{1/2}})(5) + 0.6(0.1y+5)^{1/2}[\frac{15}{(3t)^{1/2}}] = (-\frac{60}{x^{1/2}}) + 9(0.1y+5)^{1/2}[\frac{1}{(3t)^{1/2}}]$. When $t = 3$, $x = 144$, and $y = 110$, then $\frac{dz}{dt} = -\frac{60}{12} + 9(11+5)^{1/2}(\frac{1}{3}) = -5 + 12 = 7$ bicycles per month.

35. Since the output is $Q(x,y) = 0.08x^2 + 0.12xy + 0.03y^2$, the change in output is $\Delta Q \approx dQ = \frac{\partial Q}{\partial x}\Delta x + \frac{\partial Q}{\partial y}\Delta y = (0.16x + 0.12y)\Delta x + (0.12x + 0.06y)\Delta y$. When $x = 80$, $y = 200$, $\Delta x = \frac{1}{2}$, and $\Delta y = 2$, then $\Delta Q = [0.16(80) + 0.12(200)](\frac{1}{2}) + [0.12(80) + 0.06(200)](2) = 61.60$ units.

37. Since the profit is $P(x,y) = (x-30)(70-5x+4y) + (y-40)(80+6x-7y) = 70x - 5x^2 + 4xy - 2{,}100 + 150x - 120y + 80y + 6xy - 7y^2 - 3{,}200 - 240x + 280y = -20x - 5x^2 + 10xy - 5{,}300 + 240y - 7y^2$, the change in profit is $\Delta P \approx dP = \frac{\partial P}{\partial x}\Delta x + \frac{\partial P}{\partial y}\Delta y = (-20 - 10x + 10y)\Delta x + (10x + 240 - 14y)\Delta y$. When $x = 50$, $y = 52$, $\Delta x = 1$, and $\Delta y = 2$, then $\Delta P = (-20 - 500 + 520)(1) + (500 + 240 - 728)(2) = 24$ cents.

39. The area of the inner rectangle is $A(x,y) = xy$. The area of the concrete is the change in area of ΔA due to an increase in x of $\Delta x = 1.6$ and a change in y of $\Delta y = 1.6$. Hence, the area of the concrete is $\Delta A \approx dA = \frac{\partial A}{\partial x}\Delta x + \frac{\partial A}{\partial y}\Delta y = y\Delta x + x\Delta y$ and when $x = 40$, $y = 30$, $\Delta x = 1.6$, and $\Delta y = 1.6$, the area of the concrete is $30(1.6) + 40(1.6) = 112$ square yards.

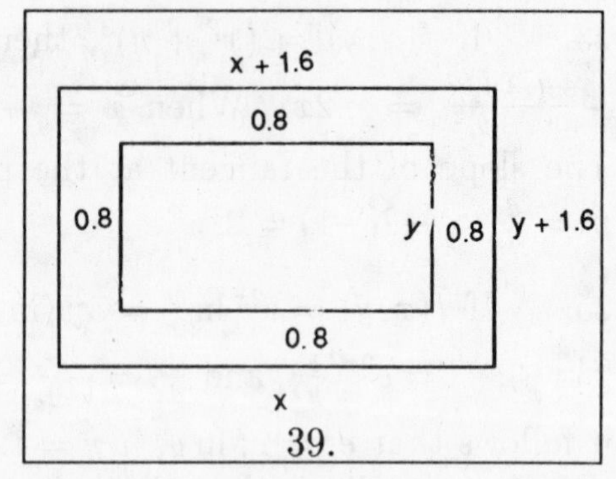

39.

41. The volume of a cylinder is $V(r,h) = \pi r^2 h$. Apply the approximation formula for the percentage change with $\Delta r = 0.01r$ and $\Delta h = -0.015h$ to get the percentage change in $V \approx 100\frac{\frac{\partial A}{\partial r}\Delta r + \frac{\partial A}{\partial h}\Delta h}{V} = 100\frac{2\pi rh(0.01r)+\pi r^2(-0.015h)}{\pi r^2 h} = 100\frac{0.02\pi r^2 h - 0.015\pi r^2 h}{\pi r^2 h} = 100\frac{0.005\pi r^2 h}{\pi r^2 h} = 0.5$ %.

43. The output function is $Q(K,L) = AK^{\alpha}L^{1-\alpha}$. Apply the approximation formula for percentage change with $\Delta K = 0.01K$ $\Delta L = 0.01L$ to get the percentage change in $Q \approx 100\frac{\frac{\partial Q}{\partial K}\Delta K + \frac{\partial Q}{\partial L}\Delta L}{Q} = 100\frac{A\alpha K^{\alpha-1}L^{1-\alpha}(0.01K)+A(1-\alpha)K^{\alpha}L^{1-\alpha-1}(0.01L)}{AK^{\alpha}L^{1-\alpha}}$
$= 100\frac{(0.01)A\alpha K^{\alpha}L^{1-\alpha}+(0.01)(1-\alpha)AK^{\alpha}L^{1-\alpha}}{AK^{\alpha}L^{1-\alpha}} = 100\frac{0.01AK^{\alpha}L^{1-\alpha}}{AK^{\alpha}L^{1-\alpha}} = 1$ %.

45. If $z = f(x,y)$, where $x = at$ and $y = bt$, then $\frac{\partial z}{\partial x} = f_x$, $\frac{\partial z}{\partial y} = f_y$, $\frac{dx}{dt} = a$, and $\frac{dy}{dt} = b$. Hence, by the chain rule, $\frac{dz}{dt} = \frac{\partial z}{\partial x}\frac{dx}{dt} + \frac{\partial z}{\partial y}\frac{dy}{dt} = af_x + bf_y$. Now, $\frac{d^2y}{dt^2} = \frac{d}{dt}(\frac{dz}{dt}) = \frac{d}{dx}(af_x + bf_y) = a\frac{d}{dx}(f_x) + b\frac{d}{dx}(f_y)$. But f_x and f_y are both functions of x and y, where $x = at$ and $y = bt$. Hence the chain rule must be applied again to get $\frac{d^2z}{dt^2} = a[\frac{\partial}{\partial x}(f_x)\frac{dx}{dt} + \frac{\partial}{\partial y}(f_x)\frac{dy}{dt}] + b[\frac{\partial}{\partial x}(f_y)\frac{dx}{dt} + \frac{\partial}{\partial y}(f_y)\frac{dy}{dt}] = a(af_{xx}) + a(bf_{xy}) + b(af_{yx}) + b(bf_{yy}) = a^2f_{xx} + abf_{xy} + abf_{yx} + b^2f_{yy} = a^2f_{xx} + 2abf_{xy} + b^2f_{yy}$. (Note that $f_{xy} = f_{yx}$ is assumed.)

9.4 Relative Maxima and Minima

1. If $f(x,y) = 5 - x^2 - y^2$, then $f_x = -2x$ and $f_y = -2y$ which are both equal to 0 only if $x = 0$ and $y = 0$. Hence, $(0,0)$ is the only critical point. Since $f_{xx} = -2$, $f_{yy} = -2$, and $f_{xy} = 0$, $D = f_{xx}f_{yy} - (f_{xy})^2 = (-2)(-2) - 0 = 4$. Since $D(0,0) = 4 > 0$ and $f_{xx}(0,0) = -2 < 0$, it follows that f has a relative maximum at (0,0).

3. If $f(x,y) = xy$, then $f_x = y$ and $f_y = x$ which are 0 only when $x = 0$ and $y = 0$. Hence,

$(0,0)$ is the only critical point. Since $f_{xx} = 0$, $f_{yy} = 0$, and $f_{xy} = 1$, $D = f_{xx}f_{yy} - (f_{xy})^2 = (0)(0) - 1 = -1$. Since $D(0,0) = -1 >< 0$ It follows that f has a saddle point at (0,0).

5. If $f(x,y) = 2x^3 + y^3 + 3x^2 - 3y - 12x - 4$, then $f_x = 6x^2 + 6x - 12 = 6(x+2)(x-1)$ and $f_y = 3y^2 - 3 = 3(y+1)(y-1)$ which indicates that $f_x = 0$ when $x = -2$ and $x = 1$, while $f_y = 0$ when $y = -1$ and $y = 1$. Hence, the critical points of f are $(-2,-1)$, $(-2,1)$, $(1,-1)$, and $(1,1)$. $f_{xx} = 12x + 6$, $f_{yy} = 6y$, and $f_{xy} = 0$, $D = f_{xx}f_{yy} - (f_{xy})^2 = (12x+6)(6y) - 0 = 36y(2x+1)$.
Since $D(-2,-1) = 36(-1)(-3) = 108 > 0$ and $f_{xx}(-2,-1) = -24 + 6 = -18 < 0$, it follows that f has a relative maximum at $(-2,-1)$,
Since $D(-2,1) = 36(1)(-3) = -108 < 0$ it follows that f has a saddle point at $(-2,1)$,
Since $D(1,-1) = 36(-1)(3) = -108 < 0$ it follows that f has a saddle point at $(1,-1)$,
Since $D(1,1) = 36(1)(3) = 108 > 0$ and $f_{xx}(1,1) = 12 + 6 = 18 > 0$, it follows that f has a relative minimum point at $(1,1)$.

7. If $f(x,y) = x^3 + y^2 - 6xy + 9x + 5y + 2$, then $f_x = 3x^2 - 6y + 9$ and $f_y = 2y - 6x + 5$. Setting $f_x = 0$ and $f_y = 0$ gives $3x^2 - 6y + 9 = 0$ and $2y - 6x + 5 = 0$ or $x^2 - 2y + 3 = 0$ and $-6x + 2y + 5 = 0$. Adding these two equations yields $x^2 - 6x + 8 = 0$, $(x-4)(x-2) = 0$, or $x = 4$ and $x = 2$. If $x = 4$, the second equation gives $-24 + 2y + 5 = 0$ or $y = \frac{19}{2}$, and if $x = 2$, the second equation gives $-12 + 2y + 5 = 0$ or $y = \frac{7}{2}$. Hence, the critical points are $(4, \frac{19}{2})$ and $(2, \frac{7}{2})$. Since $f_{xx} = 6x$, $f_{yy} = 2$, and $f_{xy} = -6$, $D = f_{xx}f_{yy} - (f_{xy})^2 = (6x)(2) - (-6)^2 = 12x - 36 = 12(x-3)$.
Since $D(4, \frac{19}{2}) = 12(4-3) = 12 > 0$ and $f_{xx}(4, \frac{19}{2}) = 24 > 0$, it follows that f has a relative minimum at $(4, \frac{19}{2})$,
since $D(2, \frac{7}{2}) = 12(2-3) = -12 < 0$ it follows that f has a saddle point at $(2, \frac{7}{2})$.

9. If $f(x,y) = (x^2 + 2y^2)e^{1-x^2-y^2}$, then $f_x = e^{1-x^2-y^2}(-2x^3 - 4xy^2 + 2x) = -2x(x^2 + 2y^2 - 1)e^{1-x^2-y^2}$ and $f_y = e^{1-x^2-y^2}(-2x^2y - 4y^3 + 4y) = -2y(x^2 + 2y^2 - 2)e^{1-x^2-y^2}$. Setting $f_x = 0$ and $f_y = 0$ gives $x^2 + 2y^2 - 1 = 0$ and $x^2 + 2y^2 - 2 = 0$ which produces no solutions. But $x = 0$ and $y = 0$ is our first solution, so $(0,0)$ is a critical point. Since $f_{xx} = e^{1-x^2-y^2}(-6x^2 - 4y^2 + 2 + 4x^4 + 8x^2y^2 - 4x^2)$, $f_{yy} = e^{1-x^2-y^2}(-2x^2 - 12y^2 + 4 + 4x^2y^2 + 8y^4 - 8y^2)$, and $f_{xy} = e^{1-x^2-y^2}(-8xy + 4x^3y + 8xy^3 - 4xy)$, $f_{xx}(0,0) = 2e$, $f_{yy}(0,0) = 4e$, and $f_{xy}(0,0) = 0$, and since $D(0,0) = f_{xx}f_{yy} - (f_{xy})^2 = (4)(2)e^2 - 0 > 0$ and $f_{xx}(0,0) > 0$, it follows that f has a relative minimum at $(0,0)$.
$f_x = -2x(x^2 + 2y^2 - 1)e^{1-x^2-y^2} = 0$ and $f_y = -2y(x^2 + 2y^2 - 2)e^{1-x^2-y^2} = 0$ are also satisfgied if $x = 0$ and $0^2 + 2y^2 - 2 = 0$, $y = \pm 1$, so at $(0, \pm 1)$. $f_{xx}(0, \pm 1) = e^{1-0^2-(\pm 1)^2}[-6(0)^2 - 4(\pm 1)^2 + 2 + 4(0)^4 + 8(0)^2(\pm 1)^2 - 4(0)^2] = -2$, $f_{yy}(0, \pm 1) = -8$, and $f_{xy}(0, \pm 1) = 0$, and since $D(0, \pm 1) = f_{xx}f_{yy} - (f_{xy})^2 = (-2)(-8) - 0 > 0$ and $f_{xx}(0, \pm 1) > 0$, it follows that f has a relative maximum at $(0, \pm 1)$.
Similarly $y = 0$ and $x^2 + 2(0)^2 - 1 = 0$, $x = \pm 1$, so at $(\pm 1, 0)$ are critical points. $f_{xx}(\pm 1, 0) = -4$, $f_{yy}(\pm 1, 0) = 2$, and $f_{xy}(\pm 1, 0) = 0$, and since $D(\pm 1, 0) = f_{xx}f_{yy} - (f_{xy})^2 = (-4)(2) - 0 <$

0, it follows that f has a saddle point at $(\pm 1, 0)$.

11. If $f(x,y) = x^3 - 4xy + y^3$, then $f_x = 3x^2 - 4y$ and $f_y = -4x + 3y^2$. Setting $f_x = 0$ and $f_y = 0$ gives $y = \frac{3x^2}{4}$ and $3y^2 = 4x$. Solving simultaneously yields $3(\frac{3x^2}{4})^2 = 4x$, $27x^4 = 64x$, $x(27x^3 - 64) = 0$, or $x = 0$ and $x = \frac{4}{3}$. If $x = 0$, then $y = 0$, and if $x = \frac{4}{3}$, then $y = \frac{4}{3}$. Hence, the critical points are $(0,0)$ and $(\frac{4}{3}, \frac{4}{3})$. Since $f_{xx} = 6x$, $f_{yy} = 6y$, and $f_{xy} = -4$, $D = f_{xx}f_{yy} - (f_{xy})^2 = (6x)(6y) - (-4)^2 = 36xy - 16 = 4(9xy - 4)$.
Since $D(0,0) = 4(-4) < 0$ it follows that f has a saddle point at at $(0,0)$.
Since $D(\frac{4}{3}, \frac{4}{3}) = 4(16 - 4) > 0$ and $f_{xx}(\frac{4}{3}, \frac{4}{3}) > 0$, it follows that f has a relative minimum at $(\frac{4}{3}, \frac{4}{3})$.

13. Let x and y denote the sales price of a the shirt endorsed by Michael Jordan and Magic Johnson, respectively. The profit $P(x,y)$ is the sum of the number of shirts sold times the difference between the sales price and the cost per shirt. Thus $P(x,y) = (x-2)(40 - 50x + 40y) + (y-2)(20 + 60x - 70y)$. Then $P_x = (x-2)(-50) + 40 - 50x + 40y + (y-2)(60) = -50x + 100 + 40 - 50x + 40y + 60y - 120 = -100x + 100y + 20 = 20(-5x + 5y + 1)$, $P_y = (x-2)(40) + (y-2)(-70) + 20 + 60x - 70y = 40x - 80 - 70y + 140 + 20 + 60x - 70y = 100x - 140y + 80 = 20(5x - 7y + 4)$. Solving simultaneously yields $-5x + 5y + (5x - 7y) = -1 + (-4)$, $-2y = -5$, $y = \frac{5}{2}$, $-5x = -\frac{25+2}{2}$, or $x = \frac{27}{10}$. Hence, the critical point is $(2.7, 2.5)$. Since $P_{xx} = 20(-5) = -100$, $P_{yy} = -140$, and $P_{xy} = 100$, $D = P_{xx}P_{yy} - (P_{xy})^2 = (-100)(-140) - (100)^2 = 10^3(14 - 10) > 0$ and $P_{xx}(2.7, 2.5) < 0$, it follows that P has a relative maximum at $(2.7, 2.5)$. Thus, to maximize the profit, $x = 2.7$ Jordan shirts and $y = 2.5$ Johnson shirts are to be sold. (If the decimal points bothers you, sell the shirts in batches of 10 each.)

15. The profit is $P = (\frac{320y}{y+2} + \frac{160x}{x+4})(150 - 50) - 1{,}000x - 1{,}000y$. $P_x = 100 \times 160\frac{4}{(x+4)^2} - 1{,}000 = 0$ if $x = 4$. $P_y = 100 \times 320\frac{2}{(y+2)^2} - 1{,}000 = 0$ if $y = 6$. $P_{xx} = -\frac{128{,}000}{8^3} = -250$, $P_{xy} = 0$, and $P_{yy} = -\frac{128{,}000}{8^3} = -250$. $D(4,6) > 0$ and $P_{xx} < 0$, so $(4,6)$ is a relative maximum. \$4,000 should be spent on development and \$6,000 on promotion.

17. a) The profit at home is $P_h = x(150 - \frac{x}{6}) - cx = (150 - c)x - \frac{x^2}{6}$ where c is the cost of manufacturing a machine. $(P_h)_x = 150 - c - \frac{x}{3} = 0$ if $x = 450 - 3c$. $(P_h)_{xx} < 0$ so $x = 450 - 3c$ represents a maximum.
b) The profit abroad is $P_f = y(100 - \frac{y}{10}) - cy = (100 - c)y - \frac{y^2}{10}$. $(P_f)_y = 100 - c - \frac{y}{5} = 0$ if $y = 500 - 5c$. $(P_f)_{yy} < 0$ so $y = 500 - 5c$ represents a maximum.
c) The total profit is $P_h + P_f$ which is maximized when $x = 450 - 3c$ and $y = 500 - 5c$.
d) P_h and P_f are completely independent (one is a function of x alone and the other of y exclusively.)

19. Let x denote the number of gallons of whole milk and y the number of gallons of skim milk. The price of whole milk is $p(x) = 20 - 5x$ and that of skim milk $q(y) = 4 - 2y$, while the cost of producing milk is $C(x,y) = 2xy + 4$. Now the profit is $P(x,y) =$(number of gallons

of whole milk)(price per gallon)+(number of gallons of skim milk)(price per gallon)−cost= $(20-5x)x+(4-2y)y-(2xy+4) = 20x-5x^2+4y-2y^2-2xy-4$. Then $P_x = 20-10x-2y$ and $P_y = 4-4y-2x$. Setting each to 0 and solving simultaneously leads to $20-10x-2y-5(4-4y-2x) = 20-10x-2y-20+20y+10x = -2y+20y = 18y = 0$ or $y=0$, so $x=2$. Also $P_{xx}(2,0) = -10$, $P_{xy}(2,0) = -2$, and $P_{yy} = -4$. $D(2,0) = (-10)(-4)-(-2)^2 > 0$ and $P_{xx} < 0$, so $(2,0)$ is a relative maximum. This means that the dairy ought to sell 2 gallons of whole milk and none of the skim milk.

21. The distance from the the origin to the point (x,y,z) is $d = \sqrt{x^2+y^2+z^2}$. We want to minimize $f(x,y) = d^2 = x^2+y^2+z^2$ if $z^2 = 3+xy$, so $f(x,y) = x^2+y^2+xy+3$. Now $f_x = 2x+y$ and $f_y = 2y+x$. Both vanish when $x=0$ and $y=0$. $f_{xx} = 2$, $f_{yy} = 2$, $f_{xy} = 1$, and so $D(0,0) = (2)(2)-(1)^2 > 0$. Since $f_{xx}(0,0) > 0$, $(0,0,3)$ is a relative minimum indicating that the shortest distance is $d = \sqrt{3}$.

23. Let x, y, and z denote the three numbers. Their sum is $x+y+z = 20$, so $z = 20-x-y$ and the product is $P = xy(20-x-y) = 20xy-x^2y-xy^2$. Now $P_x = 20y-2xy-y^2 = y(20-2x-y)$ and $P_y = 20x-x^2-2xy = x(20-x-2y)$. Both vanish when $x=0$, $y=0$, leading to a minimum, namely $P=0$. Thus we'll consider $xy \neq 0$.
Now $2x+y-20 = 0$ and $x+2y-20 = 0$. $2x+y-20-2(x+2y-20) = 2x+y-20-2x-4y+40 = -3y+20 = 0$, or $y = \frac{20}{3}$. Now $x = 20-2(\frac{20}{3}) = \frac{20}{3}$. $P_{xx} = -2y$, $P_{yy} = -2x$, $P_{xy} = 20-2x$, and so $D(\frac{20}{3},\frac{20}{3}) = [-2(\frac{20}{3})][-2(\frac{20}{3})]-[20-2(\frac{20}{3})]^2 = (\frac{40}{3})(\frac{40}{3})-\frac{400}{9} > 0$. Since $P_{xx}(\frac{20}{3},\frac{20}{3}) < 0$, $(\frac{20}{3},\frac{20}{3},\frac{20}{3})$ is a relative maximum.

25. $f(x,y) = x^2+y^2-4xy$, $f_x = 2x-4y = 0$ when $y = \frac{x}{2}$. $f_y = 2y-4x = 0$ when $y = 2x$. Thus $(0,0)$ is a critical point. $f_{xx} = 2$, $f_{xy} = -4$, and $f_{yy} = 2$, so $D(0,0) = 4-(-4)^2 < 0$ and $(0,0)$ is a saddle point.
The above is true but not asked for. If $x=0$, $f(0,y) = y^2$ which is a parabola with a minimum at $(0,0)$ (in the vertical $yz-$plane). If $y=0$, $f(x,0) = x^2$ which is a parabola with a minimum at $(0,0)$ (in the vertical $xz-$plane). If $y=x$, $f(x,x) = -2x^2$ which is a parabola with a maximum at $(0,0)$ (in the vertical plane passing through the $z-$axis and the line $y=x$ in the xy plane).

9.5 Lagrange Multipliers

1. For $f(x,y) = xy$ subject to the constraint that $g(x,y) = x+y = 1$, the partial derivatives are $f_x = y$, $f_y = x$, $g_x = 1$, and $g_y = 1$. Hence, the three Lagrange equations are $y = \lambda$, $x = \lambda$, and $x+y = 1$. From the first two equations, $x = y$ which, when substituted

into the third equation gives $2x = 1$ or $x = \frac{1}{2}$. Since $x = y$, the corresponding value for y is $y = \frac{1}{2}$. Thus, the constrained maximum is $f(\frac{1}{2}, \frac{1}{2}) = \frac{1}{4}$.

3. For $f(x, y) = x^2 + y^2$ subject to the constraint that $g(x, y) = xy = 1$, the partial derivatives are $f_x = 2x$, $f_y = 2y$, $g_x = y$, and $g_y = x$. Hence, the three Lagrange equations are $2x = \lambda y$, $2y = \lambda x$, and $xy = 1$. Multiply the first equation by y and the second by x to get $2xy = \lambda y^2$ and $2xy = \lambda x^2$. Set the two expressions for $2xy$ equal to each other to get $\lambda y^2 = \lambda x^2$, $y^2 = x^2$, or $x = \pm y$. (Note that another solution of the equation $\lambda y^2 = \lambda x^2$ is $\lambda = 0$, which implies that $x = 0$ and $y = 0$, which is not consistent with the third equation.) If $y = x$, the third equation becomes $x^2 = 1$, which implies that $x = \pm 1$ and $y = \pm 1$. If $y = -x$, the third equation becomes $-x^2 = 1$, which has no solutions. Thus, the two points at which the constrained extrema can occur are $(1, 1)$ and $(-1, -1)$. Since $f(1, 1) = 2$ and $f(-1, -1) = 2$, it follows that the minimum value is 2 and it is attained at the two points $(1, 1)$ and $(-1, -1)$.

5. For $f(x, y) = x^2 - y^2$ subject to the constraint that $g(x, y) = x^2 + y^2 = 4$, the partial derivatives are $f_x = 2x$, $f_y = -2y$, $g_x = 2x$, and $g_y = 2y$. Hence, the three Lagrange equations are $2x = 2\lambda x$, $-2y = 2\lambda y$, and $x^2 + y^2 = 4$. From the first equation, either $\lambda = 1$ or $x = 0$. If $x = 0$, the third equation becomes $y^2 = 4$ or $y = \pm 2$. From the second equation, either $\lambda = -1$ or $y = 0$. If $y = 0$, the third equation becomes $x^2 = 4$ or $x = \pm 2$. If neither $x = 0$ nor $y = 0$, the first equation implies $\lambda = 1$ while the second equation implies $\lambda = -1$, which is impossible. Hence, the only points at which the constrained extrema can occur are $(0, -2)$, $(0, 2)$, $(-2, 0)$, and $(2, 0)$. Since $f(0, -2) = -4$, $f(0, 2) = -4$, $f(-2, 0) = 4$, and $f(2, 0) = 4$, it follows that the constrained minimum is -4 and is attained at the two points $(0, -2)$ and $(0, 2)$.

7. For $f(x, y) = x^2 - y^2 - 2y$ subject to the constraint that $g(x, y) = x^2 + y^2 = 1$, the partial derivatives are $f_x = 2x$, $f_y = -2y - 2$, $g_x = 2x$, and $g_y = 2y$. Hence, the three Lagrange equations are $2x = 2\lambda x$, $-2y - 2 = 2\lambda y$, and $x^2 + y^2 = 1$. From the first equation, either $\lambda = 1$ or $x = 0$.
If $\lambda = 1$, the second equation becomes $-2y - 2 = 2y$, $4y = -2$, or $y = -\frac{1}{2}$. From the third equation, $x^2 + (\frac{-1}{2})^2 = 1$, or $x = \pm\frac{\sqrt{3}}{2}$.
If $x = 0$, the third equation becomes $0^2 + y^2 = 1$ or $y = \pm 1$ (λ, obtained from the second equation, is immaterial).
Hence, the only points at which the constrained extrema can occur are $(-\frac{\sqrt{3}}{2}, \frac{1}{2})$, $(\frac{\sqrt{3}}{2}, \frac{1}{2})$, $(0, -1)$, and $(0, 1)$. Since $f(\frac{\sqrt{3}}{2}, \frac{1}{2}) = f(-\frac{\sqrt{3}}{2}, \frac{1}{2}) = \frac{3}{4} - \frac{1}{4} - 2(\frac{-1}{2}) = \frac{1}{2} + 1 = \frac{3}{2}$, $f(0, -1) = 0^2 - (-1)^2 - 2(-1) = -1 + 2 = 1$, and $f(0, 1) = -1 - 2 = -3$, it follows that the constrained maximum is $\frac{3}{2}$ and the constrained minimum is -3.

9. Let f denote the amount of fencing needed to enclose the pasture, x the side parallel to the river and y the sides perpendicular to the river. Then, $f(x, y) = x + 2y$. The goal is to minimze this function subject to the constraint that the area $g(x, y) = xy = 3{,}200$.

The partial derivatives are $f_x = 1$, $f_y = 2$, $g_x = y$, and $g_y = x$. Hence, the three Lagrange equations are $1 = \lambda y$, $2 = \lambda x$, and $xy = 3,200$. From the first equation, $\lambda = \frac{1}{y}$. From the second equation $\lambda = \frac{2}{x}$. Setting the two expressions for λ equal to each other gives $\frac{1}{y} = \frac{2}{x}$ or $x = 2y$, and substituting this into the third equation yields $2y^2 = 3,200$, $y^2 = 1,600$, or $y = \pm 40$.
Only the positive value is meaningful in the context of this problem. Hence, $y = 40$, and (since $x = 2y$), $x = 80$. That is, to minimize the amount of fencing, the dimensions of the field should be 40 meters by 80 meters.

11. Let f denote the volume of the parcel. Then, $f(x, y) = x^2y$. The girth $4x$ plus the length y can be at most 108 inches. The goal is to maximze this function $f(x, y)$ subject to the constraint $g(x, y) = 4x + y = 108$. The partial derivatives are $f_x = 2xy$, $f_y = x^2$, $g_x = 4$, and $g_y = 1$. Hence, the three Lagrange equations are $2xy = 4\lambda$, $x^2 = \lambda$, and $4x + y = 108$. From the first equation, $\lambda = \frac{xy}{2}$, which, combined with the second equation, gives $\frac{xy}{2} = x^2$ or $y = 2x$. (Another solution is $x = 0$, which is impossible in the context of this problem.) Substituting $y = 2x$ into the third equation gives $6x = 108$ or $x = 18$, and since $y = 2x$, the corresponding value of y is $y = 36$. Hence, the largest volume is $f(18, 36) = (18)^2(36) = 11,664$ cubic inches.

13. Let f denote the cost of constructing the cylindrical can. The area of the top is πx^2, the area of the base is πx^2, and the area of the cardboard side is $2\pi xy$, where x is the radius and y is the height. Let k denote the cost per square inch of constructing the cardboard side. Then, $2k$ is the cost per square inch of constructing the base. The goal is to minimize the total cost function $f(x, y) = 2k(2\pi x^2) + k(2\pi xy) = 4\pi kx^2 + 2k\pi xy$ subject to the constraint that the volume is to be $g(x, y) = \pi x^2 y = 4\pi$. The partial derivatives are $f_x = 8k\pi x + 2k\pi y$, $f_y = 2k\pi x$, $g_x = 2\pi xy$, and $g_y = \pi x^2$. Hence, the three Lagrange equations are $8k\pi x + 2k\pi y = 2\lambda\pi xy$, $2k\pi x = \lambda\pi x^2$, and $\pi x^2 y = 4\pi$. From the first equation, $\lambda = \frac{4k}{y} + \frac{k}{x}$ (since $x \neq 0$ and $y \neq 0$), and from the second equation, $\lambda = \frac{2k}{x}$ (since $x \neq 0$). Setting the two expressions for λ equal to each other gives $\frac{4k}{y} + \frac{k}{x} = \frac{2k}{x}$, $4x + y = 2y$, or $y = 4x$. Substituting this in the third equation yields $\pi x^2(4x) = 4\pi$ or $x = 1$. Since $y = 4x$, the corresponding value of y is $y = 4$. Hence, to minimze cost, the radius of the can should be 1 inch and the height should be 4 inches.

15. The goal is to maximze $Q(x, y) = 60x^{1/3}y^{2/3}$ subject to the constraint $g(x, y) = x + y = 120$ (thousand). The partial derivatives are $Q_x = 20x^{-2/3}y^{2/3}$, $Q_y = 40x^{1/3}y^{-1/3}$, $g_x = 1$, and $g_y = 1$. Hence, the three Lagrange equations are $20x^{-2/3}y^{2/3} = \lambda$, $40x^{1/3}y^{-1/3} = \lambda$, and $x + y = 120$. It follows from the first two equations that $20x^{-2/3}y^{2/3} = 40x^{1/3}y^{-1/3}$, or $y = 2x$. Substituting this into the third equation gives $x + 2x = 120$ or $x = 40$, and (since $y = 2x$), the corresponding value of y is $y = 80$. Hence, to generate maximal output, \$40,000 should be spent on labor and \$80,000 should be spent on equipment.

17. From problem 15, the three Lagrange equations were $20x^{-2/3}y^{2/3} = \lambda$, $40x^{1/3}y^{-1/3} =$

$\lambda\pi$, and $x+y=120$ from which it was determined that the maximal output occurs when $x=40$ and $y=80$. Substituting these values in the first equation gives $\lambda = 20(40)^{-2/3}(80)^{2/3} = 31.75$ which implies that the maximal output will increase by approximately 31.75 units if the available money is increased by one thousand dollars and allocated optimally.

19. Let P denote the profit (in units of \$1,000). Then, P =(number of units)(price per unit−cost per unit)−total amount spent on development and promotion. The number of units is $\frac{320y}{y+2} + \frac{160x}{x+4}$, where x thousand is spent on development and y thousand is spent on promotion. The price per unit is \$150 and the cost per unit is \$50, so the price per unit minus the cost per unit is \$100, or $\frac{1}{10}$ thousand dollars. Hence, $P(x,y) = \frac{1}{10}\left(\frac{320y}{y+2} + \frac{160x}{x+4}\right) - (x+y) = \frac{32y}{y+2} + \frac{16x}{x+4} - x - y$. The partial derivatives of P are $P_x = \frac{(x+4)(16)-(16x)(1)}{(x+4)^2} - 1 = \frac{64}{(x+4)^2} - 1$ and $P_y = \frac{(y+2)(32)-(32y)(1)}{(y+2)^2} - 1 = \frac{64}{(y+2)^2} - 1$.
a) To maximize profit when unlimited funds are available is to maximize $P(x,y)$ without constraints. To do this, find the critical points by setting $P_x = 0$ and $P_y = 0$, that is, $P_x = \frac{64}{(x+4)^2} - 1 = 0$, $64 = (x+4)^2$, $x+4 = 8$, or $x = 4$. Similarly $P_y = \frac{64}{(y+2)^2} - 1 = 0$ $64 = (y+2)^2$, $y+2 = 8$, or $y = 6$. Thus, \$4,000 should be spent on development and \$6,000 should be spent on promotion to maximize profit.
b) If there were a restriction on the amount spent on development and promotion, then constraints would be $g(x,y) = x+y = k$ for some positive constant k. The corresponding Lagrange equations would be $\frac{64}{(x+4)^2} - 1 = \lambda$, $\frac{64}{(y+2)^2} - 1 = \lambda$, and $x+y = k$. To get the answer in part a), $\lambda = 0$. To see this from another point of view, recall that $\lambda = \frac{dM}{dk}$, where M is the maximum profit if k thousand dollars is available. This maximum profit will be greatest when its derivative $\frac{dM}{dk} = 0$, that is $\lambda = 0$.
c) Beginning with the Lagrange equations from part b), set $\lambda = 0$ to get $\frac{64}{(x+4)^2} - 1 = 0$, $64 = (x+4)^2$, $x+4 = 8$, or $x = 4$. Similarly $\frac{64}{(y+2)^2} - 1 = 0$, $64 = (y+2)^2$, $y+2 = 8$, or $y = 6$, just as in part a).

21. The goal is to maximize the utility function $U(x,y) = x^\alpha y^\beta$ subject to the budgetary constraint that $ax+by = c$. The three Lagrange equations are $\alpha x^{\alpha-1}y^\beta = a\lambda$, $\beta x^\alpha y^{\beta-1} = b\lambda$, and $ax+by = c$. From the first two equations, $\frac{\alpha x^{\alpha-1}y^\beta}{a} = \frac{\beta x^\alpha y^{\beta-1}}{b}$, $\frac{\alpha y}{a} = \frac{\beta x}{b}$, or $y = \frac{a\beta x}{b\alpha}$. Substituting this into the third equation gives $ax + b\left(\frac{a\beta x}{b\alpha}\right) = c$, $\left(a + \frac{a\beta}{\alpha}\right)x = c$, $\frac{a(\alpha+\beta)x}{\alpha} = c$, $x = \frac{c\alpha}{a(\alpha+\beta)} = \frac{c\alpha}{a}$ (since $\alpha+\beta = 1$). Finally, since $y = \frac{a\beta x}{b\alpha}$, it follows that $y = \frac{a\beta}{b\alpha}\left(\frac{c\alpha}{a}\right) = \frac{c\beta}{b}$.

23. Let x be the number of units of labor and y the number of units of capital. With the unit cost of labor and capital p and q respectively, the cost is $C(x,y) = px+qy$. The goal is to minimize cost subject to the fixed production function $Q(x,y) = c$. Since $C_x = p$ and $C_y = q$, the three Lagrange equations are $C_x = p = \lambda Q_x$, $C_y = q = \lambda Q_y$, and $Q(x,y) = c$. Solving the first two equations for $\frac{1}{\lambda}$ leads to $\frac{Q_x}{p} = \frac{Q_y}{q}$.

25. With $P = Ax^\alpha y^\beta = k$, $\alpha+\beta = 1$, and $C(x,y) = px+qy$, $P_x = \lambda A\alpha x^{\alpha-1}y^\beta$,

$P_y = \lambda A\beta x^{\alpha}y^{\beta-1}$, $C_x = p$, and $C_y = q$. The three Lagrange equations are $\lambda A\alpha x^{\alpha-1}y^{\beta} = p$, $\lambda A\beta x^{\alpha}y^{\beta-1} = q$, and $Ax^{\alpha}y^{\beta} = k$. To eliminate A divide the first equation by the second to get $\frac{p}{q} = \frac{\alpha y}{\beta x}$ or $y = \frac{p\beta x}{q\alpha}$. Substituting into the third equation yields $Ax^{\alpha}(\frac{p\beta x}{q\alpha})^{1-\alpha} = k$, $Ax(\frac{p\beta}{q\alpha})^{1-\alpha} = k$, $x = \frac{k}{A}(\frac{\alpha q}{\beta p})^{\beta}$ (since $\alpha + \beta = 1$) and $y = \frac{\beta p}{\alpha q}\frac{k}{A}\frac{(\alpha q)^{\beta}}{(\beta p)^{\beta}} = \frac{k}{A}(\frac{\beta p}{\alpha q})^{1-\beta} = \frac{k}{A}(\frac{\beta p}{\alpha q})^{\alpha}$

9.6 The Method of Least Squares

1. The sum $S(m,b)$ of the squares of the vertical distances from the three given points is $S(m,b) = d_1^2 + d_2^2 + d_3^2 = (b-1)^2 + (2m+b-3)^2 + (4m+b-2)^2$.

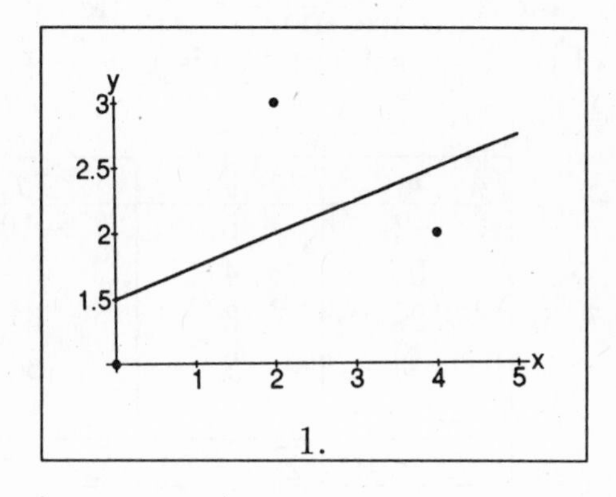

1.

To minimize $S(m,b)$, set the partial derivatives $\frac{\partial S}{\partial m} = 0$ and $\frac{\partial S}{\partial b} = 0$, namely $\frac{\partial S}{\partial m} = 2(2m+b-3)(2) + 2(4m+b-2)(4) = 40m + 12b - 28 = 0$ and $\frac{\partial S}{\partial b} = 2(b-1) + 2(2m+b-3) + 2(4m+b-2) = 12m + 6b - 12 = 0$. Solve the resulting simplified equations $10m + 3b = 7$ and $6m + 3b = 6$ to get $m = \frac{1}{4}$ and $b = \frac{3}{2}$. Hence, the equation of the least-squares line is $y = \frac{x}{4} + \frac{3}{2}$.

3. The sum $S(m,b)$ of the squares of the vertical distances from the four given points is $S(m,b) = (m+b-2)^2 + (2m+b-4)^2 + (4m+b-4)^2 + (5m+b-2)^2$.

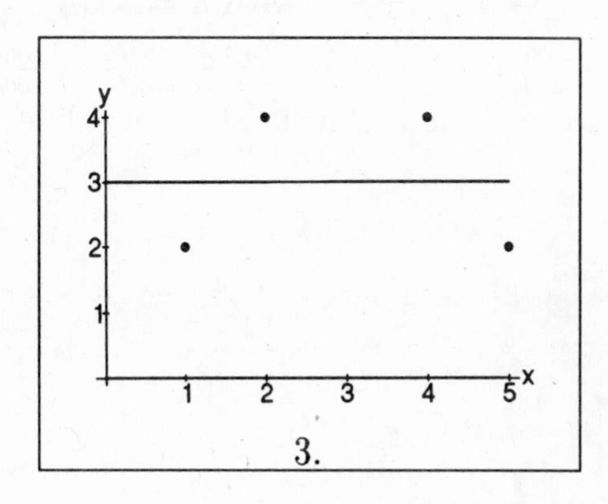

3.

To minimize $S(m,b)$, set the partial derivatives $\frac{\partial S}{\partial m} = 0$ and $\frac{\partial S}{\partial b} = 0$, namely $\frac{\partial S}{\partial m} = 2(m+b-2) + 2(2m+b-4)(2) + 2(4m+b-4)(4) + 2(5m+b-2)(5) = 92m + 24b - 72 = 0$ and $\frac{\partial S}{\partial b} = 2(m+b-2) + 2(2m+b-4) + 2(4m+b-4) + 2(5m+b-2) = 24m + 8b - 24 = 0$. Solve the resulting simplified equations $23m + 6b = 18$ and $3m + b = 3$ to get $m = 0$ and $b = 3$. Hence, the equation of the least-squares line is $y = 3$.

5.

x	y	xy	x^2
1	2	2	1
2	2	4	4
2	3	6	4
5	5	25	25
$\sum x = 10$	$\sum y = 12$	$\sum xy = 37$	$\sum x^2 = 34$

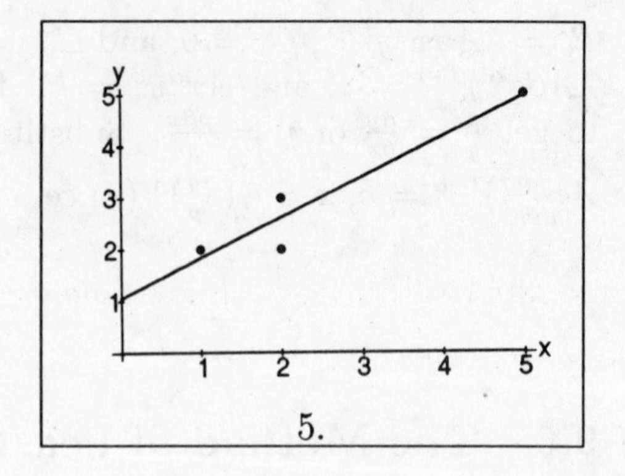

5.

From the formulas $m = \frac{n\sum xy - \sum x \sum y}{n\sum x^2 - (\sum x)^2}$ and $b = \frac{\sum x^2 \sum y - \sum x \sum xy}{n\sum x^2 - (\sum x)^2}$, with $n = 4$, $m = \frac{4(37)-10(12)}{4(34)-(10)^2} = \frac{7}{9}$ and $b = \frac{34(12)-10(37)}{4(34)-(10)^2} = \frac{19}{18}$. Hence, the equation of the least-squares line is $y = \frac{7x}{9} + \frac{19}{18} = 0.78x + 1.06$.

7.

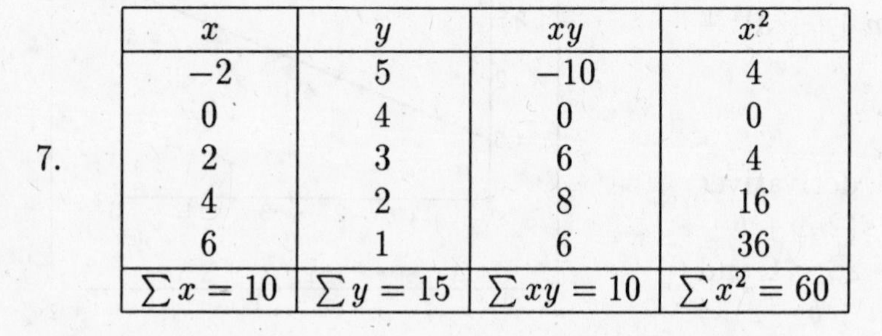

x	y	xy	x^2
−2	5	−10	4
0	4	0	0
2	3	6	4
4	2	8	16
6	1	6	36
$\sum x = 10$	$\sum y = 15$	$\sum xy = 10$	$\sum x^2 = 60$

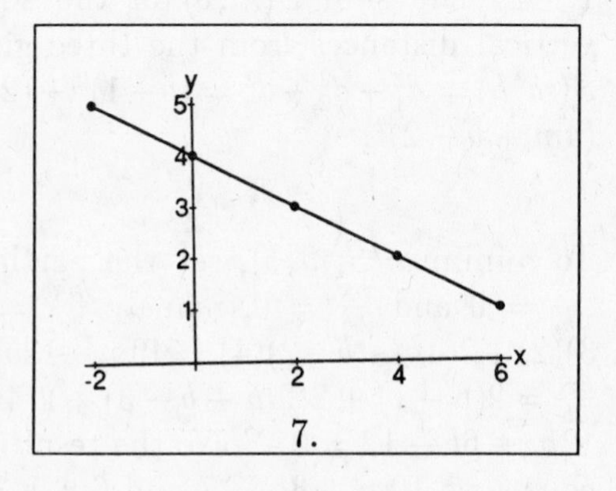

7.

From the formulas $m = \frac{n\sum xy - \sum x \sum y}{n\sum x^2 - (\sum x)^2}$ and $b = \frac{\sum x^2 \sum y - \sum x \sum xy}{n\sum x^2 - (\sum x)^2}$, with $n = 5$, $m = \frac{5(10)-10(15)}{5(60)-(10)^2} = -\frac{100}{200} = -\frac{1}{2}$ and $b = \frac{60(15)-10(10)}{5(60)-(10)^2} = \frac{800}{200} = 4$. Hence, the equation of the least-squares line is $y = -\frac{x}{2}+4$.

9. From the formulas $m = \frac{n\sum xy - \sum x \sum y}{n\sum x^2 - (\sum x)^2}$ and $b = \frac{\sum x^2 \sum y - \sum x \sum xy}{n\sum x^2 - (\sum x)^2}$, with $n = 5$, $m = \frac{5(40.29)-14.5(10.3)}{5(31.45)-(10.3)^2} = \frac{52.10}{51.16} = 1.0184$ and $b = \frac{31.45(14.5)-10.3(40.29)}{5(31.45)-(10.3)^2} = \frac{41.038}{51.16} = 0.8022$. Hence, the equation of the least-squares line is $y = 1.0184x + 0.8022$.

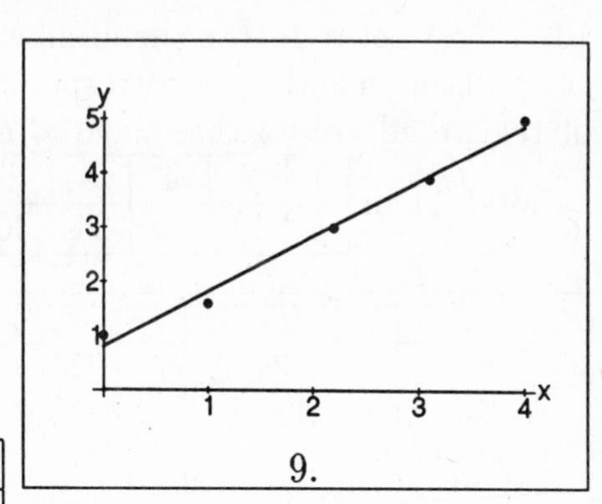

9.

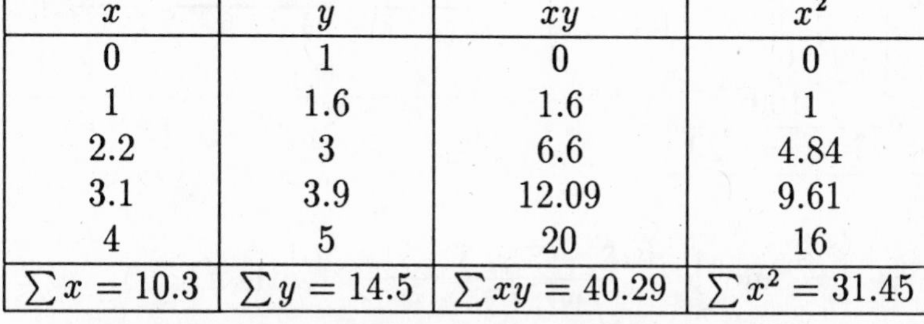

x	y	xy	x^2
0	1	0	0
1	1.6	1.6	1
2.2	3	6.6	4.84
3.1	3.9	12.09	9.61
4	5	20	16
$\sum x = 10.3$	$\sum y = 14.5$	$\sum xy = 40.29$	$\sum x^2 = 31.45$

11. a) Let x be the number of catalogs requested and y the number of applications received (both in units of 1,000). The given points (x, y) are plotted on the accompanying graph.

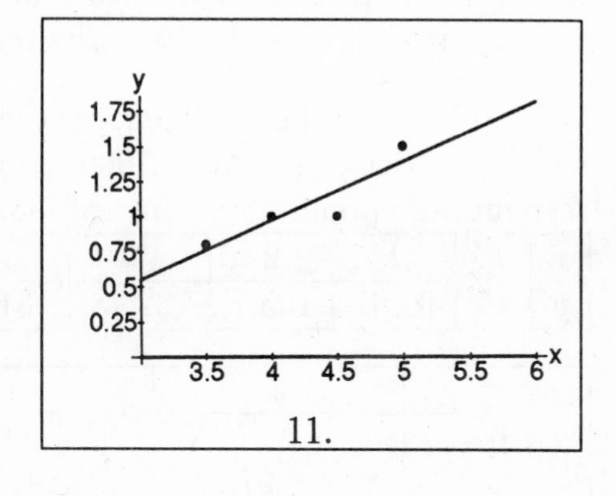

11.

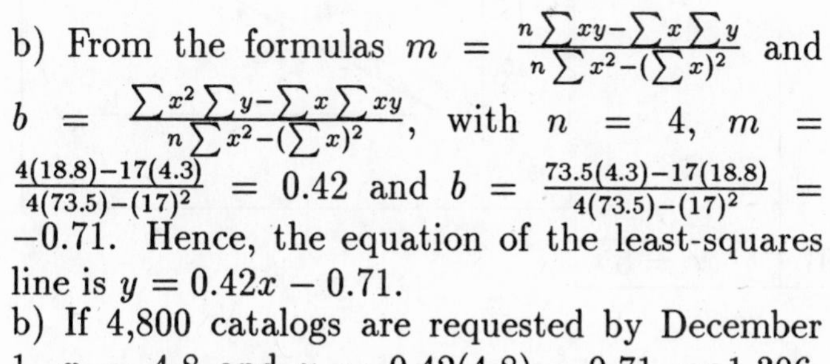

x	y	xy	x^2
4.5	1.0	4.5	20.25
3.5	0.8	2.8	12.25
4.0	1.0	4.0	16.00
5.0	1.5	7.5	25.00
$\sum x = 17.0$	$\sum y = 4.3$	$\sum xy = 18.8$	$\sum x^2 = 73.50$

b) From the formulas $m = \frac{n\sum xy - \sum x \sum y}{n\sum x^2 - (\sum x)^2}$ and $b = \frac{\sum x^2 \sum y - \sum x \sum xy}{n\sum x^2 - (\sum x)^2}$, with $n = 4$, $m = \frac{4(18.8)-17(4.3)}{4(73.5)-(17)^2} = 0.42$ and $b = \frac{73.5(4.3)-17(18.8)}{4(73.5)-(17)^2} = -0.71$. Hence, the equation of the least-squares line is $y = 0.42x - 0.71$.

b) If 4,800 catalogs are requested by December 1, $x = 4.8$ and $y = 0.42(4.8) - 0.71 = 1.306$, which means that approximately 1,306 completed applications will be received by March 1.

13. a) Let x denote the number of hours after the polls open and y the corresponding percentage of registered voters that have already cast their ballots. Then

x	2	4	6	8	10
y	12	19	24	30	37

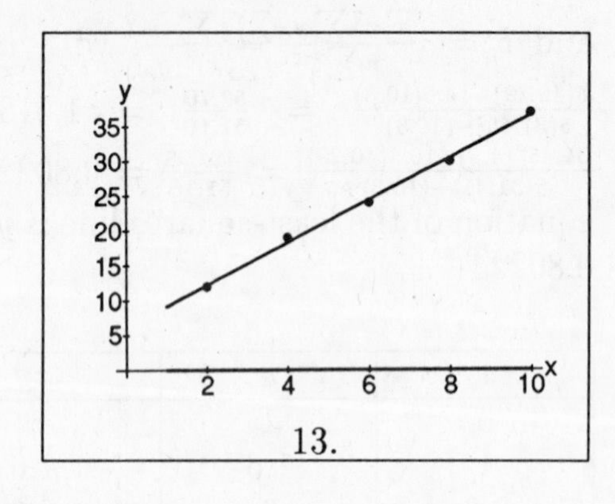

13.

x	y	xy	x^2
2	12	24	4
4	19	76	16
6	24	144	36
8	30	240	64
10	37	370	100
$\sum x = 30$	$\sum y = 122$	$\sum xy = 854$	$\sum x^2 = 220$

b) From the formulas $m = \frac{n\sum xy - \sum x \sum y}{n\sum x^2 - (\sum x)^2}$ and $b = \frac{\sum x^2 \sum y - \sum x \sum xy}{n\sum x^2 - (\sum x)^2}$, with $n = 5$, $m = \frac{5(854)-30(122)}{5(220)-(30)^2} = \frac{610}{200} = 3.05$ and $b = \frac{220(122)-30(854)}{5(220)-(30)^2} = \frac{1{,}220}{200} = 6.10$. Hence, the equation of the least-squares line is $y = 3.05x + 6.10$.
c) When the polls close at 8:00 p.m., $x = 12$ and so $y = 3.05(12) + 6.1 = 42.7$, which means that approximately 42.7 % of the registered voters can be expected to vote.

15. a) Let x denote the number of years (in decades) after 1900 and y the corresponding population (in millions). Then

x	0	1	2	3	4	5
y	75	91.97	105.7	122.78	131.7	178.5

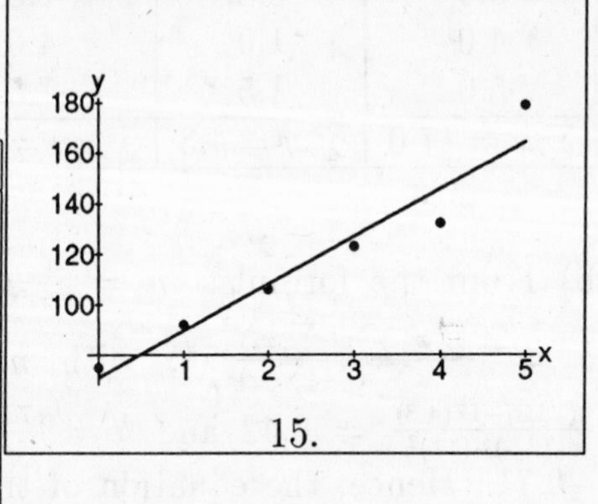

15.

x	y	xy	x^2
0	75	0	0
1	91.97	91.97	1
2	105.7	211.4	4
3	122.78	368.34	9
4	131.7	526.8	16
5	178.5	892.5	25
$\sum x = 15$	$\sum y = 705.65$	$\sum xy = 2{,}091.01$	$\sum x^2 = 55$

From the formulas $m = \frac{n\sum xy - \sum x \sum y}{n\sum x^2 - (\sum x)^2}$ and $b = \frac{\sum x^2 \sum y - \sum x \sum xy}{n\sum x^2 - (\sum x)^2}$, with $n = 6$, $m = \frac{6(2{,}091.01)-15(705.65)}{6(55)-(15)^2} = \frac{1{,}961.31}{105} = 18.68$ and $b = \frac{55(705.65)-15(2{,}091.01)}{6(55)-(15)^2} = \frac{7{,}445.6}{105} = 70.91$. Hence, the equation of the least-squares line is $y = 18.68x + 70.91$. In the year 1970, $x = 7$ and $y = 18.68(7) + 70.91 = 201.67$. Thus the population is estimated to be 201.67 million people.

b) In the year 1990, $x = 9$ and $y = 18.68(9) + 70.91 = 239.03$. Thus the population is estimated to be 239.03 million people.
c) In the year 2000, $x = 10$ and $y = 18.68(10) + 70.91 = 257.7$. Thus the population is estimated to be 257.7 million people.

Review Problems

1. a) $f(x,y) = 2x^3y + 3xy^2 + \frac{y}{x}$. Then $f_x = 6x^2y + 3y^2 - yx^{-2} = 6x^2y + 3y^2 - \frac{y}{x^2}$ and $f_y = 2x^3 + 6xy + \frac{1}{x}$.
b) $f(x,y) = (xy^2+1)^5$. Then $f_x = 5(xy^2+1)^4(y^2)$ and $f_y = 5(xy^2+1)^4(2xy) = 10xy(xy^2+1)^4$.
c) $f(x,y) = xye^{xy}$. Then $f_x = xye^{xy}(y) + e^{xy}(y) = y(xy+1)e^{xy}$ and $f_y = xye^{xy}(x) + e^{xy}(x) = x(xy+1)e^{xy}$.
d) $f(x,y) = \frac{x^2-y^2}{2x+y}$. Then $f_x = \frac{(2x+y)(2x)-(x^2-y^2)(2)}{(2x+y)^2} = \frac{4x^2+2xy-2x^2+2y^2}{(2x+y)^2} = \frac{2(x^2+xy+y^2)}{(2x+y)^2}$ and $f_y = \frac{(2x+y)(-2y)-(x^2-y^2)(1)}{(2x+y)^2} = \frac{-4xy-2y^2-x^2+y^2}{(2x+y)^2} = \frac{-4xy-y^2-x^2}{(2x+y)^2}$.
e) $f(x,y) = \ln\frac{xy}{x+3y} = \ln x + \ln y - \ln(x+3y)$. Then $f_x = \frac{1}{x} - \frac{1}{x+3y}$ and $f_y = \frac{1}{y} - \frac{3}{x+3y}$.

2. a) $f(x,y) = x^2 + y^3 - 2xy^2$. Then $f_x = 2x - 2y^2$ and $f_y = 3y^2 - 4xy$. Now $f_{xx} = 2$, $f_{yy} = 6y - 4x$, and $f_{xy} = f_{yx} = -4y$.
b) $f(x,y) = e^{x^2+y^2}$. Then $f_x = 2xe^{x^2+y^2}$ and $f_y = 2ye^{x^2+y^2}$. Now $f_{xx} = 2xe^{x^2+y^2}(2x) + e^{x^2+y^2}(2) = 2(2x^2+1)e^{x^2+y^2}$, $f_{yy} = 2ye^{x^2+y^2}(2y) + e^{x^2+y^2}(2) = 2(2y^2+1)e^{x^2+y^2}$, and $f_{xy} = f_{yx} = 2xe^{x^2+y^2}(2y) = 4xye^{x^2+y^2}$.
c) $f(x,y) = x\ln y$. Then $f_x = \ln y$ and $f_y = \frac{x}{y}$. Now $f_{xx} = 0$, $f_{yy} = -\frac{x}{y^2}$, and $f_{xy} = f_{yx} = \frac{1}{y}$.

3. Let $Q = 40K^{1/3}L^{1/2}$ denote the total output, where K denotes the capital investment and L the size of the labor force. The marginal product of capital is $\frac{\partial Q}{\partial K} = \frac{40}{3}K^{-2/3}L^{1/2} = \frac{40L^{1/2}}{3K^{2/3}}$ which is approximately the change ΔQ in output due to one (thousand dollar) unit increase in capital. When $K = 125$ (thousand) and $L = 900$, $\Delta Q \approx \frac{\partial Q}{\partial K} = \frac{40(900)^{1/2}}{3(125)^{2/3}} = = 16$ units.

4. The marginal product of labor is the partial derivative $\frac{\partial Q}{\partial L}$. To say that this partial derivative increases as K increases is to say that its derivative with respect to K is positive, that is, $\frac{\partial^2 Q}{\partial K \partial L} > 0$.

5. a) If $z = x^3 - 3xy^2$, $x = 2t$, and $y = t^2$, then $\frac{\partial z}{\partial x} = 3x^2 - 3y^2$, $\frac{\partial z}{\partial y} = -6xy$, $\frac{dx}{dt} = 2$, and $\frac{dy}{dt} = 2t$. Hence, by the chain rule, $\frac{dz}{dt} = \frac{\partial z}{\partial x}\frac{dx}{dt} + \frac{\partial z}{\partial y}\frac{dy}{dt} = (3x^2 - 3y^2)(2) + (-6xy)(2t) =$

$[3(2t)^2 - 3(t^2)^2](2) + [-6(2t)(t^2)](2t) = 24t^2 - 6t^4 - 24t^4 = 6t^2(4 - 5t^2)$.
b) If $z = x \ln y$, $x = 2t$, and $y = e^t$, then $\frac{\partial z}{\partial x} = \ln y$, $\frac{\partial z}{\partial y} = \frac{x}{y}$, $\frac{dx}{dt} = 2$, and $\frac{dy}{dt} = e^t$. Hence, by the chain rule, $\frac{dz}{dt} = \frac{\partial z}{\partial x}\frac{dx}{dt} + \frac{\partial z}{\partial y}\frac{dy}{dt} = (\ln y)(2) + (\frac{x}{y})(e^t) = 2(\ln e^t) + (\frac{2t}{e^t})(e^t) = 2t + 2t = 4t$.

6. Since the demand is $Q(x,y) = 240 + 0.1y^2 - 0.2x^2$, the change in demand is $\Delta Q \approx dQ = \frac{\partial Q}{\partial x}\Delta x + \frac{\partial Q}{\partial y}\Delta y = (-0.4x)\Delta x + (0.2y)\Delta y$. When $x = 45$, $y = 48$, $\Delta x = 2$, and $\Delta y = -1$, $\Delta Q = (-0.4)(45)(2) + (0.2)(48)(-1) = -45.6 = -46$, that is the demand will drop by approximately 46 cans per week.

7. Output $Q(K,L) = 120K^{1/3}L^{2/3}$. Apply the approximation formula for percentage change with $\Delta K = 0.02K$ $\Delta L = 0.01L$ to get the percentage change in $Q \approx$
$100\frac{\frac{\partial Q}{\partial K}\Delta K + \frac{\partial Q}{\partial L}\Delta L}{Q} = 100\frac{40\alpha K^{-2/3}L^{2/3}(0.02K) + (80)K^{1/3}L^{-1/3}(0.01L)}{120K^{1/3}L^{2/3}} = 100\frac{0.8K^{1/3}L^{2/3} + 0.8K^{1/3}L^{2/3}}{120K^{1/3}L^{2/3}}$
$= 100\frac{1.6K^{1/3}L^{2/3}}{120K^{1/3}L^{2/3}} = 1.33\ \%$.

8. The price of apple pies is $p(x,y) = \frac{1}{2}x^{1/3}y^{1/2}$ dollars per pie. The price of the apples t months from now will be $x = 23 + \sqrt{8t}$ cents per pound, and bakers' wages t months from now will be $y = 3.96 + 0.02t$ dollars per hour. The weekly demand for the pies is $Q(p) = \frac{3{,}600}{p}$. The goal of the problem is to find $\frac{\partial Q}{\partial t}$ when $t = 2$. $\frac{dQ}{dt} = \frac{\partial Q}{\partial p}\frac{dp}{dt} = -\frac{3{,}600}{p^2}\frac{dp}{dt}$. Hence, by the chain rule, $\frac{dp}{dt} = \frac{\partial p}{\partial x}\frac{dx}{dt} + \frac{\partial p}{\partial y}\frac{dy}{dt} = (\frac{1}{6}x^{-2/3}y^{1/2})(\frac{1}{2})(8t)^{-1/2}(8) + \frac{1}{4}x^{1/3}y^{-1/2}(0.02) = \frac{2}{3}x^{-2/3}y^{1/2}(8t)^{-1/2} + 0.005x^{1/3}y^{-1/2}$. When $t = 2$, $p = 3$, $x = 27$, and $y = 4$. Hence, putting it all together. $\frac{dQ}{dt} = -\frac{3{,}600}{9}[\frac{2}{3}(\frac{1}{9})(2)(\frac{1}{4}) + 0.005(3)(\frac{1}{2})] \approx -17.81$, that is, the weekly demand for pies will be decreasing at the rate of approximately 18 pies per week.

9. a) If $f(x,y) = x^2 - y$, then when $f = 2$, $x^2 - y = 2$ or $y = x^2 - 2$ which is a parabola opening upward with vertex at $(0,-2)$, and when $f = -2$, which is a parabola opening upward with vertex at $(0,2)$.

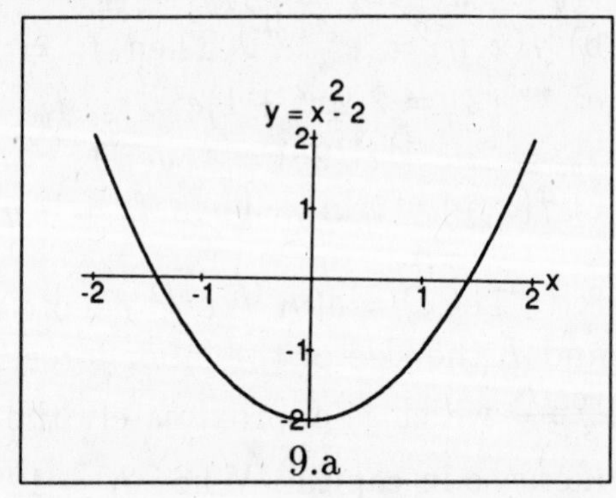

9.a

b) If $f(x,y) = 6x+2y$, then when $f = 0$, $6x+2y = 0$ or $y = -3x$ which is a straight line through the origin with slope -3; when $f = 1$, $6x + 2y = 1$ or $y = -3x + \frac{1}{2}$ which is a straight line with y–intercept $\frac{1}{2}$ and slope -3; and when $f = 2$, $6x + 2y = 2$ or $y = -3x + 1$ which is a straight line with y–intercept 1 and slope -3

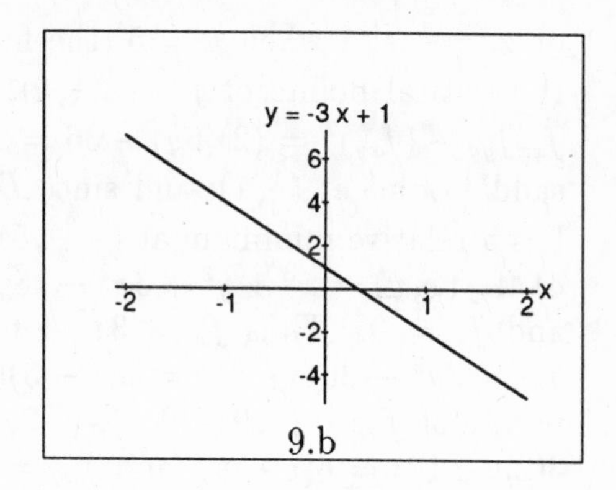

9.b

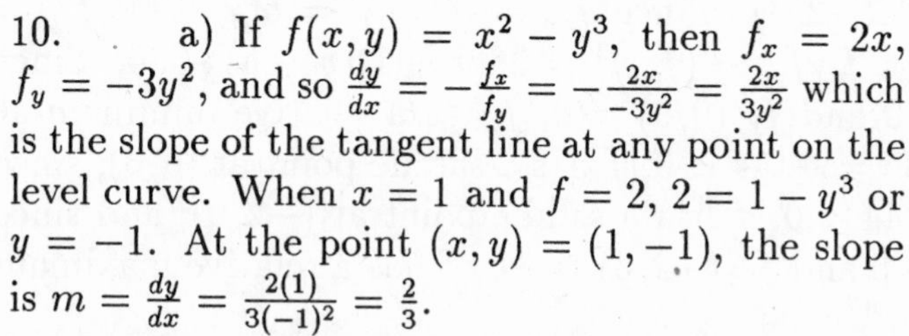
10. a) If $f(x,y) = x^2 - y^3$, then $f_x = 2x$, $f_y = -3y^2$, and so $\frac{dy}{dx} = -\frac{f_x}{f_y} = -\frac{2x}{-3y^2} = \frac{2x}{3y^2}$ which is the slope of the tangent line at any point on the level curve. When $x = 1$ and $f = 2$, $2 = 1 - y^3$ or $y = -1$. At the point $(x,y) = (1,-1)$, the slope is $m = \frac{dy}{dx} = \frac{2(1)}{3(-1)^2} = \frac{2}{3}$.

b) If $f(x,y) = xe^y$, then $f_x = e^y$, $f_y = xe^y$, and so $\frac{dy}{dx} = -\frac{f_x}{f_y} = -\frac{e^y}{xe^y} = -\frac{1}{x}$ which isd the slope of the tangent line at any point on the level curve. When $x = 2$ and $f = 2$, $2 = 2e^y$ or $y = 0$. At the point $(x,y) = (2,0)$, the slope is $m = \frac{dy}{dx} = -\frac{1}{2}$.

11. Output is $Q(x,y) = 60x^{1/3}y^{2/3}$, where x denotes the number of skilled workers. The combination of x and y for which output will remain at the current level are the coordinates of the points (x,y) that lie on the constant-production curve $Q = k$, where k is the current level of output. For any value of x, the slope of this constant-production curve is an approximation to the change in unskilled labor y that should be made to offset a one-unit increase in skilled labor x so that the level of output will remain constant. Thus, ΔQ =change in unskilled labor$\approx \frac{dQ}{dx} = -\frac{Q_x}{Q_y} = -\frac{20x^{-2/3}y^{2/3}}{40x^{1/3}y^{-1/3}} = -\frac{y}{2x}$. When $x = 10$ and $y = 40$, $\Delta Q \approx \frac{dQ}{dx} = -\frac{40}{2(10)} = -2$, that is, the level of unskilled labor should be decreased by approximately 2 workers.

12. a) If $f(x,y) = x^3+y^3+3x^2-18y^2+81y+5$, then $f_x = 3x^2+6x$ and $f_y = 3y^2-36y+81$. To find the critical points, set $f_x = 0$ and $f_y = 0$. Thus $3x^2 + 6x = 3x(x+2) = 0$ or $x = 0$ and $x = -2$. Similarly, $3y^2 - 36y + 81 = 3(y-3)(y-9)$ or $y = 3$ and $y = 9$. Hence, the critical points of f are $(0,3)$, $(0,9)$, $(-2,3)$, and $(-2,9)$. Since $f_{xx} = 6x+6$, $f_{yy} = 6y - 36$, and $f_{xy} = 0$, $D = f_{xx}f_{yy} - (f_{xy})^2 = (6x+6)(6y-36) - 0 = 36(x+1)(y-6)$. Since $D(0,3) = 36(1)(-3) = -108 < 0$, f has a saddle point at $(0,3)$; since $D(0,9) = 36(1)(3) = 108 > 0$, and $f_{xx}(0,9) = 6 > 0$, f has a relative minimum at $(0,9)$; since $D(-2,3) = 36(-1)(-3) = 108 > 0$, and $f_{xx}(-2,3) = -6 < 0$, f has a relative maximum at $(-2,3)$; and since $D(-2,9) = 36(-1)(3) = -108 < 0$, f has a saddle point at $(-2,9)$.

b) If $f(x,y) = x^2 + y^3 + 6xy - 7x - 6y$, then $f_x = 2x + 6y - 7$ and $f_y = 3y^2 + 6x - 6$. To find the critical points, set $f_x = 0$ and $f_y = 0$. Thus $2x + 6y - 7 = 0$ and $3y^2 + 6x - 6 = 0$ or $2x + 6y - 7 = 0$ and $2x + y^2 - 2 = 0$. Subtracting the two equations gives $y^2 - 6y + 5 = 0$, $(y-1)(y-5) = 0$, or $y = 1$ and $y = 5$. When $y = 1$, the first equation gives $2x + 6 - 7 = 0$

or $x = \frac{1}{2}$ and when $y = 5$, the first equation gives $2x + 30 - 7 = 0$ or $x = -\frac{23}{2}$. Hence, the critical points of f are $(\frac{1}{2}, 1)$, $(-\frac{23}{2}, 5)$. Since $f_{xx} = 2$, $f_{yy} = 6y$, and $f_{xy} = 6$, $D = f_{xx}f_{yy} - (f_{xy})^2 = (2)(6y) - 36 = 12(y - 3)$. Since $D(\frac{1}{2}, 1) = 12(-2) = -24 < 0$, f has a saddle point at $(\frac{1}{2}, 1)$; and since $D(-\frac{23}{2}, 5) = 12(2) = 24 > 0$, and $f_{xx}(-\frac{23}{2}, 5) = 2 > 0$, f has a relative minimum at $(-\frac{23}{2}, 5)$.
c) If $f(x,y) = x^3 + y^3 + 3x^2 - 18y^2 + 81y + 5$, then, to find the critical points, set $f_x = 0$ and $f_y = 0$. Thus $f_x = 3x^2 + 6x = 3x(x+2) = 0$ at $x = 0$ and $x = -2$. Similarly $f_y = 3y^2 - 36y + 81 = 3(y-3)(y-9) = 0$ at $y = 3$ and $y = 9$. Hence, the critical points of f are $(0,9)$, $(0,3)$, $(-2,9)$, and $(-2,3)$. Since $f_{xx} = 6x + 6 = 6(x+1)$, $f_{yy} = 3(2y - 12) = 6(y-6)$, and $f_{xy} = 0$, $D = f_{xx}f_{yy} - (f_{xy})^2 = 36(x+1)(y-6) - 36$. Since $D(0,9) = 36(0+1)(9-6) - 36 = 72 > 0$ and $f_{xx}(0,9) = 6$, f has a relative minimum at $(0,9)$; since $D(0,3) = 36(0+1)(3-6) - 36 = -144 < 0$, f has a saddle point at $(0,3)$; since $D(-2,9) = 36(-2+1)(9-6) - 36 = -144 < 0$, f has a saddle point at $(-2,9)$; and since $D(-2,3) = 36(-2+1)(3-6) - 36 = 72 > 0$ and $f_{xx}(-2,3) = -6$, f has a relative maximum at $(0,9)$.

13. For $f(x,y) = x^2 + 2y^2 + 2x + 3$ subject to the constraint that $g(x,y) = x^2 + y^2 = 4$, the partial derivatives are $f_x = 2x + 2$, $f_y = 4y$, $g_x = 2x$, and $g_y = 2y$. Hence, the three Lagrange equations are $2x + 2 = 2\lambda x$, $4y = 2\lambda y$, and $x^2 + y^2 = 4$. From the first equation $\lambda = 1 + \frac{1}{x}$. From the second equation, $\lambda = 2$ or $y = 0$. If $y = 0$, the third equation gives $x = \pm 2$. If $y \neq 0$, setting the two expressions for λ equal to each other yields $1 + \frac{1}{x} = 2$ or $x = 1$. From the third equation, $y = \pm\sqrt{3}$. Hence, the points at which the constrained extrema can occur are $(-2,0)$, $(2,0)$, $(1,\sqrt{3})$, and $(1,-\sqrt{3})$. Since $f(-2,0) = 3$, $f(2,0) = 11$, and $f(1,-\sqrt{3}) = f(1,\sqrt{3}) = 12$, it follows that the constrained maximum is 12 which is attained at $(1,-\sqrt{3})$ as well as $(1,\sqrt{3})$, and the constrained minimum is 3 which is attained at $(-2,0)$.

14. Let x denote the length of the rectangle, y the width, and $f(x,y)$ the corresponding area. Then $f(x,y) = xy$. Since the rectangle is to have a fixed perimieter, the goal is to maximize $f(x,y)$ subject to the constraint that $g(x,y) = x + y = k$, for some constant k. The partial derivatives are $f_x = y$, $f_y = x$, $g_x = 1$, and $g_y = 1$. The three Lagrange equations are $y = \lambda$, $x = \lambda$, and $x + y = k$. From the first two equations, $x = y$, which implies that the rectangle of greatest area is a square.

15. Let x denote the amount spent on development and y the amount spent on promotion in thousand dollars. The profit $P(x,y)$=(number of units sold)(price per unit−cost per unit)−total amount spent on development and promotion. The number of units sold is $\frac{250y}{y+2} + \frac{100x}{x+5}$. The selling price is \$350 per unit and the cost \$150 per unit. Hence, the price per unit minus the cost per unit is \$200 or $\frac{1}{5}$ of a thousand dollars. Putting it all together, $P(x,y) = (\frac{1}{5})(\frac{250y}{y+2} + \frac{100x}{x+5}) - (x+y) = \frac{50y}{y+2} + \frac{20x}{x+5} - x - y$. Then $P_x = \frac{(x+5)(20)-(20x)(1)}{(x+5)^2} - 1 = \frac{100}{(x+5)^2} - 1$, $P_y = \frac{(y+2)(50)-(50y)(1)}{(y+2)^2} - 1 = \frac{100}{(y+2)^2} - 1$. To find the critical points, set $P_x = 0$ and

$P_y = 0$. Thus $\frac{100}{(x+5)^2} - 1 = 0$, $(x+5)^2 = 100$, $x+5 = 10$, or $x = 5$, and $\frac{100}{(y+2)^2} - 1 = 0$, $(y+2)^2 = 100$, $y+2 = 10$, or $y = 8$. Hence, the critical point is $(5,8)$. Since $P_{xx} = -\frac{200}{(x+5)^3}$, $P_{yy} = -\frac{200}{(y+2)^3}$, and $P_{xy} = 0$, $D(5,8) = P_{xx}P_{yy} - (P_{xy})^2 = \frac{40{,}000}{(10)^3(10)^3}$ and $P_{xx}(5,8) = -\frac{200}{1{,}000} < 0$, it follows that $P(x,y)$ has a relative maximum at $(5,8)$. Assuming that the absolute maximum and the relative maximum are the same, it follows that to maximize the profit, \$5,000 should be spent on development and \$8,000 should be spent on promotion.

16. From problem 15, the profit function is $P(x,y) = \frac{50y}{y+2} + \frac{20x}{x+5} - x - y$. The constraint is $g(x,y) = x + y = 11$ thousand dollars. Hence, the partial derivatives are $P_x = \frac{100}{(x+5)^2} - 1$, $P_y = \frac{100}{(y+2)^2} - 1$, $g_x = 1$, and $g_y = 1$. The three Lagrange equations are $\frac{100}{(x+5)^2} - 1 = \lambda$, $\frac{100}{(y+2)^2} - 1 = \lambda$, and $x+y = 11$. From the first two equations, $\frac{100}{(x+5)^2} = \frac{100}{(y+2)^2}$, $(x+5)^2 = (y+2)^2$, or $y = x+3$. From the third equation, $x + (x+3) = 11$ or $x = 4$. Since $y = x+3$, the corresponding value of y is $y = 7$. Hence, to maximize profit, \$4,000 should be spent on development and \$7,000 should be spent on promotion.
Note that if $y + 2 = -(x+5)$, $y = -x - 7$, $x + (-x-7) = 11$, which is impossible.

17. The increase in axnimal profit M resulting from an increase in available money by one thousand dollars is $\Delta M \approx \frac{dM}{dk} = \lambda$ where, from problem 16, $\lambda = \frac{100}{(x+5)^2} - 1$. Since the optional allocation of 11 thousand dollars is $x = 4$ and $y = 7$, the increase in maximal profit resulting from the decision to spend 12 thousand dollars is $\Delta M \approx \lambda = \frac{100}{(x+5)^2} - 1 = \frac{100}{9^2} - 1 = \frac{100}{81} - 1 = 0.235$ thousand or \$235.

18. $f(x,y) = \frac{12}{x} + \frac{18}{y} + xy$. Suppose y is fixed, (say at $y = 1$), then f is very large when x is quite small. f is also large when x is large, with a dip in the values of f between these extremes. The same reasoning applies to y when x is fixed.
$f_x = -\frac{12}{x^2} + y$ and $f_y = -\frac{18}{y^2} + x$. To find the critical points, set $f_x = 0$ and $f_y = 0$. Thus $y = \frac{12}{x^2} > 0$ and $x = \frac{18}{y^2} > 0$. Substituting leads to $y = \frac{12}{x^2} = \frac{12}{(\frac{18}{y^2})^2} = \frac{12y^4}{18^2}$ or $y = 0$ (which is not in the domain of the function) and $12y^3 = 18^2$, $y^3 = 27$, $y = 3$. The corresponding value for x is $x = \frac{18}{3^2} = 2$. Hence, the critical points of f is $(2,3)$. Since $f_{xx} = \frac{24}{x^3}$, $f_{yy} = \frac{36}{y^3}$, and $f_{xy} = 1$, $D = f_{xx}f_{yy} - (f_{xy})^2 = (\frac{24}{x^3})(\frac{36}{y^3}) - 1 > 0$ for some (x,y). Since $D(2,3) = \frac{(24)(36)}{(2^3)(3^3)} > 0$ and $f_{xx}(2,3) > 0$, f has a relative minimum at $(2,3)$.

19. With $h = \frac{kV^{1/3}}{D^{2/3}} = kV^{1/3}D^{-2/3}$, $\frac{\partial h}{\partial V} = \frac{1}{3}kV^{-2/3}D^{-2/3}$ and $\frac{\partial h}{\partial D} = \frac{-2}{3}kV^{1/3}D^{-5/3}$. The desired ratio is $\frac{\frac{\partial h}{\partial V}}{\frac{\partial h}{\partial D}} = \frac{\frac{1}{3}kV^{-2/3}D^{-2/3}}{\frac{-2}{3}kV^{1/3}D^{-5/3}} = -\frac{D}{2V}$.

20. With $Q = x^a y^b$, $Q_x = ax^{a-1}y^b$ and $Q_x = bx^a y^{b-1}$. Now $xQ_x + yQ_y = x(ax^{a-1}y^b) + y(bx^a y^{b-1}) = (a+b)x^a y^b = (a+b)Q$. If $a + b = 1$ then $xQ_x + yQ_y = (a+b)Q = Q$.

21. Let r denote the radius of the cylinder and h its height. The volume of the cylinder

is $V = \pi r^2 h$. With $r = 2$ cm and $h = 6$ cm, $\Delta r = 0.02$, $\Delta h = 0.04$, $\Delta V \approx dV = \frac{\partial V}{\partial r}\Delta r + \frac{\partial V}{\partial h}\Delta h = \pi(2rhdr + r^2dh) = \pi[24(0.02) + 4(0.04)] = \frac{\pi}{100}(48+16) = 2.0106$ or roughly 2 cubic centimeters. The calculations were performed as though the tube were inflated instead of being deflated.

22. The sum $S(m,b)$ of the squares of the vertical distances from the four given points is $S(m,b) = (m+b-1)^2 + (m+b-2)^2 + (3m+b-2)^2 + (4m+b-3)^2$. To minimize $S(m,b)$, set the partial derivatives $\frac{\partial S}{\partial m} = 0$ and $\frac{\partial S}{\partial b} = 0$, namely $\frac{\partial S}{\partial m} = 2(m+b-1) + 2(m+b-2) + 2(3m+b-2)(3) + 2(4m+b-3)(4) = 2(27m+9b-21) = 0$ and $\frac{\partial S}{\partial b} = 2(m+b-1) + 2(m+b-2) + 2(3m+b-2) + 2(4m+b-3) = 18m + 8b - 16 = 0$. Solve the resulting simplified equations $9m + 3b = 7$ and $9m + 4b = 8$ to get $m = \frac{4}{9}$ and $b = 1$. Hence, the equation of the least-squares line is $y = \frac{4x}{9} + 1$.

23. a) Let x denote the monthly advertising expenditure and y the corresponding sales (both measured in units of $1,000). Then

x	3	4	7	9	10
y	78	86	138	145	156

x	y	xy	x^2
3	78	234	9
4	86	344	16
7	138	966	49
9	145	1305	81
10	156	1560	100
$\sum x = 33$	$\sum y = 603$	$\sum xy = 4,409$	$\sum x^2 = 255$

b) From the formulas $m = \frac{n\sum xy - \sum x \sum y}{n\sum x^2 - (\sum x)^2}$ and $b = \frac{\sum x^2 \sum y - \sum x \sum xy}{n\sum x^2 - (\sum x)^2}$, with $n = 5$, $m = \frac{5(4,409) - 33(603)}{5(255) - (33)^2} = 11.54$ and $b = \frac{255(603) - 33(4,409)}{5(255)-(33)^2} = 44.45$. Hence, the equation of the least-squares line is $y = 11.54x + 44.45$.

c) If the monthly advertising expenditure is $5,000, then $x = 5$ and $y = 11.54(5) + 44.45 = 102.15$. Thus the monthly sales will be approximately $102,150.

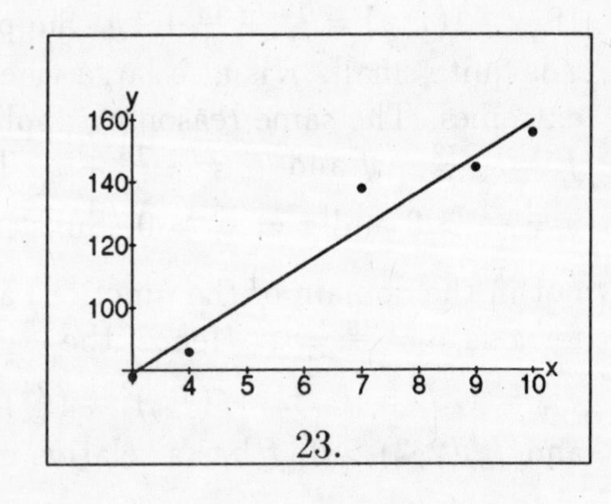

23.

Chapter 10

Double Integrals

10.1 Double Integrals

1. $\int_{y=0}^{y=1}\int_{x=1}^{x=2} x^2 y dx dy = \int_{y=0}^{y=1}[\int_{x=1}^{x=2} x^2 y dx]dy = \int_{y=0}^{y=1}[\frac{x^3}{3}y|_{x=1}^{x=2}]dy = \int_{y=0}^{y=1}[\frac{8}{3}y - \frac{1}{3}y]dy = \frac{7}{3}\int_{y=0}^{y=1} y dy = \frac{7}{6}y^2|_{y=0}^{y=1}y^2 = \frac{7}{6}$

3. $\int_{y=0}^{y=\ln 2}\int_{x=-1}^{x=0} 2xe^y dx dy = \int_{y=0}^{y=\ln 2}[\int_{x=-1}^{x=0} 2xe^y dx]dy = \int_{y=0}^{y=\ln 2}[x^2 e^y|_{x=-1}^{x=0}]dy = \int_{y=0}^{y=\ln 2}[-e^y]dy = -e^y|_{y=0}^{y=\ln 2} = -e^{\ln 2} + 1 = -2 + 1 = -1.$

5. $\int_{y=1}^{y=3}\int_{x=0}^{x=1} \frac{2xy}{x^2+1} dx dy = \int_{y=1}^{y=3}[\int_{x=0}^{x=1} \frac{2xy}{x^2+1} dx]dy = \int_{y=1}^{y=3}[y\ln(x^2+1)|_{x=0}^{x=1}]dy = \int_{y=1}^{y=3}[y(\ln 2 - \ln 1)]dy = \int_{y=1}^{y=3} y\ln 2 dy = \ln 2(\frac{1}{2})y^2|_{y=1}^{y=3} = \ln 2(\frac{9}{2} - \frac{1}{2}) = 4\ln 2$

7. $\int_{x=0}^{x=4}\int_{y=0}^{y=\sqrt{x}} x^2 y dy dx = \int_{x=0}^{x=4}[\int_{y=0}^{y=\sqrt{x}} x^2 y dy]dx = \int_{x=0}^{x=4}[\frac{y^2}{2}x^2|_{y=0}^{y=\sqrt{x}}]dx = \int_{x=0}^{x=4}[\frac{x^3}{2}]dx = \frac{x^4}{8}|_{x=0}^{x=4} = 32$

9. $\int_{y=0}^{y=1}\int_{x=y-1}^{x=1-y}(2x+y)dx dy = \int_{y=0}^{y=1}[\int_{x=y-1}^{x=1-y}(2x+y)dx]dy = \int_{y=0}^{y=1}[(x^2+xy)|_{x=y-1}^{x=1-y}]dy = \int_{y=0}^{y=1}\{[(1-y)^2 + (1-y)y] - [(y-1)^2 + (y-1)y]\}dy = \int_{y=0}^{y=1}[(1-y)y - (y-1)y]dy = \int_{y=0}^{y=1}(-2y^2+2y)dy = -2(\frac{y^3}{3} - \frac{y^2}{2})|_{y=0}^{y=1} = 2(\frac{1}{3} - \frac{1}{2}) = \frac{1}{3}.$

11. $\int_{x=0}^{x=2}\int_{y=x^2}^{y=4} xe^y dy dx = \int_{x=0}^{x=2}[\int_{y=x^2}^{y=4} xe^y dy]dx = \int_{x=0}^{x=2}[xe^y|_{y=x^2}^{y=4}]dx = \int_{x=0}^{x=2}[xe^4 - xe^{x^2}]dx = e^4\int_{x=0}^{x=2} x dx - \int_{x=0}^{x=2} xe^{x^2} dx = e^4(\frac{x^2}{2})|_{x=0}^{x=2} - \frac{e^{x^2}}{2}|_{x=0}^{x=2} = \frac{e^4}{2}(4-0) - \frac{1}{2}(e^4 - 1) = \frac{3e^4}{2} + \frac{1}{2}.$

13. $\int_{x=0}^{x=1}\int_{y=0}^{y=x} \frac{2y}{x^3+1} dy dx = \int_{x=0}^{x=1}[\int_{y=0}^{y=x} \frac{2y}{x^3+1} dy]dx = \int_{x=0}^{x=1} \frac{y^2}{x^3+1}|_{y=0}^{y=x}]dx = \int_{x=0}^{x=1} \frac{x^2}{x^3+1} dx = \frac{1}{3}\ln|x^3+1||_{x=0}^{x=1} = \frac{1}{3}(\ln 2 - \ln 1) = \frac{\ln 2}{3}.$

15. $\int_{x=0}^{x=1}\int_{y=0}^{y=x} xe^y dydx = \int_{x=0}^{x=1}[\int_{y=0}^{y=x} xe^y dy]dx = \int_{x=0}^{x=1} xe^y|_{y=0}^{y=x}]dx = \int_{x=0}^{x=1}(xe^x - x)dx = \int_{x=0}^{x=1} xe^x dx - \int_{x=0}^{x=1} xdx = (xe^x - e^x)|_{x=0}^{x=1} - \frac{x^2}{2}|_{x=0}^{x=1} = [(e-e)+1] - \frac{1}{2} = \frac{1}{2}$.

17. $\int_{y=0}^{y=1}\int_{x=y}^{x=1} ye^{x+y}dxdy = \int_{y=0}^{y=1}[\int_{x=y}^{x=1} ye^y e^x dx]dy = \int_{y=0}^{y=1} ye^y e^x|_{x=y}^{x=1}dy = \int_{y=0}^{y=1}(ye^y e^1 - ye^y e^y)dy = e\int_{y=0}^{y=1} ye^y dy - \int_{y=0}^{y=1} ye^{2y}dy = e(ye^y - e^y)|_{y=0}^{y=1} - [\frac{ye^{2y}}{2} - \frac{e^{2y}}{4}]|_{y=0}^{y=1} = e[(e-e)+1] - [(\frac{e^2}{2} - \frac{e^2}{4}) + \frac{1}{4}] = e - \frac{e^2}{4} - \frac{1}{4}$.

19. $\int_{x=0}^{x=1}\int_{y=0}^{y=4} \sqrt{xy}dydx = \int_{x=0}^{x=1}[\int_{y=0}^{y=4} x^{1/2}y^{1/2}dy]dx = \int_{x=0}^{x=1} x^{1/2}\frac{2y^{3/2}}{3}|_{y=0}^{y=4}]dx = \int_{x=0}^{x=1} \frac{16x^{1/2}}{3}dx = \frac{16}{3}\frac{2x^{3/2}}{3}|_{x=0}^{x=1} = \frac{32}{9}$.

21. $\int_{x=1}^{x=e}\int_{y=0}^{y=\ln x} xydydx = \int_{x=1}^{x=e}[\int_{y=0}^{y=\ln x} xydy]dx = \int_{x=1}^{x=e} x\frac{y^2}{2}|_{y=0}^{y=\ln x}dx = \int_{x=1}^{x=e} \frac{x}{2}(\ln x)^2 dx = \frac{1}{2}[\frac{x^2(\ln x)^2}{2}|_{x=1}^{x=e} - \int_1^e x\ln x dx] = \frac{1}{2}[\frac{e^2}{2} - \int_1^e x\ln x dx] = \frac{1}{2}(\frac{e^2}{2} - \frac{x^2\ln x}{2}|_1^e + \frac{x^2}{4}|_1^e) = \frac{e^2-1}{8}$.

10.2 Finding Limits of Integration

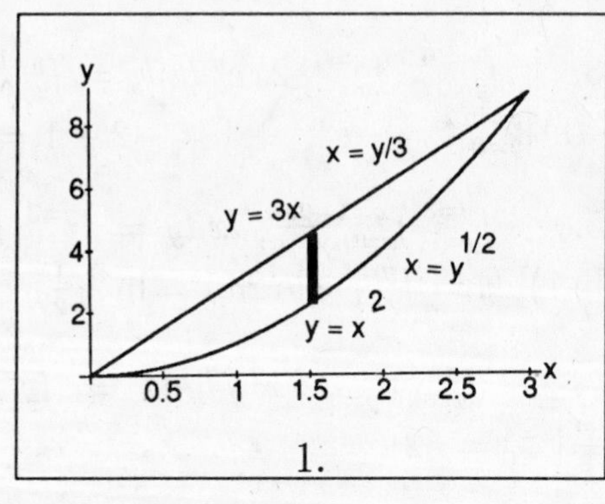

1.

1. R can be described in terms of vertical cross sections by $0 \le x \le 3$ and $x^2 \le y \le 3x$ and in terms of horizontal cross sections by $0 \le y \le 9$ and $\frac{y}{3} \le x \le \sqrt{y}$.

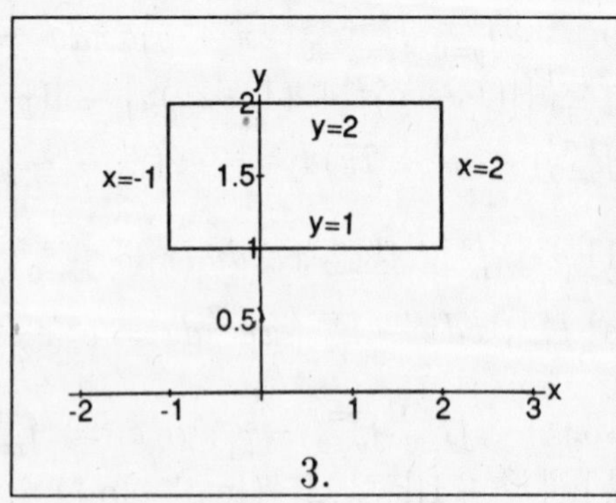

3.

3. R can be described in terms of vertical cross sections by $-1 \le x \le 2$ and $1 \le y \le 2$ and in terms of horizontal cross sections by $1 \le y \le 2$ and $-1 \le x \le 2$.

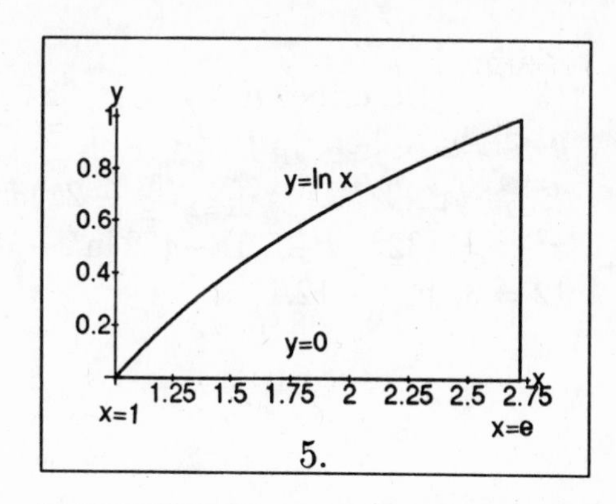

5.

5. R can be described in terms of vertical cross sections by $1 \le x \le e$ and $0 \le y \le \ln x$ and in terms of horizontal cross sections by $0 \le y \le 1$ and $e^y \le x \le e$.

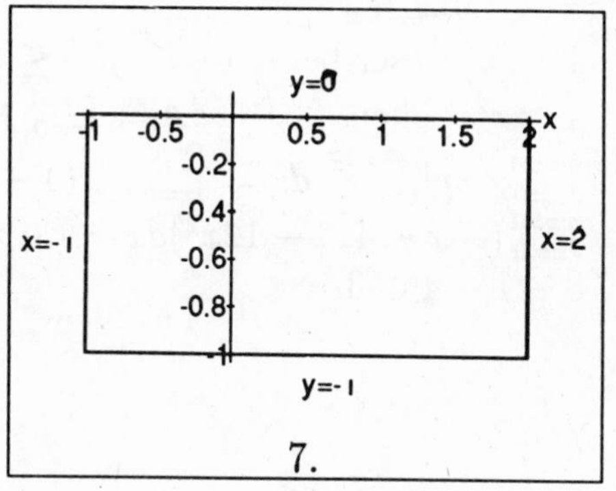

7.

7. Describe R by $-1 \le x \le 2$ and $-1 \le y \le 0$. Then $\int_R\int 3xy^2 dA = \int_{y=-1}^{y=0}\int_{x=-1}^{x=2} 3xy^2 dx dy = \int_{y=-1}^{y=0} \frac{3x^2y^2}{2}\Big|_{x=-1}^{x=2} dy = \int_{y=-1}^{y=0}(6y^2 - \frac{3y^2}{2})dy = \int_{y=-1}^{y=0} \frac{9y^2}{2} dy = \frac{3y^3}{2}\Big|_{y=-1}^{y=0} = 0 - \frac{3(-1)}{2} = \frac{3}{2}$.

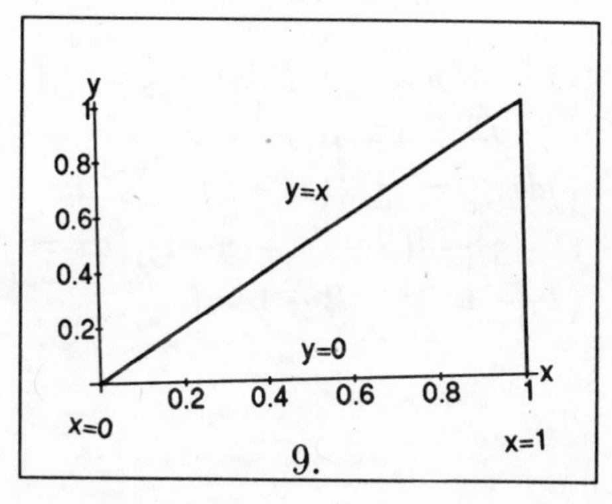

9.

9. Describe R by $0 \le x \le 1$ and $0 \le y \le x$. Then $\int_R\int xe^y dA = \int_{x=0}^{x=1}\int_{y=0}^{y=x} xe^y dy dx = \int_{x=0}^{x=1} xe^y\Big|_{y=0}^{y=x} dx = \int_{x=0}^{x=1}(xe^x - x)dx = \int_0^1 xe^x dx - \int_0^1 x dx = (xe^x - e^x)\Big|_0^1 - \frac{x^2}{2}\Big|_0^1 = [(e-e)+1] - \frac{1}{2} = \frac{1}{2}$.

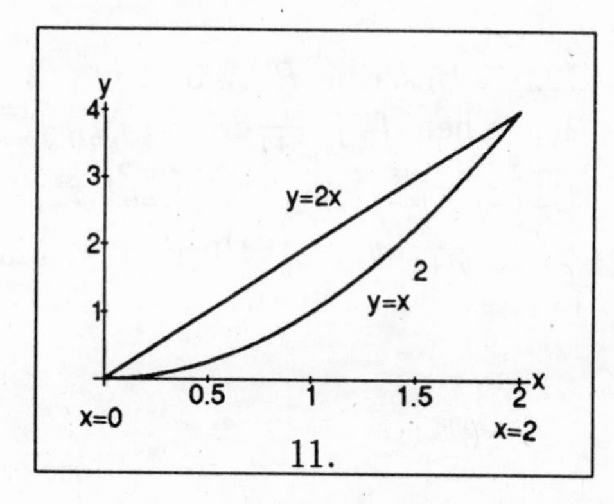

11.

11. Describe R by $0 \le x \le 2$ and $x^2 \le y \le 2x$. Then $\int_R\int(2y - x)dA = \int_{x=0}^{x=2}\int_{y=x^2}^{y=2x}(2y - x)dy dx = \int_{x=0}^{x=2}(y^2 - xy)\Big|_{y=x^2}^{y=2x} dx = \int_{x=0}^{x=2}[(4x^2 - 2x^2) - (x^4 - x^3)]dx = \int_0^2(2x^2 - x^4 + x^3)dx = (\frac{2x^3}{3} - \frac{x^5}{5} + \frac{x^4}{4})\Big|_0^2 = \frac{16}{3} - \frac{32}{5} + \frac{16}{4} = \frac{44}{15}$.

13. Describe R by $2 \le x \le 4$ and $x \le y \le \frac{16}{x}$. Then $\int_R \int 2dA = \int_{x=2}^{x=4} \int_{y=x}^{y=16/x} 2dydx = \int_{x=2}^{x=4} 2y|_{y=x}^{y=16/x} dx = \int_{x=2}^{x=4} (\frac{32}{x} - 2x)dx = (32 \ln|x| - x^2)|_2^4 = (32 \ln 4 - 16) - (32 \ln 2 - 4) = 32 \ln(\frac{4}{2}) - 12 = 32 \ln 2 - 12$.

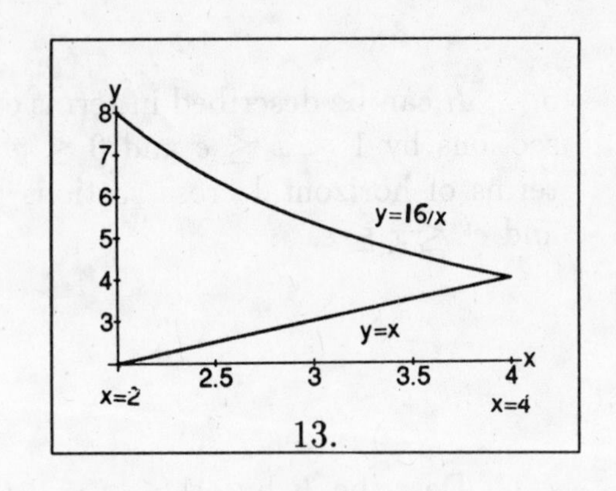

13.

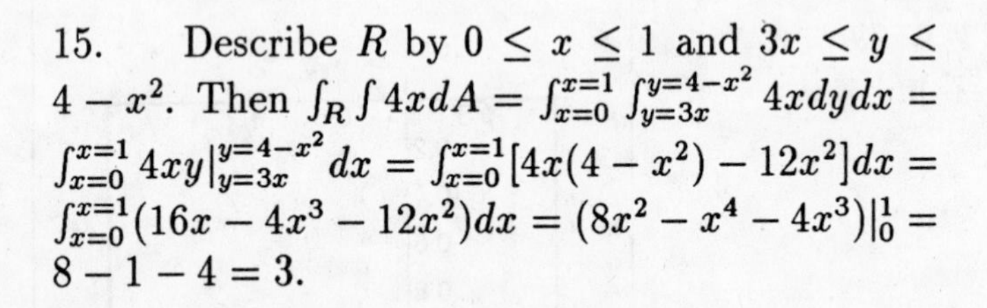

15. Describe R by $0 \le x \le 1$ and $3x \le y \le 4 - x^2$. Then $\int_R \int 4xdA = \int_{x=0}^{x=1} \int_{y=3x}^{y=4-x^2} 4xdydx = \int_{x=0}^{x=1} 4xy|_{y=3x}^{y=4-x^2} dx = \int_{x=0}^{x=1} [4x(4 - x^2) - 12x^2]dx = \int_{x=0}^{x=1} (16x - 4x^3 - 12x^2)dx = (8x^2 - x^4 - 4x^3)|_0^1 = 8 - 1 - 4 = 3$.

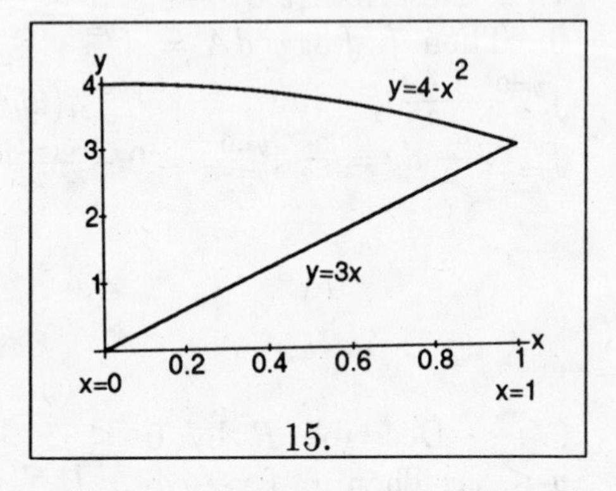

15.

17. Describe R by $0 \le y \le 1$ and $y - 1 \le x \le 1 - y$. Then $\int_R \int (2x + 1)dA = \int_{y=0}^{y=1} \int_{x=y-1}^{x=1-y} (2x + 1)dxdy = \int_{y=0}^{y=1} (x^2 + x)|_{x=1-y}^{x=y-1} dy = \int_{y=0}^{y=1} \{[(1 - y)^2 + (1 - y)] - [(y - 1)^2 + (y - 1)]\}dx = \int_{y=0}^{y=1} (2 - 2y)dx = (2y - y^2)|_0^1 = 2 - 1 = 1$.

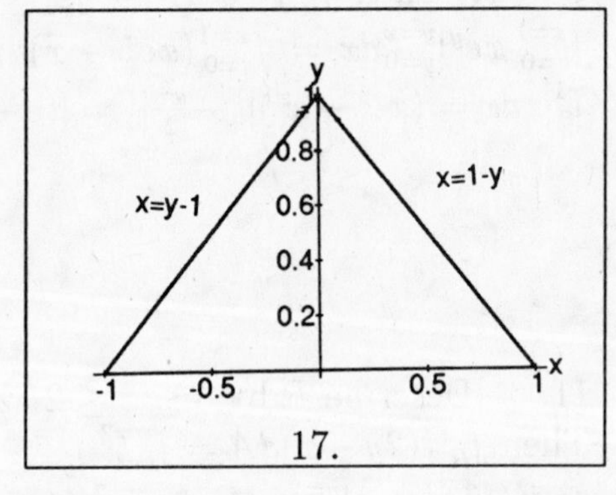

17.

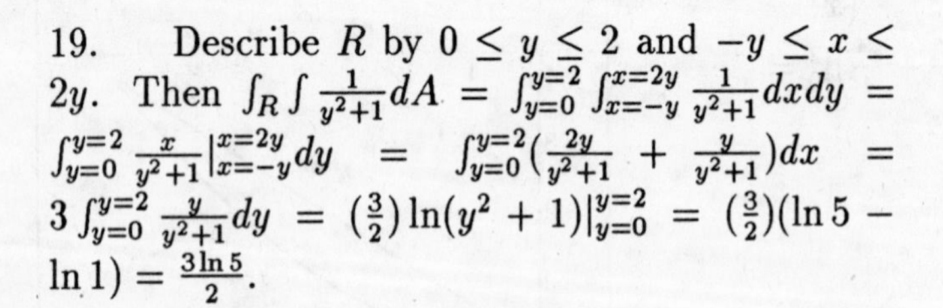

19. Describe R by $0 \le y \le 2$ and $-y \le x \le 2y$. Then $\int_R \int \frac{1}{y^2+1} dA = \int_{y=0}^{y=2} \int_{x=-y}^{x=2y} \frac{1}{y^2+1} dxdy = \int_{y=0}^{y=2} \frac{x}{y^2+1}|_{x=-y}^{x=2y} dy = \int_{y=0}^{y=2} (\frac{2y}{y^2+1} + \frac{y}{y^2+1})dx = 3 \int_{y=0}^{y=2} \frac{y}{y^2+1} dy = (\frac{3}{2}) \ln(y^2 + 1)|_{y=0}^{y=2} = (\frac{3}{2})(\ln 5 - \ln 1) = \frac{3 \ln 5}{2}$.

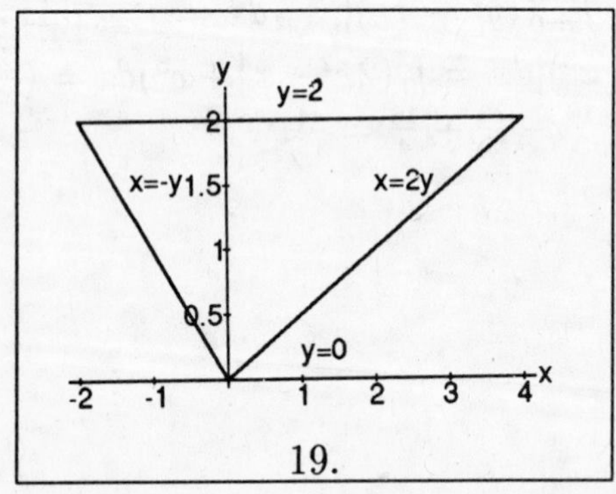

19.

21. Describe R by $0 \leq y \leq 1$ and $y \leq x \leq y^{1/3}$. Then $\int_R\int 12x^2e^{y^2}dA = \int_{y=0}^{y=1}\int_{x=y}^{x=y^{1/3}} 12x^2e^{y^2}dxdy = \int_{y=0}^{y=1} 4x^3e^{y^2}|_{x=y}^{x=y^{1/3}} dy = \int_{y=0}^{y=1}(4ye^{y^2} - 4y^3e^{y^2})dy = 4\int_{y=0}^{y=1} ye^{y^2}dy - 4\int_{y=0}^{y=1} y^2[ye^{y^2}]dy = 2e^{y^2}|_0^1 - 4(\frac{y^2e^{y^2}}{2} - \frac{e^{y^2}}{2})|_0^1 = 2(e-1) - 2[(e-e)+1] = 2e-4.$

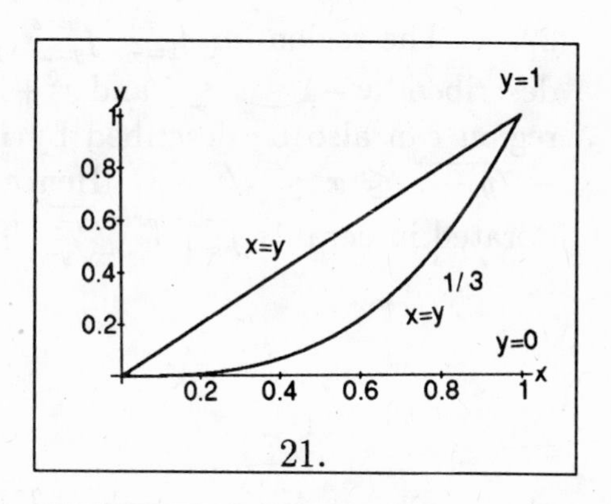

21.

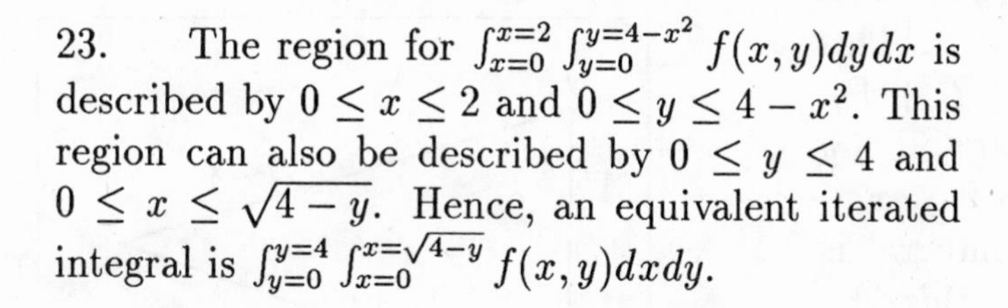
23. The region for $\int_{x=0}^{x=2}\int_{y=0}^{y=4-x^2} f(x,y)dydx$ is described by $0 \leq x \leq 2$ and $0 \leq y \leq 4-x^2$. This region can also be described by $0 \leq y \leq 4$ and $0 \leq x \leq \sqrt{4-y}$. Hence, an equivalent iterated integral is $\int_{y=0}^{y=4}\int_{x=0}^{x=\sqrt{4-y}} f(x,y)dxdy$.

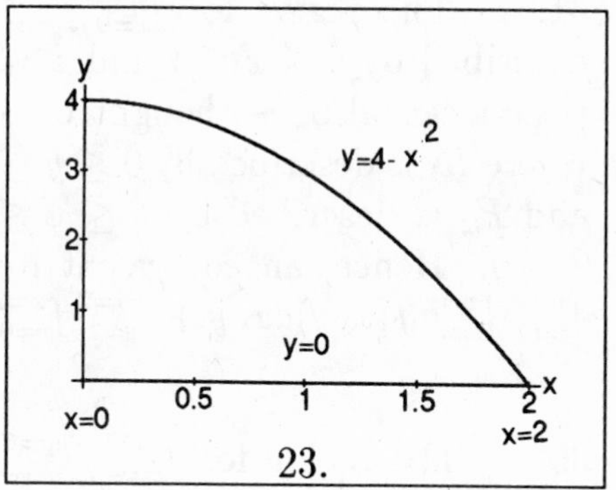

23.

25. The region for $\int_{x=0}^{x=1}\int_{y=x^3}^{y=\sqrt{x}} f(x,y)dydx$ is described by $0 \leq x \leq 1$ and $x^3 \leq y \leq \sqrt{x}$. This region can also be described by $0 \leq y \leq 1$ and $y^2 \leq x \leq y^{1/3}$. Hence, an equivalent iterated integral is $\int_{y=0}^{y=1}\int_{x=y^2}^{x=y^{1/3}} f(x,y)dxdy$

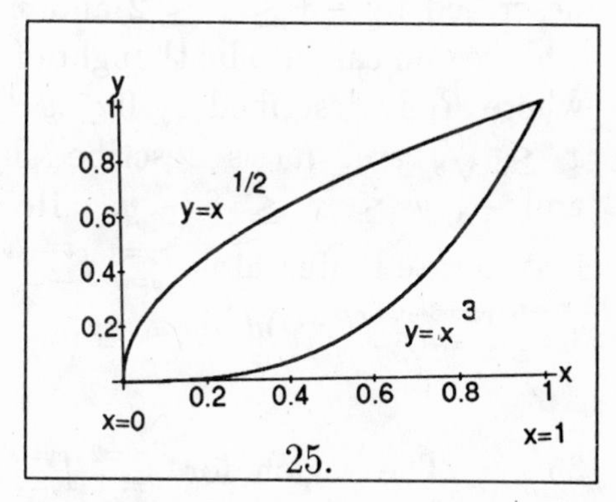

25.

27. The region for $\int_{x=1}^{x=e^2}\int_{y=\ln x}^{y=2} f(x,y)dydx$ is described by $1 \leq x \leq e^2$ and $\ln x \leq y \leq 2$. This region can also be described by $0 \leq y \leq 2$ and $1 \leq x \leq e^y$. Hence, an equivalent iterated integral is $\int_{y=0}^{y=2}\int_{x=1}^{x=e^y} f(x,y)dxdy$

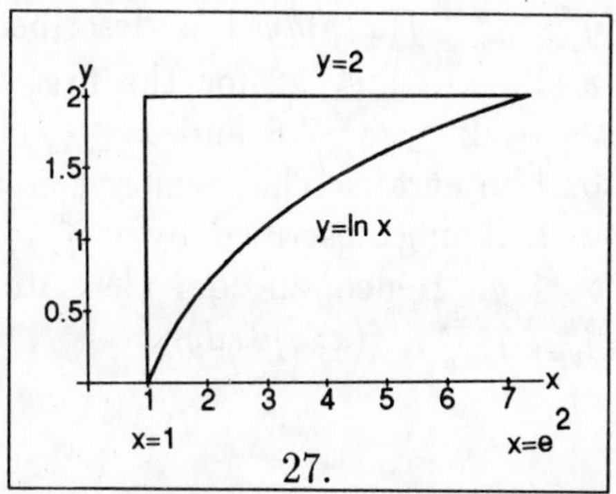

27.

29. The region for $\int_{x=-1}^{x=1}\int_{y=x^2+1}^{y=2} f(x,y)dydx$ is described by $-1 \le x \le 1$ and $x^2+1 \le y \le 2$. This region can also be described by $1 \le y \le 2$ and $-\sqrt{y-1} \le x \le \sqrt{y-1}$. Hence, an equivalent iterated integral is $\int_{y=1}^{y=2}\int_{x=-\sqrt{y-1}}^{x=\sqrt{y-1}} f(x,y)dxdy$

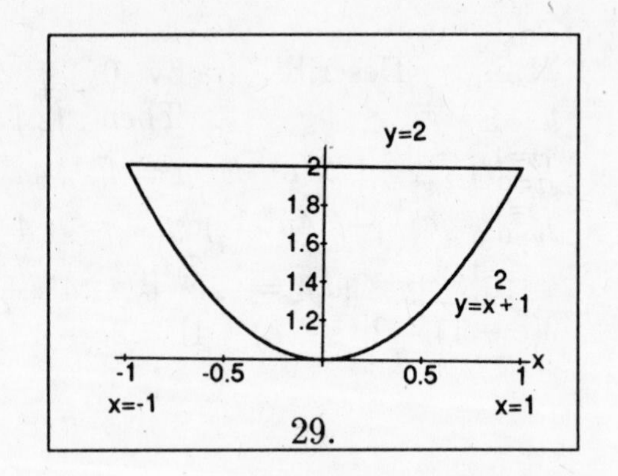

29.

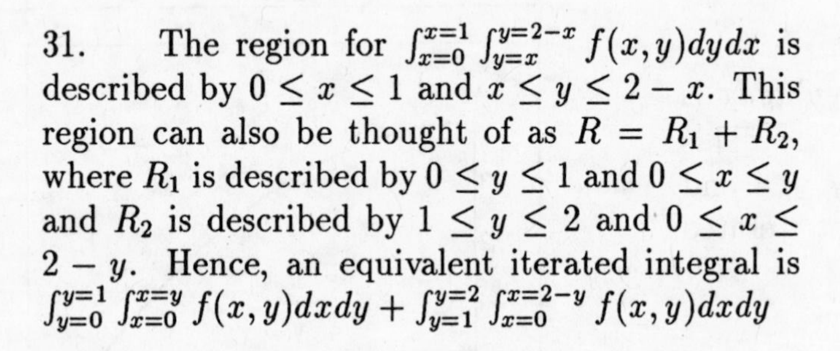

31. The region for $\int_{x=0}^{x=1}\int_{y=x}^{y=2-x} f(x,y)dydx$ is described by $0 \le x \le 1$ and $x \le y \le 2-x$. This region can also be thought of as $R = R_1 + R_2$, where R_1 is described by $0 \le y \le 1$ and $0 \le x \le y$ and R_2 is described by $1 \le y \le 2$ and $0 \le x \le 2-y$. Hence, an equivalent iterated integral is $\int_{y=0}^{y=1}\int_{x=0}^{x=y} f(x,y)dxdy + \int_{y=1}^{y=2}\int_{x=0}^{x=2-y} f(x,y)dxdy$

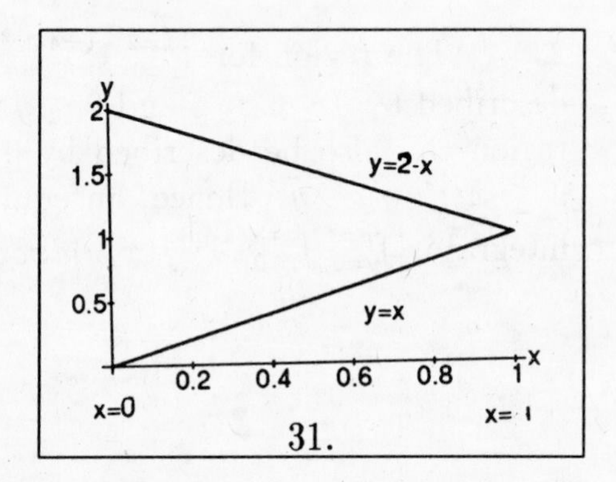

31.

33. The region for $\int_{x=-3}^{x=2}\int_{y=x^2}^{y=6-x} f(x,y)dydx$ is described by $-3 \le x \le 2$ and $x^2 \le y \le 6-x$. This region can also be thought of as $R = R_1 + R_2$, where R_1 is described by $0 \le y \le 4$ and $-\sqrt{y} \le x \le \sqrt{y}$ and R_2 is described by $4 \le y \le 9$ and $-\sqrt{y} \le x \le 6-y$. Hence, an equivalent iterated integral is $\int_{y=0}^{y=4}\int_{x=-\sqrt{x}}^{x=\sqrt{y}} f(x,y)dxdy + \int_{y=4}^{y=9}\int_{x=-\sqrt{y}}^{x=6-y} f(x,y)dxdy$

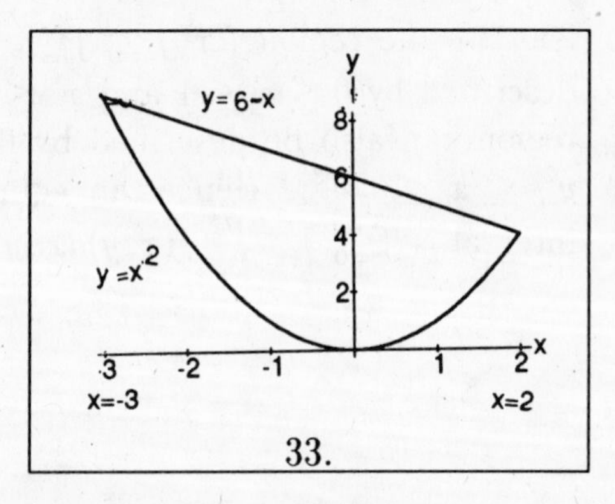

33.

35. The region for $\int_{x=1}^{x=2}\int_{y=x}^{y=x^3} f(x,y)dydx + \int_{x=2}^{x=8}\int_{y=x}^{y=8} f(x,y)dydx$ is described by $1 \le x \le 2$ and $x \le y \le x^3$ for the first integral, as well as by $2 \le x \le 8$ and $x \le y \le 8$ for the second integral. This region can also be thought of as being described by $1 \le y \le 8$ and $\sqrt[3]{y} \le x \le y$. Hence, an equivalent iterated integral is $\int_{y=1}^{y=8}\int_{x=y^{1/3}}^{x=y} f(x,y)dxdy$.

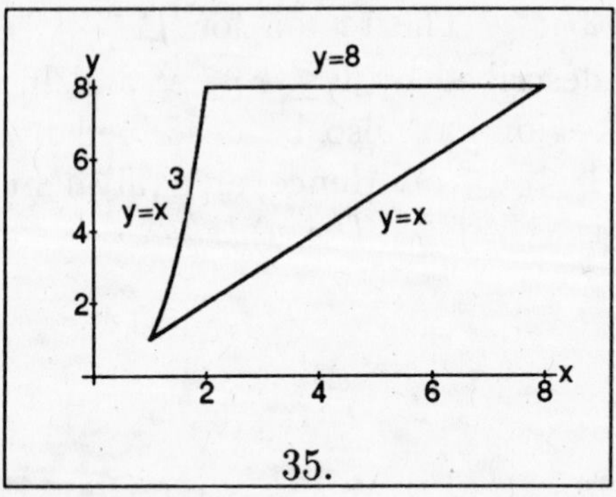

35.

10.3 Applications of Double Integrals

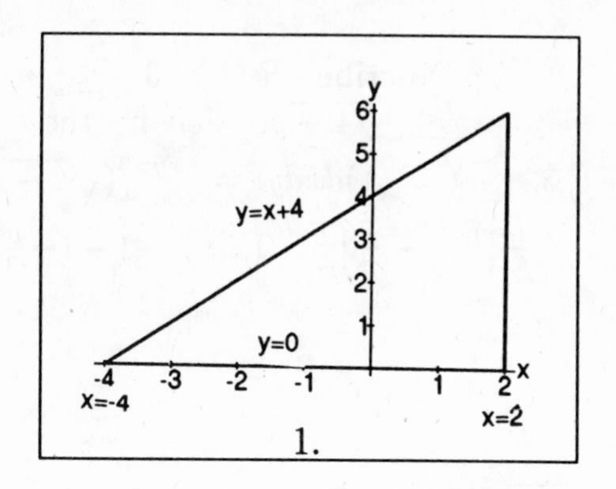

1.

1. Describe R by $-4 \le x \le 2$ and $0 \le y \le x+4$. Then, the area of R is $\int_{x=-4}^{x=2}\int_{y=0}^{y=x+4}(1)dydx = \int_{x=-4}^{x=2} y|_{y=0}^{y=x+4}dx = \int_{x=-4}^{x=2}(x+4)dx = (\frac{x^2}{2}+4x)|_{-4}^{2} = (2+8)-(8-16) = 18$.

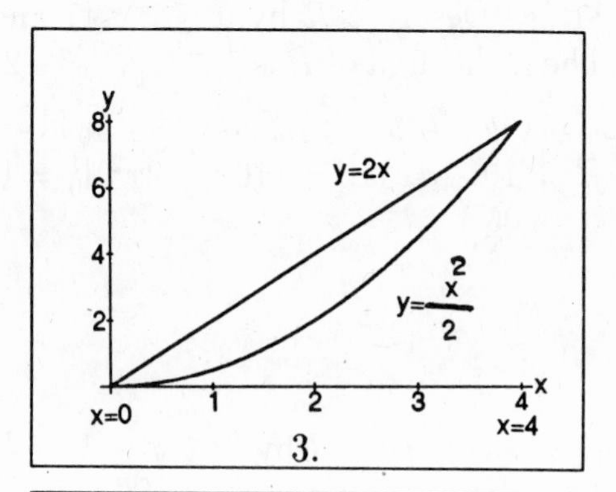

3.

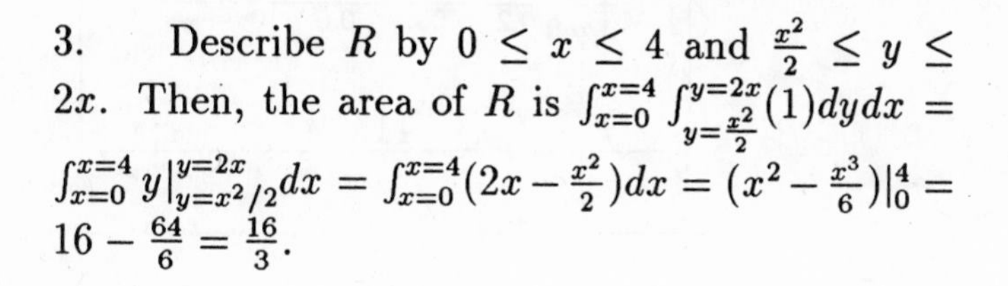

3. Describe R by $0 \le x \le 4$ and $\frac{x^2}{2} \le y \le 2x$. Then, the area of R is $\int_{x=0}^{x=4}\int_{y=\frac{x^2}{2}}^{y=2x}(1)dydx = \int_{x=0}^{x=4} y|_{y=x^2/2}^{y=2x}dx = \int_{x=0}^{x=4}(2x-\frac{x^2}{2})dx = (x^2-\frac{x^3}{6})|_0^4 = 16-\frac{64}{6} = \frac{16}{3}$.

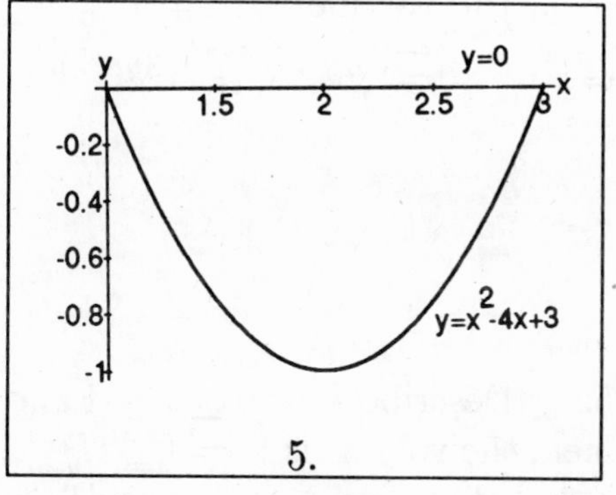

5.

5. Describe R by $1 \le x \le 3$ and $x^2-4x+3 \le y \le 0$. Then, the area of R is $\int_{x=1}^{x=3}\int_{y=x^2-4x+3}^{y=0}(1)dydx = \int_{x=1}^{x=3} y|_{y=x^2-4x+3}^{y=0}dx = \int_{x=1}^{x=3}(-x^2+4x-3)dx = (-\frac{x^3}{3}+2x^2-3x)|_1^3 = (-9+18-9)-(-\frac{1}{3}+2-3) = \frac{4}{3}$.

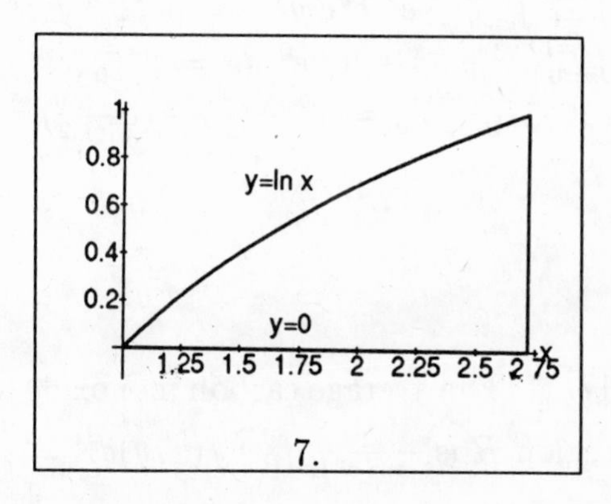

7.

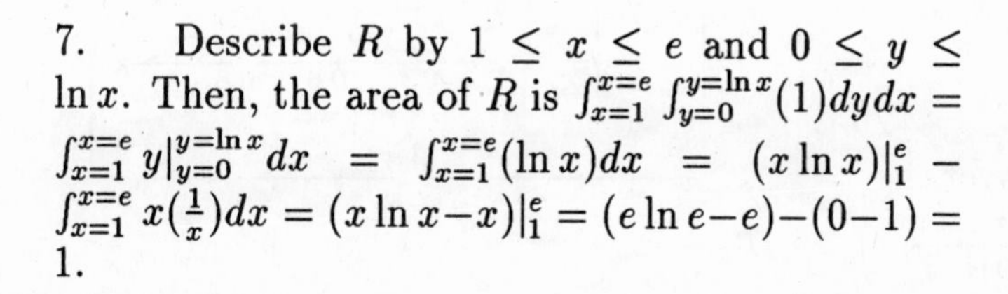

7. Describe R by $1 \le x \le e$ and $0 \le y \le \ln x$. Then, the area of R is $\int_{x=1}^{x=e}\int_{y=0}^{y=\ln x}(1)dydx = \int_{x=1}^{x=e} y|_{y=0}^{y=\ln x}dx = \int_{x=1}^{x=e}(\ln x)dx = (x\ln x)|_1^e - \int_{x=1}^{x=e} x(\frac{1}{x})dx = (x\ln x - x)|_1^e = (e\ln e - e)-(0-1) = 1$.

9. Describe R by $0 \le y \le 3$ and $\frac{y}{3} \le x \le \sqrt{4-y}$. Then, the area of R is $\int_{y=0}^{y=3}\int_{x=\frac{y}{3}}^{x=\sqrt{4-y}}(1)dxdy = \int_{y=0}^{y=3}(\sqrt{4-y}-\frac{y}{3})dy = [-\frac{2(4-y)^{3/2}}{3}-\frac{y^2}{6}]|_0^3 = (-\frac{2}{3}-\frac{9}{6})-(-\frac{16}{3}-0) = \frac{19}{6}$.

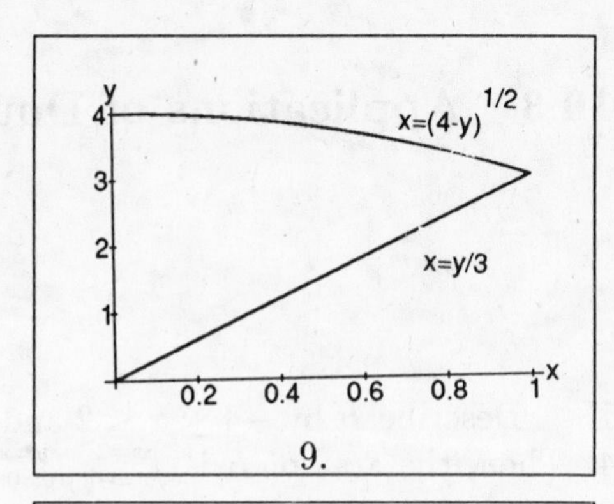

9.

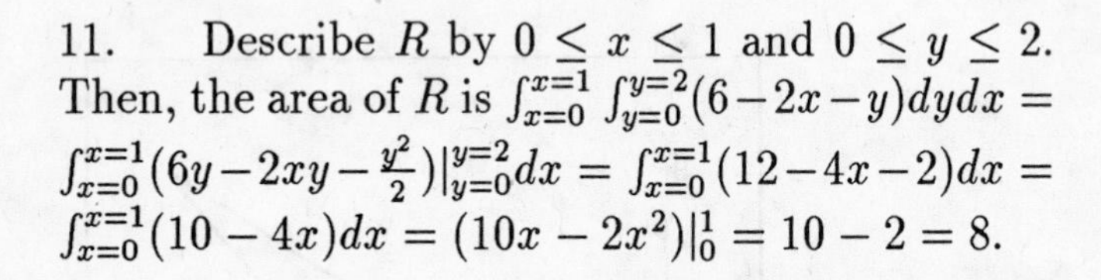
11. Describe R by $0 \le x \le 1$ and $0 \le y \le 2$. Then, the area of R is $\int_{x=0}^{x=1}\int_{y=0}^{y=2}(6-2x-y)dydx = \int_{x=0}^{x=1}(6y-2xy-\frac{y^2}{2})|_{y=0}^{y=2}dx = \int_{x=0}^{x=1}(12-4x-2)dx = \int_{x=0}^{x=1}(10-4x)dx = (10x-2x^2)|_0^1 = 10-2 = 8$.

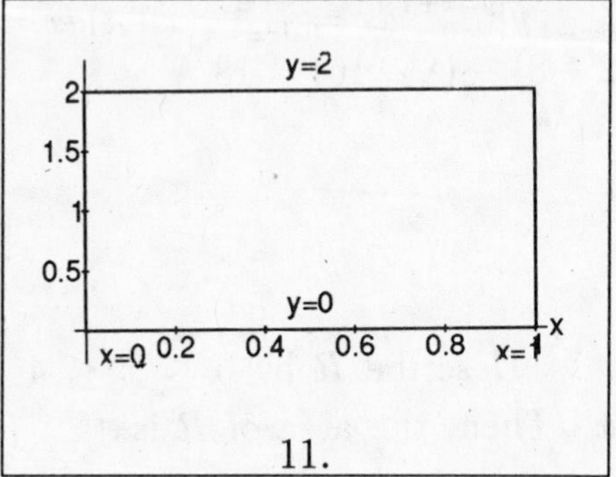

11.

13. Describe R by $0 \le y \le 1$ and $0 \le x \le 2y$. Then, the volume is $V = \int_{y=0}^{y=1}\int_{x=0}^{x=2y}(e^{y^2})dxdy = \int_{y=0}^{y=1}(xe^{y^2})|_{x=0}^{x=2y}dy = \int_{y=0}^{y=1}(2ye^{y^2})dy = e^{y^2}|_0^1 = e - 1$.

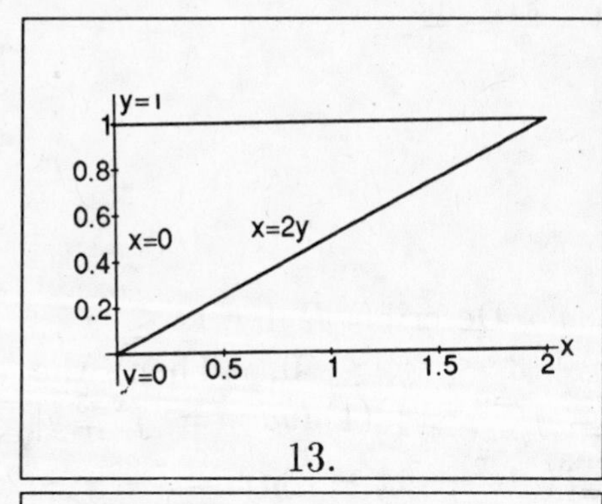

13.

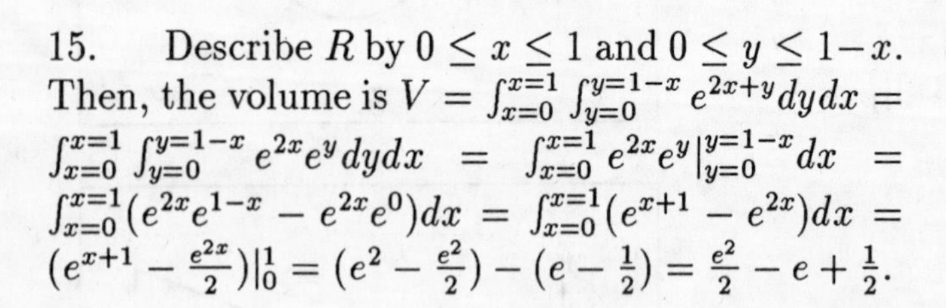
15. Describe R by $0 \le x \le 1$ and $0 \le y \le 1-x$. Then, the volume is $V = \int_{x=0}^{x=1}\int_{y=0}^{y=1-x} e^{2x+y}dydx = \int_{x=0}^{x=1}\int_{y=0}^{y=1-x} e^{2x}e^{y}dydx = \int_{x=0}^{x=1} e^{2x}e^{y}|_{y=0}^{y=1-x}dx = \int_{x=0}^{x=1}(e^{2x}e^{1-x}-e^{2x}e^{0})dx = \int_{x=0}^{x=1}(e^{x+1}-e^{2x})dx = (e^{x+1}-\frac{e^{2x}}{2})|_0^1 = (e^2-\frac{e^2}{2})-(e-\frac{1}{2}) = \frac{e^2}{2}-e+\frac{1}{2}$.

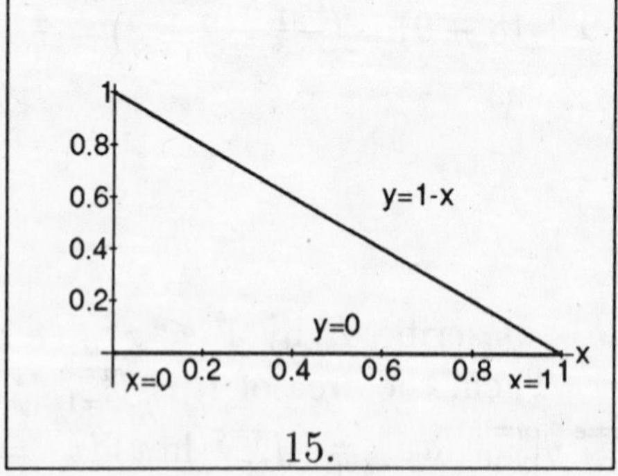

15.

17. The average carbon monoxide level over the region R is $\frac{1}{\text{area of R}}\int_R\int f(x,y)dA = \frac{\int_R\int f(x,y)dA}{\int_R\int(1)dA}$.

19. Describe R by $0 \le x \le 1$ and $0 \le y \le x$. The area is $\frac{1}{2}$(base)(height)$= \frac{1}{2}(1)(1) = \frac{1}{2}$. Then, the average value is $\frac{1}{\text{area of R}} \int_R \int f(x,y)dA = \int_{x=0}^{x=1} \int_{y=0}^{y=x} 400xe^{-y}dydx = 2\int_{x=0}^{x=1}(-400xe^{-y})|_{y=0}^{y=x}dx = 2\int_{x=0}^{x=1}(-400xe^{-x} + 400x)dx = -800\int_{x=0}^{x=1}(xe^{-x} - e^{-2x})dx + 800\int_{x=0}^{x=1} xdx = -800(-xe^{-x} - e^{-x})|_0^1 + 800(\frac{x^2}{2})|_0^1 = -800[(-e^{-1} - e^{-1}) + 1] + \frac{800}{2}) = -800(1 - 2e^{-1}) + 400 = \$188.61$ per acre.

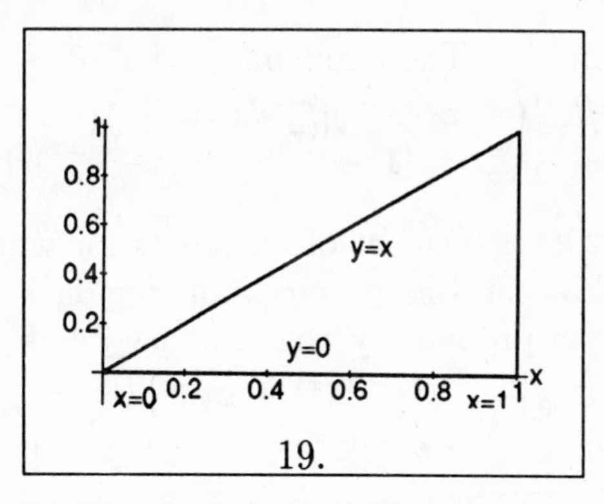

19.

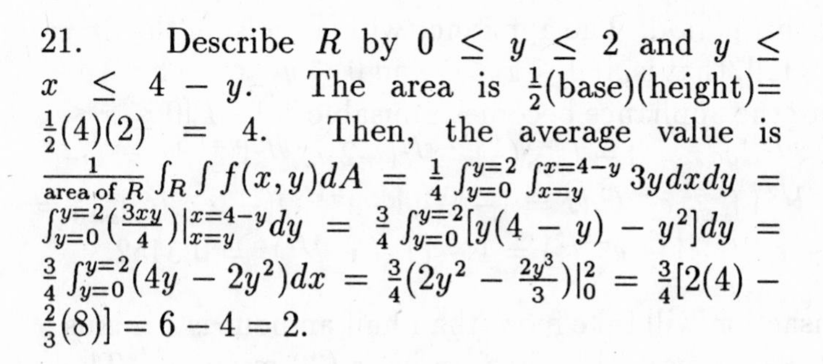

21. Describe R by $0 \le y \le 2$ and $y \le x \le 4 - y$. The area is $\frac{1}{2}$(base)(height)$= \frac{1}{2}(4)(2) = 4$. Then, the average value is $\frac{1}{\text{area of } R} \int_R \int f(x,y)dA = \frac{1}{4}\int_{y=0}^{y=2} \int_{x=y}^{x=4-y} 3ydxdy = \int_{y=0}^{y=2}(\frac{3xy}{4})|_{x=y}^{x=4-y}dy = \frac{3}{4}\int_{y=0}^{y=2}[y(4-y) - y^2]dy = \frac{3}{4}\int_{y=0}^{y=2}(4y - 2y^2)dx = \frac{3}{4}(2y^2 - \frac{2y^3}{3})|_0^2 = \frac{3}{4}[2(4) - \frac{2}{3}(8)] = 6 - 4 = 2.$

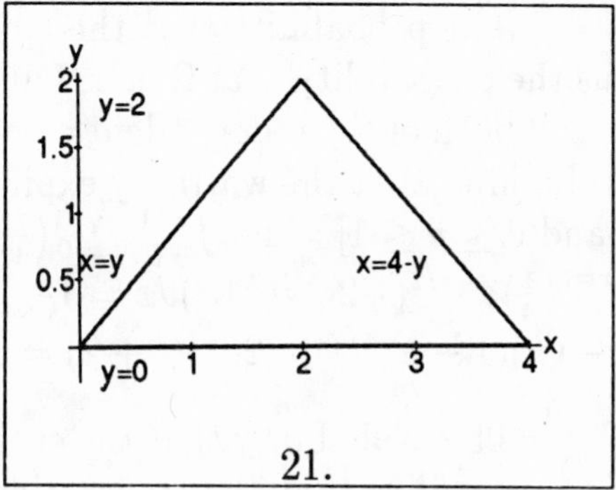

21.

23. Describe R by $-2 \le x \le 2$ and $0 \le y \le 4 - x^2$. Then, $\int_R \int f(x,y)dA = \int_{x=-2}^{x=2} \int_{y=0}^{y=4-x^2} xdydx = \int_{x=-2}^{x=2} xy|_{y=0}^{y=4-x^2} dx = \int_{x=-2}^{x=2} x(4 - x^2)dx = \int_{x=-2}^{x=2}(4x - x^3)dx = (2x^2 - \frac{x^4}{4})|_{x=-2}^{x=2} = (8 - \frac{16}{4}) - (8 - \frac{16}{4}) = 0$. Hence, the average value is $\frac{1}{\text{area of } R} \int_R \int f(x,y)dA = 0$.

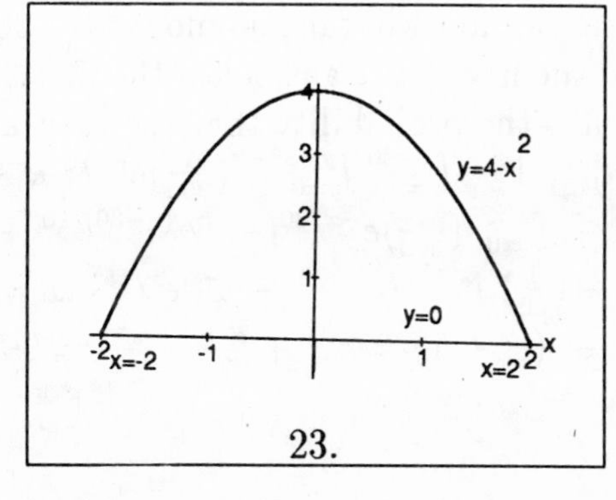

23.

25. Describe R by $0 \le x \le 1$ and $x^2 \le y \le 1$. Then, the area of R is $\int_R \int(1)dA = \int_{x=0}^{x=1} \int_{y=x^2}^{y=1} dydx = \int_{x=0}^{x=1} y|_{y=x^2}^{y=1}dx = \int_{x=0}^{x=1}(1 - x^2)dx = (x - \frac{x^3}{3})|_{x=0}^{x=1} = 1 - \frac{1}{3} = \frac{2}{3}$ and $\int_R \int f(x,y)dA = \int_{x=0}^{x=1} \int_{y=x^2}^{y=1} e^x y^{-1/2}dydx = \int_{x=0}^{x=1} 2e^x y^{1/2}|_{y=x^2}^{y=1}dx = \int_{x=0}^{x=1}(2e^x - 2xe^x)dx = 2\int_{x=0}^{x=1} e^x dx - 2\int_{x=0}^{x=1} xe^x dx = 2e^x|_0^1 - 2(xe^x - e^x)|_0^1 = (4e^x - 2xe^x)|_0^1 = (4e - 2e) - 4 = 2e - 4$. Hence, the average value is $\frac{1}{\text{area of } R} \int_R \int f(x,y)dA = \frac{3(2e-4)}{2} = 3(e-2)$.

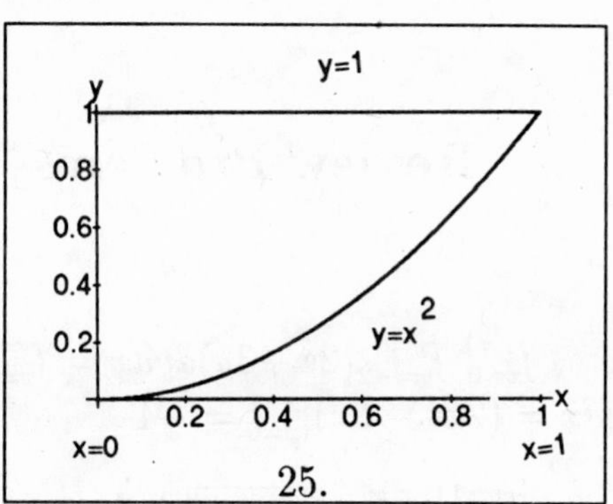

25.

27. The probability that $0 \le x \le 1$ and $0 \le y \le 2$ is $\int_{x=0}^{x=1}\int_{y=0}^{y=2} xe^{-x}e^{-y}dydx = \int_{x=0}^{x=1}(-xe^{-x}e^{-y})|_{y=0}^{y=2}dx = \int_{x=0}^{x=1}(-xe^{-x}e^{-2}+xe^{-x})dx = (1-e^{-2})\int_{x=0}^{x=1} xe^{-x}dx = (1-e^{-2})(-xe^{-x}-e^{-x})|_{x=0}^{x=1} = (1-e^{-2})[(-e^{-1}-e^{-1})+1] = (1-e^{-2})(1-2e^{-1}) = 0.2285.$

29. The set of all points for which $x+y \le 1$ is the region R in the first quadrant below the line $x+y$. This region is described by $0 \le x \le 1$ and $0 \le y \le 1-x$. Hence, the probability that $x+y \le 1$ is $\int_{x=0}^{x=1}\int_{y=0}^{y=1-x} xe^{-x}e^{-y}dydx = \int_{x=0}^{x=1}(-xe^{-x}e^{-y})|_{y=0}^{y=1-x}dx = \int_{x=0}^{x=1}[-xe^{-x}e^{-(1-x)} + xe^{-x}]dx = \int_{x=0}^{x=1}(xe^{-x} - xe^{-1})dx = \int_{x=0}^{x=1} xe^{-x}dx - e^{-1}\int_{x=0}^{x=1} xdx = (-xe^{-x}-e^{-x})|_{x=0}^{x=1} - e^{-1}\frac{x^2}{2}|_{x=0}^{x=1} = [(-e^{-1}-e^{-1})+1] - e^{-1}(\frac{1}{2}) = -\frac{5}{2}e^{-1}+1 = 0.0803.$

31. The probability that the appliance does not fail during the first year is equal to 1 minus the probability that it does fail during this period. The appliance will fail during the first year if both of the independent components fail, that is, if $0 \le x \le 1$ and $0 \le y \le 1$. Thus the probability that the warranty expires before the appliance becomes unusable is $1-P[0 \le x \le 1$ and $0 \le y \le 1] = 1-\int_{x=0}^{x=1}\int_{y=0}^{y=1}(\frac{1}{4})e^{-x/2}e^{-y/2}dydx = 1-\int_{x=0}^{x=1}(\frac{1}{4})e^{-x/2}(-2)e^{-y/2}|_{y=0}^{y=1}dx = 1-\int_{x=0}^{x=1}(\frac{1}{4})e^{-x/2}(-2e^{-1/2}+2)dx = 1-\frac{1}{4}(2-2e^{-1/2})\int_{x=0}^{x=1} e^{-x/2}dx = 1-(\frac{1}{2})(1-e^{-1/2})(-2e^{-x/2})|_{x=0}^{x=1} = 1-(\frac{1}{2})(1-e^{-1/2})(-2e^{-1/2}+2) = 1-(1-e^{-1/2})(1-e^{-1/2}) = 1-(1-e^{-1/2})^2 = 0.8452.$

33. The probability that the entire transaction will take more than half an hour is 1 minus the probability that the transaction will take no more than half an hour (30 minutes). The transaction will take no more than 30 minutes if $x+y \le 30$, that is, if (x,y) is in the region R in the first quadrant below the line $x+y=30$. This region is $0 \le x \le 30$ and $0 \le y \le 30-x$. Thus the probability that the transaction will take more than half an hour is $1-P[(x,y)$ in $R] = 1-\int_{x=0}^{x=30}\int_{y=0}^{y=30-x}(\frac{1}{300})e^{-x/30}e^{-y/10}dydx = 1-\int_{x=0}^{x=30}(\frac{1}{300}e^{-x/30})(-10e^{-y/10})|_{y=0}^{y=30-x}dx = 1-\int_{x=0}^{x=30}(\frac{1}{300})e^{-x/30}(-10e^{(x-30)/10}+10)dx = 1-(\frac{1}{300})\int_{x=0}^{x=30} e^{-x/30}(10-10e^{-3}e^{x/10})dx = 1-(\frac{1}{30})\int_{x=0}^{x=30}(e^{-x/30}-e^{-3}e^{x/15})dx = 1-(-e^{-x/30})|_{x=0}^{x=30}+(\frac{e^{-3}}{2})e^{x/15}|_{x=0}^{x=30} = 1-(-e^{-1}+1)+(\frac{e^{-3}}{2})(e^2-1) = e^{-1}+\frac{e^{-1}}{2}-\frac{e^{-3}}{2} = 0.5269.$

Review Problems

1. $\int_{x=0}^{x=1}\int_{y=-2}^{y=0}(2x+3y)dydx = \int_{x=0}^{x=1}[2xy+\frac{3y^2}{2}]|_{y=-2}^{y=0}dx = \int_{x=0}^{x=1}[0-(-4x+6)]dx = \int_{x=0}^{x=1}(4x-6)dx = (2x^2-6x)|_{x=0}^{x=1} = -4$

2. $\int_{y=0}^{y=1}\int_{x=0}^{x=y} x\sqrt{1-y^3}dxdy = \int_{y=0}^{y=1}\frac{x^2}{2}\sqrt{1-y^3}|_{x=0}^{x=y}dy = \frac{1}{2}\int_{y=0}^{y=1} y^2\sqrt{1-y^3}dy = (-\frac{1}{9})(1-y^3)^{3/2}|_{y=0}^{y=1} = \frac{1}{9}$

3. $\int_{x=0}^{x=1}\int_{y=-x}^{y=x}\frac{6xy^2}{x^5+1}dydx = \int_{x=0}^{x=1}\frac{2xy^3}{x^5+1}\Big|_{y=-x}^{y=x}dx = \int_{x=0}^{x=1}(\frac{2x}{x^5+1})[x^3-(-x^3)]dx = 4\int_{x=0}^{x=1}\frac{x^4}{x^5+1}dx = \frac{4}{5}\ln|x^5+1|\,|_{x=0}^{x=1} = (\frac{4}{5})(\ln 2 - \ln 1) = \frac{4}{5}\ln 2.$

4. $\int_{x=0}^{x=1}\int_{y=0}^{y=2}e^{-x-y}dydx = \int_{x=0}^{x=1}\int_{y=0}^{y=2}e^{-x}e^{-y}dydx = \int_{x=0}^{x=1}(-e^{-x}e^{-y})|_{y=0}^{y=2}dx = \int_{x=0}^{x=1}(-e^{-x}e^{-2}+e^{-x})dx = (1-e^{-2})\int_{x=0}^{x=1}e^{-x}dx = (1-e^{-2})(-e^{-x})|_{x=0}^{x=1} = (1-e^{-2})(-e^{-1}+1) = 0.5466$

5. $\int_{x=0}^{x=1}\int_{y=0}^{y=x}xe^{2y}dydx = \int_{x=0}^{x=1}(\frac{1}{2})xe^{2y}|_{y=0}^{y=x}dx = \int_{x=0}^{x=1}[\frac{xe^{2x}}{2}-\frac{x}{2}]dx = \frac{1}{2}\int_{x=0}^{x=1}xe^{2x}dx - \frac{1}{2}\int_{x=0}^{x=1}xdx = (\frac{1}{2})[\frac{x}{2}e^{2x}|_{x=0}^{x=1} - \frac{1}{2}\int_{x=0}^{x=1}e^{2x}dx] - \frac{1}{2}\int_{x=0}^{x=1}xdx = (\frac{1}{2})[\frac{x}{2}e^{2x} - \frac{1}{4}e^{2x}]|_{x=0}^{x=1} - \frac{x^2}{4}|_{x=0}^{x=1}xdx = (\frac{1}{2})[\frac{e^2}{2} - \frac{e^2}{4} + \frac{1}{4}] - \frac{1}{4} = (\frac{1}{8})(e^2-1)$

6. Describe R by $-1 \le x \le 2$ and $0 \le y \le 3$. Then, the area of R is $\int_R\int(6x^2y)dA = \int_{x=-1}^{x=2}\int_{y=0}^{y=3}6x^2ydydx = \int_{x=-1}^{x=2}3x^2y^2|_{y=0}^{y=3}dx = \int_{x=-1}^{x=2}27x^2dx = 9x^3|_{x=-1}^{x=2} = 9(8+1) = 81.$

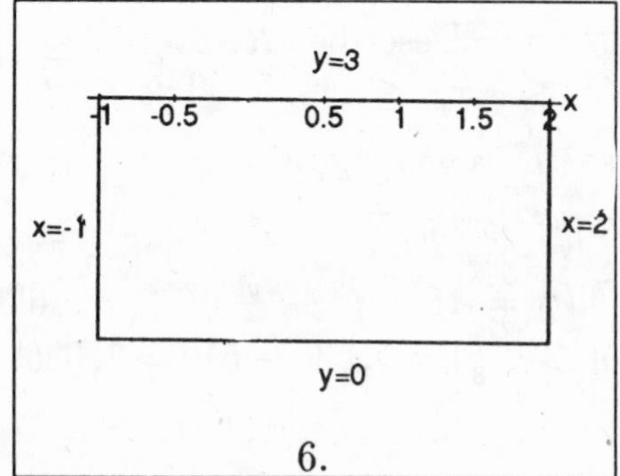

6.

7. Describe R by $0 \le x \le 1$ and $x \le y \le 3-2x$. Then, the area of R is $\int_R\int(x+2y)dA = \int_{x=0}^{x=1}\int_{y=x}^{y=3-2x}(x+2y)dydx = \int_{x=0}^{x=1}(xy+y^2)|_{y=x}^{y=3-2x}dx = \int_{x=0}^{x=1}[3x-2x^2+(3-2x)^2-(x^2+x^2)]dx = \int_{x=0}^{x=1}(9-9x)dx = (9x-\frac{9x^2}{2})|_0^1 = 9-\frac{9}{2} = \frac{9}{2}.$

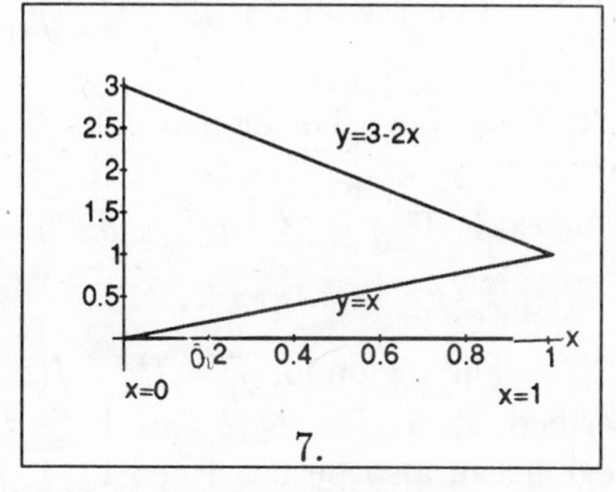

7.

8. Describe R by $0 \le x \le 4$ and $\frac{x}{2} \le y \le \sqrt{x}$. Then, the area of R is $\int_R\int 40x^2ydA = \int_{x=0}^{x=4}\int_{y=x/2}^{y=\sqrt{x}}40x^2ydydx = \int_{x=0}^{x=4}20x^2y^2|_{y=x/2}^{y=\sqrt{x}}dx = \int_{x=0}^{x=4}(20x^3-5x^4)dx = (5x^4-x^5)|_{x=0}^{x=4} = 256.$

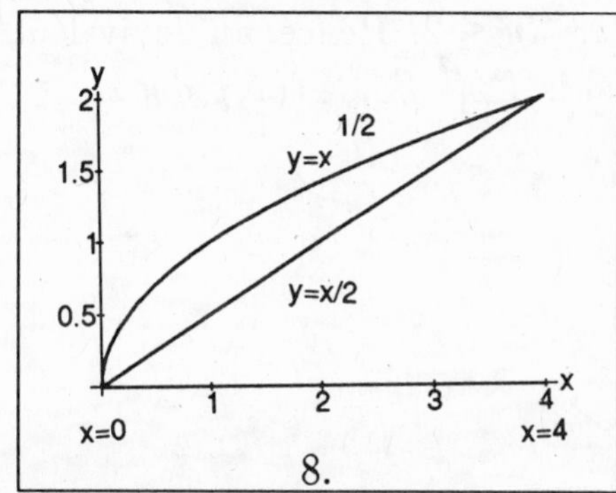

8.

9. Describe R by $0 \leq x \leq 1$ and $x \leq y \leq \sqrt[3]{x}$. Then, the area of R is $\int_R \int 6y^2 e^{x^2} dA = \int_{x=0}^{x=1} \int_{y=x}^{y=\sqrt[3]{x}} 6y^2 e^{x^2} dy dx = \int_{x=0}^{x=1} 2y^3 e^{x^2} |_{y=x}^{y=\sqrt[3]{x}} dx = \int_{x=0}^{x=1} (2xe^{x^2} - 2x^3 e^{x^2}) dx = \int_{x=0}^{x=1} 2xe^{x^2} dx - \int_{x=0}^{x=1} x^2 (2xe^{x^2}) dx = e^{x^2}|_{x=0}^{x=1} - [x^2 e^{x^2}|_{x=0}^{x=1} - \int_{x=0}^{x=1} 2xe^{x^2} dx] = (e^{x^2} - x^2 e^{x^2} + e^{x^2})|_{x=0}^{x=1} = (2 - x^2) e^{x^2}|_{x=0}^{x=1} = e - 2$.

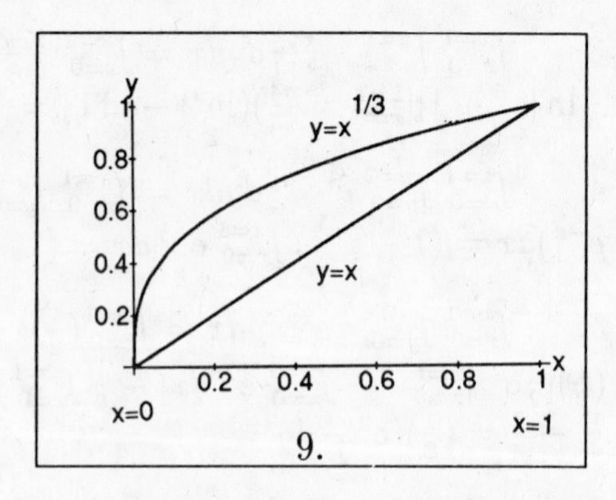

9.

10. Describe R by $1 \leq y \leq 2$ and $y^2 \leq x \leq \frac{8}{y}$. Then, $\int_R \int 32xy^3 dA = \int_{y=1}^{y=2} \int_{x=y^2}^{x=\frac{8}{y}} 32xy^3 dx dy = \int_{y=1}^{y=2} 16x^2 y^3 |_{x=y^2}^{x=8/y} dy = 16 \int_{y=1}^{y=2} [(\frac{64}{y^2})(y^3) - y^4(y^3)] dy = 16 \int_{y=1}^{y=2} (64y - y^7) dy = 16(32y^2 - \frac{y^8}{8})|_{y=1}^{y=2} = 16[32(4) - \frac{256}{8}] - 16(32 - \frac{1}{8}) = 1,536 - 510 = 1,026$.

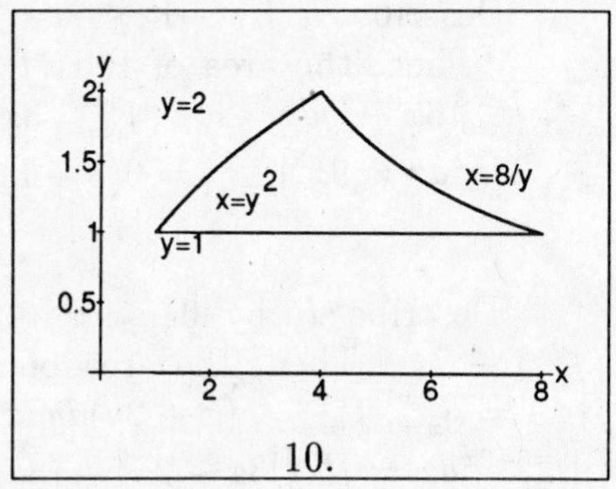

10.

11. The region for $\int_{x=0}^{x=4} \int_{y=x^2/8}^{y=\sqrt{x}} f(x,y) dy dx$ is described by $0 \leq x \leq 4$ and $\frac{x^2}{8} \leq y \leq \sqrt{x}$. This region can also be described by $0 \leq y \leq 2$ and $y^2 \leq x \leq \sqrt{8y}$. Hence, an equivalent iterated integral is $\int_{y=0}^{y=2} \int_{x=y^2}^{x=\sqrt{8y}} f(x,y) dx dy$.

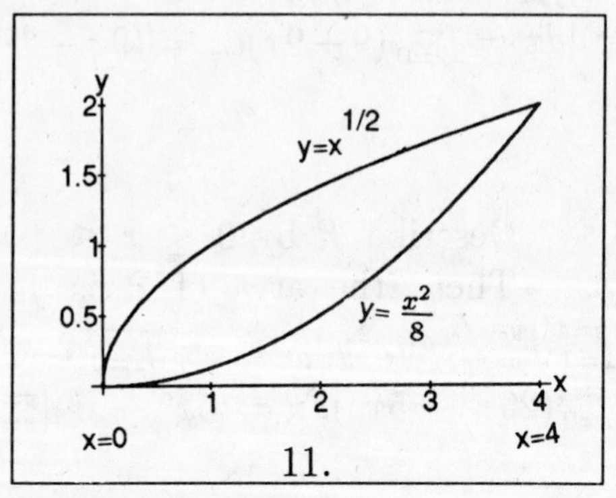

11.

12. The region for $\int_{y=0}^{y=2} \int_{x=1}^{x=e^y} f(x,y) dx dy$ is described by $0 \leq y \leq 2$ and $1 \leq x \leq e^y$. This region can also be described by $1 \leq x \leq e^2$ and $\ln x \leq y \leq 2$. Hence, an equivalent iterated integral is $\int_{x=1}^{x=e^2} \int_{y=\ln x}^{y=2} f(x,y) dy dx$.

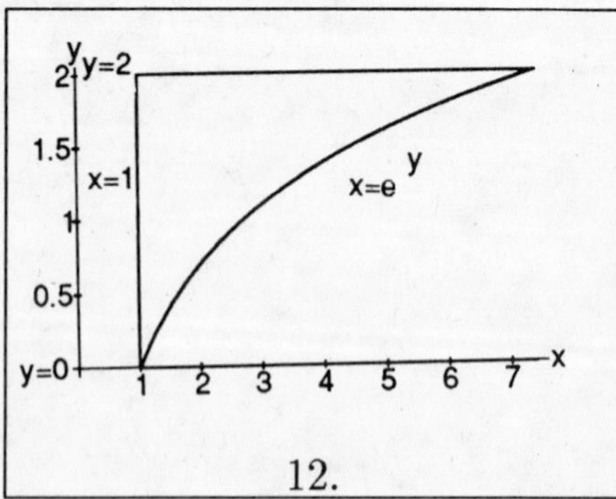

12.

13. The region for $\int_{x=0}^{x=2}\int_{y=x^2}^{y=8-2x} f(x,y)dydx$ is described by $0 \leq x \leq 2$ and $x^2 \leq y \leq 8-2x$. This region can also be described as $R = R_1 + R_2$, where R_1 is described by $0 \leq y \leq 4$ and $0 \leq x \leq \sqrt{y}$, while R_2 is described by $4 \leq y \leq 8$ and $0 \leq x \leq 4 - \frac{y}{2}$. Hence, an equivalent iterated integral is $\int_{y=0}^{y=4}\int_{x=0}^{x=\sqrt{y}} f(x,y)dxdy + \int_{y=4}^{y=8}\int_{x=0}^{x=4-y/2} f(x,y)dxdy$.

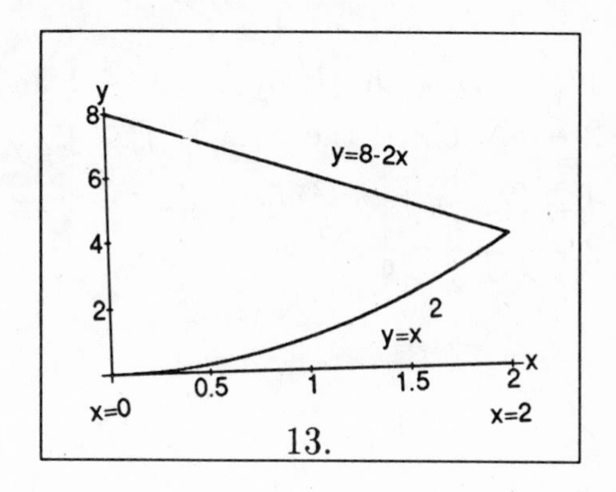

13.

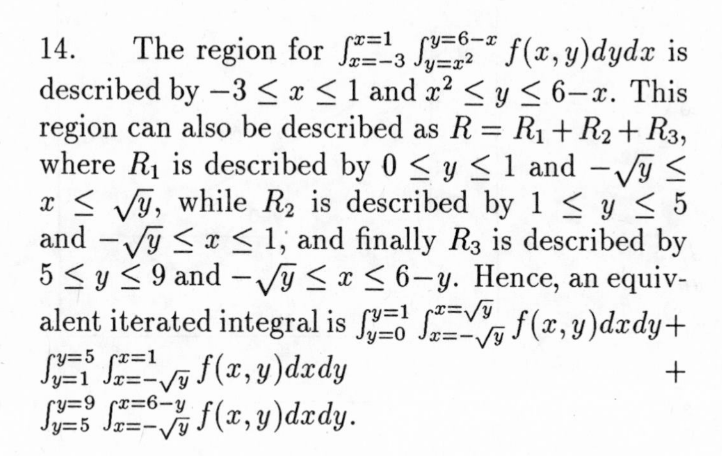

14. The region for $\int_{x=-3}^{x=1}\int_{y=x^2}^{y=6-x} f(x,y)dydx$ is described by $-3 \leq x \leq 1$ and $x^2 \leq y \leq 6-x$. This region can also be described as $R = R_1 + R_2 + R_3$, where R_1 is described by $0 \leq y \leq 1$ and $-\sqrt{y} \leq x \leq \sqrt{y}$, while R_2 is described by $1 \leq y \leq 5$ and $-\sqrt{y} \leq x \leq 1$, and finally R_3 is described by $5 \leq y \leq 9$ and $-\sqrt{y} \leq x \leq 6-y$. Hence, an equivalent iterated integral is $\int_{y=0}^{y=1}\int_{x=-\sqrt{y}}^{x=\sqrt{y}} f(x,y)dxdy + \int_{y=1}^{y=5}\int_{x=-\sqrt{y}}^{x=1} f(x,y)dxdy + \int_{y=5}^{y=9}\int_{x=-\sqrt{y}}^{x=6-y} f(x,y)dxdy$.

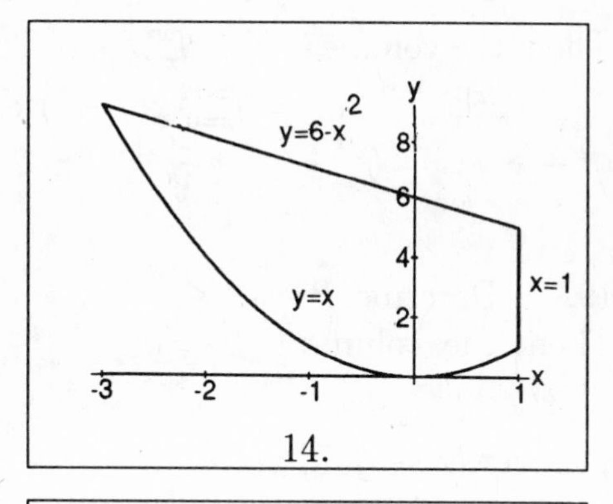

14.

15. Describe R by $-1 \leq x \leq 2$ and $x^2 - 4 \leq y \leq x - 2$. Then, the area of R is $\int_{x=-1}^{x=2}\int_{y=x^2-4}^{y=x-2}(1)dydx = \int_{x=-1}^{x=2} y|_{y=x^2-4}^{y=x-2}dx = \int_{x=-1}^{x=2}(x-2-x^2+4)dx = \int_{x=-1}^{x=2}(x+2-x^2)dx = (\frac{x^2}{2}+2x-\frac{x^3}{3})|_{x=-1}^{x=2} = (2+4-\frac{8}{3})-(\frac{1}{2}-2+\frac{1}{3}) = \frac{9}{2}$.

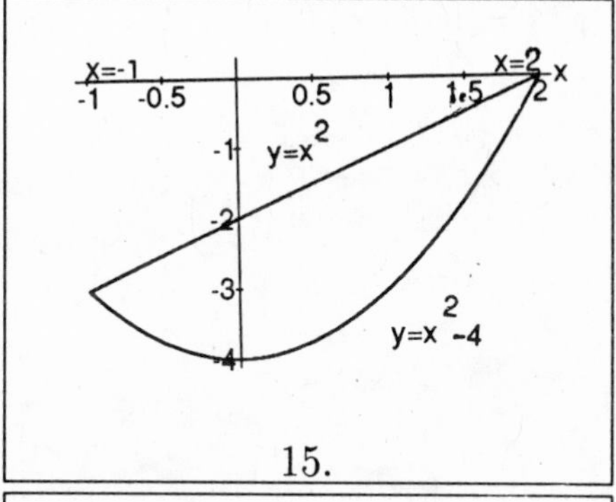

15.

16. Describe R by $0 \leq y \leq 1$ and $0 \leq x \leq e^y$. Then, the area is $\int_{y=0}^{y=1}\int_{x=0}^{x=e^y}(1)dxdy = \int_{y=0}^{y=1} x|_{x=0}^{x=e^y}dy = \int_{y=0}^{y=1} e^y dy = e^y|_{y=0}^{y=1} = e-1$.

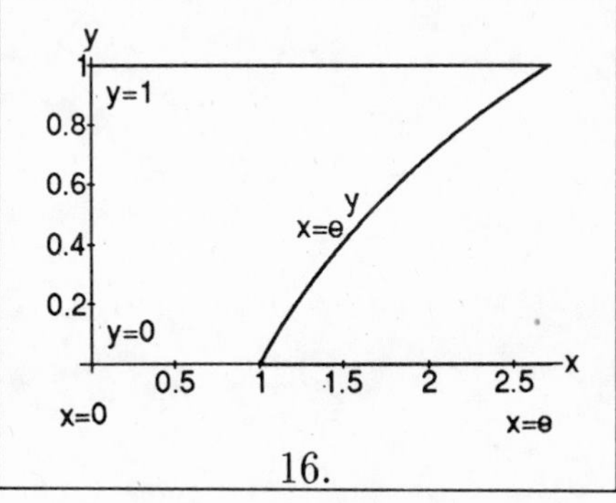

16.

17. Describe R by $0 \leq y \leq 1$ and $\sqrt{y} \leq x \leq 2 - y$. Then, the area is $\int_{y=0}^{y=1} \int_{x=\sqrt{y}}^{x=2-y} (1)dxdy = \int_{y=0}^{y=1}(2 - y - y^{1/2})dy = (2y - \frac{y^2}{2} - \frac{2y^{3/2}}{3})|_{y=0}^{y=1} = 2 - \frac{1}{2} - \frac{2}{3} = \frac{5}{6}$.

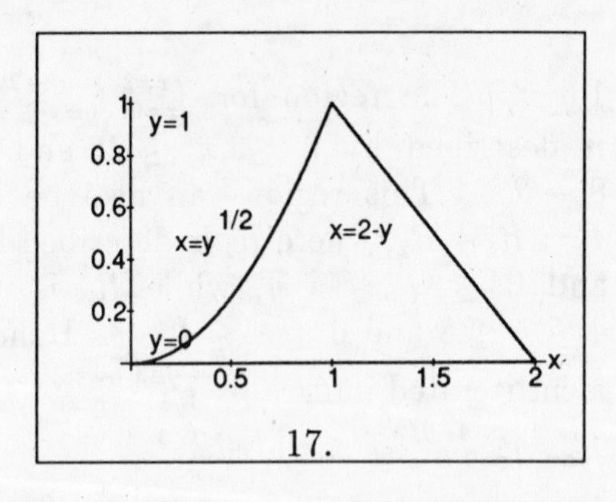

17.

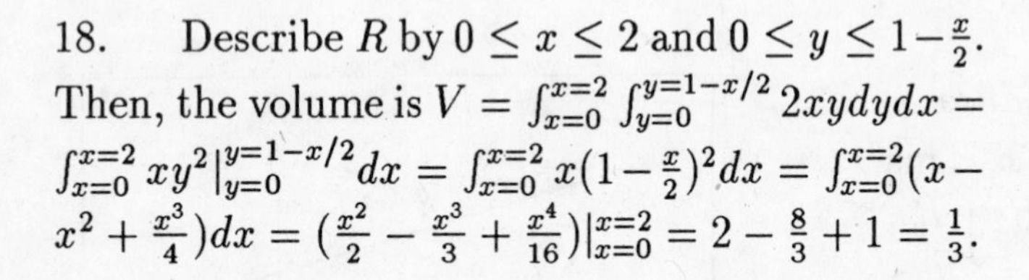

18. Describe R by $0 \leq x \leq 2$ and $0 \leq y \leq 1 - \frac{x}{2}$. Then, the volume is $V = \int_{x=0}^{x=2} \int_{y=0}^{y=1-x/2} 2xydydx = \int_{x=0}^{x=2} xy^2|_{y=0}^{y=1-x/2}dx = \int_{x=0}^{x=2} x(1 - \frac{x}{2})^2 dx = \int_{x=0}^{x=2}(x - x^2 + \frac{x^3}{4})dx = (\frac{x^2}{2} - \frac{x^3}{3} + \frac{x^4}{16})|_{x=0}^{x=2} = 2 - \frac{8}{3} + 1 = \frac{1}{3}$.

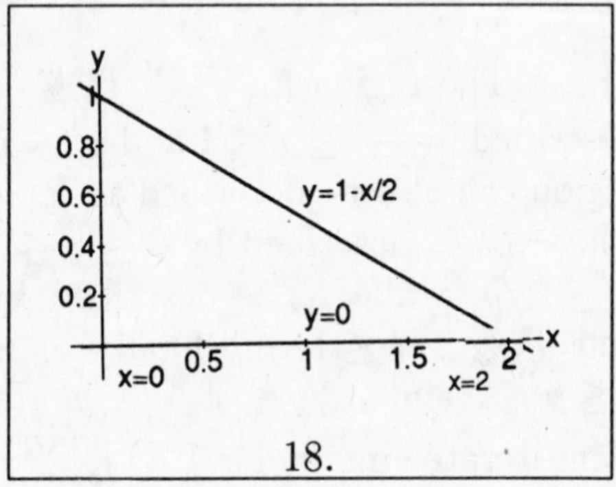

18.

19. Describe R by $0 \leq x \leq 1$ and $x^2 \leq y \leq x$. Then, the volume is $V = \int_{x=0}^{x=1} \int_{y=x^2}^{y=x} xe^y dydx = \int_{x=0}^{x=1}(xe^y)|_{y=x^2}^{y=x}dx = \int_{x=0}^{x=1}(xe^x - xe^{x^2})dx = \int_{x=0}^{x=1} xe^x dx - \int_{x=0}^{x=1} xe^{x^2}dx = (xe^x - e^x)|_{x=0}^{x=1} - \frac{e^{x^2}}{2}|_{x=0}^{x=1} = [(e - e) + 1] - (\frac{e}{2} - \frac{1}{2}) = \frac{3}{2} - \frac{e}{2} = \frac{3-e}{2}$.

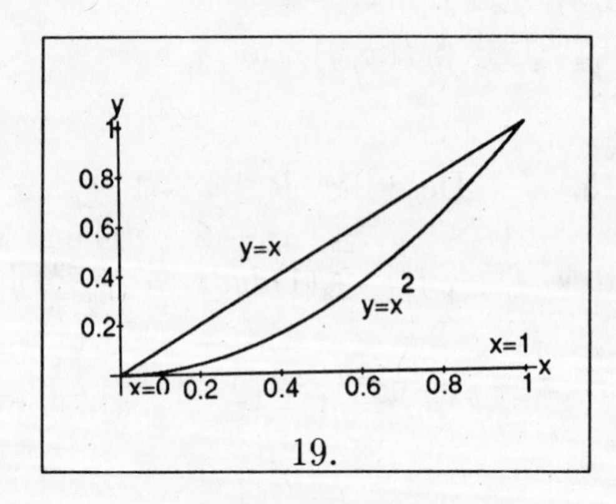

19.

20. Describe R by $0 \le x \le 1$ and $x^2 \le y \le x$. Then, the area of R is $\int_{x=0}^{x=1}\int_{y=x^2}^{y=x}(1)dydx = \int_{x=0}^{x=1} y|_{y=x^2}^{y=x}dx = \int_{x=0}^{x=1}(x - x^2)dx = (\frac{x^2}{2} - \frac{x^3}{3})|_{x=0}^{x=1} = \frac{1}{2} - \frac{1}{3} = \frac{1}{6}$ and $\int_R\int f(x,y)dA = \int_{x=0}^{x=1}\int_{y=x^2}^{y=x} 2xydydx = \int_{x=0}^{x=1} xy^2|_{y=x^2}^{y=x}dx = \int_{x=0}^{x=1}(x^3 - x^5)dx = (\frac{x^4}{4} - \frac{x^6}{6})|_0^1 = \frac{1}{4} - \frac{1}{6} = \frac{1}{12}$. Hence, the average value is $\frac{1}{\text{area of } R}\int_R\int f(x,y)dA = \frac{6}{12} = \frac{1}{2}$.

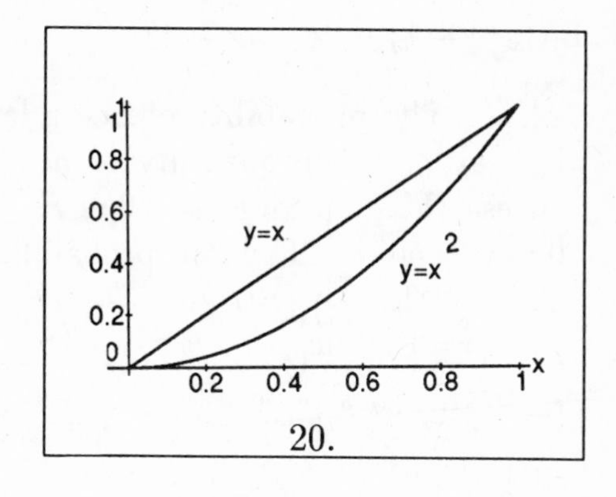

20.

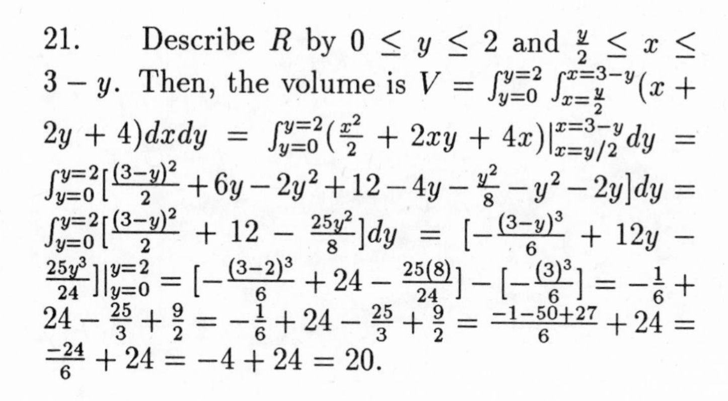
21. Describe R by $0 \le y \le 2$ and $\frac{y}{2} \le x \le 3 - y$. Then, the volume is $V = \int_{y=0}^{y=2}\int_{x=\frac{y}{2}}^{x=3-y}(x + 2y + 4)dxdy = \int_{y=0}^{y=2}(\frac{x^2}{2} + 2xy + 4x)|_{x=y/2}^{x=3-y}dy = \int_{y=0}^{y=2}[\frac{(3-y)^2}{2} + 6y - 2y^2 + 12 - 4y - \frac{y^2}{8} - y^2 - 2y]dy = \int_{y=0}^{y=2}[\frac{(3-y)^2}{2} + 12 - \frac{25y^2}{8}]dy = [-\frac{(3-y)^3}{6} + 12y - \frac{25y^3}{24}]|_{y=0}^{y=2} = [-\frac{(3-2)^3}{6} + 24 - \frac{25(8)}{24}] - [-\frac{(3)^3}{6}] = -\frac{1}{6} + 24 - \frac{25}{3} + \frac{9}{2} = -\frac{1}{6} + 24 - \frac{25}{3} + \frac{9}{2} = \frac{-1-50+27}{6} + 24 = \frac{-24}{6} + 24 = -4 + 24 = 20.$

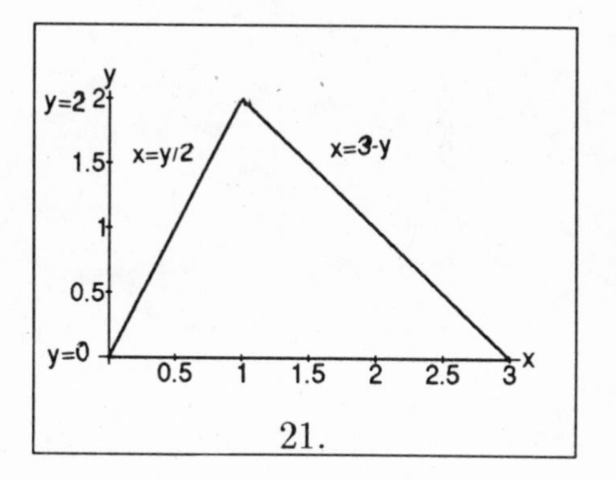

21.

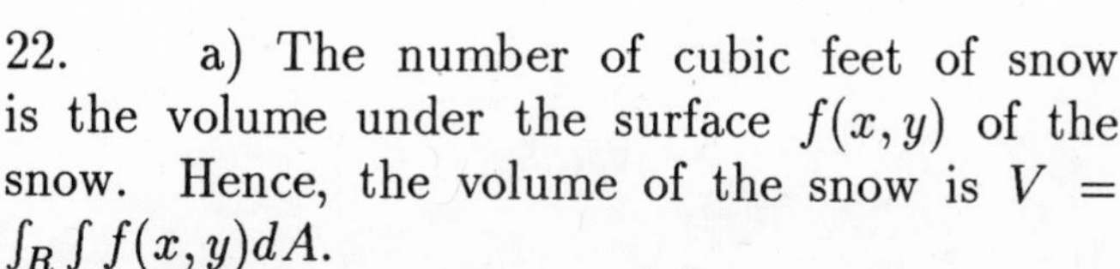
22. a) The number of cubic feet of snow is the volume under the surface $f(x,y)$ of the snow. Hence, the volume of the snow is $V = \int_R\int f(x,y)dA$.

b) The average depth is $\frac{1}{\text{area of } R}\int_R\int f(x,y)dA = \frac{\int_R\int f(x,y)dA}{\int_R\int (1)dA}$.

23. The joint probability density function is $f(x,y) = 6e^{-2x}e^{-3y}$.

a) $P(0 \le x \le 1 \text{ and } 0 \le y \le 2) = \int_{y=0}^{y=2}\int_{x=0}^{x=1} 6e^{-2x}e^{-3y}dxdy = \int_{y=0}^{y=2}(-3e^{-2x}e^{-3y})|_{x=0}^{x=1}dy = \int_{y=0}^{y=2}(-3e^{-2}e^{-3y} + 3e^{-3y})dy = 3(1 - e^{-2})\int_{y=0}^{y=2} e^{-3y}dy = -(1 - e^{-2})e^{-3y}|_{y=0}^{y=2} = -(1 - e^{-2})(e^{-6} - 1) = 0.8625.$

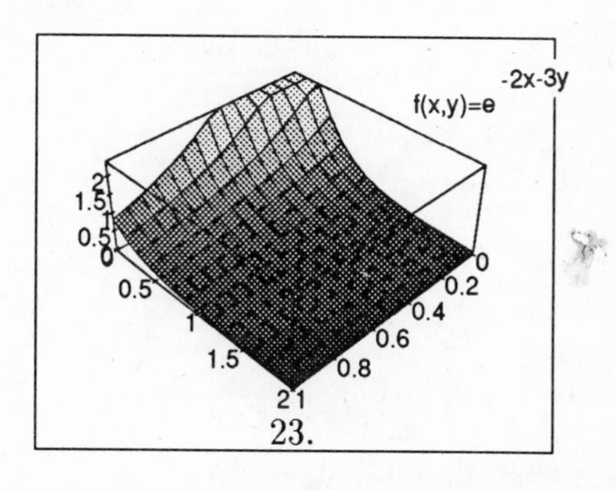

23.

b) The region R defined by $x + y \le 2$ is described by by $0 \le x \le 2$ and $0 \le y \le 2 - x$. Hence, $P(x + y \le 2) = \int_{x=0}^{x=2}\int_{y=0}^{y=2-x} 6e^{-2x}e^{-3y}dydx = \int_{x=0}^{x=2}(-2e^{-2x}e^{-3y})|_{y=0}^{y=2-x}dx =$

$\int_{x=0}^{x=2}(-2e^{-2x}e^{-6+3x}+2e^{-2x})dx = 2\int_{x=0}^{x=2} e^{-2x}dx - 2e^{-6}\int_{x=0}^{x=2} e^{x}dx = -e^{-2x}|_{x=0}^{x=2} - 2e^{-6}e^{x}|_{x=0}^{x=2} = -(e^{-4}-1) - 2e^{-6}(e^{2}-1) = 1 - 3e^{-4} + 2e^{-6} = 0.9500.$

24. The probability of being late for the meeting is 1 minus the probability of not being late. The probability of not being late is the probability that the total time $x+y$ is less than or equal to 50. The corresponding region is described by $0 \le x \le 50$ and $0 \le y \le 50-x$. Thus the probability of being late is $1 - \int_{x=0}^{x=50}\int_{y=0}^{y=50-x}(\frac{1}{500})e^{-x/10}e^{-y/50}dydx = 1 - \int_{x=0}^{x=50}(-\frac{1}{10})(e^{-x/10}e^{-y/50})|_{y=0}^{y=50-x}dx = 1 - \int_{x=0}^{x=50}(-\frac{1}{10})e^{-x/10}[e^{-(50-x)/50} - 1]dx = 1 + (\frac{1}{10})\int_{x=0}^{x=50} e^{-x/10}(e^{-1}e^{x/50} - 1)dx = 1 + (\frac{1}{10}e^{-1})\int_{x=0}^{x=50} e^{-2x/25}dx - (\frac{1}{10})\int_{x=0}^{x=50} e^{-x/10}dx = 1 + (-\frac{5}{4}e^{-1}e^{-2x/25})|_{x=0}^{x=50} + e^{-x/10}|_{x=0}^{x=50} = 1 - \frac{5}{4}e^{-1}(e^{-4}-1) + e^{-5} - 1 = \frac{5e^{-1}}{4} - \frac{e^{-5}}{4} = 0.4582.$

Chapter 11

Infinite Series and Taylor Approximations

11.1 Infinite Series

1. $\frac{1}{3}+\frac{1}{9}+\frac{1}{27}+\frac{1}{81}+\cdots=\frac{1}{3^1}+\frac{1}{3^2}+\frac{1}{3^3}+\frac{1}{3^4}+\cdots=\sum_{n=1}^{\infty}\frac{1}{3^n}$. This is not unique. One could write $\sum_{k=1}^{\infty}\frac{1}{3^k}$ or $\sum_{n=0}^{\infty}\frac{1}{3^{n+1}}$ or $\frac{1}{3}+\sum_{n=2}^{\infty}\frac{1}{3^n}$ for instance.

3. $\frac{1}{2}+\frac{2}{3}+\frac{3}{4}+\frac{4}{5}+\cdots=\frac{2-1}{2}+\frac{3-1}{3}+\frac{4-1}{4}+\frac{5-1}{5}+\cdots=\sum_{n=2}^{\infty}\frac{n-1}{n}$ or, equivalently, $\frac{1}{2}+\frac{2}{3}+\frac{3}{4}+\frac{4}{5}+\cdots=\frac{1}{1+1}+\frac{2}{2+1}+\frac{3}{3+1}+\frac{4}{4+1}+\cdots=\sum_{n=1}^{\infty}\frac{n}{n+1}$.

5. $\frac{1}{2}-\frac{4}{3}+\frac{9}{4}-\frac{16}{5}+\cdots=\frac{1^2}{1+1}-\frac{2^2}{2+1}+\frac{3^2}{3+1}-\frac{4^2}{4+1}+\cdots=\sum_{n=1}^{\infty}(-1)^{n+1}\frac{n^2}{n+1}$.

7. For $\sum_{n=1}^{\infty}\frac{1}{2^n}$, $S_4=\frac{1}{2^1}+\frac{1}{2^2}+\frac{1}{2^3}+\frac{1}{2^4}=\frac{1}{2}+\frac{1}{4}+\frac{1}{8}+\frac{1}{16}=\frac{15}{16}$.

9. For $\sum_{n=1}^{\infty}\frac{(-1)^n}{n}$, $S_4=\frac{(-1)^1}{1}+\frac{(-1)^2}{2}+\frac{(-1)^3}{3}+\frac{(-1)^4}{4}=-1+\frac{1}{2}-\frac{1}{3}+\frac{1}{4}=-\frac{7}{12}$.

11. For $\sum_{n=1}^{\infty}(\frac{1}{n+3}-\frac{1}{n+4})$, $S_n=(\frac{1}{4}-\frac{1}{5})+(\frac{1}{5}-\frac{1}{6})+(\frac{1}{6}-\frac{1}{7})+(\frac{1}{7}-\frac{1}{8})+\cdots+(\frac{1}{n+3}-\frac{1}{n+4})=\frac{1}{4}-\frac{1}{n+4}$. Hence, $\lim_{n\to\infty}S_n=\lim_{n\to\infty}(\frac{1}{4}-\frac{1}{n+4})=\frac{1}{4}$ and so $\sum_{n=1}^{\infty}(\frac{1}{n+3}-\frac{1}{n+4})=\frac{1}{4}$.

13. For $\sum_{n=1}^{\infty}\frac{1}{(n+1)(n+2)}=\sum_{n=1}^{\infty}(\frac{1}{n+1}-\frac{1}{n+2})$. $S_n=(\frac{1}{2}-\frac{1}{3})+(\frac{1}{3}-\frac{1}{4})+(\frac{1}{4}-\frac{1}{5})+(\frac{1}{5}-\frac{1}{6})+\cdots+(\frac{1}{n+1}-\frac{1}{n+2})=\frac{1}{2}-\frac{1}{n+2}$. Hence, $\lim_{n\to\infty}S_n=\lim_{n\to\infty}(\frac{1}{2}-\frac{1}{n+2})=\frac{1}{2}$ and so $\sum_{n=1}^{\infty}\frac{1}{(n+1)(n+2)}=\frac{1}{2}$.

15. For $\sum_{n=1}^{\infty}\frac{6}{10^n}$. $S_n=\frac{6}{10}+\frac{6}{10^2}+\frac{6}{10^3}+\cdots+\frac{6}{10^n}$ and $\frac{1}{10}S_n=\frac{6}{10^2}+\frac{6}{10^3}+\frac{6}{10^4}+\cdots+\frac{6}{10^{n+1}}$. Hence, $S_n-\frac{1}{10}S_n=\frac{9}{10}S_n=\frac{6}{10}-\frac{6}{10^{n+1}}=\frac{6}{10}(1-\frac{1}{10^n})$. Now $S_n=(\frac{10}{9})(\frac{6}{10})(1-\frac{1}{10^n})=(\frac{2}{3})(1-\frac{1}{10^n})$.

Hence, $\lim_{n\to\infty} S_n = \lim_{n\to\infty}(\frac{2}{3})(1-\frac{1}{10^n}) = \frac{2}{3}$ and so $\sum_{n=1}^{\infty} \sum_{n=1}^{\infty} \frac{6}{10^n} = \frac{2}{3}$.

17. For $\sum_{n=1}^{\infty} \frac{4}{3^n}$. $S_n = \frac{4}{3} + \frac{4}{3^2} + \frac{4}{3^3} + \cdots + \frac{4}{3^n}$ and $\frac{1}{3}S_n = \frac{4}{3^2} + \frac{4}{3^3} + \frac{4}{3^4} + \cdots + \frac{4}{3^{n+1}}$. Hence, $S_n - \frac{1}{3}S_n = \frac{2}{3}S_n = \frac{4}{3} - \frac{4}{3^{n+1}} = \frac{4}{3}(1-\frac{1}{3^n})$. Now $S_n = (\frac{3}{2})(\frac{4}{3})(1-\frac{1}{3^n}) = 2(1-\frac{1}{3^n})$. Hence, $\lim_{n\to\infty} S_n = \lim_{n\to\infty} 2(1-\frac{1}{3^n}) = 2$ and so $\sum_{n=1}^{\infty} \sum_{n=1}^{\infty} \frac{4}{3^n} = 2$.

19. The series $\sum_{n=1}^{\infty} \frac{n}{2n+1}$ diverges since $\lim_{n\to\infty} a_n = \lim_{n\to\infty} \frac{n}{2n+1} = \lim_{n\to\infty} \frac{1}{2+1/n} = \frac{1}{2} \neq 0$. The necessary condition for convergence is not satisfied.

21. The series $\sum_{n=1}^{\infty}(\frac{1}{n+1} - \frac{1}{n})$ converges to -1 since $\lim_{n\to\infty} S_n = \lim_{n\to\infty}[(\frac{1}{2}-1)+(\frac{1}{3}-\frac{1}{2})+(\frac{1}{4}-\frac{1}{3})+(\frac{1}{5}-\frac{1}{4})+\cdots+(\frac{1}{n+1}-\frac{1}{n})] = \lim_{n\to\infty}(-1+\frac{1}{n+1}) = -1$.

23. The series $\sum_{n=1}^{\infty}(-\frac{3}{2})^n$ converges to $-\frac{2}{3}$ since $S_n = (-\frac{2}{3})+(-\frac{2}{3})^2+(-\frac{2}{3})^3+\cdots+(-\frac{2}{3})^n$ and $-\frac{2}{3}S_n = (-\frac{2}{3})^2+(-\frac{2}{3})^3+(-\frac{2}{3})^4+\cdots+(-\frac{4}{3})^{n+1}$. Hence, $S_n+\frac{2}{3}S_n = \frac{5}{3}S_n = -\frac{2}{3}-(-\frac{2}{3})^{n+1} = -\frac{2}{3}[1-(-\frac{2}{3})^n]$. Now $S_n = (-\frac{2}{5})[1-(-\frac{2}{3})^n]$. Hence, $\lim_{n\to\infty} S_n = \lim_{n\to\infty}(-\frac{2}{5})[1-(-\frac{2}{3})^n] = -\frac{2}{5}$.

25. The series $\sum_{n=1}^{\infty} \frac{1}{\sqrt{n}}$ diverges since $S_n = \frac{1}{\sqrt{1}} + \frac{1}{\sqrt{2}} + \frac{1}{\sqrt{3}} + \cdots + \frac{1}{\sqrt{n}} > \frac{1}{\sqrt{n}} + \frac{1}{\sqrt{n}} + \frac{1}{\sqrt{n}} + \cdots + \frac{1}{\sqrt{n}} = \frac{n}{\sqrt{n}} = \sqrt{n}$ which increases without bound as $n \to \infty$.

11.2 Geometric Series

1. $\sum_{n=0}^{\infty}(\frac{4}{5})^n = \frac{1}{1-4/5} = 5$

3. $\sum_{n=0}^{\infty} \frac{2}{3^n} = 2\sum_{n=0}^{\infty}(\frac{1}{3})^n = 2(\frac{1}{1-1/3}) = 3$

5. For $\sum_{n=1}^{\infty}(\frac{3}{2})^n$, the ratio is $r = \frac{3}{2} > 1$, and so the geometric series diverges.

7. $\sum_{n=2}^{\infty}(\frac{3}{-4})^n = 3[(-\frac{1}{4})^2 + (-\frac{1}{4})^3 + (-\frac{1}{4})^4 + \cdots] = 3(-\frac{1}{4})^2[(-\frac{1}{4}) + (-\frac{1}{4})^2 + \cdots] = (\frac{3}{16})[\frac{1}{1-(-1/4)}] = \frac{3}{20}$.

9. $\sum_{n=1}^{\infty} 5(0.9)^n = 5[0.9 + (0.9)^2 + (0.9)^3 + \cdots] = 4.5(10) = 45$.

11. $\sum_{n=1}^{\infty} \frac{3^n}{4^{n+2}} = \sum_{n=1}^{\infty} \frac{3^n}{4^n 4^2} = (\frac{1}{16})\sum_{n=1}^{\infty}(\frac{3}{4})^n = (\frac{1}{16})[(\frac{3}{4}) + (\frac{3}{4})^2 + (\frac{3}{4})^3 + \cdots] = (\frac{1}{16})(\frac{3}{4})[1 + \frac{3}{4} + (\frac{3}{4})^2 + \cdots] = (\frac{3}{64})[\frac{1}{1-3/4}] = (\frac{3}{64})(4) = \frac{3}{16}$.

13. $\sum_{n=0}^{\infty} \frac{4^{n+1}}{5^{n-1}} = \sum_{n=0}^{\infty} \frac{4(4^n)}{5^n 5^{-1}} = (\frac{4}{5^{-1}})\sum_{n=0}^{\infty}(\frac{4}{5})^n = 20(\frac{1}{1-4/5}) = 20(5) = 100$.

15. Write the decimal as a geometric series as follows: $0.3333\cdots = \frac{3}{10} + \frac{3}{100} + \frac{3}{1{,}000} + \frac{3}{10{,}000} + \cdots = (\frac{3}{10})(1+\frac{1}{10}+\frac{1}{100}+\frac{1}{1{,}000}+\cdots) = (\frac{3}{10})[1+\frac{1}{10}+(\frac{1}{10})^2+(\frac{1}{10})^3+\cdots] = (\frac{3}{10})(\frac{1}{1-1/10}) = (\frac{3}{10})(\frac{10}{9}) = \frac{1}{3}$.

17. Write the decimal as a geometric series as follows: $0.252525\cdots = \frac{25}{100}+\frac{25}{10{,}000}+\frac{25}{1{,}000{,}000}+\cdots = (\frac{25}{100})(1+\frac{1}{100}+\frac{1}{10{,}000}+\cdots) = (\frac{25}{100})[1+\frac{1}{100}+(\frac{1}{100})^2+\cdots] = (\frac{25}{100})(\frac{1}{1-1/100}) = (\frac{25}{100})(\frac{100}{99}) = \frac{25}{99}$.

19. Since 92 percent of all income is spent, the amount (in billions) spent by beneficiaries of the 50 billion dollar tax cut is $0.92(50)$. This becomes new income, of which 92 percent or $0.92[0.92(50)] = (0.92)^2(50)$ is spent. This, in turn, generates additional spending of $0.92[(0.92)^2(50)] = (0.92)^3(50)$ and so on. The total amount spent if this process continues indefinitely is $(0.92)(50) + (0.92)^2(50) + (0.92)^3(50) + \cdots = (0.92)(50)[1 + (0.92) + (0.92)^2 + \cdots] = (0.92)(50)(\frac{1}{1-0.92}) = 46(\frac{1}{0.08}) = 575$ billion dollars.

21. The present value of the investment is the sum of present values of the individual payments. Thus, the present value is $2{,}000e^{-0.15} + 2{,}000(e^{-0.15})^2 + 2{,}000(e^{-0.15})^3 + \cdots = 2{,}000e^{-0.15}[1 + e^{-0.15} + (e^{-0.15})^2 + \cdots = 2{,}000e^{-0.15}(\frac{1}{1-e^{-0.15}}) = \$12{,}358.32$.

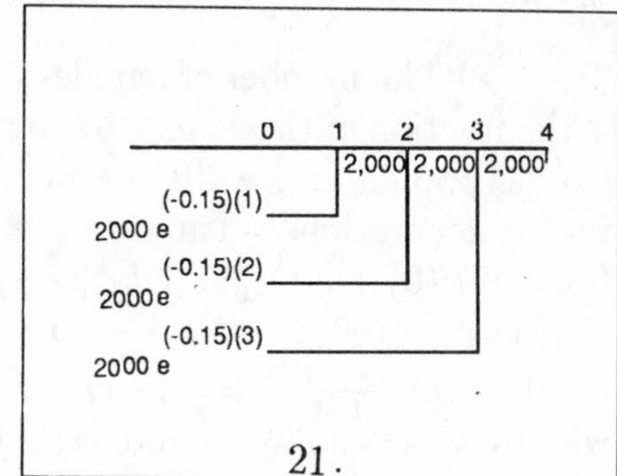

21.

23. Since d is the daily dosage of toxins consumed by the body and q percent is excreted daily, the amount left in the body after the first day is $(1 - 0.01q)d$.
On the second day, the body contains what is left from the first day plus the new amount that is consumed, namely $d + (1 - 0.01q)d$. The amount left in the body on the second day is $[d + (1 - 0.01q)d](1 - 0.01q) = d[(1 - 0.01q) + (1 - 0.01q)^2]$.
On the third day, the body contains what is left from the second day plus the new amount that is consumed, namely $d + d[(1 - 0.01q) + (1 - 0.01q)^2]$. The amount left in the body on the third day is $\{d+d[(1-0.01q)+(1-0.01q)^2]\}(1-0.01q) = d[(1-0.01q)+(1-0.01q)^2 + (1-0.01q)^3]$, and so on.
The total amount left if this process continues indefinitely is $d[(1-0.01q)+(1-0.01q)^2+(1-0.01q)^3+\cdots] = d(1-0.01q)[1+(1-0.01q)+(1-0.01q)^2+\cdots] = d(1-0.01q)[\frac{1}{1-(1-0.01q)}] = d(1-0.01q)(\frac{1}{0.01q}) = d(100-q)(\frac{1}{q})$.

25. Let T be the time it takes to run the first half of the course. Then it takes $\frac{T}{2}$ units of time to run one half of the remainder of the course. The total time to this point is $T + \frac{T}{2}$. One-half of the remainder is covered in $(\frac{1}{2})(\frac{T}{2}) = \frac{T}{4}$ units of time for a total time of $T + \frac{T}{2} + \frac{T}{4}$, and so on. The total time if this process continues indefinitely is $T + \frac{T}{2} + \frac{T}{4} + \cdots = T(1 + \frac{1}{2} + \frac{1}{4} + \cdots) = T(\frac{1}{1-1/2}) = 2T$.

27. Let x denote the number of the trial on which the first old penny is found. Then, $P(x \geq 100) = \sum_{n=100}^{\infty} P(x = n) = \sum_{n=100}^{\infty}(1-p)^{n-1}p$, where $p = 0.002$ is the probability of finding an old penny and $1 - p = 0.998$. Hence, $P(x \geq 100) = \sum_{n=100}^{\infty}(0.998)^{n-1}(0.002) = (0.998)^{99}(0.002) + (0.998)^{100}(0.002) + \cdots = (0.998)^{99}(0.002)[1 + 0.998 + (0.998)^2 + \cdots] = (0.998)^{99}(0.002)(\frac{1}{1-0.998}) = (0.998)^{99}(0.002)(\frac{1}{0.002}) = (0.998)^{99} = 0.8202$.

29. Since 7.3 percent of the rare gas is consumed yearly and 1.7×10^9 is the quantity consumed in the first year (1990), the quantity consumed in the second year is $1.073(1.7 \times 10^9)$. During the third year the consumption is $(1.073)^2(1.7 \times 10^9)$. During the nth year, the consumption is $(1.073)^{n-1}(1.7 \times 10^9)$. The total consumption during these n years is $1.7 \times 10^9 \frac{1-(1.073)^n}{1-1.073}$. With the total reserves being 3×10^{11}, $[1 - (1.073)^n](1.7 \times 10^9) = 3 \times 10^{11}(-0.073)$, $1 - (1.073)^n = \frac{(-0.073)3\times10^{11}}{1.7\times10^9} = -12.8824$, $(1.073)^n = 13.8824$, $n \ln 1.073 = \ln 13.8824$, $n = \frac{\ln 13.8824}{\ln 1.073} = 37.34$, which brings us to the year $1990 + 37$ or 2027.

31. a) The number of females in the $(n+1)^{st}$ generation will be $(1+b)P(n) - d$, because b is the fraction of the generation by which the females are increased, thus added to the present female population, while d will be removed, say after they had a chance to contribute to future generations. Thus we use $c = 1 + b$ and $r = -d$ in problem 30, so that $P(n) = (1+b)^n P(0) + [\sum_{k=0}^{n-1}(1+b)^k](-d) = 100(1+b)^n - d\sum_{k=0}^{n-1}(1+b)^k$.
b) $P(n) = 100(1+0.2)^n - 15\sum_{k=0}^{n-1}(1+0.2)^k$, so $P(10) = 100(1.2)^{10} - 15\sum_{k=0}^{9}(1.2)^k = 619.17 - 15[\frac{1-(1.2)^{10}}{1-1.2}] = 619.17 - 389.38 \approx 229.8$. In the long run, as $n \to \infty$, the population will grow beyond all bounds as the ratio $r = 1.2 > 1$.

11.3 Taylor Approximations

1. For $f(x) = e^{3x}$ about $x = 0$, compute the Taylor coefficients as follows: $f(x) = e^{3x}$, $f'(x) = 3e^{3x}$, $f''(x) = 3^2e^{3x}$, $f^{(3)}(x) = (3^3)e^{3x}, \ldots, f^{(n)}(x) = 3^n e^{3x}$. Thus $f(0) = 1$, $f'(0) = 3$, $f''(0) = 3^2$, $f^{(3)}(0) = 3^3, \ldots, f^{(n)}(0) = 3^n$, and so $a_n = \frac{f^{(n)}(0)}{n!} = \frac{3^n}{n!}$, which leads to the corresponding Taylor series $\sum_{n=0}^{\infty}\frac{3^n}{n!}x^n$. To find the general pattern, note the relation between the exponents and the order of the derivatives.

3. For $f(x) = \ln(x+1)$ about $x = 0$, compute the Taylor coefficients as follows: $f(x) = \ln(x+1)$, $f'(x) = (1+x)^{-1}$, $f''(x) = -(1+x)^{-2}$, $f^{(3)}(x) = 2(1+x)^{-3}$, $f^{(4)}(x) = -(2)(3)(1+x)^{-4}, \ldots, f^{(n)}(x) = (-1)^{n+1}\frac{(n-1)!}{(1+x)^n}$. To find the general pattern, note the relation between the exponents and the order of the derivatives. Thus $f(0) = 0$, $f'(0) = 1$, $f''(0) = -1$, $f^{(3)}(0) =$

$2!$, $f^{(4)}(0) = -3!, \ldots, f^{(n)}(0) = (-1)^{n+1}(n-1)!$, and so $a_n = \frac{f^{(n)}(0)}{n!} = \frac{(-1)^{n+1}(n-1)!}{n!} = \frac{(-1)^{n+1}}{n}$, which leads to the corresponding Taylor series $\sum_{n=0}^{\infty} \frac{(-1)^{n+1}}{n} x^n$.

5. For $f(x) = \frac{e^x+e^{-x}}{2}$ about $x = 0$, compute the Taylor coefficients as follows: $f(x) = \frac{e^x+e^{-x}}{2}$, $f'(x) = \frac{e^x-e^{-x}}{2}$, $f''(x) = \frac{e^x+e^{-x}}{2}$, $f^{(3)}(x) = \frac{e^x-e^{-x}}{2}, \ldots, f^{(n)}(x) = \frac{e^x+(-1)^n e^{-x}}{2}$. Thus $f(0) = 1$, $f'(0) = 0$, $f''(0) = 1$, $f^{(3)}(0) = 0, \ldots, f^{(n)}(0) = \frac{1+(-1)^n}{2}$, and so $a_n = \frac{f^{(n)}(0)}{n!} = \frac{1}{n!}$ if n is even (0 otherwise), or $a_{2n} = \frac{1}{(2n)!}$. This leads to the corresponding Taylor series $\sum_{n=0}^{\infty} \frac{1}{(2n)!} x^{2n}$.

7. For $f(x) = (1+x)e^x$ about $x = 0$, compute the Taylor coefficients as follows: $f(x) = (1+x)e^x$, $f'(x) = (2+x)e^x$, $f''(x) = (3+x)e^x$, $f^{(3)}(x) = (4+x)e^x, \ldots, f^{(n)}(x) = (n+1+x)e^x$. Thus $f(0) = 1$, $f'(0) = 2$, $f''(0) = 3$, $f^{(3)}(0) = 4, \ldots, f^{(n)}(0) = n+1$, and so $a_n = \frac{f^{(n)}(0)}{n!} = \frac{n+1}{n!}$, which leads to the corresponding Taylor series $\sum_{n=0}^{\infty} \frac{n+1}{n!} x^n$.

9. For $f(x) = e^{2x}$ about $x = 1$, compute the Taylor coefficients as follows: $f(x) = e^{2x}$, $f'(x) = 2e^{2x}$, $f''(x) = 2^2 e^{2x}$, $f^{(3)}(x) = 2^3 e^{2x}, \ldots, f^{(n)}(x) = 2^n e^{2x}$. Thus $f(0) = e^2$, $f'(1) = 2e^2$, $f''(1) = 2^2e^2$, $f^{(3)}(1) = 2^3e^2, \ldots, f^{(n)}(1) = 2^n e^2$, and so $a_n = \frac{f^{(n)}(1)}{n!} = \frac{2^n e^2}{n!}$, which leads to the corresponding Taylor series $\sum_{n=0}^{\infty} \frac{2^n e^2}{n!}(x-1)^n$.

11. For $f(x) = \frac{1}{x}$ about $x = 1$, compute the Taylor coefficients as follows: $f(x) = \frac{1}{x} = x^{-1}$, $f'(x) = -x^{-2}$, $f^{(2)}(x) = 2x^{-3}$, $f^{(3)}(x) = -(2)(3)x^{-4}, \ldots, f^{(n)}(x) = (-1)^n \frac{n!}{x^{n+1}}$. Thus $f(1) = 1$, $f'(1) = -1$, $f''(1) = 2!$, $f^{(3)}(1) = -3!$, $f^{(4)}(1) = 4!, \ldots, f^{(n)}(1) = (-1)^n n!$, and so $a_n = \frac{f^{(n)}(1)}{n!} = \frac{(-1)^n n!}{n!} = (-1)^n$, which leads to the corresponding Taylor series $\sum_{n=0}^{\infty}(-1)^n x^n$.

13. For $f(x) = \frac{1}{2-x}$ about $x = 1$, compute the Taylor coefficients as follows: $f(x) = \frac{1}{2-x} = (2-x)^{-1}$, $f'(x) = -(2-x)^{-2}(-1)$, $f^{(2)}(x) = 2(2-x)^{-3}$, $f^{(3)}(x) = (2)(3)(2-x)^{-4}$, $\ldots, f^{(n)}(x) = \frac{n!}{(2-x)^{n+1}}$. Thus $f(1) = 1$, $f'(1) = 1$, $f''(1) = 2!$, $f^{(3)}(1) = 3!$, $f^{(4)}(1) = 4!$, $\ldots, f^{(n)}(1) = n!$, and so $a_n = \frac{f^{(n)}(1)}{n!} = \frac{n!}{n!} = 1$, which leads to the corresponding Taylor series $\sum_{n=0}^{\infty}(x-1)^n$.

15. Start with the fact that $\frac{1}{1-x} = \sum_{n=0}^{\infty} x^n$ for $|x| < 1$. Replace x by $5x$ to get $\frac{1}{1-5x} = \sum_{n=0}^{\infty}(5x)^n$ for $|5x| < 1$ or $|x| < \frac{1}{5}$. Now multiply by x for $\frac{x}{1-5x} = \sum_{n=0}^{\infty} 5^n x^{n+1}$ for $|x| < \frac{1}{5}$.

17. Start with the fact that $e^x = \sum_{n=0}^{\infty} \frac{x^n}{n!}$ for all x. Replace x by $-\frac{x}{2}$ to get $e^{-x/2} = \sum_{n=0}^{\infty} \frac{1}{n!}(-\frac{x}{2})^n = \sum_{n=0}^{\infty} \frac{x^n}{n!(-2)^n}$ for all x.

19. Start with the fact that $\frac{1}{1-x} = \sum_{n=0}^{\infty} x^n$ for all $|x| < 1$. Replace x by $1-x$ to get $\frac{1}{1-x} = \sum_{n=0}^{\infty}(1-x)^n$ for all $|1-x| < 1$, $-1 < 1-x < 1$, $-2 < -x < 0$, or $0 < x < 2$.

21. From example 3.6, the Taylor polynomial of degree 3 for $f(x) = \sqrt{x}$ about $x = 4$ is $P_3(x) = 2+\frac{x-4}{4}-\frac{(x-4)^2}{64}+\frac{(x-4)^3}{512}$ and so $\sqrt{3.8} \approx P_3(3.8) = 2+\frac{3.8-4}{4}-\frac{(3.8-4)^2}{64}+\frac{(3.8-4)^3}{512} = 1.94936$.

23. From example 3.4, the Taylor polynomial of degree 5 for $f(x) = \ln x$ about $x = 1$ is $P_5(x) = (x-1) - \frac{(x-1)^2}{2} + \frac{(x-1)^3}{3} - \frac{(x-4)^4}{4} + \frac{(x-1)^5}{5}$ and so $\ln 1.1 \approx P_5(1.1) = (1.1-1) - \frac{(1.1-1)^2}{2} + \frac{(1.1-1)^3}{3} - \frac{(1.1-4)^4}{4} + \frac{(1.1-1)^5}{5} = 0.09531$.

25. From example 3.5, the Taylor polynomial of degree 4 for $f(x) = e^x$ about $x = 0$ is $P_4(x) = 1 + x + \frac{x^2}{2!} + \frac{x^3}{3!} + \frac{x^4}{4!}$ and so $e^{0.3} \approx P_4(0.3) = 1 + 0.3 + \frac{0.3^2}{2!} + \frac{0.3^3}{3!} + \frac{0.3^4}{4!} = 1.34984$.

27. Since $e^x = \sum_{n=0}^{\infty} \frac{x^n}{n!}$, it follows that $e^{-x^2} = \sum_{n=0}^{\infty} \frac{(-x^2)^n}{n!} = \sum_{n=0}^{\infty} \frac{(-1)^n}{n!}x^{2n}$ and so $P_6(x) = 1 - x^2 + \frac{x^4}{2} - \frac{x^6}{6}$. Hence, $\int_0^{1/2} e^{-x^2}dx \approx \int_0^{1/2} P_6(x)dx = \int_0^{1/2}(1 - x^2 + \frac{x^4}{2} - \frac{x^6}{6})dx = (x - \frac{x^3}{3} + \frac{x^5}{10} - \frac{x^7}{42})|_0^{1/2} = (0.5 - \frac{0.5^3}{3} + \frac{0.5^5}{10} - \frac{0.5^7}{42})|_0^{1/2} = 0.46127$

29. Start with the fact that $\frac{1}{1-x} = \sum_{n=0}^{\infty} x^n$ for all $|x| < 1$. Replace x by $-x^2$ to get $\frac{1}{1+x^2} = \sum_{n=0}^{\infty}(-x^2)^n = \sum_{n=0}^{\infty}(-1)^n x^{2n}$ for all $|x^2| < 1$ so that $P_2(x) = 1 - x^2 + x^4$. Hence, $\int_0^{0.1} \frac{1}{1+x^2}dx \approx \int_0^{0.1} P_2(x) = \int_0^{0.1}(1-x^2+x^4)dx = (x-\frac{x^3}{3}+\frac{x^5}{5})|_0^{0.1} = (0.1-\frac{0.1^3}{3}+\frac{0.1^5}{5}) = 0.09967$.

Review Problems

1. For $\sum_{n=1}^{\infty}(\frac{1}{n+1} - \frac{1}{n+3})$, $S_n = (\frac{1}{2} - \frac{1}{4}) + (\frac{1}{3} - \frac{1}{5}) + (\frac{1}{4} - \frac{1}{6}) + (\frac{1}{5} - \frac{1}{7}) + \cdots + (\frac{1}{n-1} - \frac{1}{n+1}) + (\frac{1}{n} - \frac{1}{n+2}) + (\frac{1}{n+1} - \frac{1}{n+3}) = \frac{1}{2} + \frac{1}{3} - \frac{1}{n+2} - \frac{1}{n+3} = \frac{5}{6} - \frac{1}{n+2} - \frac{1}{n+3}$. Hence, $\lim_{n\to\infty} S_n = \lim_{n\to\infty} \frac{5}{6} - \frac{1}{n+2} - \frac{1}{n+3} = \frac{5}{6}$.

2. The series $\sum_{n=2}^{\infty} \frac{1}{n(n-1)}$ can be rewritten $\sum_{n=2}^{\infty}(\frac{1}{n-1} - \frac{1}{n})$. $S_n = (1 - \frac{1}{2}) + (\frac{1}{2} - \frac{1}{3}) + (\frac{1}{3} - \frac{1}{4}) + (\frac{1}{4} - \frac{1}{5}) + \cdots + (\frac{1}{n-2} - \frac{1}{n-1}) + (\frac{1}{n-1} - \frac{1}{n}) = 1 - \frac{1}{n}$. Hence, $\lim_{n\to\infty} S_n = \lim_{n\to\infty}(1 - \frac{1}{n}) = 1$.

3. For the geometric series $\sum_{n=1}^{\infty} \frac{2}{(-3)^n} = 2\sum_{n=1}^{\infty}(-\frac{1}{3})^n$, $S_n = 2[-\frac{1}{3}+(-\frac{1}{3})^2+(-\frac{1}{3})^3+\cdots+(-\frac{1}{3})^n]$. Multiply S_n by $-\frac{1}{3}$ to get $-\frac{1}{3}S_n = 2[(-\frac{1}{3})^2 + (-\frac{1}{3})^3 + (-\frac{1}{3})^4 + \cdots + (-\frac{1}{3})^{n+1}]$ and so $S_n + \frac{1}{3}S_n = \frac{4}{3}S_n = 2[-\frac{1}{3} - (-\frac{1}{3})^{n+1}] = -\frac{2}{3}[1 - (-\frac{1}{3})^n]$. Hence, $S_n = \frac{3}{4}(-\frac{2}{3})[1 - (-\frac{1}{3})^n] = -\frac{1}{2}[1 - (-\frac{1}{3})^n]$ and the sum of the series is $\lim_{n\to\infty} S_n = \lim_{n\to\infty}(-\frac{1}{2})[1 - (-\frac{1}{3})^n] = -\frac{1}{2}$.

4. $\sum_{n=1}^{\infty} \frac{3}{(-5)^n} = 3\sum_{n=1}^{\infty}(-\frac{1}{5})^n = -\frac{3}{5}\sum_{n=0}^{\infty}(-\frac{1}{5})^n = -\frac{1}{1-(-1/5)} = -\frac{3}{5}(\frac{5}{6}) = -\frac{1}{2}$.

5. The series $\sum_{n=0}^{\infty}(-\frac{3}{2})^n$ is a geometric series with ratio $|-\frac{3}{2}| > 1$ and hence diverges.

6. $\sum_{n=1}^{\infty} e^{-0.5n} = \sum_{n=1}^{\infty}(e^{-0.5})^n = e^{-0.5}\sum_{n=0}^{\infty}(e^{-0.5})^n = e^{-0.5}\frac{1}{1-e^{-0.5}} = 1.5415.$

7. $\sum_{n=2}^{\infty}\frac{2^{n+1}}{3^{n-3}} = \sum_{n=2}^{\infty}\frac{(2)2^n}{(3^{-3})3^n} = \frac{2}{3^{-3}}\sum_{n=2}^{\infty}\frac{2^n}{3^n} = 54[(\frac{2}{3})^2 + (\frac{2}{3})^3 + (\frac{2}{3})^4 + \cdots] = 54(\frac{2}{3})^2[1 + \frac{2}{3} + (\frac{2}{3})^2 + (\frac{2}{3})^3 + \cdots] = 54(\frac{2}{3})^2(\frac{1}{1-2/3}) = 54(\frac{4}{9})(3) = 72.$

8. Write the decimal as a geometric series as follows: $1.545454\cdots = 1 + \frac{54}{100} + \frac{54}{10{,}000} + \frac{54}{1{,}000{,}000} + \cdots = 1 + (\frac{54}{100})(1 + \frac{1}{100} + \frac{1}{10{,}000} + \cdots) = 1 + (\frac{54}{100})[1 + \frac{1}{100} + (\frac{1}{100})^2\cdots] = 1 + (\frac{54}{100})(\frac{1}{1-1/100}) = 1 + (\frac{54}{100})(\frac{100}{99}) = 1 + \frac{54}{99} = \frac{17}{11}.$
Alternate method involving no power series: Let $N = 1.5454\cdots54\cdots$, then $100N = 154.5454\cdots54\cdots$ and $100N - N = 154.5454\cdots54\cdots - 1.5454\cdots54\cdots$ or $99N = 153.0000\cdots00\cdots$ from which we conclude that $N = \frac{153}{99} = \frac{17}{11}.$

9. The ball drops from a height of 6 feet. It rises and then falls to/from a height of 6(0.8) feet. Then it rises and the falls to/from a height of $6(0.8)^2$ feet. Thus, the total distance travelled is $6 + 2(6)(0.8) + 2(6)(0.8)^2 + 2(6)(0.8)^3 + \ldots = 6 + 12(0.8)[1 + 0.8 + (0.8)^2 + \cdots] = 6 + 12(0.8)[\frac{1}{1-0.8}] = 54$ feet.

10. The present value of the first withdrawal of \$500 invested at 12 percent compounded continuously is $P_1 = 500e^{(-0.12)(1)}$. The present value of the second withdrawal is $P_2 = 500e^{(-0.12)(2)}$, and in general, the present value of the nth withdrawal of \$500 is $P_n = 500e^{(-0.12)(n)}$. Hence, the amount to be invested is $\sum_{n=1}^{\infty} P_n = \sum_{n=1}^{\infty} 500e^{(-0.12)(n)} = 500\sum_{n=1}^{\infty}(e^{-0.12})^n = 500e^{-0.12}\sum_{n=0}^{\infty}(e^{-0.12})^n = 500e^{-0.12}(\frac{1}{1-e^{-0.12}}) = \$3,921.67.$

11. Of the original dose of 10 units, only $10e^{-0.8}$ units are left in the patient's body just prior to the second injection one day later. Hence, the number of units in the patient's body after one day (just after the next injection) is $S_1 = 10 + 10e^{-0.8}$. The medication in the patient's body after two days (immediately following the injection) consists of the 10 units just injected plus what remains from the first two injections. Of the original dose, only $10e^{(-0.8)(2)}$ units are left, and of the second dose, only $10e^{-0.8}$ units are left. Hence, the number of units in the patient's body after 2 days is $S_2 = 10 + 10e^{-0.8} + 10e^{(-0.8)(2)}$. In general, the number of units in the patient's body after n days is $S_n = 10 + 10e^{-0.8} + 10e^{(-0.8)(2)} + \cdots + 10e^{(-0.8)(n)}$. The number S of units in the patient's body in the long run is the limit S_n as n increases without bound. That is, $S = \lim_{n\to\infty} S_n = \sum_{n=0}^{\infty} 10e^{(-0.8)(n)} = 10\sum_{n=0}^{\infty}(e^{-0.8})^n = 10(\frac{1}{1-e^{-0.8}}) = 18.16$ units.

12. Let x denote the number of the roll on which the first six comes up. If you roll second, you will get the first six if $x = 2, 5, 8, \cdots$. Since the probability of rolling a six is $\frac{1}{6}$ and the probability of not rolling a six is $\frac{5}{6}$, the probability that you win is $P(x = 2) + P(x = 5) + P(x = 8) + \cdots = (\frac{1}{6})(\frac{5}{6}) + (\frac{1}{6})(\frac{5}{6})^4 + (\frac{1}{6})(\frac{5}{6})^7 + \cdots = (\frac{1}{6})(\frac{5}{6})[1 + (\frac{5}{6})^3 + (\frac{5}{6})^6 + \cdots] = (\frac{5}{36})\{1+(\frac{5}{6})^3+[(\frac{5}{6})^3]^2+\cdots\} = (\frac{5}{36})[1+\frac{125}{216}+(\frac{125}{216})^2+\cdots] = (\frac{5}{36})[\frac{1}{1-(125/216)}] = (\frac{5}{36})(\frac{216}{91}) = 0.3297.$

13. For $f(x) = \frac{1}{x+3}$ about $x = 0$, compute the Taylor coefficients as follows: $f(x) = \frac{1}{x+3} = (x+3)^{-1}$, $f'(x) = -(x+3)^{-2}$, $f^{(2)}(x) = 2(x+3)^{-3}$, $f^{(3)}(x) = -(2)(3)(x+3)^{-4}, \ldots, f^{(n)}(x) = (-1)^n \frac{n!}{(x+3)^{n+1}}$. Thus $f(0) = \frac{1}{3}$, $f'(0) = -\frac{1}{3^2}$, $f''(0) = \frac{2!}{3^3}$, $f^{(3)}(0) = -\frac{3!}{3^4}, \ldots, f^{(n)}(0) = (-1)^n \frac{n!}{3^{n+1}}$, and so $a_n = \frac{f^{(n)}(1)}{n!} = \frac{(-1)^n n!}{3^{n+1} n!} = \frac{(-1)^n}{3^{n+1}}$, which leads to the corresponding Taylor series $\sum_{n=0}^{\infty} \frac{(-1)^n}{3^{n+1}} x^n$.

14. For $f(x) = e^{-3x}$ about $x = 0$, compute the Taylor coefficients as follows: $f(x) = e^{-3x}$, $f'(x) = -3e^{-3x}$, $f''(x) = 3^2 e^{-3x}$, $f^{(3)}(x) = -3^3 e^{-3x}, \ldots, f^{(n)}(x) = (-1)^n 3^n e^{-3x}$. Thus $f(0) = 1$, $f'(0) = -3$, $f''(0) = 3^2$, $f^{(3)}(0) = -3^3, \ldots, f^{(n)}(0) = (-1)^n 3^n$, and so $a_n = \frac{f^{(n)}(0)}{n!} = \frac{(-3)^n}{n!}$, which leads to the corresponding Taylor series $\sum_{n=0}^{\infty} \frac{(-3)^n}{n!} x^n$.

15. For $f(x) = (1+2x)e^x$ about $x = 0$, compute the Taylor coefficients as follows: $f(x) = (1+2x)e^x$, $f'(x) = (3+2x)e^x$, $f''(x) = (5+2x)e^x$, $f^{(3)}(x) = (7+2x)e^x, \ldots, f^{(n)}(x) = (2n+1+2x)e^x$. Thus $f(0) = 1$, $f'(0) = 3$, $f''(0) = 5$, $f^{(3)}(0) = 7, \ldots, f^{(n)}(0) = 2n+1$, and so $a_n = \frac{f^{(n)}(0)}{n!} = \frac{2n+1}{n!}$, which leads to the corresponding Taylor series $\sum_{n=0}^{\infty} \frac{2n+1}{n!} x^n$.

16. For $f(x) = \frac{1}{(1+x)^2} = (1+x)^{-2}$ about $x = -2$, compute the Taylor coefficients as follows: $f(x) = (1+x)^{-2}$, $f'(x) = -2(1+x)^{-3}$, $f^{(2)}(x) = 3!(1+x)^{-4}$, $f^{(3)}(x) = -4!(1+x)^{-5}$, $\ldots, f^{(n)}(x) = (-1)^n \frac{(n+1)!}{(1+x)^{n+2}}$. Thus $f(-2) = 1$, $f'(-2) = 2$, $f''(-2) = 3!$, $f^{(3)}(-2) = 4!$, $\ldots, f^{(n)}(-2) = (-1)^n \frac{(n+1)!}{(-1)^{n+2}} = (n+1)!$, and so $a_n = \frac{f^{(n)}(-2)}{n!} = \frac{(n+1)!}{n!} = n+1$, which leads to the corresponding Taylor series $\sum_{n=0}^{\infty}(n+1)(x+2)^n$.

17. Start with the fact that $\frac{1}{1-x} = \sum_{n=0}^{\infty} x^n$ for $|x| < 1$. Replace x by $-4x^2$ to get $\frac{1}{1+4x^2} = \sum_{n=0}^{\infty}(-4x^2)^n = \sum_{n=0}^{\infty}(-1)^n 4^n x^{2n}$ for $|-4x^2| < 1$ or $|x| < \frac{1}{2}$. Now multiply by $4x$ for $\frac{4x}{1+4x^2} = \sum_{n=0}^{\infty}(-1)^n 4^{n+1} x^{2n+1} = \sum_{n=0}^{\infty} 2(2x)^{2n+1}$ for $|x| < \frac{1}{2}$.

18. For $f(x) = \sqrt{x} = x^{1/2}$ about $x = 1$, compute the Taylor coefficients as follows: $f(x) = x^{1/2}$, $f'(x) = \frac{1}{2}x^{-1/2}$, $f^{(2)}(x) = -\frac{1}{4}x^{-3/2}$, $f^{(3)}(x) = \frac{3}{8}x^{-5/2}$. Thus $f(1) = 1$, $f'(1) = \frac{1}{2}$, $f''(1) = -\frac{1}{4}$, $f^{(3)}(1) = \frac{3}{8}$, and so $P_3(x) = 1 + \frac{1}{2}(x-1) - \frac{1}{8}(x-1)^2 + \frac{1}{16}(x-1)^3$, and $\sqrt{0.9} \approx P_3(0.9) = 1 + \frac{1}{2}(0.9-1) - \frac{1}{8}(0.9-1)^2 + \frac{1}{16}(0.9-1)^3 = 0.94869$

19. For $f(x) = \frac{x}{1+x^3}$, it would be time-consumming to calculate the Taylor coefficients directly. Instead, start with $\frac{1}{1-x} = \sum_{n=0}^{\infty} x^n$ for $|x| < 1$. Replace x by $-x^3$ to get $\frac{1}{1+x^3} = \sum_{n=0}^{\infty}(-x^3)^n = \sum_{n=0}^{\infty}(-1)^n x^{3n}$ for $|-x^3| < 1$ or $|x| < 1$. Now multiply by x for $\frac{x}{1+x^3} = \sum_{n=0}^{\infty}(-1)^n x^{3n+1}$ for $|x| < 1$. Hence, the Taylor polynomial of degree 10 is $P_{10}(x) = x - x^4 + x^7 - x^{10}$ and $\int_0^{1/2} \frac{x}{1+x^3} dx \approx \int_0^{1/2} P_{10}(x) dx = \int_0^{1/2}(x - x^4 + x^7 - x^{10}) dx = \left(\frac{x^2}{2} - \frac{x^5}{5} + \frac{x^8}{8} - \frac{x^{11}}{11}\right)\Big|_0^{1/2} = \frac{1}{2^3} - \frac{1}{5(2^5)} + \frac{1}{8(2^8)} - \frac{1}{11(2^{11})} = 0.11919$

Chapter 12

Trigonometric Functions

12.1 The Trigonometric Functions

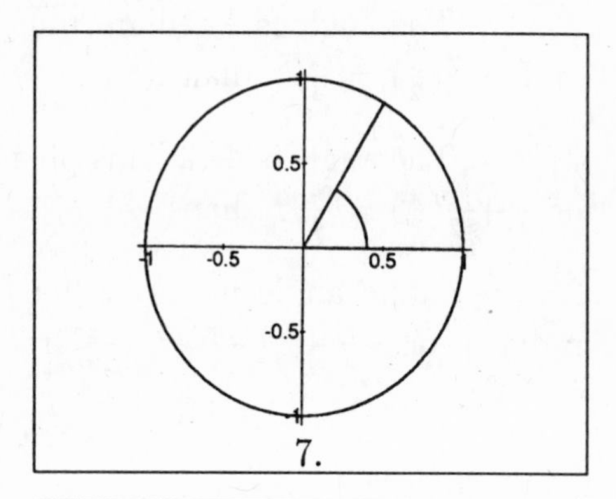

7.

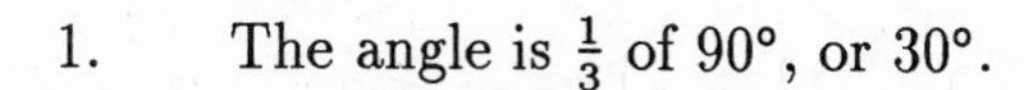

1. The angle is $\frac{1}{3}$ of $90°$, or $30°$.

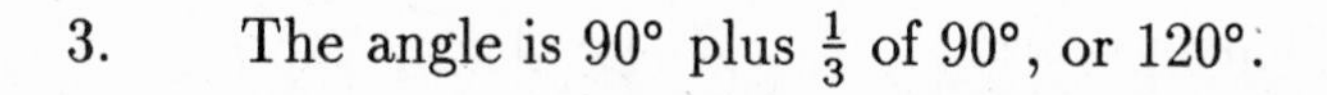

3. The angle is $90°$ plus $\frac{1}{3}$ of $90°$, or $120°$.

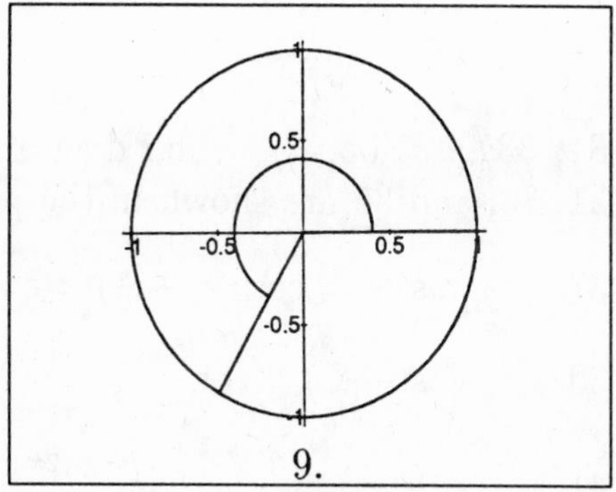

9.

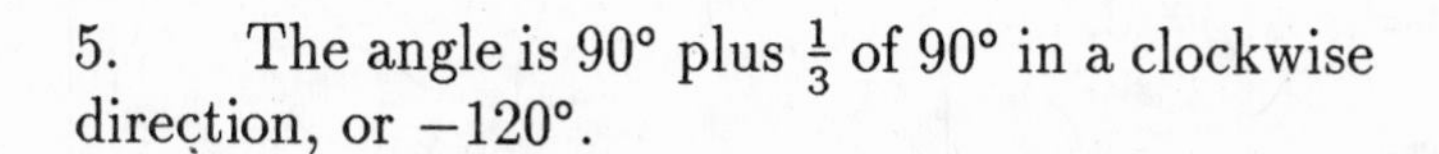

5. The angle is $90°$ plus $\frac{1}{3}$ of $90°$ in a clockwise direction, or $-120°$.

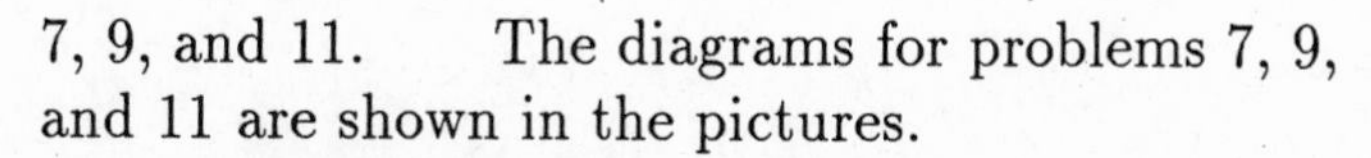

7, 9, and 11. The diagrams for problems 7, 9, and 11 are shown in the pictures.

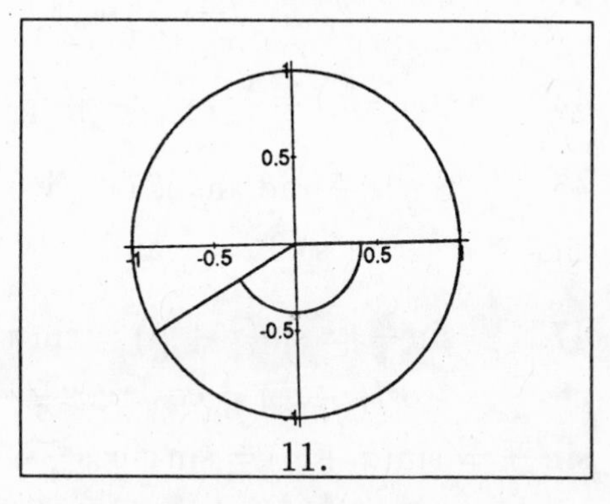

11.

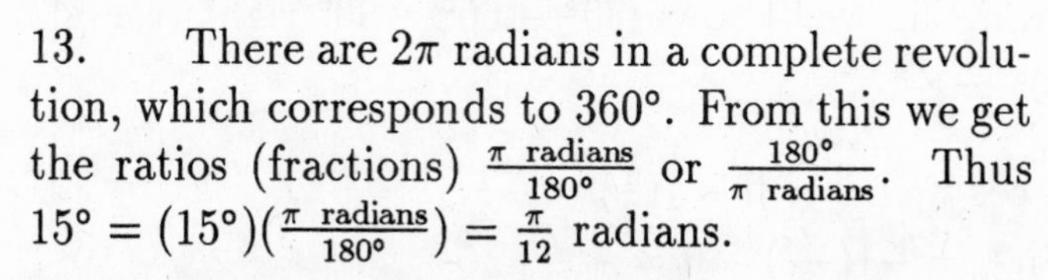

13. There are 2π radians in a complete revolution, which corresponds to $360°$. From this we get the ratios (fractions) $\frac{\pi \text{ radians}}{180°}$ or $\frac{180°}{\pi \text{ radians}}$. Thus $15° = (15°)(\frac{\pi \text{ radians}}{180°}) = \frac{\pi}{12}$ radians.

15. $-150° = (-150°)(\frac{\pi \text{ radians}}{180°}) = -\frac{5\pi}{6}$ radians.

17. $540° = (540°)(\frac{\pi \text{ radians}}{180°}) = 3\pi$ radians.

19. There are 2π radians in a complete revolution, which corresponds to $360°$. From this we get the ratios (fractions) $\frac{\pi \text{ radians}}{180°}$ or $\frac{180°}{\pi \text{ radians}}$. Thus $\frac{5\pi}{6} = \frac{5\pi \text{ radians}}{6}\frac{180°}{\pi \text{ radians}} = 150°$.

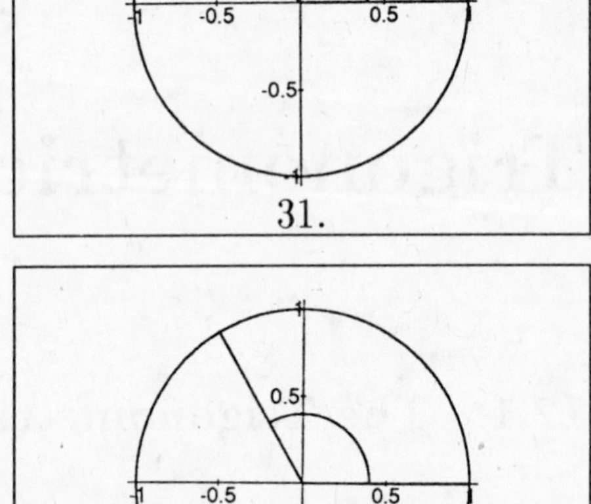

31.

21. $\frac{3\pi}{2} = \frac{3\pi \text{ radians}}{2}\frac{180°}{\pi \text{ radians}} = 270°$.

23. $-\frac{3\pi}{4} = -\frac{3\pi \text{ radians}}{4}\frac{180°}{\pi \text{ radians}} = -135°$.

25. The angle is $\frac{\pi}{2}$ radians plus $\frac{2}{3}$ of $\frac{\pi}{2}$ radians, or $\frac{\pi}{2} + \frac{2}{3}(\frac{\pi}{2}) = \frac{5\pi}{6}$ radians.

27. The angle is $\frac{\pi}{2}$ radians plus $\frac{1}{2}$ of $\frac{\pi}{2}$ radians, or $\frac{\pi}{2} + \frac{1}{2}(\frac{\pi}{2}) = \frac{3\pi}{4}$ radians.

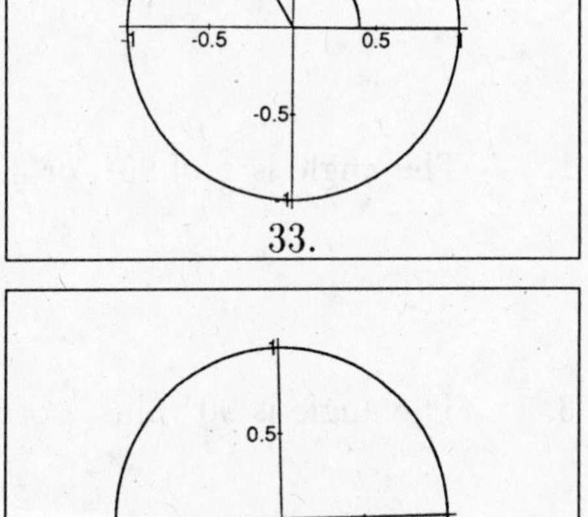

33.

29. The angle is $-\frac{\pi}{2}$ radians plus $\frac{1}{3}$ of $-\frac{\pi}{2}$ radians, or $-\frac{\pi}{2} + \frac{1}{3}(-\frac{\pi}{2}) = -\frac{2\pi}{3}$ radians.

31, 33, and 35. The diagrams for problems 31, 33, and 35 are shown in the pictures.

35.

37. $\cos\frac{7\pi}{2} = \cos(4\pi - \frac{\pi}{2}) = \cos(-\frac{\pi}{2}) = \cos\frac{\pi}{2} = 0$.

39. $\cos(-\frac{5\pi}{2}) = \cos\frac{5\pi}{2} = \cos(2\pi + \frac{\pi}{2}) = \cos\frac{\pi}{2} = 0$.

41. $\cot(\frac{5\pi}{2}) = \frac{\cos(\frac{5\pi}{2})}{\sin(\frac{5\pi}{2})} = \frac{\cos(2\pi+\frac{\pi}{2})}{\sin(2\pi+\frac{\pi}{2})} = \frac{\cos\frac{\pi}{2}}{\sin\frac{\pi}{2}} = \frac{0}{1} = 0$.

43. $\tan(-\pi) = \frac{\sin(-\pi)}{\cos(-\pi)} = \frac{-\sin\pi}{\cos\pi} = \frac{0}{-1} = 0$.

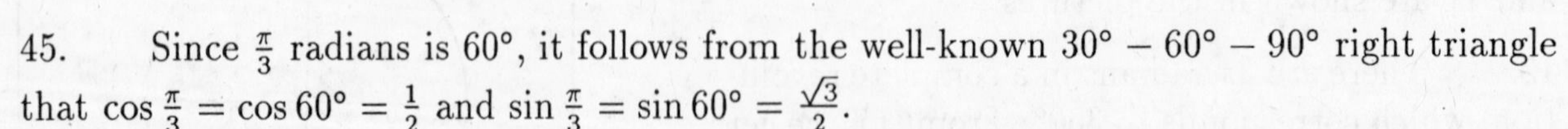

45. Since $\frac{\pi}{3}$ radians is $60°$, it follows from the well-known $30° - 60° - 90°$ right triangle that $\cos\frac{\pi}{3} = \cos 60° = \frac{1}{2}$ and $\sin\frac{\pi}{3} = \sin 60° = \frac{\sqrt{3}}{2}$.

47. $\sin\frac{7\pi}{6} = \sin(\pi + \frac{\pi}{6}) = \sin\pi\cos\frac{\pi}{6} + \cos\pi\sin\frac{\pi}{6} = 0 - 1(\frac{1}{2}) = -\frac{1}{2}$
$\cos\frac{7\pi}{6} = \cos(\pi + \frac{\pi}{6}) = \cos\pi\cos\frac{\pi}{6} - \sin\pi\sin\frac{\pi}{6} = -1(\frac{\sqrt{3}}{2}) - 0 = -\frac{\sqrt{3}}{2}$
$\sin\frac{5\pi}{4} = \sin(\pi + \frac{\pi}{4}) = \sin\pi\cos\frac{\pi}{4} + \cos\pi\sin\frac{\pi}{4} = 0 - 1(\frac{\sqrt{2}}{2}) = -\frac{\sqrt{2}}{2}$
$\cos\frac{5\pi}{4} = \cos(\pi + \frac{\pi}{4}) = \cos\pi\cos\frac{\pi}{4} - \sin\pi\sin\frac{\pi}{4} = -1(\frac{\sqrt{2}}{2}) - 0 = -\frac{\sqrt{2}}{2}$

$\sin\frac{4\pi}{3} = \sin(\pi+\frac{\pi}{3}) = \sin\pi\cos\frac{\pi}{3} + \cos\pi\sin\frac{\pi}{3} = 0 - 1(\frac{\sqrt{3}}{2}) = -\frac{\sqrt{3}}{2}$
$\cos\frac{4\pi}{3} = \cos(\pi+\frac{\pi}{3}) = \cos\pi\cos\frac{\pi}{3} - \sin\pi\sin\frac{\pi}{3} = -1(\frac{1}{2}) - 0 = -\frac{1}{2}$
$\sin\frac{3\pi}{2} = -1$ and $\cos\frac{3\pi}{2} = 0$
$\sin\frac{5\pi}{3} = \sin(\pi+\frac{2\pi}{3}) = \sin\pi\cos\frac{2\pi}{3} + \cos\pi\sin\frac{2\pi}{3} = 0 - 1(\frac{\sqrt{3}}{2}) = -\frac{\sqrt{3}}{2}$
$\cos\frac{5\pi}{3} = \cos(\pi+\frac{2\pi}{3}) = \cos\pi\cos\frac{2\pi}{3} - \sin\pi\sin\frac{2\pi}{3} = -1(-\frac{1}{2}) - 0 = \frac{1}{2}$
$\sin\frac{7\pi}{4} = \sin(\pi+\frac{3\pi}{4}) = \sin\pi\cos\frac{3\pi}{4} + \cos\pi\sin\frac{3\pi}{4} = 0 - 1(\frac{\sqrt{2}}{2}) = -\frac{\sqrt{2}}{2}$
$\cos\frac{7\pi}{4} = \cos(\pi+\frac{3\pi}{4}) = \cos\pi\cos\frac{3\pi}{4} - \sin\pi\sin\frac{3\pi}{4} = -1(-\frac{\sqrt{2}}{2}) - 0 = \frac{\sqrt{2}}{2}$
$\sin\frac{11\pi}{6} = \sin(\pi+\frac{5\pi}{6}) = \sin\pi\cos\frac{5\pi}{6} + \cos\pi\sin\frac{5\pi}{6} = 0 - 1(\frac{1}{2}) = -\frac{1}{2}$
$\cos\frac{11\pi}{6} = \cos(\pi+\frac{5\pi}{6}) = \cos\pi\cos\frac{5\pi}{6} - \sin\pi\sin\frac{5\pi}{6} = -(-\frac{\sqrt{3}}{2}) - 0 = \frac{\sqrt{3}}{2}$
$\sin 2\pi = 0$ and $\cos 2\pi = 1$.

49. $\sin(-\frac{\pi}{6}) = -\sin\frac{\pi}{6} = -\frac{1}{2}$.

51. $\cos\frac{7\pi}{3} = \cos(2\pi+\frac{\pi}{3}) = \cos\frac{\pi}{3} = \frac{1}{2}$.

53. $\tan\frac{\pi}{6} = \frac{\sin(\pi/6)}{\cos(\pi/6)} = \frac{1/2}{\sqrt{3}/2} = \frac{1}{\sqrt{3}} = \frac{\sqrt{3}}{3}$.

55. $\sec\frac{\pi}{3} = \frac{1}{\cos(\pi/3)} = \frac{1}{1/2} = 2$.

57. $\tan(-\frac{\pi}{4}) = \frac{\sin(-\pi/4)}{\cos(-\pi/4)} = \frac{-\sin(\pi/4)}{\cos(\pi/4)} = \frac{-\sqrt{2}/2}{\sqrt{2}/2} = -1$.

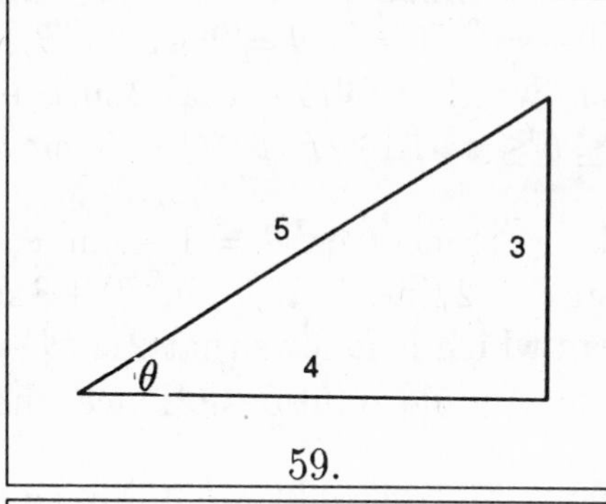

59.

59. Since $\sec\theta = \frac{5}{4}$, the corresponding right triangle shows that $\tan\theta = \frac{3}{4}$.

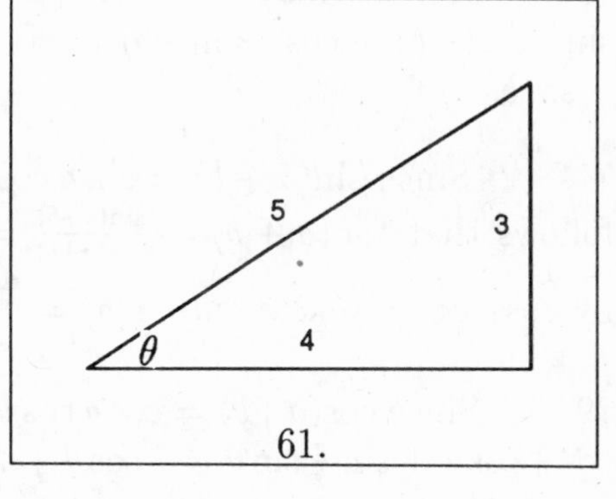

61.

61. Since $\csc\theta = \frac{1}{\sin\theta} = \frac{5}{3}$, the corresponding right triangle shows that $\tan\theta = \frac{3}{4}$.

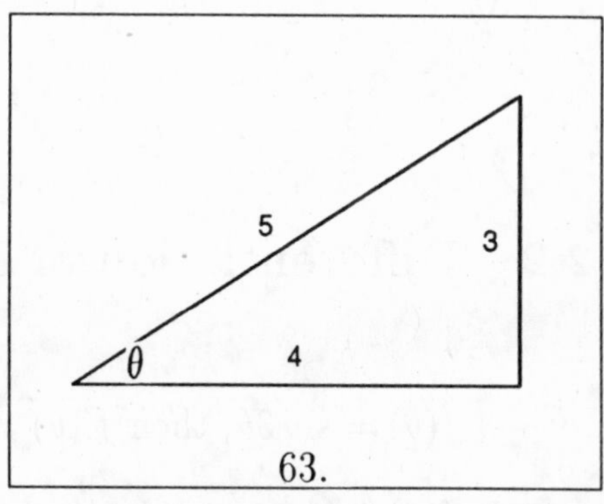

63.

63. Since $\tan\theta = \frac{3}{4}$, the corresponding right triangle shows that $\cos\theta = \frac{4}{5}$.

65. Since $\sin 2\theta = 2\sin\theta\cos\theta$, the equation can be rewritten as $0 = \cos\theta - \sin 2\theta = \cos\theta - 2\sin\theta\cos\theta = \cos\theta(1-2\sin\theta)$ from which it follows that $\cos\theta = 0$ or $\theta = \frac{\pi}{2}$ and $\theta = \frac{3\pi}{2}$. Also $1-2\sin\theta = 0$, $2\sin\theta = 1$, $\sin\theta = \frac{1}{2}$, or $\theta = \frac{\pi}{6}$ and $\theta = \frac{5\pi}{6}$.

Hence, the solutions are $\theta = \frac{\pi}{6}$, $\theta = \frac{\pi}{2}$, $\theta = \frac{5\pi}{6}$, $\theta = \frac{3\pi}{2}$.

67. Since $\sin 2\theta = 2\sin\theta\cos\theta$, the equation can be rewritten as $0 = \sin 2\theta - \sqrt{3}\cos\theta = 2\sin\theta\cos\theta - \sqrt{3}\cos\theta = \cos\theta(2\sin\theta - \sqrt{3})$ from which it follows that $\cos\theta = 0$ or $\theta = \frac{\pi}{2}$. Also $2\sin\theta - \sqrt{3} = 0$, $\sin\theta = \frac{\sqrt{3}}{2}$, or $\theta = \frac{\pi}{3}$ and $\theta = \frac{2\pi}{3}$. Hence, the solutions are $\theta = \frac{\pi}{3}$, $\theta = \frac{\pi}{2}$, and $\theta = \frac{2\pi}{3}$.

69. Since $\sin 2\theta = 2\sin\theta\cos\theta$, the equation can be rewritten as $0 = 2\cos^2\theta - \sin 2\theta = 2\cos^2\theta - 2\sin\theta\cos\theta = 2\cos\theta(\cos\theta - \sin\theta)$ from which it follows that $2\cos\theta = 0$ or $\theta = \frac{\pi}{2}$. Also $\cos\theta - \sin\theta = 0$, $\cos\theta = \sin\theta$, or $\theta = \frac{\pi}{4}$. Hence, the solutions are $\theta = \frac{\pi}{4}$ and $\theta = \frac{\pi}{2}$.

71. Since $\cos^2\theta = 1 - \sin^2\theta$, the equation can be rewritten as $1 = 2\cos^2\theta + \sin\theta = 2(1-\sin^2\theta) + \sin\theta = 2 - 2\sin^2\theta + \sin\theta$, $2\sin^2\theta - \sin\theta - 1 = 0$ or $(2\sin\theta + 1)(\sin\theta - 1) = 0$ from which it follows that $2\sin\theta + 1 = 0$ or $\sin\theta = -\frac{1}{2}$, which has no solution in the interval $0 \le \theta \le \pi$. Also $\sin\theta - 1 = 0$, $\sin\theta = 1$, or $\theta = \frac{\pi}{2}$. Hence, the only solution is $\theta = \frac{\pi}{2}$.

73. Since $\cos^2\theta = 1 - \sin^2\theta$, the equation can be rewritten as $1 = 2\sin^2\theta - \cos^2\theta + 3\sin\theta = 2\sin^2\theta - (1 - \sin^2\theta) + 3\sin\theta$, $2\sin^2\theta + 3\sin\theta - 2 = 0$, or $(2\sin\theta - 1)(\sin\theta + 2) = 0$ from which it follows that $2\sin\theta - 1 = 0$, $\sin\theta = \frac{1}{2}$, or $\theta = \frac{\pi}{6}$ and $\theta = \frac{5\pi}{6}$. Also $\sin\theta + 2 = 0$, which has no solutions. Hence, the solutions are $\theta = \frac{\pi}{6}$ and $\theta = \frac{5\pi}{6}$.

75. Replace b by $-b$ in the identity $\sin(a+b) = \sin a\cos b + \cos a\sin b$ to get $\sin(a-b) = \sin a\cos(-b) + \cos a\sin(-b) = \sin a\cos b - \cos a\sin b$ since $\cos(-b) = \cos b$ and $\sin(-b) = -\sin b$.

77. Since $\sin(a+b) = \sin a\cos b + \cos a\sin b$ and $\cos(a+b) = \cos a\cos b - \sin a\sin b$, it follows that $\tan(a+b) = \frac{\sin(a+b)}{\cos(a+b)} = \frac{\sin a\cos b + \cos a\sin b}{\cos a\cos b - \sin a\sin b}$. Now divide each term in the quotient by $\cos a\cos b$ to get $\tan(a+b) = \frac{\frac{\sin a\cos b}{\cos a\cos b} + \frac{\cos a\sin b}{\cos a\cos b}}{\frac{\cos a\cos b}{\cos a\cos b} - \frac{\sin a\sin b}{\cos a\cos b}} = \frac{\frac{\sin a}{\cos a} - \frac{\sin b}{\cos b}}{1 - \frac{\sin a}{\cos a}\frac{\sin b}{\cos b}} = \frac{\tan a - \tan b}{1 - \tan a\tan b}$.

79. Since $\cos(a+b) = \cos a\cos b - \sin a\sin b$, $\cos(\frac{\pi}{2} - \theta) = \cos\frac{\pi}{2}\cos(-\theta) - \sin(\frac{\pi}{2})\sin(-\theta) = \cos\frac{\pi}{2}\cos\theta + \sin\frac{\pi}{2}\sin\theta = 0\cos\theta + 1\sin\theta = \sin\theta$.

12.2 Differentiation of Trigonometric Functions

1. If $f(\theta) = \sin 3\theta$, then $f'(\theta) = \cos 3\theta\frac{d}{d\theta}(3\theta) = 3\cos 3\theta$.

3. If $f(\theta) = \sin(1 - 2\theta)$, then $f'(\theta) = \cos(1 - 2\theta)\frac{d}{d\theta}(1 - 2\theta) = -2\cos(1 - 2\theta)$.

5. If $f(\theta) = \cos(\theta^3 + 1)$, then $f'(\theta) = -\sin(\theta^3+1)\frac{d}{d\theta}(\theta^3+1) = -3\theta^2\sin(\theta^3+1)$.

7. If $f(\theta) = \cos^2(\frac{\pi}{2} - \theta) = [\cos(\frac{\pi}{2} - \theta)]^2$, then $f'(\theta) = 2\cos(\frac{\pi}{2} - \theta)\frac{d}{d\theta}[\cos(\frac{\pi}{2} - \theta)] = 2\cos(\frac{\pi}{2}-\theta)[-\sin(\frac{\pi}{2}-\theta)]\frac{d}{d\theta}(\frac{\pi}{2}-\theta) = 2\cos(\frac{\pi}{2}-\theta)[-\sin(\frac{\pi}{2}-\theta)](-1) = 2\cos(\frac{\pi}{2}-\theta)\sin(\frac{\pi}{2}-\theta) = \sin 2(\frac{\pi}{2}-\theta) = \sin(\pi - 2\theta)$ (since $\sin 2a = 2\sin a \cos a$).

9. If $f(\theta) = \cos(1+3\theta)^2$ then $f'(\theta) = -\sin(1+3\theta)^2\frac{d}{d\theta}(1+3\theta)^2 = -\sin(1+3\theta)^2[2(1+3\theta)(3)] = -6(1+3\theta)\sin(1+3\theta)^2$.

11. If $f(\theta) = e^{-\theta/2}\cos 2\pi\theta$, then, by the product rule, $f'(\theta) = e^{-\theta/2}\frac{d}{d\theta}\cos 2\pi\theta + \cos 2\pi\theta\frac{d}{d\theta}e^{-\theta/2} = e^{-\theta/2}[-\sin 2\pi\theta\frac{d}{d\theta}(2\pi\theta)] + \cos 2\pi\theta[e^{-\theta/2}\frac{d}{d\theta}(-\theta/2)] = e^{-\theta/2}(-2\pi\sin 2\pi\theta) + \cos 2\pi\theta[-\frac{1}{2}e^{-\theta/2}] = -e^{-\theta/2}(2\pi\sin 2\pi\theta + \frac{1}{2}\cos 2\pi\theta)$.

13. If $f(\theta) = \frac{\sin\theta}{1+\sin\theta}$, then, by the quotient rule, $f'(\theta) = \frac{(1+\sin\theta)\frac{d}{d\theta}\sin\theta - (\sin\theta)\frac{d}{d\theta}(1+\sin\theta)}{(1+\sin\theta)^2} = \frac{(1+\sin\theta)(\cos\theta)-(\sin\theta)(\cos\theta)}{(1+\sin\theta)^2} = \frac{\cos\theta}{(1+\sin\theta)^2}$.

15. If $f(\theta) = \tan(1-\theta^5)$, then $f'(\theta) = \sec^2(1-\theta^5)\frac{d}{d\theta}(1-\theta^5) = -5\theta^4\sec^2(1-\theta^5)$.

17. If $f(\theta) = \tan^2(\frac{\pi}{2} - 2\pi\theta) = [\tan(\frac{\pi}{2} - 2\pi\theta)]^2$, then $f'(\theta) = 2[\tan(\frac{\pi}{2} - 2\pi\theta)]\frac{d}{d\theta}[\tan(\frac{\pi}{2} - 2\pi\theta)] = 2[\tan(\frac{\pi}{2} - 2\pi\theta)][\sec^2(\frac{\pi}{2} - 2\pi\theta)]\frac{d}{d\theta}(\frac{\pi}{2} - 2\pi\theta) = 2\tan(\frac{\pi}{2} - 2\pi\theta)\sec^2(\frac{\pi}{2} - 2\pi\theta)(-2\pi) = -4\pi\tan(\frac{\pi}{2} - 2\pi\theta)\sec^2(\frac{\pi}{2} - 2\pi\theta)$.

19. If $f(\theta) = \ln\sin^2\theta = \ln(\sin\theta)^2 = 2\ln\sin\theta$, then $f'(\theta) = \frac{2}{\sin\theta}\frac{d}{d\theta}(\sin\theta) = \frac{2\cos\theta}{\sin\theta} = 2\cot\theta$.

21. If $f(\theta) = \sec\theta = \frac{1}{\cos\theta} = (\cos\theta)^{-1}$, then $f'(\theta) = -(\cos\theta)^{-2}\frac{d}{d\theta}(\cos\theta)$ $= -(\cos\theta)^{-2}(-\sin\theta) = \frac{\sin\theta}{\cos^2\theta} = (\frac{\sin\theta}{\cos\theta})(\frac{1}{\cos\theta}) = \tan\theta\sec\theta$. Warning: $(\cos\theta)^{-1}$ is not the same as $\cos^{-1}\theta$. This latter symbol refers to the inverse cosine, which is beyond the scope of this course.

23. If $f(\theta) = \cot\theta = \frac{\cos\theta}{\sin\theta}$, then $f'(\theta) = \frac{\sin\theta\frac{d}{d\theta}(\cos\theta) - \cos\theta\frac{d}{d\theta}(\sin\theta)}{\sin^2\theta} = \frac{\sin\theta(-\sin\theta)-\cos\theta(\cos\theta)}{\sin^2\theta} = \frac{-\sin^2\theta-\cos^2\theta}{\sin^2\theta} = -\frac{\sin^2\theta+\cos^2\theta}{\sin^2\theta} = \frac{1}{\sin^2\theta} = -\csc^2\theta$.

25. If $f(\theta) = \cos\theta$, then $f'(\theta) = \lim_{\Delta\theta\to 0}\frac{\cos(\theta+\Delta\theta)-\cos\theta}{\Delta\theta}$ (from the definition of the derivative). Thus, $f'(\theta) = \lim_{\Delta\theta\to 0}\frac{\cos(\theta)\cos\Delta\theta - \sin\theta\sin\Delta\theta - \cos\theta}{\Delta\theta} = \lim_{\Delta\theta\to 0}[\frac{\cos(\theta)\cos\Delta\theta-\cos\theta}{\Delta\theta} - \frac{\sin\theta\sin\Delta\theta}{\Delta\theta}] = \lim_{\Delta\theta\to 0}\cos\theta(\frac{\cos\Delta\theta-1}{\Delta\theta}) - \lim_{\Delta\theta\to 0}\sin\theta\frac{\sin\Delta\theta}{\Delta\theta} = \cos\theta\lim_{\Delta\theta\to 0}\frac{\cos\Delta\theta-1}{\Delta\theta} - \sin\theta\lim_{\Delta\theta\to 0}\frac{\sin\Delta\theta}{\Delta\theta} = (\cos\theta)(0) - (\sin\theta)(1) = -\sin\theta$.

12.3 Applications of Trigonometric Functions

1. Let x denote the horizontal distance (in miles) between the plane and the observer, let t denote time (in hours), and draw a diagram representing the situation. It is given that $\frac{dx}{dt} = -500$ $\frac{\text{miles}}{\text{hour}}$. From the right triangle in the first figure, $\tan\theta = \frac{3}{x}$. Differentiating both sides of this equation with respect to t yields $\sec^2\theta \frac{d\theta}{dt} = -\frac{3}{x^2}\frac{dx}{dt}$ or $\frac{d\theta}{dt} = -\frac{3}{x^2}\frac{1}{\sec^2\theta}\frac{dx}{dt} = -\frac{3\cos^2\theta}{x^2}\frac{dx}{dt}$. When $x = 4$, the corresponding $3-4-5$ right triangle gives $\cos\theta = \frac{4}{5}$. Hence, $\frac{d\theta}{dt} = -\frac{3}{4^2}(\frac{4}{5})^2(-500) = 60$ $\frac{\text{radians}}{\text{hour}}$.

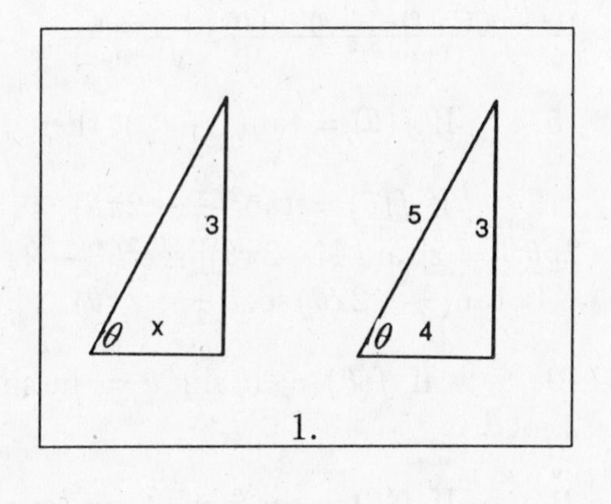

1.

3. Let x denote the length of the rope (from the rowboat to the end of the pier), let t denote time (in minutes), and draw a diagram representing the situation. It is given that $\frac{dx}{dt} = -4$ $\frac{\text{feet}}{\text{minute}}$. From the right triangle in the first figure, $\sin\theta = \frac{12}{x}$. Differentiating both sides of this equation with respect to t yields $\cos\theta\frac{d\theta}{dt} = -\frac{12}{x^2}\frac{dx}{dt}$ or $\frac{d\theta}{dt} = -\frac{12}{x^2}\frac{1}{\cos\theta}\frac{dx}{dt} = -\frac{12\sec\theta}{x^2}\frac{dx}{dt}$. When the distance from the boat to the pier is 16 feet, the corresponding $12-16-20$ right triangle gives $\sec\theta = \frac{20}{16} = \frac{5}{4}$. Hence, $\frac{d\theta}{dt} = -\frac{12}{(20)^2}(\frac{5}{4})(-4) = 0.15$ $\frac{\text{radians}}{\text{minute}}$.

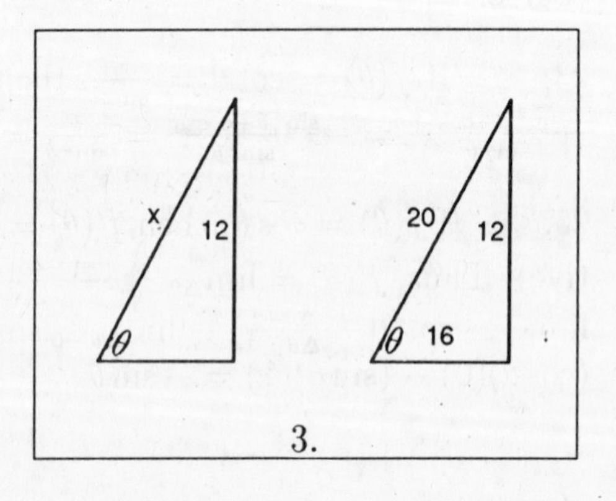

3.

5. Let x denote the volume of the trough. Then $V =$(area of triangular cross section)(length of trough). The area A of the triangle is $\frac{1}{2}$(base)(height)$= \frac{bh}{2}$. From the right triangle ABD $\cos\frac{\theta}{2} = \frac{h}{3}$ or $h = 3\cos\frac{\theta}{2}$ and $\sin\frac{\theta}{2} = \frac{b/2}{3}$ or $b = 6\sin\frac{\theta}{2}$. Hence, $A = \frac{bh}{2} = \frac{1}{2}(6\sin\frac{\theta}{2})(3\cos\frac{\theta}{2}) = 9\sin\frac{\theta}{2}\cos\frac{\theta}{2}$. By the double-angle formula $\sin 2\alpha = 2\sin\alpha\cos\alpha$ with $\alpha = \frac{\theta}{2}$, so $\sin\frac{\theta}{2}\cos\frac{\theta}{2} = \frac{1}{2}\sin\theta$. Thus $A = \frac{9}{2}\sin\theta$. Hence, $V(\theta) = (\frac{9}{2}\sin\theta)(20) = 90\sin\theta$. The goal is to maximize $V(\theta)$ for $0 \le \theta \le \pi$. The derivative is $V'(\theta) = 90\cos\theta$ which is 0 when $\theta = \frac{\pi}{2}$. Since $V(0) = 0$, $V(\pi) = 0$, and $V(\frac{\pi}{2}) > 0$, it follows that the volume is greatest when $\theta = \frac{\pi}{2}$ radians.

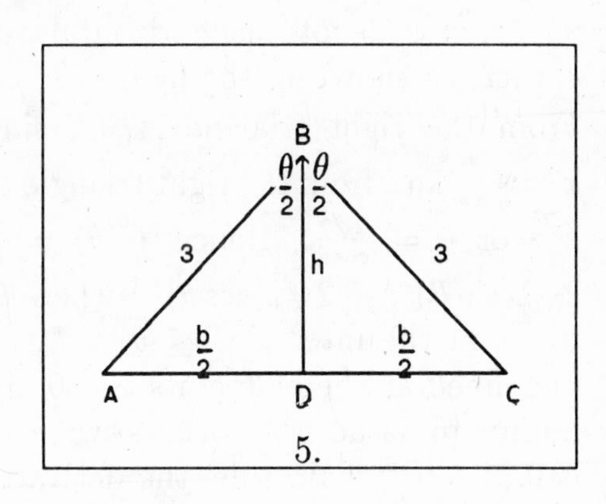

5.

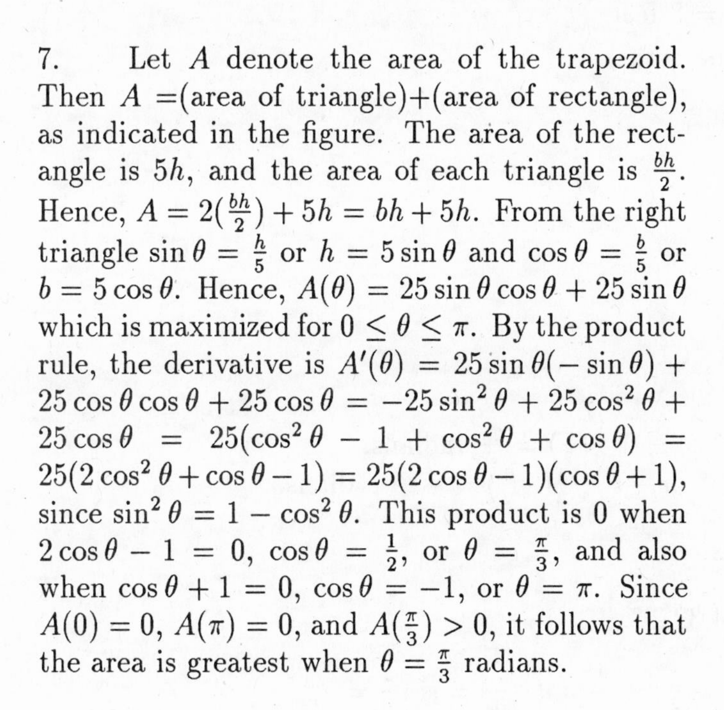

7. Let A denote the area of the trapezoid. Then $A =$(area of triangle)+(area of rectangle), as indicated in the figure. The area of the rectangle is $5h$, and the area of each triangle is $\frac{bh}{2}$. Hence, $A = 2(\frac{bh}{2}) + 5h = bh + 5h$. From the right triangle $\sin\theta = \frac{h}{5}$ or $h = 5\sin\theta$ and $\cos\theta = \frac{b}{5}$ or $b = 5\cos\theta$. Hence, $A(\theta) = 25\sin\theta\cos\theta + 25\sin\theta$ which is maximized for $0 \le \theta \le \pi$. By the product rule, the derivative is $A'(\theta) = 25\sin\theta(-\sin\theta) + 25\cos\theta\cos\theta + 25\cos\theta = -25\sin^2\theta + 25\cos^2\theta + 25\cos\theta = 25(\cos^2\theta - 1 + \cos^2\theta + \cos\theta) = 25(2\cos^2\theta + \cos\theta - 1) = 25(2\cos\theta - 1)(\cos\theta + 1)$, since $\sin^2\theta = 1 - \cos^2\theta$. This product is 0 when $2\cos\theta - 1 = 0$, $\cos\theta = \frac{1}{2}$, or $\theta = \frac{\pi}{3}$, and also when $\cos\theta + 1 = 0$, $\cos\theta = -1$, or $\theta = \pi$. Since $A(0) = 0$, $A(\pi) = 0$, and $A(\frac{\pi}{3}) > 0$, it follows that the area is greatest when $\theta = \frac{\pi}{3}$ radians.

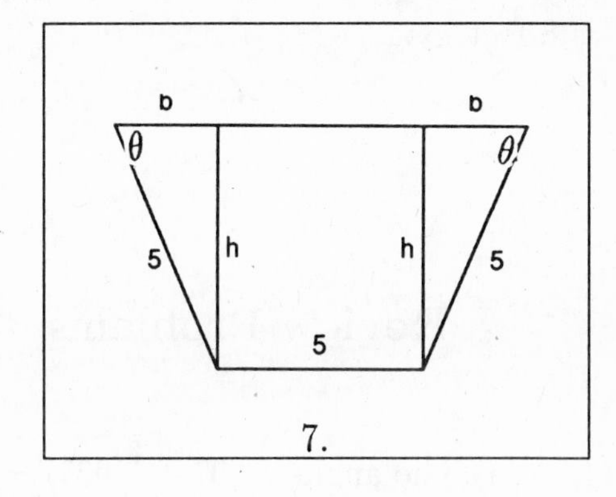

7.

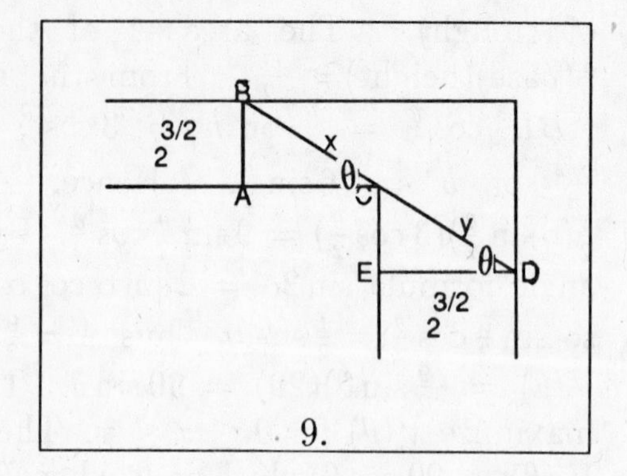

9.

9. Let C denote the horizontal clearance at the corner, as shown in the figure. Then $C = x + y$. From the right triangle ABC, $\sin\theta = \frac{2\sqrt{2}}{x}$ or $x = \frac{2\sqrt{2}}{\sin\theta}$ and from the right triangle CED, $\cos\theta = \frac{2\sqrt{2}}{y}$ or $y = \frac{2\sqrt{2}}{\cos\theta}$. Hence, $C(\theta) = \frac{2\sqrt{2}}{\sin\theta} + \frac{2\sqrt{2}}{\cos\theta} = 2\sqrt{2}(\sin\theta)^{-1} + 2\sqrt{2}(\cos\theta)^{-1}$ where θ is restricted to be in the interval $0 \le \theta \le \frac{\pi}{2}$. Since $C(\theta)$ is undefined at the endpoints $\theta = 0$ and $\theta = \frac{\pi}{2}$, the minimum value of C occurs when the derivative is 0. By the chain rule, the derivative is $C'(\theta) = -2\sqrt{2}(\sin\theta)^{-2}(\cos\theta) - 2\sqrt{2}(\cos\theta)^{-2}(-\sin\theta) = 2\sqrt{2}(\frac{\sin\theta}{\cos^2\theta} - \frac{\cos\theta}{\sin^2\theta}) = 2\sqrt{2}\frac{\sin^3\theta - \cos^3\theta}{\cos^2\theta\sin^2\theta} = 0$ when $\cos^3\theta = \sin^3\theta$, $\cos\theta = \sin\theta$, or $\theta = \frac{\pi}{4}$ radians. Thus the minimal clearance is $C(\frac{\pi}{4}) = \frac{2\sqrt{2}}{\sin(\pi/4)} + \frac{2\sqrt{2}}{\cos(\pi/4)} = \frac{2\sqrt{2}}{\sqrt{2}/2} + \frac{2\sqrt{2}}{\sqrt{2}/2} = 4 + 4 = 8$ feet. Hence, the longest pipe that can clear the corner is 8 feet long.

Review Problems

1. a) The angle is $90° + \frac{1}{3}(90°) = 120°$ or $\frac{\pi}{2} + \frac{1}{3}(\frac{\pi}{2}) = \frac{2\pi}{3}$ radians.
b) The angle is $-180° + \frac{1}{2}(-90°) = -225°$ or $-\pi + \frac{1}{2}(-\frac{\pi}{2}) = -\frac{5\pi}{4}$ radians.

2. $50° = 50°\frac{\pi \text{ radians}}{180°} = 0.8727$ radians.

3. $0.25 = (0.25 \text{ radians})\frac{180°}{\pi \text{ radians}} = 14.3239°$.

4. a) $\sin(-\frac{5\pi}{3}) = -\sin\frac{5\pi}{3} = -\sin(2\pi - \frac{\pi}{3}) = -\sin(-\frac{\pi}{3}) = \sin\frac{\pi}{3} = \frac{\sqrt{3}}{2}$
b) $\cos\frac{15\pi}{4} = \cos(4\pi - \frac{\pi}{4}) = \cos(-\frac{\pi}{4}) = \cos\frac{\pi}{4} = \frac{\sqrt{2}}{2}$
c) $\sec\frac{7\pi}{3} = \sec(2\pi + \frac{\pi}{3}) = \sec(\frac{\pi}{3}) = \frac{1}{\cos(\pi/3)} = 2$
d) $\cot\frac{2\pi}{3} = \frac{\cos(2\pi/3)}{\sin(2\pi/3)} = \frac{-1/2}{\sqrt{3}/2} = -\frac{\sqrt{3}}{3}$.

5. If $\sin\theta = \frac{4}{5}$, the corresponding $3-4-5$ right triangle shows that $\tan\theta = \frac{4}{3}$.

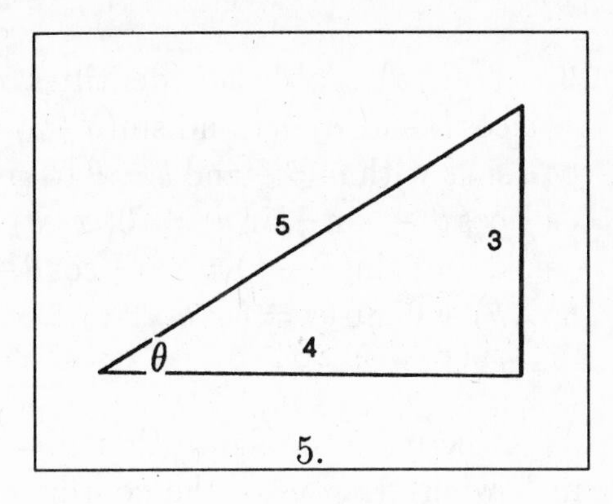

5.

6. If $\cot\theta = \frac{\sqrt{5}}{2}$, the corresponding right triangle shows that $\csc\theta = \frac{3}{2}$.

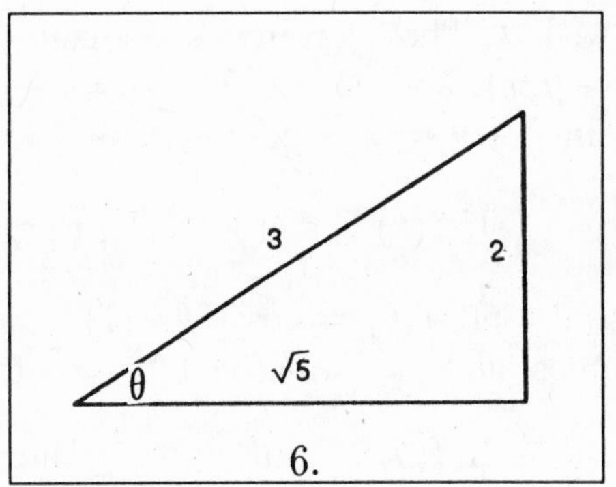

6.

7. Since $\sin 2\theta = 2\sin\theta\cos\theta$, the equation can be rewritten as $0 = 2\cos\theta + \sin 2\theta = 2\cos\theta + 2\sin\theta\cos\theta = 2\cos\theta(1+\sin\theta) = 0$ if $\cos\theta = 0$ or $\theta = \frac{\pi}{2}$ and $\theta = \frac{3\pi}{2}$ or $1 + \sin\theta = 0$, $\sin\theta = -1$, or $\theta = \frac{3\pi}{2}$. Hence the solutions are $\theta = \frac{\pi}{2}$ and $\frac{3\pi}{2}$.

8. Since $\cos^2\theta = 1 - \sin^2\theta$, the equation can be rewritten as $2 = 3\sin^2\theta - \cos^2\theta = 3\sin^2\theta - (1 - \sin^2\theta) = 4\sin^2\theta - 1$, and so $\sin^2\theta = \frac{3}{4}$ or $\sin\theta = \pm\frac{\sqrt{3}}{2}$. On the interval $0 \le \theta \le \pi$, the equation $\sin\theta = \frac{\sqrt{3}}{2}$ implies $\theta = \frac{\pi}{3}$ and $\theta = \frac{2\pi}{3}$, while the equation $\sin\theta = -\frac{\sqrt{3}}{2}$ has no solution on this interval. Hence the only solutions are $\theta = \frac{\pi}{3}$ and $\frac{2\pi}{3}$.

9. Since $\cos 2\theta = \cos^2\theta - \sin^2\theta = 1 - \sin^2\theta - \sin^2\theta = 1 - 2\sin^2\theta$, the equation can be rewritten as $0 = 2\sin^2\theta - (1 - 2\sin^2\theta) = 4\sin^2\theta - 1$ or $\sin\theta = \pm\frac{1}{2}$. On the interval $0 \le \theta \le \pi$, the equation $\sin\theta = \frac{1}{2}$ implies $\theta = \frac{\pi}{6}$ or $\theta = \frac{5\pi}{6}$, while the equation $\sin\theta = -\frac{1}{2}$ has no solution. Hence the solutions are $\theta = \frac{\pi}{6}$ and $\frac{5\pi}{6}$.

10. Since $\cos^2\theta = 1 - \sin^2\theta$, the equation can be rewritten as $5\sin\theta - 2(1 - \sin^2\theta) = 1$, $2\sin^2\theta + 5\sin\theta - 3 = 0$, or $(2\sin\theta - 1)(\sin\theta + 3) = 0$ which implies that $2\sin\theta - 1 = 0$, $\sin\theta = \frac{1}{2}$, or $\theta = \frac{\pi}{6}$ and $\theta = \frac{5\pi}{6}$. The other factor $\sin\theta + 3 = 0$ has no solution. Hence the solutions are $\theta = \frac{\pi}{6}$ and $\frac{5\pi}{6}$.

11. Start with $\sin^2\theta + \cos^2\theta = 1$, and divide by $\sin^2\theta$ to get $1 + \frac{\cos^2\theta}{\sin^2\theta} = \frac{1}{\sin^2\theta}$, $1 + (\frac{\cos\theta}{\sin\theta})^2 = (\frac{1}{\sin\theta})^2$, or $1 + \cot^2\theta = \csc^2\theta$.

12. Since $\sin 2\theta = 2\sin\theta\cos\theta$ and $\cos 2\theta = \cos^2\theta - \sin^2\theta$, $\tan 2\theta = \frac{\sin 2\theta}{\cos 2\theta} = \frac{2\sin\theta\cos\theta}{\cos^2\theta - \sin^2\theta} = \frac{\frac{2\sin\theta\cos\theta}{\cos^2\theta}}{\frac{\cos^2\theta}{\cos^2\theta} - \frac{\sin^2\theta}{\cos^2\theta}} = \frac{\frac{2\sin\theta}{\cos\theta}}{1 - (\frac{\sin\theta}{\cos\theta})^2} = \frac{2\tan\theta}{1 - \tan^2\theta}$.

13. a)Apply the identities $\cos(a+b) = \cos a \cos b - \sin a \sin b$ and $\sin(a+b) = \sin a \cos b + \cos a \sin b$ with $a = \frac{\pi}{2}$ and $b = \theta$ to get $\cos(\frac{\pi}{2}+\theta) = \cos\frac{\pi}{2}\cos\theta - \sin\frac{\pi}{2}\sin\theta = 0(\cos\theta) - 1(\sin\theta) = -\sin\theta$ and $\sin(\frac{\pi}{2}+\theta) = \sin\frac{\pi}{2}\cos\theta + \cos\frac{\pi}{2}\sin\theta = 1(\cos\theta) + 0(\sin\theta) = \cos\theta$

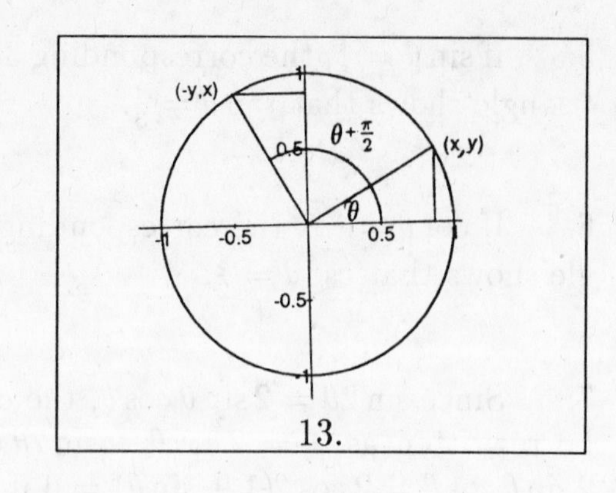

13.

b) Notice from the figure that $x = \cos\theta$ and $y = \sin\theta$, where (x, y) are the coordinates associated with θ. The coordinates associated with $\theta + \frac{\pi}{2}$ are $(-y, x)$, and so $\cos(\frac{\pi}{2}+\theta) = -y = -\sin\theta$ and $\sin(\frac{\pi}{2}+\theta) = x = \cos\theta$.

14. If $f(\theta) = \sin(3\theta+1)^2$, then $f'(\theta) = \cos(3\theta+1)^2\frac{d}{d\theta}(3\theta+1)^2 = 6(3\theta+1)\cos(3\theta+1)^2$.

15. If $f(\theta) = \cos^2(3\theta+1) = [\cos(3\theta+1)]^2$, then $f'(\theta) = 2\cos(3\theta+1)\frac{d}{d\theta}\cos(3\theta+1) = 2\cos(3\theta+1)[-3\sin(3\theta+1)] = -6\cos(3\theta+1)\sin(3\theta+1) = -3\cos(6\theta+2)$.

16. If $f(\theta) = \tan(3\theta^2+1)$, then $f'(\theta) = \sec^2(3\theta^2+1)\frac{d}{d\theta}(3\theta^2+1) = 6\theta\sec^2(3\theta^2+1)$.

17. If $f(\theta) = \tan^2(3\theta^2+1) = [\tan(3\theta^2+1)]^2$, then $f'(\theta) = 2\tan(3\theta^2+1)\frac{d}{d\theta}\tan(3\theta^2+1) = 2\tan(3\theta^2+1)\sec^2(3\theta^2+1)(6\theta) = 12\theta\tan(3\theta^2+1)\sec^2(3\theta^2+1)$

18. If $f(\theta) = \frac{\sin\theta}{1-\cos\theta}$, then, by the quotient rule, $f'(\theta) = \frac{(1-\cos\theta)\frac{d}{d\theta}(\sin\theta) - \sin\theta\frac{d}{d\theta}(1-\cos\theta)}{(1-\cos\theta)^2} = \frac{(1-\cos\theta)(\cos\theta)-(\sin\theta)(\sin\theta)}{(1-\cos\theta)^2} = \frac{\cos\theta-\cos^2\theta-\sin^2\theta}{(1-\cos\theta)^2} = \frac{\cos\theta-1}{(1-\cos\theta)^2} = -\frac{1}{1-\cos\theta}$

19. If $f(\theta) = \ln\cos^2\theta = 2\ln\cos\theta$, then, $f'(\theta) = 2\frac{1}{\cos\theta}\frac{d}{d\theta}(\cos\theta) = -\frac{2\sin\theta}{\cos\theta} = -2\tan\theta$.

20. By the product rule $\frac{d}{d\theta}(\sin\theta\cos\theta) = (\sin\theta)\frac{d}{d\theta}\cos\theta + (\cos\theta)\frac{d}{d\theta}\sin\theta = (\sin\theta)(-\sin\theta) + (\cos\theta)(\cos\theta) = \cos^2\theta - \sin^2\theta = \cos 2\theta$.

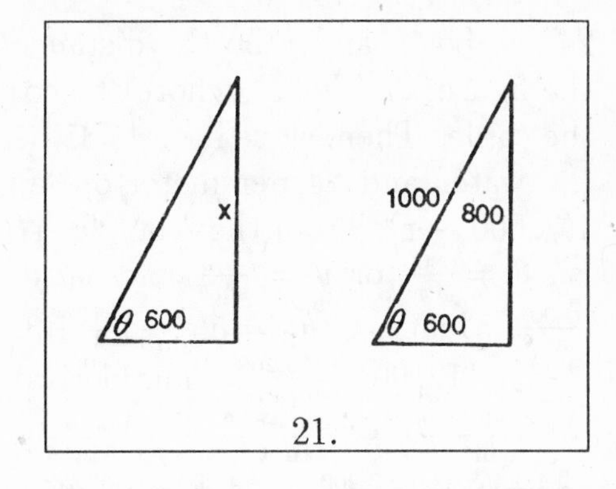

21.

21. Let x denote the distance (in feet) from the ground to the ball, let t denote time (in minutes), and draw a diagram representing the situation.
It is given that $\frac{dx}{dt} = -20 \frac{\text{feet}}{\text{minute}}$. From the right triangle in the first figure, $\tan\theta = \frac{x}{600}$. Differentiating both sides of this equation with respect to t yields $\sec^2\theta \frac{d\theta}{dt} = \frac{1}{600}\frac{dx}{dt}$ or $\frac{d\theta}{dt} = \frac{1}{600}\frac{1}{\sec^2\theta}\frac{dx}{dt} = \frac{\cos^2\theta}{600}\frac{dx}{dt}$. When $x = 800$, the corresponding $600-800-1,000$ triangle gives $\cos\theta = \frac{600}{1,000} = \frac{3}{5}$. Hence, $\frac{d\theta}{dt} = \frac{1}{600}(\frac{3}{5})^2(-20) = -0.012 \frac{\text{radians}}{\text{minute}}$.

22. Let V denote the volume of the trough. Then, V =(area of trapezoidal cross section)(length of trough). The length of the trough is 9 meters. As in problem 7 of section 12.3 (with 5 replaced by 4), the area A of the cross section is $A = (4\cos\theta)(4\sin\theta) + 4(4\sin\theta) = 16\cos\theta\sin\theta + 16\sin\theta$. Hence, $V(\theta) = 9(16)(\cos\theta\sin\theta + \sin\theta)$. The derivative is $V'(\theta) = 144(\cos^2\theta) - \sin^2\theta + \cos\theta) = 144[\cos^2\theta - (1-\cos^2\theta) + \cos\theta] = 144(2\cos^2\theta + \cos\theta - 1) = 144(2\cos\theta - 1)(\cos\theta + 1)$ which equals 0 if $2\cos\theta - 1 = 0$, $\cos\theta = \frac{1}{2}$, or $\theta = \frac{\pi}{3}$, and $\cos\theta = -1$ or $\theta = \pi$. The relevant interval is $0 \le \theta \le \pi$. Since $V(0) = 0$, $V(\pi) = 0$, and $V(\frac{\pi}{3}) > 0$, it follows that the maximum volume is achieved when $\theta = \frac{\pi}{3}$ radians.

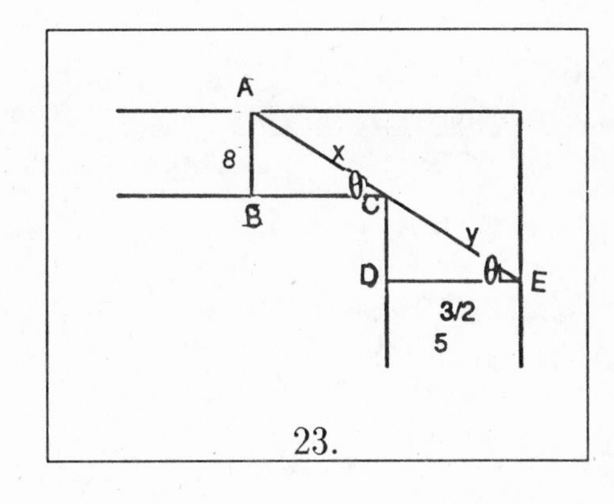

23.

23. Let C denote the horizontal clearance at the corner, as shown in the figure. Then From the right triangle ABC, $\sin\theta = \frac{8}{x}$ or $x = \frac{8}{\sin\theta}$ and from the right triangle CED, $\cos\theta = \frac{5\sqrt{5}}{y}$ or $y = \frac{5\sqrt{5}}{\cos\theta}$. Hence, $C(\theta) = \frac{8}{\sin\theta} + \frac{5\sqrt{5}}{\cos\theta} = 8(\sin\theta)^{-1} + 5\sqrt{5}(\cos\theta)^{-1}$ where the relevant interval is $0 \le \theta \le \frac{\pi}{2}$. Since the function is undefined at the endpoints, the minimum occurs when the derivative is 0. By the chain rule, $C'(\theta) = -8(\sin\theta)^{-2}(\cos\theta) - 5\sqrt{5}(\cos\theta)^{-2}(-\sin\theta) = 5\sqrt{5}(\frac{\sin\theta}{\cos^2\theta}) - 8(\frac{\cos\theta}{\sin^2\theta}) = \frac{5\sqrt{5}\sin^3\theta - 8\cos^3\theta}{\cos^2\theta\sin^2\theta} = 0$ when $8\cos^3\theta = 5\sqrt{5}\sin^3\theta$, $\frac{\sin^3\theta}{\cos^3\theta} = \frac{8}{5\sqrt{5}}$, $\tan^3\theta = \frac{8}{5^{3/2}}$, or $\tan\theta = \frac{8^{1/3}}{(5^{3/2})^{1/3}} = \frac{2}{\sqrt{5}}$.

From the corresponding right triangle, $\sin\theta = \frac{2}{3}$ and $\cos\theta = \frac{\sqrt{5}}{3}$. Hence, the minimal clearance is $C = \frac{8}{\sin\theta} + \frac{5\sqrt{5}}{\cos\theta} = 8(\frac{3}{2}) + 5\sqrt{5}(\frac{3}{\sqrt{5}}) = 12 + 15 = 27$ feet.

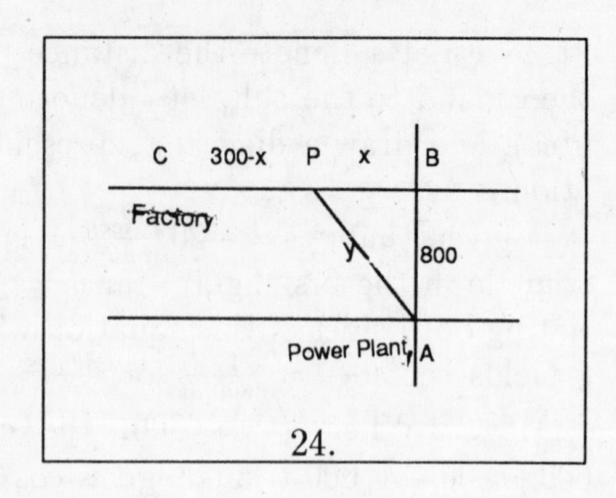

24.

24. Let x and y be the distances indicated in the figure, and let C denote the cost of installing the cable. Then, since it costs \$5 per meter under the water and \$4 per meter on land, $C = 5y + 4(3,000 - x)$. From the triangle ABP, $\sin\theta = \frac{800}{y}$ or $y = \frac{800}{\sin\theta}$ and $\tan\theta = \frac{800}{x}$ or $x = \frac{800}{\tan\theta}$. Hence, $C(\theta) = 5(\frac{800}{\sin\theta}) + 4(3,000 - \frac{800}{\tan\theta}) = \frac{4,000}{\sin\theta} + 12,000 - \frac{3,200}{\tan\theta}$. The derivative is $C'(\theta) = -(\frac{4,000\cos\theta}{\sin^2\theta}) + \frac{3,200\sec^2\theta}{\tan^2\theta} = -\frac{4,000\cos\theta}{\sin^2\theta} + \frac{3,200/\cos^2\theta}{\sin^2\theta/\cos^2\theta} = -\frac{4,000\cos\theta}{\sin^2\theta} + \frac{3,200}{\sin^2\theta} = \frac{-4,000\cos\theta + 3,200}{\sin^2\theta}$ which is 0 (and hence C is minimal) if $-4,000\cos\theta + 3,200 = 0$ or $\cos\theta = \frac{3,200}{4,000} = \frac{4}{5}$.

Appendix: Algebra Review

Section A: Algebra Review

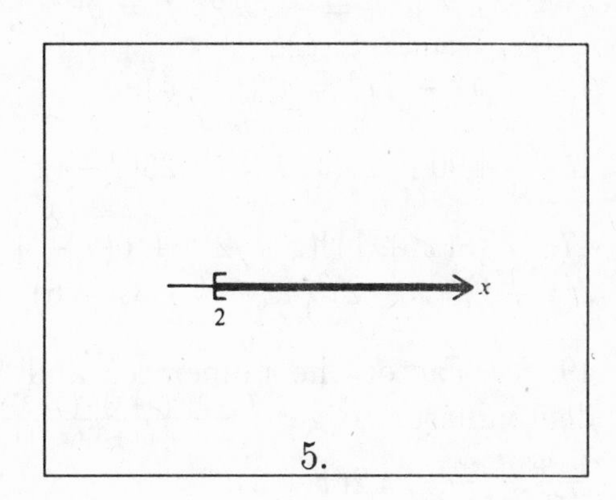

5.

1. 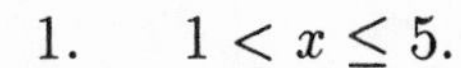 $1 < x \leq 5$.

3. $-5 < x$.

5. See the picture.

7. See the picture.

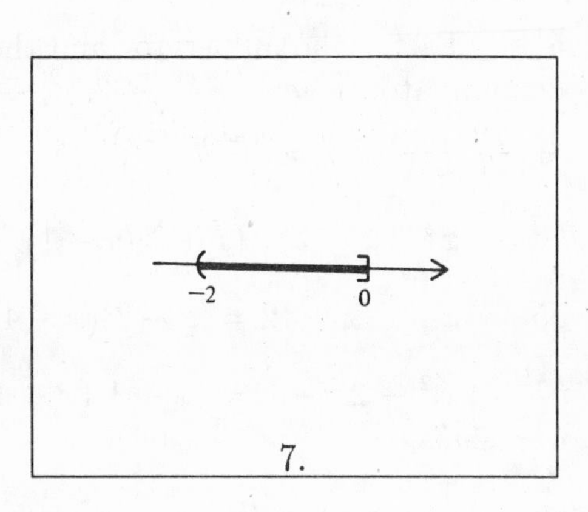

7.

9. The distance between 0 and -4 is $|0-(-4)| = 4$.

11. The distance between -2 and 3 is $|3-(-2)| = 5$.

13. If $|x| \leq 3$, then $-3 \leq x \leq 3$.

15. If $|x+4| \leq 2$, then $-2 \leq x+4 \leq 2$, $-2-4 \leq x \leq 2-4$, or $-6 \leq x \leq -2$.

17. If $|x+2| \geq 5$, then $x+2 \geq 5$ or $x+2 \leq -5$, which means $3 \leq x$ or $x \leq -7$.

19. $5^3 = (5)(5)(5) = 125$.

21. $16^{1/2} = \sqrt{16} = 4$ (not -4).

23. $8^{2/3} = (8^{1/3})^2 = 2^2 = 4$.

25. $(\frac{1}{4})^{1/2} = \sqrt{\frac{1}{4}} = \frac{1}{2}$.

27. $\frac{2^5 2^2}{2^8} = \frac{2^{5+2}}{2^8} = \frac{2^7}{2^8} = \frac{1}{2}$.

29. $\frac{2^{4/3}2^{5/3}}{2^5} = \frac{2^{4/3+5/3}}{2^5} = \frac{2^{9/3}}{2^5} = \frac{2^3}{2^5} = \frac{1}{2^2} = \frac{1}{4}$.

31. $\frac{2(16^{3/4})}{2^3} = \frac{2(16^{1/4})^3}{2^3} = \frac{2(2^3)}{2^3} = 2$.

33. $[\sqrt{8}(2^{5/2})]^{-1/2} = [(2^3)^{1/2}(2)^{5/2}]^{-1/2} = [2^{3/2+5/2}]^{-1/2} = [2^{8/2}]^{-1/2} = 2^{-2} = \frac{1}{4}$.

35. $a^3a^7 = a^n$, $a^{3+7} = a^n$, $a^{10} = a^n$, $n = 10$.

37. $a^4a^{-3} = a^n$, $a^{4-3} = a^n$, $a^1 = a^n$, $n = 1$.

39. $(a^3)^n = a^{12}$, $a^{3n} = a^{12}$, $3n = 12$, $n = 4$.

41. $a^{3/5}a^{-n} = \frac{1}{a^2}$, $a^{3/5-n} = a^{-2}$, $\frac{3}{5} - n = -2$, $n = \frac{3}{5} + 2 = \frac{13}{5}$.

43. $x^5 - 4x^4 = x^4(x-4)$.

45. $100 - 25(x-3) = 25[4 - (x-3)] = 25(4 - x + 3) = 25(7 - x)$.

47. $8(x+1)^3(x-2)^2 + 6(x+1)^2(x-2)^3 = 2(x+1)^2(x-2)^2[4(x+1) + 3(x-2)] = 2(x+1)^2(x-2)^2(4x+4+3x-6) = 2(x+1)^2(x-2)^2(7x-2)$.

49. Factor the numerator and "cancel" the common factor from the numerator and denominator to get $\frac{(x+3)^3(x+1)-(x+3)^2(x+1)^2}{(x+3)(x+1)} = \frac{(x+3)^2(x+1)[(x+3)-(x+1)]}{(x+3)(x+1)} = \frac{(x+3)^2(x+1)(x+3-x-1)}{(x+3)(x+1)} = \frac{2(x+3)^2(x+1)}{(x+3)(x+1)} = 2(x+3)$.

51. Factor the numerator and then "cancel" the common factor from the numerator and denominator to get $\frac{4(1-x)^2(x+3)^3+2(1-x)(x+3)^4}{(1-x)^4} = \frac{2(1-x)(x+3)^3[2(1-x)+(x+3)]}{(1-x)^4} = \frac{2(1-x)(x+3)^3(2-2x+x+3)}{(1-x)^4} = \frac{2(1-x)(x+3)^3(5-x)}{(1-x)^4} = \frac{2(x+3)^3(5-x)}{(1-x)^3}$.

53. $x^2 + x - 2 = (x+2)(x-1)$.

55. $x^2 - 7x + 12 = (x-3)(x-4)$.

57. $x^2 - 2x + 1 = (x-1)(x-1) = (x-1)^2$. Note the important identity $(a \pm b)2 = a^2 \pm 2ab + b^2$.

59. $x^2 - 4 = (x+2)(x-2)$. Note the important identity $a^2 - b^2 = (a+b)(a-b)$.

61. $x^3 - 1 = (x-1)(x^2+x+1)$. Note the important identity $a^3 - b^3 = (a-b)(a^2+ab+b^2)$.

63. $x^7 - x^5 = x^5(x^2-1) = x^5(x+1)(x-1)$.

65. $2x^3 - 8x^2 - 10x = 2x(x^2 - 4x - 5) = 2x(x-5)(x+1)$.

67. Factor the equation to get $2x^2 - 2x - 8 = (x-4)(x+2) = 0$ or $x = 4$ and $x = -2$.

69. Factor the equation to get $x^2 + 10x + 25 = (x+5)^2 = 0$ or $x = -5$. Note the important identity $(a \pm b)2 = a^2 \pm 2ab + b^2$.

71. Factor the equation to get $x^2 - 16 = (x+4)(x-4) = 0$ or $x = -4$ and $x = 4$. Note the important identity $a^2 - b^2 = (a+b)(a-b)$.

73. Factor the equation to get $2x^2+3x+1=(2x+1)(x+1)=0$ or $x=-\frac{1}{2}$ and $x=-1$.

75. Factor the equation to get $4x^2+12x+9=(2x+3)(2x+3)=(x+3)^2=0$ or $x=-\frac{3}{2}$. Note the important identity $(a\pm b)2=a^2\pm 2ab+b^2$.

77. Rewrite the equation as $1+\frac{4}{x}-\frac{5}{x^2}=\frac{x^2+4x-5}{x^2}=\frac{(x+5)(x-1)}{x^2}=0$ or $x=-5$ and $x=1$.

79. Rewrite the equation as $2+\frac{2}{x}-\frac{4}{x^2}=\frac{2x^2+2x-4}{x^2}=\frac{2(x^2+x-2)}{x^2}=\frac{2(x+2)(x-1)}{x^2}=0$ or $x=-2$ and $x=1$.

81. For the equation $2x^2+3x+1=0$, use the quadratic formula with $a=2$, $b=3$, and $c=1$ to get $x=\frac{-3\pm\sqrt{3^2-4(2)(1)}}{2(2)}=\frac{-3\pm\sqrt{1}}{4}$ or $x=\frac{-3+1}{4}=-\frac{1}{2}$ and $x=\frac{-3-1}{4}=-1$.

83. For the equation $x^2-2x+3=0$, use the quadratic formula with $a=1$, $b=-2$, and $c=3$ to get $x=\frac{-(-2)\pm\sqrt{(-2)^2-4(1)(3)}}{2(1)}=\frac{2\pm\sqrt{-8}}{2}$. Since $\sqrt{-8}$ is not a real number, the equation has no real solutions.

85. For the equation $4x^2+12x+9=0$, use the quadratic formula with $a=4$, $b=12$, and $c=9$ to get $x=\frac{-12\pm\sqrt{(12)^2-4(4)(9)}}{2(4)}=\frac{-12\pm\sqrt{0}}{8}=-\frac{3}{2}$.

87. Starting with the system $x+5y=13$ and $3x-10y=-11$, multiply the first equation by 2 to get $2x+10y=26$ and add it to $3x-10y=11$ to form $5x=15$ or $x=3$. Substitute this into the first equation to $3+5y=13$ or $y=2$. Hence, the solution of the system is $x=3$ and $y=2$.

89. To eliminate x in the system $5x-4y=12$ and $2x-3y=2$, multiply the first equation by 2 and the second by -5 to get $10x-8y=24$ and $-10x+15y=-10$. Add together to form $0+7y=14$ or $y=2$. Substitute this into the second equation to $2x-3(2)=2$, $2x=8$, or $x=4$. Hence, the solution of the system is $x=4$ and $y=2$.

91. To solve the system $2y^2-x^2=1$ and $x-2y=3$, solve the second equation for x to get $x=3+2y$. Substitute into the first equation to get $2y^2-(3+2y)^2=1$, $2y^2-(9+12y+4y^2)=1$, $2y^2-9-12y-4y^2=1$, $2y^2+12y+10=0$, $y^2+6y+5=0$, $(y+5)(y+1)=0$, or $y=-5$ and $y=-1$. If $y=-5$, the second equation gives $x-2(-5)=3$, or $x=-7$. If $y=-1$, the second equation gives $x-2(-1)=3$, or $x=1$. Hence, the solutions of the system are $x=-7$ and $y=-5$ or $x=1$ and $y=-1$.

93. $\sum_{j=1}^{4}(3j+1)=[3(1)+1]+[3(2)+1]+[3(3)+1]+[3(4)+1]=4+7+10+13=34$.

95. $\sum_{j=1}^{10}(-1)^j=(-1)^1+(-1)^2+(-1)^3+(-1)^4+(-1)^5+(-1)^6+(-1)^7+(-1)^8+(-1)^9+(-1)^{10}=-1+1-1+1-1+1-1+1-1+1=0$

97. $1 + \frac{1}{2} + \frac{1}{3} + \frac{1}{4} + \frac{1}{5} + \frac{1}{6} = \sum_{j=1}^{6} \frac{1}{j}$.

99. $2x_1 + 2x_2 + 2x_3 + 2x_4 + 2x_5 + 2x_6 = \sum_{j=1}^{6} 2x_j$

101. $1 - 2 + 3 - 4 + 5 - 6 + 7 - 8 = \sum_{j=1}^{8} (-1)^{j+1} j$ since $(-1)^{2n} = 1$ and $(-1)^{2n+1} = -1$.